Mechanics

Masud Chaichian · Ioan Merches
Anca Tureanu

Mechanics

An Intensive Course

Springer

Prof. Dr. Masud Chaichian
Department of Physics
University of Helsinki
PO Box 64
00014 Helsinki
Finland
e-mail: masud.chaichian@helsinki.fi

Prof. Dr. Ioan Merches
Faculty of Physics
Al. I. Cuza University
Carol I Boulevard 11
700506 Iasi
Romania
e-mail: merches@uaic.ro

Doc. Dr. Anca Tureanu
Department of Physics
University of Helsinki
PO Box 64
00014 Helsinki
Finland
e-mail: anca.tureanu@helsinki.fi

ISBN 978-3-642-42986-6
ISBN 978-3-642-17234-2 (eBook)
DOI 10.1007/978-3-642-17234-2
Springer Heidelberg Dordrecht London New York

Softcover reprint of the hardcover 1st edition 2012

Printed on acid-free paper

Springer is part of Springer Science+Business Media (www.springer.com)

He who knows nothing, loves nothing. But he who understands also loves, notices, sees. The more knowledge is involved in a thing, the greater the love.

Paracelsus

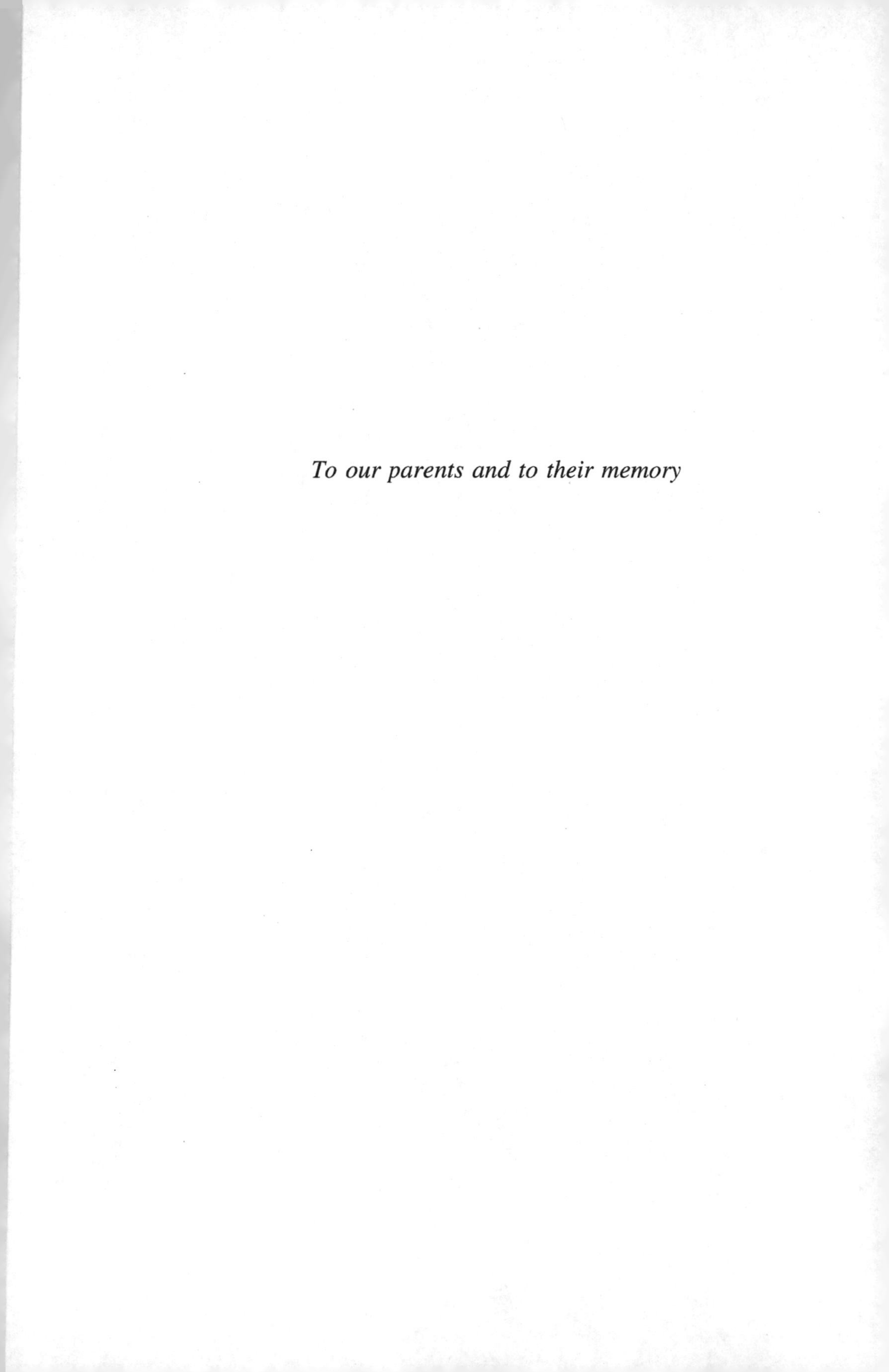

To our parents and to their memory

Preface

Mechanics is the oldest discipline among the fundamental natural sciences. The name comes from the Greek word "mechanike", which means "mechanism". The subject of mechanics as a science is the investigation of the motion of bodies and their equilibrium under the action of applied forces. Depending on the nature of the bodies, mechanics can be divided into three branches: (a) general mechanics, dealing with the mechanical behaviour of material points and rigid bodies; (b) fluid mechanics (or the mechanics of continuous media), which is concerned with ideal and viscous fluids and (c) mechanics of deformable media, which studies the deformation of solid bodies under applied external forces.

The knowledge of mechanical motion or displacement of bodies can be accomplished by a very general procedure based on a system of basic axioms, called principles. These principles are the core of what is known as Newtonian mechanics, relativistic mechanics, quantum mechanics and so forth. During the eighteenth century, after the huge success achieved by the mechanics of Galileo Galilei (1564–1642) and Isaac Newton (1643–1727), there appeared the tendency of making mechanics more abstract and general. This tendency leads to what nowadays is called analytical mechanics. Among the founders of analytical mechanics are: Pierre-Louis Moreau de Maupertuis (1698–1759), Leonhard Euler (1707–1783), Jean Baptiste le Rond D'Alembert (1717–1783), Joseph-Louis Lagrange (1736–1813), Carl Friedrich Gauss (1777–1855) and William Rowan Hamilton (1805–1865). Analytical mechanics has proved to be a very useful tool of investigation not only in Newtonian mechanics, but also in other disciplines of Physics: electrodynamics, quantum field theory, theory of relativity, magnetofluid dynamics – to mention a few.

Classical mechanics has undergone an important revival during the last few decades, due to the progress in non-linear dynamics, stochastic processes and various applications of Noether's theorem in the study of both discrete and continuous systems. We recall that there are no exactly linear processes in Nature, but only approximately. All linear models studied in any science are only approximations of reality.

This book is dedicated to the principles and applications of classical mechanics, written for undergraduate and graduate students in physics and related subjects. Its main purpose is to make the students familiar with the fundamentals of the theory, to stimulate them in the use of applications and to contribute to the formation of their background as specialists.

The first two chapters are dedicated to the basic notions and principles of both Newtonian and analytical mechanics, as different approaches to the same purpose: the investigation of mechanical behaviour of both discrete and continuous systems. A special emphasis is put on the large applicability of analytical formalism in various branches of physics.

In the third chapter, the Lagrangian formalism is applied to the study of some classic mechanical systems, as the harmonic oscillator and the gravitational pendulum, as well as to the investigation of some non-mechanical systems, like electric circuits.

The fourth chapter is concerned with the mechanics of the rigid body. The derivation of velocity and acceleration distributions in relative motion makes possible to study the motion of a rigid body about a fixed point. The chapter ends with some applications, such as the physical pendulum and the symmetrical top, together with some mechanical–electromagnetic analogies.

The aim of the fifth chapter is to make the reader familiar with the Hamiltonian formalism. The derivation of the canonical equations is followed by several applications and extensions in mechanics and electrodynamics. The canonical transformations, integral invariants and the Hamilton–Jacobi formalism are also described. They are very useful for students for their further studies of thermodynamics, statistics and quantum theory.

The sixth, final, chapter deals with the mechanics of continuous deformable media. Here, both the Lagrangian and Hamiltonian formalisms are applied in order to study some well-known models of continuous media: the elastic medium, the ideal and viscous fluids. Special attention is paid to the extension of Noether's theorem to continuous media and its applications to the fundamental theorems of ideal fluids.

Since classical mechanics has undergone a considerable evolution during the last century, the authors have tried to draw the attention of the reader to three main directions of development of post-classical mechanics: theory of relativity, quantum mechanics and stochastic processes. These three basic orientations in post-classical mechanics are very briefly exposed in three addenda, which conclude the main substance of the book. At the end of the book, for the convenience of readers, two appendices are provided, which contain the most frequently used formulas on vector and tensor algebra, as well as on vector calculus.

The present book is an outcome of the authors' teaching experience over many years in different countries and with different students studying diverse fields of physics and engineering. The authors believe that the presentation and the distribution of the topics, the various applications in several branches of physics and the set of more than 100 proposed problems make this book a comprehensive and useful tool for students, teachers and researchers.

During the preparation of this book the authors have benefited from discussing various questions with many of their colleagues and students. It is a pleasure to express gratitude to all of them and to acknowledge the stimulating discussions and their useful advice. Our special thanks go to Professor Peter Presnajder for valuable suggestions and for his considerable help in improving the manuscript.

Helsinki, Iasi, October 2011

M. Chaichian
I. Merches
A. Tureanu

Contents

1 Foundations of Newtonian Mechanics 1
1.1 Notions, Principles and Fundamental Theorems of Newtonian Mechanics. 1
1.1.1 Velocity. Acceleration 2
1.1.2 Analytical Expressions for Velocity and Acceleration in Different Coordinate Systems 4
1.2 Principles of Newtonian Mechanics 6
1.2.1 The Principle of Inertia (Newton's First Law) 6
1.2.2 The Law of Force (Newton's Second Law). 7
1.2.3 The Principle of Action and Reaction (Newton's Third Law) 8
1.3 General Theorems of Newtonian Mechanics 9
1.3.1 Integration of the Equations of Motion 9
1.3.2 First Integrals 11
1.3.3 General Theorems of One-Particle Mechanics 12
1.3.4 General Theorems for Systems of Particles 15
1.4 Problems. 23

2 Principles of Analytical Mechanics 27
2.1 Constraints 27
2.1.1 One-Particle Systems 27
2.1.2 Many-Particle Systems 39
2.2 Elementary Displacements 40
2.3 Principle of Virtual Work 42
2.3.1 Free Particle 42
2.3.2 Particle Subject to Constraints 42
2.3.3 System of Free Particles 43
2.3.4 System of Particles Subject to Constraints 43
2.3.5 Application 46

2.4 Generalized Coordinates . . . 47
2.4.1 Configuration Space . . . 48
2.4.2 Generalized Forces . . . 49
2.4.3 Kinetic Energy in Generalized Coordinates . . . 51
2.5 Differential and Integral Principles in Analytical Mechanics . . . 52
2.5.1 D'Alembert's Principle . . . 53
2.5.2 Lagrange Equations for Holonomic Systems . . . 53
2.5.3 Velocity-Dependent Potential . . . 57
2.5.4 Non-potential Forces . . . 60
2.6 Elements of Calculus of Variations . . . 62
2.6.1 Shortest Distance Between Two Points in a Plane . . . 66
2.6.2 Brachistochrone Problem . . . 67
2.6.3 Surface of Revolution of Minimum Area . . . 68
2.6.4 Geodesics . . . 69
2.7 Hamilton's Principle . . . 74
2.7.1 Euler–Lagrange Equations for the Action Integral . . . 77
2.7.2 Criteria for the Construction of Lagrangians . . . 79
2.8 Symmetry Properties and Conservation Theorems . . . 79
2.8.1 First Integrals as Constants of Motion . . . 80
2.8.2 Symmetry Transformations . . . 82
2.8.3 Noether's Theorem . . . 82
2.9 Principle of Least Action . . . 86
2.10 Problems . . . 93

3 Applications of the Lagrangian Formalism in the Study of Discrete Particle Systems . . . 97
3.1 Central Force Fields . . . 97
3.1.1 Two-Body Problem . . . 97
3.1.2 General Properties of Motion in Central Field . . . 99
3.1.3 Discussion of Trajectories . . . 101
3.1.4 Bertrand's Theorem . . . 105
3.2 Kepler's Problem . . . 108
3.2.1 Determination of Trajectories . . . 109
3.2.2 Law of Motion . . . 113
3.2.3 Runge–Lenz Vector . . . 116
3.2.4 Artificial Satellites of the Earth. Cosmic Velocities . . . 118
3.3 Classical Theory of Collisions Between Particles . . . 121
3.3.1 Collisions Between Two Particles . . . 121
3.3.2 Effective Scattering Cross Section . . . 125
3.3.3 Scattering on a Spherical Potential Well . . . 129
3.3.4 Rutherford's Formula . . . 131
3.4 Periodical Motion of a Particle Under the Influence of Gravity . . . 132
3.4.1 Simple Pendulum . . . 133
3.4.2 Cycloidal Pendulum . . . 136
3.4.3 Spherical Pendulum . . . 137

3.5 Motion of a Particle Subject to an Elastic Force . . . 140
3.5.1 Harmonic Linear Oscillator . . . 140
3.5.2 Space Oscillator . . . 141
3.5.3 Non-linear Oscillations . . . 143
3.6 Small Oscillations About a Position of Stable Equilibrium . . . 144
3.6.1 Equations of Motion. Normal Coordinates . . . 146
3.6.2 Small Oscillations of Molecules . . . 149
3.7 Analogy Between Mechanical and Electric Systems . . . 153
3.7.1 Kirchhoff's Rule Relative to the Loops of an Electric Circuit . . . 153
3.7.2 Kirchhoff's Rule Relative to the Junction Points of an Electric Circuit . . . 157
3.8 Problems . . . 159

4 Rigid Body Mechanics . . . 163
4.1 General Considerations . . . 163
4.2 Distribution of Velocities and Accelerations in a Rigid Body . . . 165
4.3 Inertial Forces . . . 168
4.3.1 Action of the Coriolis Force on the Motion of Bodies at the Surface of the Earth . . . 170
4.3.2 Foucault's Pendulum . . . 172
4.4 Euler's Angles . . . 175
4.5 Motion of a Rigid Body About a Fixed Point . . . 177
4.5.1 Kinematic Preliminaries . . . 178
4.5.2 Angular Momentum . . . 179
4.5.3 Kinetic Energy . . . 180
4.5.4 Ellipsoid of Inertia . . . 182
4.5.5 Euler's Equations of Motion . . . 183
4.6 Applications . . . 188
4.6.1 Physical Pendulum . . . 188
4.6.2 Symmetrical Top . . . 189
4.6.3 Fast Top. The Gyroscope . . . 197
4.6.4 Motion of a Rigid Body Relative to a Non-inertial Frame. The Gyrocompass . . . 198
4.6.5 Motion of Rigid Bodies in Contact . . . 202
4.6.6 Mechanical–Electromagnetic Analogies . . . 205
4.7 Problems . . . 208

5 Hamiltonian Formalism . . . 211
5.1 Hamilton's Canonical Equations . . . 211
5.1.1 Motion of a Particle in a Plane . . . 216
5.1.2 Motion of a Particle Relative to a Non-inertial Frame . 218
5.1.3 Motion of a Charged Particle in an Electromagnetic Field . . . 219
5.1.4 Energy of a Magnetic Dipole in an External Field . . . 222

5.2 Routh's Equations 222
5.3 Poisson Brackets 225
5.3.1 Poisson Brackets for Angular Momentum 228
5.3.2 Poisson Brackets and Commutators 231
5.3.3 Lagrange Brackets 234
5.4 Canonical Transformations 236
5.4.1 Extensions and Applications 241
5.4.2 Mechanical–Thermodynamical Analogy. Thermodynamic Potentials 247
5.5 Infinitesimal Canonical Transformations 249
5.5.1 Total Momentum as Generator of Translations 251
5.5.2 Total Angular Momentum as Generator of Rotations 251
5.5.3 Hamiltonian as Generator of Time-Evolution 252
5.6 Integral Invariants 253
5.6.1 Integral Invariants of the Canonical Equations 255
5.6.2 The Relative Universal Invariant of Mechanics 256
5.6.3 Liouville's Theorem 258
5.6.4 Pfaff Forms 261
5.6.5 Quantum Mechanical Harmonic Oscillator 263
5.7 Hamilton–Jacobi Formalism 265
5.7.1 Hamilton–Jacobi Equation 265
5.7.2 Methods for Solving the Hamilton–Jacobi Equation . . . 268
5.7.3 Applications 273
5.7.4 Action–Angle Variables 278
5.7.5 Adiabatic Invariants 287
5.8 Problems 289

6 Mechanics of Continuous Deformable Media 293
6.1 General Considerations 293
6.2 Kinematics of Continuous Deformable Media 294
6.2.1 Lagrange's Method 294
6.2.2 Euler's Method 295
6.3 Dynamics of Continuous Media 297
6.3.1 Equation of Continuity 297
6.3.2 Forces Acting upon a Continuous Deformable Medium 300
6.3.3 General Theorems 301
6.3.4 Equations of Motion of a CDM. Cauchy Stress Tensor 302
6.4 Deformation of a Continuous Deformable Medium About a Point. Linear Approximation 306
6.4.1 Rotation Tensor and Small-Strain Tensor 306
6.4.2 Saint-Venant Compatibility Conditions 309
6.4.3 Finite-Strain Tensor 310

6.5 Elastic Medium 311
6.5.1 Hooke's Generalized Law 311
6.5.2 Equations of Motion of an Elastic Medium 315
6.5.3 Plane Waves in Isotropic Elastic Media 317
6.6 Perfect Fluid 319
6.6.1 Equation of Motion of a Perfect Fluid 319
6.6.2 Particular Types of Motion of an Ideal Fluid 320
6.6.3 Fundamental Conservation Theorems 328
6.6.4 Magnetodynamics of Ideal Fluids 338
6.7 Viscous Fluid 339
6.8 Lagrangian Formalism 344
6.8.1 Euler–Lagrange Equations for Continuous Systems . . . 344
6.8.2 Applications 347
6.9 Hamiltonian Formalism 357
6.9.1 Hamilton's Canonical Equations for Continuous Systems 357
6.9.2 Applications 362
6.10 Noether's Theorem for Continuous Systems 373
6.10.1 Hamilton's Principle and the Equations of Motion 373
6.10.2 Symmetry Transformations 377
6.10.3 Energy Conservation Theorem 380
6.10.4 Momentum Conservation Theorem 381
6.10.5 Angular Momentum Conservation Theorem 382
6.10.6 Centre of Mass Theorem 383
6.11 Problems 385

Addenda 387

Addendum I: Special Theory of Relativity 389

Addendum II: Quantum Theory and the Atom 399

Addendum III: Stochastic Processes and the Langevin Equation 409

Appendix A: Elements of Vector and Tensor Algebra 411

Appendix B: Elements of Vector and Tensor Analysis 427

References 447

Author Index 449

Subject Index 451

Chapter 1
Foundations of Newtonian Mechanics

1.1 Notions, Principles and Fundamental Theorems of Newtonian Mechanics

In physics, by *mechanical motion* we mean the change in time of the position of a body with respect to another body, chosen as a reference. Generally speaking, the motion of a body does not reduce to its mechanical motion, since the body can be simultaneously animated by several types of motion (mechanical, chemical, biological, etc.) depending on its complexity. For the sake of simplicity we shall, nevertheless, call mechanical motion just *motion*.

The study of the motion of a body implies the choice of another body, supposed to be fixed, with respect to which the motion of the first body is considered. The body chosen as a reference defines, by abstraction, a *reference system* or *reference frame*.

If the position of a body does not change relative to a certain reference frame, then it is *at rest* relative to that frame. There are no immobile reference frames in Nature; nevertheless, the motion of the bodies is conventionally referred to reference frames considered to be *fixed*. Any other reference frame which is immobile with respect to the first is, in its turn, *immobile*.

The aforementioned considerations show that neither *absolute reference frames* (i.e. independent of the motion of bodies) nor *absolute rest state* can exist. However, sometimes the expression *absolute motion* is used when we refer to a fixed reference frame, in order to distinguish this motion from that relative to a frame which is mobile with respect to the first, called *relative motion*.

A body whose dimensions can be neglected when studying its motion is called *point mass* or *particle*. In this case, the mass of the body is supposed to be concentrated in a geometric point (e.g. the centre of mass). Such an approximation depends, obviously, upon the concrete conditions of the mechanical model. For example, a planet moving around the Sun can be considered as a particle, but this approach is not possible when the motion of the planet around its axis is studied.

M. Chaichian et al., *Mechanics*, DOI: 10.1007/978-3-642-17234-2_1,

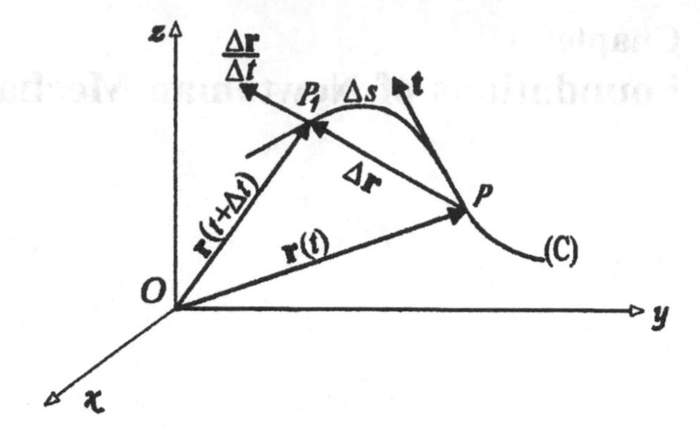

Fig. 1.1 Geometry of the trajectory described by a particle.

The *position* of a particle is defined relative to a given reference frame. To do this, we can chose either Cartesian coordinates x, y, z, spherical coordinates r, θ, φ, or cylindrical coordinates ρ, φ, z, etc. Most commonly, the position of a particle is defined by its *radius-vector* or *position vector* $\mathbf{r}$, relative to the origin of the coordinate frame.

To know the *motion* of a particle one has to know its coordinates (e.g. Cartesian coordinates $x_1 = x$, $x_2 = y$, $x_3 = z$) as functions of time

$$x_i = x_i(t) \quad (i = 1, 2, 3) \tag{1.1.1}$$

or, in vector form,

$$\mathbf{r} = \mathbf{r}(t). \tag{1.1.2}$$

The vector function $\mathbf{r}(t)$ must obey certain mathematical requirements, imposed by the physical phenomenon of the motion. It must be continuous and homogeneous in time, finite in magnitude, and at least twice differentiable. The last condition is required due to fact that the differential equations of motion are of the second-order.

When the parameter t varies, the particle describes a curve called *trajectory* (Fig. 1.1). In other words, the trajectory is the geometric locus of the successive positions occupied by the moving particle.

Equations (1.1.1) are called the *finite equations* of motion. They express the *law of motion* of the particle. At the same time, these equations are the *parametric equations* of the trajectory of the particle. If the trajectory is known, then the motion of the particle can be defined by a single scalar equation

$$s = s(t). \tag{1.1.3}$$

Here, s is the curve between the origin and the actual position of the particle, at time t.

1.1.1 Velocity. Acceleration

Let P be a particle tracing the trajectory (C) and $\mathbf{r}(t)$ the radius-vector of the particle, at time t, relative to the origin O of the Cartesian reference frame $Oxyz$.

Suppose that at time $t' = t + \Delta t$, the particle reaches the position P_1, defined by the radius-vector $\mathbf{r}(t + \Delta t)$. Then the ratio

$$\bar{\mathbf{v}} = \frac{\Delta \mathbf{r}}{\Delta t}, \tag{1.1.4}$$

where $\Delta \mathbf{r} = \mathbf{r}(t + \Delta t) - \mathbf{r}(t)$, is a vector collinear with $\Delta \mathbf{r}$, named *average velocity* of the particle on the arc of curve PP_1 (averages are customarily denoted by a bar above the corresponding quantity). If $\Delta t \to 0$, then $P_1 \to P$ and (1.1.4) yields

$$\mathbf{v} = \lim_{\Delta t \to 0} \frac{\Delta \mathbf{r}}{\Delta t} = \frac{d\mathbf{r}}{dt} = \dot{\mathbf{r}}, \tag{1.1.5}$$

which is the *instantaneous linear velocity* of the particle at time t. The vector $\mathbf{v}$ with the origin in P is tangent to the trajectory (C) and shows the direction of motion. If $\boldsymbol{\tau}$ is the unit vector of $\mathbf{v}$, we can write

$$\mathbf{v} = v\boldsymbol{\tau}, \quad v = |\mathbf{v}|. \tag{1.1.6}$$

The magnitude $|\mathbf{v}|$ of the vector $\mathbf{v}$ is called the *speed* of motion. If we denote $d\mathbf{r} = \mathbf{r}(t + dt) - \mathbf{r}(t)$ and observe that $|d\mathbf{r}| = ds$, the magnitude of $\mathbf{v}$ can be defined by

$$v = \frac{ds}{dt} = \left| \frac{d\mathbf{r}}{dt} \right| = \dot{s}. \tag{1.1.7}$$

Suppose now that the particle P is forced to remain on the curve (C) during its motion. If the curve (C) is moving, then the actual trajectory (Γ) of the particle and the arc (C) will not coincide, the velocity $\mathbf{v}$ being tangent to the trajectory (Γ) (Fig. 1.2). If $|\mathbf{v}| = \text{const.}$, the motion is called *uniform*; if $|\mathbf{v}|$ changes in time, then the motion is called *varied*.

Let the velocities corresponding to the positions P, P_1 be $\mathbf{v}$ and $\mathbf{v} + \Delta \mathbf{v}$, respectively. The vector quantity

$$\bar{\mathbf{a}} = \frac{\Delta \mathbf{v}}{\Delta t} \tag{1.1.8}$$

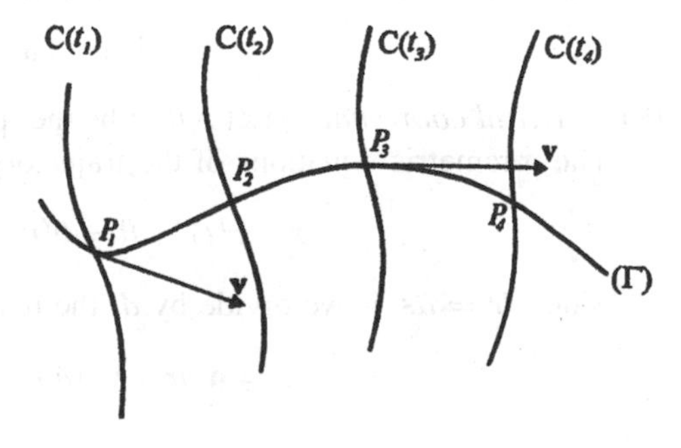

Fig. 1.2 The instantaneous velocity of a particle is tangent to its trajectory.

is the *average acceleration* of the particle in the time interval $[t, t+\Delta t]$, while

$$\mathbf{a} = \lim_{\Delta t \to 0} \frac{\Delta \mathbf{v}}{\Delta t} = \frac{d\mathbf{v}}{dt} = \frac{d^2\mathbf{r}}{dt^2} = \dot{\mathbf{v}} = \ddot{\mathbf{r}} \tag{1.1.9}$$

is the *instantaneous acceleration* of the particle at the time t. If $|\mathbf{a}| = \text{const.}$, the motion is called *uniformly varied.* The definitions (1.1.5) and (1.1.9) yield the following units for velocity and acceleration:

$$[v] = LT^{-1}, \quad [a] = LT^{-2}. \tag{1.1.10}$$

1.1.2 Analytical Expressions for Velocity and Acceleration in Different Coordinate Systems

From this point on, we shall assume that the direction of axes of the *Oxyz* reference frame is fixed, and the summation convention (Einstein's convention) for repeated indices running from 1 to 3 is used. The summation convention was introduced by *Albert Einstein* in 1916, to simplify the formulas which involved sums over coordinates. According to this notational convention, when an index appears twice in a product, that index is summed over, without the sum symbol being explicitly written. If, however, in some expression an index (for example, j) appears twice, but no sum over it has to be taken, this is customarily mentioned in brackets ("no summation over j"). Mostly we shall use Einstein's convention for coordinate indices, but at some point when the formulas will become too complicated, we shall use it also for sums over particle indices (this will be specified when it occurs). Any exceptions to these rules will be pointed out at the right time.

(a) *Orthogonal Cartesian coordinates.* If we denote (see Appendix A):

$$\mathbf{i} = \mathbf{u}_1, \quad \mathbf{j} = \mathbf{u}_2, \quad \mathbf{k} = \mathbf{u}_3, \tag{1.1.11}$$

then the radius-vector $\mathbf{r}$ can be written as

$$\mathbf{r} = x_i \mathbf{u}_i. \tag{1.1.12}$$

The first and second time derivatives of (1.1.12) give

$$\mathbf{v} = \dot{x}_i \mathbf{u}_i, \quad \mathbf{a} = \ddot{x}_i \mathbf{u}_i. \tag{1.1.13}$$

(b) *Spherical coordinates.* Let r, θ, φ be the spherical coordinates of the particle P. The parametric equations of the trajectory are then:

$$r = r(t), \quad \theta = \theta(t), \quad \varphi = \varphi(t). \tag{1.1.14}$$

Since $d\mathbf{r} = d\mathbf{s}$, if we divide by dt the relation (see Appendix B)

$$d\mathbf{s} = \mathbf{u}_r dr + \mathbf{u}_\theta r d\theta + \mathbf{u}_\varphi r \sin\theta d\varphi,$$

we obtain

$$\mathbf{v} = \dot{r}\mathbf{u}_r + r\dot{\theta}\mathbf{u}_\theta + r\sin\theta\dot{\varphi}\mathbf{u}_\varphi, \tag{1.1.15}$$

where

$$\begin{aligned} \mathbf{u}_r &= \mathbf{i}\sin\theta\cos\varphi + \mathbf{j}\sin\theta\sin\varphi + \mathbf{k}\cos\theta, \\ \mathbf{u}_\theta &= \mathbf{i}\cos\theta\cos\varphi + \mathbf{j}\cos\theta\sin\varphi - \mathbf{k}\sin\theta, \\ \mathbf{u}_\varphi &= -\mathbf{i}\sin\varphi + \mathbf{j}\cos\varphi. \end{aligned} \tag{1.1.16}$$

The acceleration vector is found by taking the time derivative of (1.1.15). Using (1.1.16) we finally get

$$\begin{aligned} \mathbf{a} = {} & (\ddot{r} - r\dot{\theta}^2 - r\dot{\varphi}^2\sin^2\theta)\mathbf{u}_r + (2\dot{r}\dot{\theta} + r\ddot{\theta} - r\dot{\varphi}^2\sin\theta\cos\varphi)\mathbf{u}_\theta \\ & + (2\dot{r}\dot{\varphi}\sin\theta + 2r\dot{\theta}\dot{\varphi}\cos\theta + r\ddot{\varphi}\sin\theta)\mathbf{u}_\varphi. \end{aligned} \tag{1.1.17}$$

(c) *Plane polar coordinates*. Suppose that the particle P moves in the xy-plane ($\theta = \frac{\pi}{2}$). The position of the particle is then determined by r and φ. Taking $\theta = \frac{\pi}{2}$ and $\dot{\theta} = 0$ in (1.1.15) and (1.1.17), we arrive at

$$\mathbf{v} = \dot{r}\mathbf{u}_r + r\dot{\varphi}\mathbf{u}_\varphi, \tag{1.1.18}$$

$$\mathbf{a} = (\ddot{r} - r\dot{\varphi}^2)\mathbf{u}_r + (2\dot{r}\dot{\varphi} + r\ddot{\varphi})\mathbf{u}_\varphi. \tag{1.1.19}$$

(d) *Cylindrical coordinates*. Observing that the cylindrical coordinate system is a combination of the plane polar coordinates ρ, φ and a Cartesian coordinate z, we have at once:

$$\mathbf{v} = \dot{\rho}\mathbf{u}_\rho + \rho\dot{\varphi}\mathbf{u}_\varphi + \dot{z}\mathbf{k}, \tag{1.1.20}$$

$$\mathbf{a} = (\ddot{\rho} - \rho\dot{\varphi}^2)\mathbf{u}_\rho + (2\dot{\rho}\dot{\varphi} + \rho\ddot{\varphi})\mathbf{u}_\varphi + \ddot{z}\mathbf{k}, \tag{1.1.21}$$

where $\mathbf{k}$ is the z-axis unit vector.

(e) *Natural coordinates* (see Appendix B). If the particle P describes a trajectory whose equation is known,

$$\mathbf{r} = \mathbf{r}(s), \tag{1.1.22}$$

then the motion of the particle is defined by (1.1.3). The time derivatives of (1.1.22) yield

$$\mathbf{v} = \dot{\mathbf{r}} = \frac{d\mathbf{r}}{ds}\frac{ds}{dt} = \dot{s}\boldsymbol{\tau} = v\boldsymbol{\tau}, \tag{1.1.23}$$

$$\mathbf{a} = \ddot{\mathbf{r}} = \dot{v}\boldsymbol{\tau} + \frac{v^2}{\rho}\boldsymbol{\nu} = \mathbf{a}_\tau + \mathbf{a}_\nu, \tag{1.1.24}$$

where $\boldsymbol{\tau} = \frac{d\mathbf{r}}{ds}$, ρ is the *radius of curvature* and $\boldsymbol{\nu}$ – the *principal normal unit vector* of the curve (C) at the point P.

1.2 Principles of Newtonian Mechanics

Newtonian mechanics is based on some fundamental statements called *principles* or *laws*. These principles were established by the generalization of a large number of particular experiments. They cannot be demonstrated, but they have not been contradicted by any known particular experiment. In his famous book "Mathematical Principles of Natural Philosophy" (1687), *Isaac Newton* gave the following fundamental principles.

1.2.1 The Principle of Inertia (Newton's First Law)

A body is in a state of rest or performs a uniform motion along a straight line, unless subjected to an external force.

To give a well-determined content to this principle, one must indicate the reference frame relative to which the motion of the body is considered. This frame is called *inertial*, and the property of a body to be at rest or in uniform straight motion with respect to such a frame is called *inertia*. That is why Newton's first law is called *the principle of inertia*. Any reference frame which is at rest or in a uniform straight motion (translation) with respect to an inertial frame is also an inertial frame.

Consider an observer in a train carriage that moves straightly and uniformly relative to the ground. Then, there is no mechanical experiment that the observer can perform inside the carriage that can show whether the carriage is at rest or in a straight uniform motion with respect to the Earth. By generalizing this mental experiment, we arrive at the classical (Galilean) principle of relativity:

No mechanical experiment can be done within an inertial frame, that can put into evidence either the rest state or the uniform straight motion of that frame.

Let $S(Oxyz)$ and $S'(O'x'y'z')$ be two inertial frames and $\mathbf{V}$ – the velocity of S' relative to S (Fig. 1.3). If Δt and $\Delta t'$ are the time intervals between two events, as determined by observers placed in O and O', respectively, then we shall assume that $\Delta t = \Delta t'$, i.e. the two observers measure *the same duration*. If we choose the same origin of the time interval in S and S', the two events are recorded

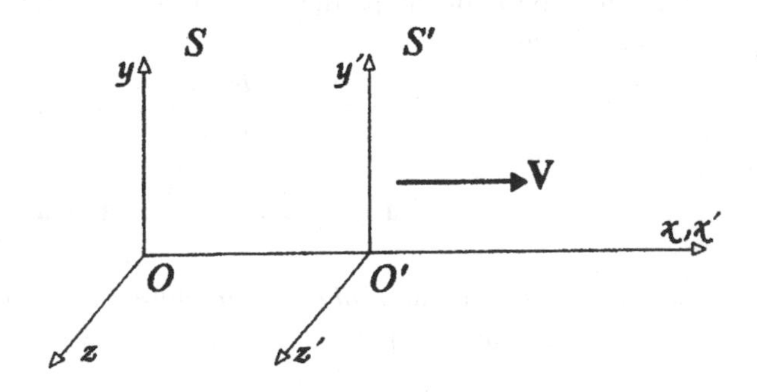

Fig. 1.3 Schematic representation of two inertial frames, S and S'.

simultaneously by the observers. This postulate is called *principle of absolute simultaneity.*

Let x, y, z, t, and x', y', z', t' be the space–time coordinates of a certain event in S and S', respectively. Then we can write

$$\mathbf{r}' = \mathbf{r} - \mathbf{V}t, \quad t' = t. \tag{1.2.1}$$

If the displacement takes place along the common axis $Ox \equiv Ox'$, the relations (1.2.1) yield

$$x' = (x - Vt), \quad y' = y, \quad z' = z, \quad t' = t. \tag{1.2.1'}$$

These transformations fulfill the properties of a *group*, called the *Galilei transformations group*. The theory of special relativity shows that the Galilei–Newton group is a limiting case ($c \to \infty$) of a more general group of transformations, the Lorentz group:

$$x' = \gamma(x - Vt), \quad y' = y, \quad z' = z, \quad t' = \gamma\left(t - \frac{V}{c^2}x\right), \tag{1.2.2}$$

where c is the speed of light in empty space and $\gamma = \sqrt{1 - \frac{V^2}{c^2}}$.

1.2.2 The Law of Force (Newton's Second Law)

The rate of change of momentum of a body is proportional to the magnitude of the external force and takes place in the direction of the acting force.

This principle (which is traditionally known as the *law of force*) introduces two fundamental notions: *mass* and *force*. It is very difficult to give comprehensive, perfectly logical definitions of these notions. We shall accept as satisfactory the following definitions:

The *mass* of a body is a scalar positive quantity, that characterizes the body; it is a measure of its inertia and gravitational interaction with other bodies.

The *force* is a measure of the mechanical interaction between a body and other bodies, characterizing the magnitude and direction of this interaction, and having the effect of an acceleration, or a deformation.

The product $m\mathbf{v}$ is called *linear momentum* and its variation is called *impulse*. The force principle can be then written in the form:

$$\mathbf{F} = \frac{d}{dt}(m\mathbf{v}). \tag{1.2.3}$$

In Newtonian mechanics the mass m is a constant quantity, i.e. it does not depend on the motion. As a result,

$$\mathbf{F} = m\frac{d\mathbf{v}}{dt} = m\mathbf{a}, \tag{1.2.4}$$

which is the *fundamental equation of Newtonian mechanics*. Equation (1.2.4) shows that the acceleration of a particle, the second time-derivative of the radius-vector, is not an independent quantity, but is given by the force acting on the particle divided by its mass m.

In classical, non-relativistic mechanics the mass occurs both as an *inertial* mass and a *gravitational* mass. The first appears in the fundamental equation (1.2.4), while the second is found for instance in the gravitational force formula

$$\mathbf{F}_{\mathrm{G}} = m\mathbf{g}. \tag{1.2.5}$$

The experiments done by *Loránd Eötvös* (1890) and *Pieter Zeeman* (1907) showed that the two masses are proportional (even equal, if the unit system is suitably chosen).

Most generally, the force $\mathbf{F}$ is a known vector function of time t, position $\mathbf{r}$, and velocity $\mathbf{v}$, i.e. $\mathbf{F} = \mathbf{F}(\mathbf{r}, \mathbf{v}, t)$. The product $m\mathbf{a}$ is then a function of the same variables, being a *polar vector*.

Observations:

(a) The first two principles can be put in a unique form, as follows:
There is at least one space–time reference frame in which the law $\mathbf{F} = m\mathbf{a}$ *is valid.*
Indeed, if $\mathbf{F} = 0$, one has $\mathbf{a} = 0$ (since $m \neq 0$), i.e. the body is either at rest, or it moves uniformly along a straight line.

(b) The aforementioned considerations lead to an equivalent formulation of the classical principle of relativity:
The laws of Newtonian mechanics are the same in any inertial frame,
or
The laws of Newtonian mechanics keep their form (are covariant) under the transformations of the Galilei group (1.2.1).

(c) The second law gives rise to another principle, given by Newton in the following form:
If two forces act on a body simultaneously and in different directions, then the body describes the diagonal of the parallelogram constructed on the forces in the same time in which it would describe the sides of the parallelogram under the separate action of the two forces. This formulation is known as the *principle of the independence of forces,* sometimes referred to as *Newton's fourth law.*

1.2.3 The Principle of Action and Reaction (Newton's Third Law)

To any force (action) corresponds an equal and directly opposed reaction. In other words, *the mutual actions of two bodies are equal and directly opposed.*

Following this principle, if a body (considered as a particle) A acts on another body (particle) B with the force $\mathbf{F}_{BA}$, then B acts in its turn on A, with the force

$$\mathbf{F}_{AB} = -\mathbf{F}_{BA}. \tag{1.2.6}$$

Let us interpret, in the light of this postulate, the fundamental equation of Newtonian mechanics (1.2.4). If $\mathbf{F} = m\mathbf{a}$ is the force acting on a body, then the body will respond with the force $-m\mathbf{a}$. The force $\mathbf{J} = -m\mathbf{a}$ is called (somewhat improperly) *force of inertia*.

The action and reaction principle lies at the foundation of Newtonian mechanics of particle systems. It also extends to the electrostatic and gravitational phenomena.

Observation: Let us agree to call *free* any particle whose motion is in no way restricted, i.e. it moves according to the law (1.2.4) and it is subject to applied forces. If the motion of the particle is restricted (e.g. the particle *must* move on a curve, or on a surface, or its velocity must obey a certain condition, etc.), we shall say that the motion is *subject to constraints*. The constraints appear in (1.2.4) as *constraint forces*, and have the character of reaction forces. This means that the constraints are subject to the principle of action and reaction.

1.3 General Theorems of Newtonian Mechanics

1.3.1 Integration of the Equations of Motion

The *fundamental problem of mechanics* is to determine the motion of a particle of given mass, knowing the force acting at any moment on the particle.

Consider a free particle of mass m, subject to the resultant force $\mathbf{F}$, and having the acceleration $\mathbf{a}$. If the motion is referred to the Cartesian frame $Oxyz$, the projection on the axes of the fundamental equation (1.2.4) yields

$$m\ddot{x}_i = F_i \quad (i = 1, 2, 3), \tag{1.3.1}$$

which is a system of three second-order differential equations. Assuming that the existence conditions for solutions are fulfilled, the *general integral* of the system (1.3.1) is

$$x_i = x_i(t, C_1, C_2, \dots, C_6) \quad (i = 1, 2, 3), \tag{1.3.2}$$

where $C_1, \dots, C_6$ are constants of integration. To know the *motion* of the particle means to know its coordinates x_i ($i = 1, 2, 3$) as functions of time t. Therefore, it is necessary to determine the six arbitrary constants $C_1, \dots, C_6$. To this end, we impose as the *initial conditions*: at the initial time $t = t_0$ (e.g. $t_0 = 0$), the coordinates x_i^0 of the particle and the components $\dot{x}_i^0$ of its velocity are given, i.e.

$$x_i^0 = x_i(t_0, C_1, \dots, C_6), \quad \dot{x}_i^0 = \dot{x}_i(t_0, C_1, \dots, C_6). \tag{1.3.3}$$

The system of six algebraic equations (1.3.3) yields the constants $C_1, \dots, C_6$. Finally, these solutions are introduced in (1.3.2) and this determines the motion of the particle uniquely.

Example. Let us find the finite equations of motion and the trajectory of a shell of mass m, thrown at an angle α relative to the horizontal plane and having the initial velocity $\mathbf{v}_0$.

Since the length of the trajectory of the body is supposed to be much longer than any of its three dimensions, we may consider the shell as a heavy particle, the only force acting on it being the force of gravity. This is a simplified model, since there are many factors contributing to the real motion of the body, like: force of friction with the air, density of the air, speed and direction of the wind, etc. The branch of mechanics which takes into account all these aspects is called *ballistics*.

Newton's second law (1.2.4) gives

$$m\ddot{\mathbf{r}} = \mathbf{F}.$$

Next, we use (1.2.5) and project the last equation on the axes of the Cartesian orthogonal frame $Oxyz$, the axis z being along the vertical line and pointing upwards. Then,

$$\ddot{x} = 0, \quad \ddot{y} = 0, \quad \ddot{z} = -g.$$

Integrating these equations twice, yields

$$\dot{x} = C_1, \quad \dot{y} = C_2, \quad \dot{z} = -gt + C_3,$$

$$x = C_1 t + C_4, \quad y = C_2 t + C_5, \quad z = -\frac{g}{2}t^2 + C_3 t + C_6.$$

To determine uniquely the motion of the body, we must find the six constants of integration $C_1, \ldots, C_6$ (otherwise, there would be ∞^6 possibilities of motion). To do this, one must know all six initial conditions. For instance, let us take, at $t = 0$:

$$x(0) = 0, \quad y(0) = 0, \quad z(0) = 0,$$

$$\dot{x}(0) = v_{0x} = 0, \quad \dot{y}(0) = v_{0y} = v_0 \cos\alpha, \quad \dot{z}(0) = v_{0z} = v_0 \sin\alpha,$$

i.e. at the initial time the shell is at the origin of the frame and its initial velocity lies in the yz-plane. Then we find:

$$C_1 = C_4 = C_5 = C_6 = 0, \quad C_2 = v_0 \cos\alpha, \quad C_3 = v_0 \sin\alpha,$$

and the finite equations of motion are

$$x = 0, \quad y = v_0 t \cos\alpha, \quad z = -\frac{g}{2}t^2 + v_0 t \sin\alpha.$$

If the initial velocity is horizontal ($\alpha = 0$), then

$$x = 0, \quad y = v_0 t, \quad z = -\frac{g}{2}t^2,$$

while in the case of ascending vertical initial velocity,

$$x = 0, \quad y = 0, \quad z = v_0 t - \frac{g}{2}t^2.$$

The trajectory is found by eliminating the time t from the parametric equations of motion:

$$z = y \tan\alpha - \frac{g}{2v_0^2 \cos^2\alpha} y^2.$$

This is a *parabola* with its concavity downwards. The points of intersection with the horizontal plane are $y_1 = 0$ and $y_2 = \frac{v_0^2}{g} \sin 2\alpha$. Taking the derivative of y_2 with respect to α and then equating to zero the result, we conclude that the maximum horizontal distance reached by the shell, for a given v_0, is for $\alpha = \frac{\pi}{4}$.

We can also find the angle α at which the projectile reaches a certain given point y_1, z_1. This is obtained by means of the last relation:

$$\tan \alpha = \frac{v_0^2}{g y_1} \left[1 \pm \sqrt{1 - \frac{2g}{v_0} \left(z_1 + \frac{g y_1^2}{2 v_0^2} \right)} \right].$$

It is clear that, if

$$z_1 + \frac{g y_1^2}{2 v_0^2} - \frac{v_0}{2g} \leq 0,$$

the sought after angle has two values.

1.3.2 First Integrals

In some cases, there exists the possibility of obtaining information about the motion of mechanical systems without the full integration of the differential equations of motion. Suppose, for instance, that we are able to find a relation between the time t, the coordinates x_i, with $i = 1, 2, 3$, of the particle, the components $\dot{x}_i$ of its velocity, and a single constant C, for *any* initial conditions, generally written as

$$f(\mathbf{r}, \mathbf{v}, t) = C. \tag{1.3.4}$$

This relation is a first-order differential equation. It is called *a first integral* of (1.2.4). The constants that occur in first integrals are determined by means of the initial conditions. In our case,

$$f(\mathbf{r}_0, \mathbf{v}_0, t_0) = C. \tag{1.3.5}$$

Two or more first integrals are called *distinct* if there is no relation between them. Since the knowledge of a first integral diminishes by one the number of unknowns, the maximum number of distinct first integrals of (1.3.1) is six. It follows then that to know six distinct first integrals of (1.3.1) means to determine the general integral of the system.

Some first integrals present a special importance, since they express the conservation of certain fundamental physical quantities. The determination of first integrals of mechanical (and, as we shall see, non-mechanical) systems is tightly related to the general theorems that express the space–time variation of fundamental quantities: linear momentum, angular momentum and energy. In the following, we shall prove these theorems for both one-particle and many-particle mechanical systems.

1.3.3 General Theorems of One-Particle Mechanics

1.3.3.1 Linear Momentum Theorem

Let us consider, as before, the motion of the particle P of mass m, relative to the fixed Cartesian orthogonal frame $Oxyz$. If we denote by

$$\mathbf{p} = m\mathbf{v} \tag{1.3.6}$$

the *linear momentum* of the particle, then Eq. (1.2.3) reads:

$$\frac{d\mathbf{p}}{dt} = \mathbf{F}, \tag{1.3.7}$$

which says that: *The time derivative of the linear momentum of a particle is equal to the vector resultant of the applied forces.* This is the *linear momentum theorem.* Note that if $\mathbf{F} = 0$, then

$$\mathbf{p} = \text{const.} \tag{1.3.8}$$

This is a vector first integral, equivalent to three distinct scalar first integrals.

1.3.3.2 Angular Momentum Theorem

By definition, the *angular momentum* or *kinetic momentum* of the particle P about the point O is the cross product

$$\mathbf{l} = \mathbf{r} \times \mathbf{p} = \mathbf{r} \times (m\mathbf{v}). \tag{1.3.9}$$

Taking the time derivative of (1.3.9) and observing that $\mathbf{v}$ and $\mathbf{p}$ are collinear, we arrive at

$$\frac{d\mathbf{l}}{dt} = \mathcal{M}, \tag{1.3.10}$$

where

$$\mathcal{M} = \mathbf{r} \times \mathbf{F} \tag{1.3.11}$$

is the *moment* of the force $\mathbf{F}$ with respect to O, also called the *torque* about the point O. Equation (1.3.10) expresses the *angular momentum theorem: The time derivative of the angular momentum of a particle is equal to the moment of the force applied to it, both momenta being taken about the same point O.* If $\mathcal{M} = 0$ (i.e. $\mathbf{F} = 0$, or $\mathbf{r} || \mathbf{F}$), then (1.3.10) leads to the vector first integral

$$\mathbf{l} = \text{const.} \tag{1.3.12}$$

1.3.3.3 Areas Theorem

If the moment $\mathcal{M}$ of the force **F** is permanently orthogonal to a fixed axis (Δ) of unit vector **u**, passing through O, then the projection of (1.3.10) on (Δ) gives $\frac{d}{dt}(\mathbf{l}\cdot\mathbf{u}) = \mathcal{M}\cdot\mathbf{u} = 0$, leading to

$$\mathbf{l}\cdot\mathbf{u} = l_\Delta = C_1(\text{const.}), \tag{1.3.13}$$

i.e. a first integral. If (Δ) coincides with the z-axis, then

$$x\dot{y} - y\dot{x} = \frac{C_1}{m} = C_2(\text{const.}). \tag{1.3.14}$$

The first integral (1.3.14) allows an interesting geometric interpretation. Let $\mathbf{r}(t)$ and $\mathbf{r}(t+dt)$ be the radius-vector of the particle P at times t and $t+dt$, respectively. Then Fig. 1.4 shows that the differential area $d\mathbf{S}$ swept by the radius-vector **r** during the time interval dt can be approximated by the area of the triangle OPP_1:

$$d\mathbf{S} = \frac{1}{2}\mathbf{r}\times d\mathbf{r}. \tag{1.3.15}$$

Projecting (1.3.15) on the z-axis and then dividing by dt gives:

$$\frac{dS_z}{dt} = \frac{1}{2}(x\dot{y} - y\dot{x}). \tag{1.3.16}$$

The quantity $\frac{dS_z}{dt}$ is called *areal velocity*. Comparing (1.3.16) and (1.3.14), it follows that

$$\frac{dS_z}{dt} = C \quad \left(C = \frac{C_2}{2}\right). \tag{1.3.17}$$

The first integral (1.3.17) expresses the *areas theorem*, in projection on the z-axis: *If the moment $\mathcal{M}$ of the force* **F** *is permanently orthogonal to the z-axis, then the motion of the particle, in the xy-plane, is performed with constant areal velocity*. In other words, *the radius-vector* **r** *sweeps equal areas in equal time intervals* (Kepler's second law).

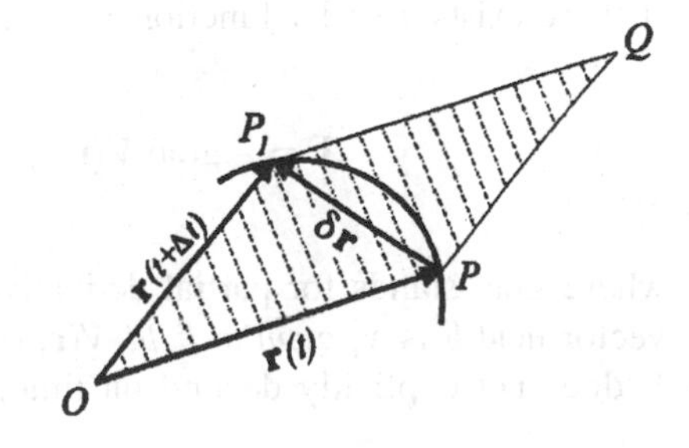

Fig. 1.4 Area swept by the radius-vector **r** during the time interval dt (see Eq. (1.3.15)).

1.3.3.4 Kinetic Energy Theorem

Let us suppose that under the action of the applied resultant force **F**, the particle P undergoes an infinitesimal displacement $d\mathbf{r}$. By definition, the scalar product

$$dW = \mathbf{F} \cdot d\mathbf{r} \tag{1.3.18}$$

is the *infinitesimal work* done by the force **F** as the particle performs the displacement $d\mathbf{r}$. We may write

$$dW = m\frac{d\mathbf{v}}{dt} \cdot \mathbf{v}dt = d\left(\frac{1}{2}m|\mathbf{v}|^2\right).$$

The scalar quantity

$$T = \frac{1}{2}m|\mathbf{v}|^2 \tag{1.3.19}$$

is the *kinetic energy* of the particle. It then follows that

$$dW = dT, \tag{1.3.20}$$

which is the *differential form* of the *kinetic energy theorem*: *The infinitesimal work of the resultant of forces acting on a particle is equal, at any time, to the differential of the kinetic energy of the particle*. Integrating (1.3.20) between t_1 and t_2, corresponding to the velocities v_1 and v_2 of the particle, we get

$$W = \int_1^2 \mathbf{F} \cdot d\mathbf{r} = \int_{v_1}^{v_2} d\left(\frac{1}{2}m|\mathbf{v}|^2\right) = \frac{1}{2}m(v_2^2 - v_1^2) = T_2 - T_1, \tag{1.3.21}$$

i.e. the *integral form* of the *kinetic energy theorem*: *The work done by the force* **F** *acting on a particle during the time interval* $(t_1,\ t_2)$ *is equal to the change in the kinetic energy of the particle during the given interval.*

1.3.3.5 Energy Conservation Theorem

If there exists a scalar function $V(\mathbf{r}, t)$, so that we may write

$$\mathbf{F} = -\text{grad } V(\mathbf{r}, t) \equiv -\left(\frac{\partial V}{\partial x_1}, \frac{\partial V}{\partial x_2}, \frac{\partial V}{\partial x_3}\right), \tag{1.3.22}$$

where grad stands for partial derivative with respect to **r**, we shall say that the vector field **F** is a *potential field*, $V(\mathbf{r}, t)$ being the *potential function* of the field. If V does not explicitly depend on time, the field **F** is called *conservative*, while

$V(\mathbf{r})$ is the *potential energy*. The work done by the particle under the action of a conservative force, when moving between any two positions P_1 and P_2, is:

$$W = \int_{P_1}^{P_2} \mathbf{F} \cdot d\mathbf{r} = -\int_{P_1}^{P_2} \frac{\partial V}{\partial x_i}\, dx_i = -\int_{P_1}^{P_2} dV = V_1 - V_2, \qquad (1.3.23)$$

where $V_1 = V(P_1), V_2 = V(P_2)$.

We also observe that the infinitesimal work done by a conservative force is an exact differential. In other words, the circulation of $\mathbf{F}$ between P_1 and P_2 does not depend on the path; it depends only on the initial and final positions of the particle. In particular, the circulation of the field $\mathbf{F} = -\mathrm{grad}\, V(\mathbf{r})$ along a closed curve is zero:

$$\oint_{(C)} \mathbf{F} \cdot d\mathbf{r} = -\oint_{(C)} dV = 0. \qquad (1.3.24)$$

Comparing (1.3.21) and (1.3.23), we deduce: $T_1 + V_1 = T_2 + V_2 = \cdots = T_n + V_n = \cdots$, or

$$T + V = \text{const.}, \qquad (1.3.25)$$

expressing the *energy conservation theorem*: *The total energy of a particle in a conservative force field is constant.*

The relation (1.3.25) is also a first integral, called *the energy first integral.*

Observation: The definition (1.3.22) does not uniquely determine the function $V(\mathbf{r}, t)$. Indeed, if we take $V'(\mathbf{r}, t) = V(\mathbf{r}, t) + \text{const.}$, we arrive at the same force $\mathbf{F}$. Hence, the choice for the zero level of V is arbitrary.

1.3.4 General Theorems for Systems of Particles

1.3.4.1 Generalities

A number of mutually interacting particles is a *system of particles*. The system is *continuous* if there is a particle in each geometrical point of the region occupied by the system, or, in other words, if at any point of the region one can define a non-zero mass density. Otherwise, the system is *discrete*. The first five chapters of this book deal with discrete and continuous particle systems having a finite number of degrees of freedom.

For compactness of writing, from now on we shall represent a sequence of consecutive positive integers between m and n (m and n being themselves positive integers, with $m<n$) by the notation $\overline{m,n}$, meaning $m, m+1, \ldots, n$. For example, the notation $i=\overline{1,N}$ means $i=1,2,\ldots,N$.

The forces acting on a N-particle system fall, as we know, into two categories: *applied* and *constraint* forces. From another point of view, these forces belong to the following two classes: *internal* and *external* forces. The internal forces act between the particles of the system and are subject to the action and reaction principle, i.e. if P_i and P_k are any two particles of the system, then (see (1.2.6))

$$\mathbf{F}_{ik} = -\mathbf{F}_{ki} \quad (i,k=\overline{1,N}). \tag{1.3.26}$$

Any force acting on the particles from outside the system is an *external force*. As an example, the solar system can be considered as a discrete system of particles; the internal forces act between the planets and the planets and Sun, while the external forces come from other celestial bodies.

1.3.4.2 Integration of the Equation of Motion

Consider a system of N free particles and let $\mathbf{r}_1, \ldots, \mathbf{r}_N$ be their respective radius-vectors relative to O. Also, let $\mathbf{F}_i^{(e)}$ and $\sum_{j=1}^{N} \mathbf{F}_{ij}$ be the resultants of external and internal forces acting on the particle P_i of mass m_i. Then the fundamental equation (1.2.4), written for this particle, is

$$m_i \ddot{\mathbf{r}}_i = \mathbf{F}_i^{(e)} + \sum_{j=1}^{N} \mathbf{F}_{ij} \quad (i=\overline{1,N}), \quad \mathbf{F}_{ii}=0, \tag{1.3.27}$$

or, in components,

$$m_i \ddot{x}_i^\alpha = F_i^{(e)\alpha} + \sum_{j=1}^{N} F_{ij}^\alpha \quad (i=\overline{1,N};\alpha=\overline{1,3}). \tag{1.3.28}$$

This is a system of $3N$ second-order differential equations. In general, the forces $\mathbf{F}_i^{(e)}$ depend on the positions of particles $\mathbf{r}_1, \ldots, \mathbf{r}_N$, their velocities $\mathbf{v}_1, \ldots, \mathbf{v}_N$, as well as the time t, while the forces $\mathbf{F}_{ij}$ are functions of the relative positions of the particles. The *general integral* of (1.3.28) is then

$$x_i^\alpha = x_i^\alpha(t, C_1, \ldots, C_{6N}) \quad (i=\overline{1,N};\alpha=\overline{1,3}). \tag{1.3.29}$$

The constants $C_1, \ldots, C_{6N}$ are determined from the initial conditions:

$$x_{i0}^\alpha = x_i^\alpha(t_0, C_1, \ldots, C_{6N}), \quad \dot{x}_{i0}^\alpha = \dot{x}_i^\alpha(t_0, C_1, \ldots, C_{6N}), \tag{1.3.30}$$

i.e. from $6N$ algebraic equations.

The integration of the differential equations (1.3.27) is facilitated, as for the one-particle mechanics, by some general theorems, which we are going to prove in the following.

1.3.4.3 Total Linear Momentum Theorem

If the summation over $i = \overline{1, N}$ in (1.3.27) is performed, we obtain

$$\frac{d}{dt}\sum_{i=1}^{N} m_i \mathbf{v}_i = \sum_{i=1}^{N} \mathbf{F}_i^{(e)} + \sum_{i=1}^{N}\sum_{j=1}^{N} \mathbf{F}_{ij}. \tag{1.3.31}$$

Let

$$\mathbf{P} = \sum_{i=1}^{N} m_i \mathbf{v}_i \tag{1.3.32}$$

be the *total linear momentum* and

$$\sum_{i=1}^{N} \mathbf{F}_i^{(e)} = \mathbf{F}^{(e)} \tag{1.3.33}$$

the resultant of exterior forces. Since, in view of (1.3.26), $\sum_i \sum_j \mathbf{F}_{ij} = 0$, we may write

$$\frac{d\mathbf{P}}{dt} = \mathbf{F}^{(e)}, \tag{1.3.34}$$

expressing the *theorem of total linear momentum*: *The time derivative of the linear momentum of a system of particles is equal to the resultant of the exterior forces acting on the system.* Once again we note that if $\mathbf{F}^{(e)} = 0$, we have the first integral $\mathbf{P} =$ const.

1.3.4.4 Total Angular Momentum Theorem

By definition, the axial vector

$$\mathbf{L} = \sum_{i=1}^{N} \mathbf{r}_i \times m_i \mathbf{v}_i \tag{1.3.35}$$

is the *total angular momentum*, or the *angular momentum of the particle system* about the point O. Taking the time derivative of (1.3.35) and using (1.3.27), we find

$$\frac{d\mathbf{L}}{dt} = \sum_{i=1}^{N} \mathbf{r}_i \times \mathbf{F}_i^{(e)} + \sum_{i=1}^{N}\sum_{j=1}^{N} \mathbf{r}_i \times \mathbf{F}_{ij}. \tag{1.3.36}$$

Due to the action and reaction principle, the last term on the r.h.s. of (1.3.36) vanishes. Indeed, for any two particles P_i and P_k $(i \neq k)$, the vectors $\mathbf{r}_{ik} = \mathbf{r}_i - \mathbf{r}_k$ and $\mathbf{F}_{ik}$ are collinear $(\mathbf{r}_{ik} = \lambda \mathbf{F}_{ik}; \lambda \neq 0)$, implying

$$\mathbf{r}_i \times \mathbf{F}_{ik} + \mathbf{r}_k \times \mathbf{F}_{ki} = (\mathbf{r}_i - \mathbf{r}_k) \times \mathbf{F}_{ik} = \mathbf{r}_{ik} \times \mathbf{F}_{ik} = 0. \tag{1.3.37}$$

It follows then from (1.3.36) that

$$\frac{d\mathbf{L}}{dt} = \mathcal{M}^{(e)}, \tag{1.3.38}$$

where

$$\mathcal{M}^{(e)} = \sum_{i=1}^{N} \mathbf{r}_i \times \mathbf{F}_i^{(e)} \tag{1.3.39}$$

is the resultant moment of exterior forces. Equation (1.3.38) expresses the *total angular momentum theorem: The time derivative of the total angular momentum is equal to the resultant moment of external forces, both momenta being taken about the same point O.* If $\mathcal{M}^{(e)} = 0$, we arrive at the first integral $\mathbf{L} = \text{const}$.

1.3.4.5 Theorem of the Total Kinetic Energy

Assume that $\mathbf{F}_i$ is the resultant of internal and external forces acting on the particle P_i of mass m_i. The work done by the system is then

$$dW = \sum_{i=1}^{N} \mathbf{F}_i \cdot d\mathbf{r}_i = \sum_{i=1}^{N} m_i \frac{d\mathbf{v}_i}{dt} \cdot \mathbf{v}_i dt = dT, \tag{1.3.40}$$

where

$$T = \frac{1}{2} \sum_{i=1}^{N} m_i |\mathbf{v}_i|^2 \tag{1.3.41}$$

is the kinetic energy of the system of particles. On the other hand, since $\mathbf{F}_i = \mathbf{F}_i^{(e)} + \sum_{j=1}^{N} \mathbf{F}_{ij}$, we may write

$$\sum_{i=1}^{N} \mathbf{F}_i \cdot d\mathbf{r}_i = \sum_{i=1}^{N} \mathbf{F}_i^{(e)} \cdot d\mathbf{r}_i + \sum_{i=1}^{N} \sum_{j=1}^{N} \mathbf{F}_{ij} \cdot d\mathbf{r}_i = dW^{(e)} + dW^{(i)}, \tag{1.3.42}$$

where $dW^{(e)}$ and $dW^{(i)}$ are the infinitesimal amounts of work done by the external and internal forces, respectively. From (1.3.40) and (1.3.42), we obtain the differential form of the *total kinetic energy theorem*,

$$dT = dW^{(e)} + dW^{(i)}, \tag{1.3.43}$$

stating that: *The differential of the total kinetic energy is equal to the sum of the infinitesimal amounts of work done by the external and internal forces.*

If one integrates (1.3.40) over a finite time interval (t_1, t_2), corresponding to two definite positions of the system, then

$$W = \sum_{i=1}^{N} \int_1^2 \mathbf{F}_i \cdot d\mathbf{r}_i = \sum_{i=1}^{N} \int_1^2 d\left(\frac{1}{2} m_i |\mathbf{v}_i|^2\right) = \left[\sum_{i=1}^{N} T_i\right]_1^2, \qquad (1.3.44)$$

which is similar to (1.3.21).

It is interesting to remark that in (1.3.43) both external and internal forces appear. Let us write $dW^{(i)}$ in a form suitable to a direct physical interpretation. Thus, since

$$\sum_{i=1}^{N}\sum_{j=1}^{N} \mathbf{F}_{ij} \cdot d\mathbf{r}_i = \sum_{i=1}^{N}\sum_{j=1}^{N} \mathbf{F}_{ji} \cdot d\mathbf{r}_j = -\sum_{i=1}^{N}\sum_{j=1}^{N} \mathbf{F}_{ij} \cdot d\mathbf{r}_j, \qquad (1.3.45)$$

we can write

$$\begin{aligned} dW^{(i)} &= \sum_{i=1}^{N}\sum_{j=1}^{N} \mathbf{F}_{ij} \cdot d\mathbf{r}_i = \frac{1}{2}\sum_{i=1}^{N}\sum_{j=1}^{N} \mathbf{F}_{ij} \cdot d\mathbf{r}_i - \frac{1}{2}\sum_{i=1}^{N}\sum_{j=1}^{N} \mathbf{F}_{ij} \cdot d\mathbf{r}_j \\ &= \frac{1}{2}\sum_{i=1}^{N}\sum_{j=1}^{N} \mathbf{F}_{ij} \cdot d\mathbf{r}_{ij}. \end{aligned} \qquad (1.3.46)$$

If the system of particles is a rigid (i.e. non-deformable) body, then $|\mathbf{r}_{ij}|^2 =$ const., i.e. $\mathbf{r}_{ij} \cdot d\mathbf{r}_{ij} = 0$. Since $\mathbf{F}_{ij}$ and $\mathbf{r}_{ij}$ are collinear, we conclude that $\mathbf{F}_{ij}$ and $d\mathbf{r}_{ij}$ are orthogonal, or, in other words: *In a rigid body, the internal forces perform no work.*

1.3.4.6 Theorem of Conservation of the Total Energy

The work done during a finite time interval (t_1, t_2) by both the internal and external forces is

$$W = W^{(e)} + W^{(i)} = \sum_{j=1}^{N} \int_1^2 \mathbf{F}_j^{(e)} \cdot d\mathbf{r}_j + \sum_{j=1}^{N}\sum_{k=1}^{N} \int_1^2 \mathbf{F}_{jk} \cdot d\mathbf{r}_j. \qquad (1.3.47)$$

If the external forces are conservative, i.e.

$$\mathbf{F}_j^{(e)} = -\mathrm{grad}_j V_j \quad (\text{no summation; } j = \overline{1,N}), \qquad (1.3.48)$$

where grad_j stands for partial derivative with respect to $\mathbf{r}_j$, then the first term of the r.h.s. of (1.3.47) reads

$$W^{(e)} = -\sum_{j=1}^{N} \int_{1}^{2} \operatorname{grad}_j V_j^{(e)} \cdot d\mathbf{r}_j = \left[\sum_{j=1}^{N} V_j^{(e)} \right]_1^2. \quad (1.3.49)$$

Suppose now that the internal forces are also conservative

$$\mathbf{F}_{jk} = -\operatorname{grad}_{jk} V_{jk}^{(i)} \quad \text{(no summation; } j,k = \overline{1,N}). \quad (1.3.50)$$

Here, grad_{jk} means partial derivative with respect to $\mathbf{r}_{jk}$, while V_{jk} is the interaction potential energy of the particles P_j and P_k. The last term on the r.h.s. of (1.3.47) then becomes

$$W^{(i)} = -\frac{1}{2} \sum_{j=1}^{N} \sum_{k=1}^{N} \int_{1}^{2} \operatorname{grad}_{jk} V_{jk}^{(i)} \cdot d\mathbf{r}_{jk} = -\frac{1}{2} \left[\sum_{j=1}^{N} \sum_{k=1}^{N} V_{jk}^{(i)} \right]_1^2. \quad (1.3.51)$$

We can define the *total potential energy*

$$V = \sum_{j=1}^{N} V_j^{(e)} + \frac{1}{2} \sum_{j=1}^{N} \sum_{k=1}^{N} V_{jk}^{(i)}, \quad (1.3.52)$$

and, using (1.3.44), (1.3.47) and (1.3.51), we finally arrive at

$$E = T + V = \text{const.}, \quad (1.3.53)$$

which is the *law of conservation and transformation of the mechanical energy*.

Observation: Our discussion refers to an ideal mechanical model. In fact, no material system is isolated in the Universe, and mechanical energy can be transformed into other forms of energy: thermal, electrical, etc. and vice versa.

1.3.4.7 Centre of Mass Theorem

Let G be the point defined by the radius-vector

$$\mathbf{r}_G = \frac{1}{M} \sum_{i=1}^{N} m_i \mathbf{r}_i, \quad (1.3.54)$$

where $M = \sum_{i=1}^{N} m_i$ is the total mass. The point G is called *centre of mass* or *centre of inertia* of the system of particles. Taking the second time derivative of (1.3.54) and using (1.3.34), we get

$$M\ddot{\mathbf{r}}_G = \mathbf{F}^{(e)}, \quad (1.3.55)$$

meaning that: *The centre of mass of a system of particles obeys the equation of motion of a point in which the entire mass of the system would be concentrated, and such that the resultant of the external forces would act there*. This is the *centre of mass theorem*. In particular, if $\mathbf{F} = 0$,

$$M\dot{\mathbf{r}}_G = \sum_{i=1}^{N} m_i \mathbf{v}_i = \text{const.}, \tag{1.3.56}$$

i.e. the centre of mass is either at rest or it moves uniformly in a straight line.

Observation: If the system of particles is continuous, the centre of mass is determined by

$$\mathbf{r}_G = \frac{1}{M} \int_V \rho(\mathbf{r}) \mathbf{r} d\tau, \tag{1.3.57}$$

where $d\tau$ is the volume element, $\rho = \frac{dm}{d\tau}$ is the *mass density* at any point of the domain occupied by the system and $M = \int_V \rho \, d\tau$ is the total mass.

1.3.4.8 König's Theorems

Now we shall establish two results due to the German mathematician Samuel König. Let us consider the motion of a system of particles relative to two reference frames $S(Oxyz)$ and $S'(Gx'y'z')$, the first being inertial and the latter non-inertial, but having its origin at the centre of mass G and its axes along fixed directions with respect to S. Then for *any* point P_i of the system is satisfied the relation

$$\mathbf{r}_i = \mathbf{r}_G + \mathbf{r}'_i. \tag{1.3.58}$$

Taking the time derivative of (1.3.58), we obtain $\mathbf{v}_i = \mathbf{v}_G + \mathbf{v}'_i$, where $\mathbf{v}'_i = d\mathbf{r}'_i/dt$ is the velocity of P_i with respect to G. Using (1.3.54) and (1.3.58), we obtain

$$\sum_{i=1}^{N} m_i \mathbf{r}'_i = 0. \tag{1.3.59}$$

In view of (1.3.58) and (1.3.59), the angular momentum of the system is (Fig. 1.5)

$$\mathbf{L} = \mathbf{r}_G \times M\mathbf{v}_G + \sum_{i=1}^{N} m_i \mathbf{r}'_i \times \mathbf{v}'_i = \mathbf{L}_G + \mathbf{L}', \tag{1.3.60}$$

known as *König's first theorem: The angular momentum of the system, relative to O, is equal to the sum of the angular momentum of the centre of mass relative to O and the angular momenta of the components of the system with respect to G.*

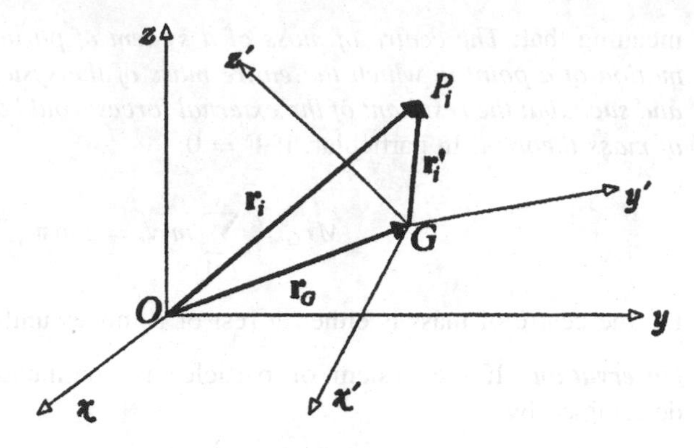

Fig. 1.5 Two reference frames: an inertial one, $S(Oxyz)$, and a non-inertial one, $S'(Gx'y'z')$, with G at the centre of mass, used to prove König's theorems (1.3.60) and (1.3.61).

In the same way, we can calculate the kinetic energy:

$$T = \frac{1}{2}M|\mathbf{v}_G|^2 + \frac{1}{2}\sum_{i=1}^{N} m_i|\mathbf{v}'_i|^2 = T_G + T', \tag{1.3.61}$$

which is *König's second theorem: The kinetic energy of the system, relative to O, is equal to the sum of the centre of mass kinetic energy with respect to O and the kinetic energy of the components of the system relative to G.*

Consequences. Using König's theorems, we shall write the angular momentum and kinetic energy theorems relative to the non-inertial frame S'. In view of (1.3.55) and (1.3.60), we have:

$$\frac{d\mathbf{L}}{dt} = \mathbf{r}_G \times \mathbf{F}^{(e)} + \frac{d\mathbf{L}'}{dt}, \quad \mathcal{M}^{(e)} = \mathbf{r}_G \times \mathbf{F}^{(e)} + \mathcal{M}'^{(e)}.$$

In this case, (1.3.38) yields

$$\frac{d\mathbf{L}'}{dt} = \mathcal{M}', \tag{1.3.62}$$

i.e. the angular momentum theorem also applies in the case of the motion relative to the centre of mass.

By virtue of (1.3.43) and (1.3.61), we have:

$$dW^{(e)} = \sum_{i=1}^{N} \mathbf{F}_i^{(e)} \cdot d\mathbf{r}_i = \mathbf{F}^{(e)} \cdot d\mathbf{r}_G + dW'^{(e)}, \tag{1.3.63}$$

$$dW^{(i)} = \sum_{i=1}^{N}\sum_{j=1}^{N} \mathbf{F}_{ij} \cdot d\mathbf{r}_i = dW'^{(i)}. \tag{1.3.64}$$

On the other hand, multiplying (1.3.55) by $d\mathbf{r}_G = \mathbf{v}_G dt$, we can also write

$$M\frac{d\mathbf{v}_G}{dt} \cdot \mathbf{v}_G dt = d\left(\frac{1}{2}M|\mathbf{v}_G|^2\right) = \mathbf{F}^{(e)} \cdot d\mathbf{r}_G, \tag{1.3.65}$$

leading to

$$dT = d\left(T' + \frac{1}{2}M|\mathbf{v}_G|^2\right) = dT' + \mathbf{F}^{(e)} \cdot d\mathbf{r}_G. \qquad (1.3.66)$$

By substituting (1.3.63)–(1.3.66) into (1.3.43), we finally get:

$$dT' = dW'^{(e)} + dW'^{(i)}, \qquad (1.3.67)$$

i.e. the kinetic energy theorem is valid also relative to G.

This analysis leads to the following conclusion: *There exists a non-inertial reference frame S' relative to which the angular momentum and kinetic energy theorems keep their form. This frame has fixed axes with respect to S and has its origin at the centre of mass of the system.* Note that, if the axes of S' are not fixed relative to S, this statement is no longer valid (see Chap. 4).

1.4 Problems

1. Study the motion of a projectile of mass m, thrown at an angle α relative to the horizontal plane, supposing that besides the force of gravity $\mathbf{G}$, a friction force $\mathbf{F}_f = -\eta\mathbf{v}$ acts on it.
2. The forces acting on a sky-diver of mass m are the force of gravity and the force of air resistance, proportional to the squared velocity. Find the diver's velocity v as a function of time and the final velocity v_f.
3. From the upper point A of a fixed sphere of radius R, a particle P of mass m begins to move without friction on the surface of the sphere. Find the distance between the lower point B of the sphere and the point C where the particle's trajectory intersects the horizontal plane.

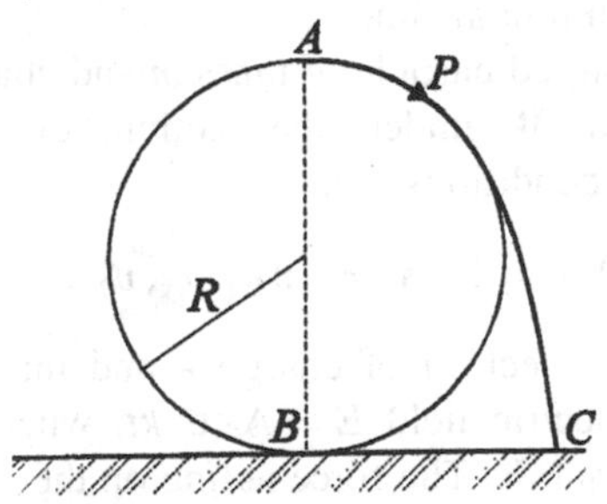

4. A rope is suspended over a massless pulley. At one end of the rope a mass m_1 is fastened, while at the other end a monkey of mass m_2 begins to climb-up, according to the law $\xi = \xi(t)$ relative to the rope. Find the motion of the monkey relative to the point O. The initial conditions are: $\xi(0) = 2l, z(0) = l$, $\dot{\xi}(0) = 0, \dot{z}(0) = 0$.

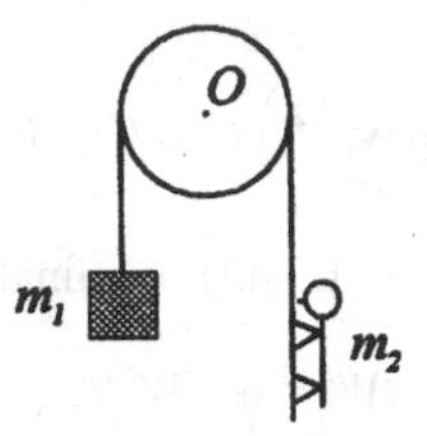

5. A particle M traces the curve given by the equations

$$x = 4\sqrt{2}\sin\theta, \quad y = \sin 2\theta.$$

(a) Determine the velocity, as well as the tangent and normal components of the acceleration of the particle with respect to the trajectory.
(b) Assuming that θ is the solution of the differential equation

$$\frac{d\theta}{dt} = \sin\theta,$$

with $\theta = \theta_0 = \frac{\pi}{2}$ at $t = 0$, give the explicit expression for θ as a function of time.
(c) Does the particle describe the entire curve if the time t varies from $-\infty$ to $+\infty$?

6. Determine the plane trajectory of a particle, whose normal and tangent components of the acceleration are constant during the motion. As initial condition, take $\theta = \theta_0$, at $t = 0$.
7. A body situated at a height h on an inclined plane of angle α is pushed downwards with the velocity $\mathbf{v}_0$ parallel to the plane. Neglecting the friction, determine the value of the angle α for which the body would arrive at the bottom of the plane in minimum time.
8. Study the motion of a charged particle of mass m and charge e, moving in the constant magnetic field $\mathbf{B}$ under the action of the Lorentz force $\mathbf{F} = e\mathbf{v} \times \mathbf{B}$. The initial conditions are:

$$\mathbf{r} = (0, R, 0), \quad \mathbf{v}_0 = (v_{0x}, v_{0y}, v_{0z}).$$

9. Find the trajectory of an electron of charge e and mass m entering in the variable homogeneous electric field $\mathbf{E} = \mathbf{A}\cos kt$, where $\mathbf{A}$ and k are constants, with the velocity $\mathbf{v}_0 \perp \mathbf{E}$. The force acting on the electron is $\mathbf{F} = -e\mathbf{E}$.
10. A particle moves without friction on the surface of a cone of angle 2α at the top, its velocity and its areal velocity being constant in the plane Ox_1x_2. Find the equations of motion of the particle.

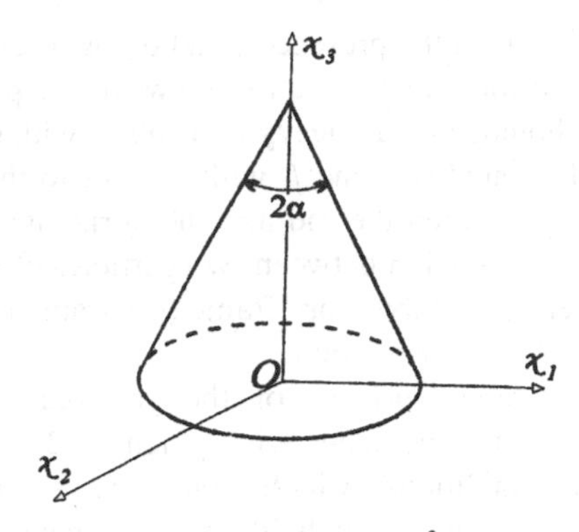

11. A particle of mass m moves on the ellipse $\frac{x^2}{a^2} + \frac{y^2}{b^2} = 1$, its acceleration being permanently directed along the y-axis. Taking as initial conditions $\mathbf{r} = (0, b)$ and $\mathbf{v} = (v_0, 0)$ at $t = 0$, determine the force acting on the particle.
12. A heavy particle moves on a vertical circle. At time $t = 0$, the particle is located at one end of the horizontal diameter, its initial velocity being zero. The velocity of the particle at the lowest point of the circle is also zero. Determine the coefficient of friction μ between the particle and the circle, knowing that the friction force is $F = \mu N$, where N is the component of the gravitational force normal to the surface and the direction of the friction force is tangent to the trajectory.
13. Solve the equation of motion of a particle for the force $F = -\frac{k}{x^3}$, if the particle is subject to the initial conditions: $x(0) = x_0 \neq 0$, $v(0) = 0$.
14. The velocity of a particle is proportional to the $(n - 1)$th power of its radius vector, while its areolar velocity is constant. Determine the acceleration of the particle and its trajectory.
15. The law of universal gravitation, giving the force between a particle of mass m and an extended object of mass $M = \int_V \rho(\mathbf{r}')d\tau'$ is

$$\mathbf{F} = -Gm \int\limits_V \frac{\rho(\mathbf{r}')(\mathbf{r} - \mathbf{r}')}{|\mathbf{r} - \mathbf{r}'|^3} d\tau',$$

where G is the gravitational constant. Find the gravitational potential energy $\varphi(\mathbf{r})$ of the object of mass m in the field of M and obtain Poisson's equation for $\varphi(\mathbf{r})$:

$$\Delta\varphi(\mathbf{r}) - 4\pi Gm\rho(\mathbf{r}) = 0.$$

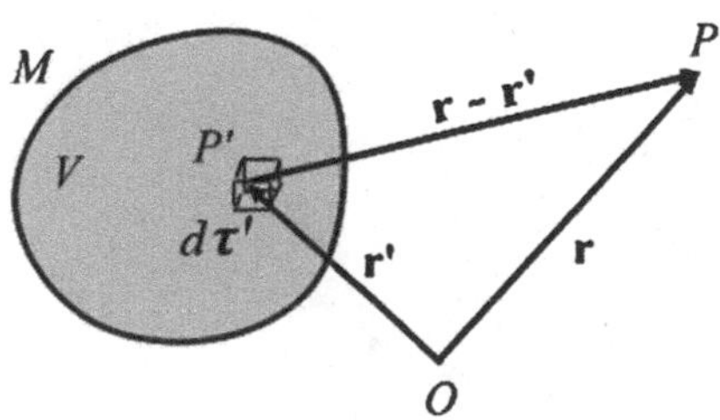

16. A point of mass m is in the presence of a homogeneous sphere of radius R and mass M. Find the force of interaction between the particle and the sphere.
17. The ends of a homogeneous heavy rod move without friction on two fixed planes, defined by angles α and β with respect to the horizontal line. Determine the angle θ between the rod and the horizontal line, at equilibrium.
18. Discuss the elastic collision between two particles of masses m_1 and m_2 in two coordinate systems: the laboratory frame (the frame in which one particle is at rest) and the centre of mass frame.
19. Find the equipotential surfaces of the gravitational field produced by a straight, finite and homogeneous wire of length $2c$ and linear density λ.
20. Determine the conditions which must be satisfied by the constants k_1, k_2, k_3, k_4, so that the force field of components $F_1 = k_1x_1 + k_2x_2$, $F_2 = k_3x_1 + k_4x_2$ (defined in the Ox_1x_2-plane) be conservative.

Chapter 2
Principles of Analytical Mechanics

2.1 Constraints

As we have already mentioned in Chap. 1, a particle (or a system of particles) is *subject to constraints* if its motion is restricted by a *constraint force* on a certain surface, or on some curve, etc. The notion of *constraint* is essential in understanding the analytical mechanics formalism, and we shall begin this chapter with a thorough analysis of this basic concept.

By definition, a constraint is a geometric or kinematic condition that limits the possibilities of motion of a mechanical system. For example, a body sliding on an inclined plane cannot leave the plane, or a pebble inside a soccer ball is compelled to move within a given volume, etc.

2.1.1 One-Particle Systems

Assuming Cartesian coordinates are used, let us begin our investigation with a single particle. If $\mathbf{r}$ is the radius-vector of the particle and $\mathbf{v}$ its velocity at time t, then a relation of the form

$$f(\mathbf{r}, \mathbf{v}, t) = 0 \tag{2.1.1}$$

is the mathematical expression for a *constraint*. One says that the particle is *subject to the constraint* (2.1.1).

We can classify the constraints according to three criteria:

(a) A constraint can be expressed either by an equality

$$f(x, y, z, t) = 0, \tag{2.1.2}$$

M. Chaichian et al., *Mechanics*, DOI: 10.1007/978-3-642-17234-2_2,

or by an inequality

$$f(x,y,z) \le 0, \quad f(x,y,z) \ge 0. \tag{2.1.3}$$

The first type of constraint is called *bilateral* and the second *unilateral*. For example, the relation

$$(x-at)^2 + (y-bt)^2 + (z-ct)^2 = R^2 \tag{2.1.4}$$

indicates that the particle is permanently on a moving sphere, with its centre at the point (*at*, *bt*, *ct*), while the inequality

$$x^2 + y^2 + z^2 - R^2 \le 0 \tag{2.1.5}$$

shows that the motion of the particle is restricted inside a fixed sphere of radius R.

(b) If the time t does not explicitly appear in the equation of the constraint, this is called a *scleronomous* or *stationary* constraint. Such a constraint is, for instance, (2.1.5). If the constraint is time-dependent, like (2.1.4), it is named a *rheonomous* or *non-stationary* constraint. An example of rheonomous constraints is provided by the system

$$f_1(x,y,z,t) = 0, \quad f_2(x,y,z,t) = 0, \tag{2.1.6}$$

meaning that the particle is forced to slide on a moving curve.

(c) A velocity-dependent constraint is called a *kinematic* or *differential* constraint, like

$$f(x,y,z,\dot{x},\dot{y},\dot{z}) = 0, \tag{2.1.7}$$

while a constraint in which the components of the velocity do not appear is named a *geometric* or *finite* constraint. For example, the constraints (2.1.2)–(2.1.6) are geometric, while (2.1.1) is kinematic. From now on, we shall consider only those differential constraints which are *linear* in the velocity components, as

$$a_i \dot{x}_i + b = 0, \tag{2.1.8}$$

where

$$a_i = a_i(\mathbf{r}, t), \quad b_i = b_i(\mathbf{r}, t) \quad (i = 1, 2, 3) \tag{2.1.9}$$

and the summation convention has been used. Taking the total time derivative of (2.1.2), we have:

$$\frac{\partial f}{\partial x_i}\dot{x}_i + \frac{\partial f}{\partial t} = 0, \tag{2.1.10}$$

meaning that a geometric constraint can be written as a linear differential constraint. Obviously, the reciprocal of this statement is not true.

Those differential constraints which can be put in a finite form are called *integrable* constraints. The geometric constraints, together with the integrable constraints, form the class of *holonomic* constraints. Such constraints are, for

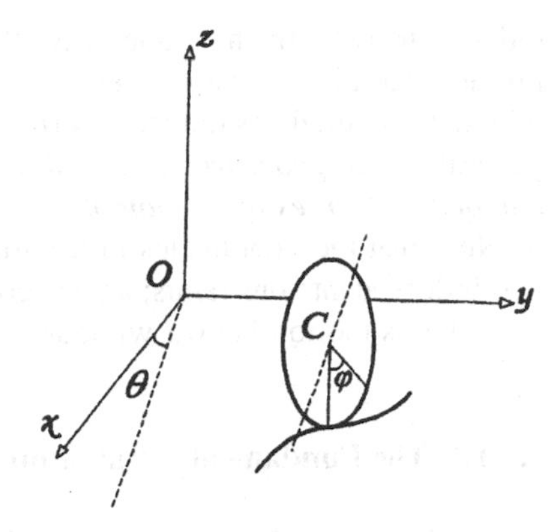

Fig. 2.1 The motion of a coin of radius a, rolling on a horizontal plane, as an example of non-integrable constraint.

example, those given by (2.1.2), (2.1.4), (2.1.6). The non-integrable constraints, together with constraints expressed by inequalities, are said to be *non-holonomic* constraints.

As an example of non-integrable constraint, let us consider a coin of radius a, rolling on a horizontal plane and keeping always a vertical position (Fig. 2.1). If θ is the angle between the x-axis and the normal to the coin, and φ is the angle of rotation of the coin, the velocity of the point C is

$$\mathbf{v} = a\dot{\varphi}\mathbf{u}, \tag{2.1.11}$$

which is permanently orthogonal to the axis of the coin. The components of the velocity are:

$$v_x = a\dot{\varphi}\cos\theta, \quad v_y = a\dot{\varphi}\sin\theta, \quad v_z = 0,$$

or, in differential form,

$$dx - a\cos\theta\, d\varphi = 0, \quad dy - a\sin\theta\, d\varphi = 0. \tag{2.1.12}$$

These two equations cannot be integrated, because their left hand sides do not represent total differentials of some functions. Consequently, they provide an example of a non-holonomic (vector) constraint.

There are no general methods of solving problems involving non-holonomic constraints. Each case must be studied separately. Fortunately, most of the problems arising in mechanics are connected with holonomic constraints.

A constraint can be characterized simultaneously upon all possible criteria. For instance, the constraint expressed by (2.1.2) is bilateral, scleronomous and geometric, while the constraint (2.1.7) is bilateral, scleronomous and differential.

There is a close relation between the number of constraints and the number of degrees of freedom of a mechanical system. The minimal number of real independent parameters that determine the position of a particle defines the *number of degrees of freedom* of that particle. A free particle, i.e. a particle subject only to applied forces, has three degrees of freedom. If the coordinates of the particle are connected by a relation of type (2.1.2), the number of its degrees of freedom reduces to two. In the same way, the existence of the two constraints (2.1.6) implies that the particle moves on a curve: the position of the particle is determined by a single parameter, corresponding to a single degree of freedom. In general, *each geometric bilateral constraint applied to a system reduces its number of degrees of freedom by one*.

Note that the coordinates of a particle cannot simultaneously obey more than two independent constraints; a third constraint would either keep the particle fixed, or make its motion known without considering the forces acting on it.

2.1.1.1 The Fundamental Equation of Motion

As we have already mentioned in Chap. 1, the existence of a constraint can be connected with a *reaction* or *constraint force*, which determines the particle to obey the constraint. If we denote by $\mathbf{F}$ and $\mathbf{L}$ the resultants of the applied and constraint forces, respectively, acting on a particle P of mass m, the differential equation of motion reads:

$$m\ddot{\mathbf{r}} = \mathbf{F} + \mathbf{L}. \tag{2.1.13}$$

The fundamental problem of mechanics of a system subject to both applied and constraint forces is: *given* $\mathbf{F}$ *and the initial conditions, consistent with the constraints, find the motion of the system and determine the reaction force* $\mathbf{L}$. The constraint force $\mathbf{L}$ is *a priori* unknown, therefore in order to use Eq. (2.1.13) one must make certain assumptions on it. The following two examples will familiarize the reader with the methods of solving problems involving constraint forces.

(1) *Motion on a curve*. First, assume that the curve, considered to be fixed, is given by its *parametric* equations:

$$x_i = x_i(q) \quad (i = 1, 2, 3), \tag{2.1.14}$$

where q is a real, time-dependent parameter. On the other hand, projecting (2.1.13) on the axes, we have:

$$m\ddot{x}_i = F_i + L_i \quad (i = 1, 2, 3), \tag{2.1.15}$$

where F_x, F_y, F_z are given as functions of $\mathbf{r}$, $\dot{\mathbf{r}}$, t or, in view of (2.1.14),

$$F_i = F_i(q, \dot{q}, t) \quad (i = 1, 2, 3). \tag{2.1.16}$$

We have arrived at a system of three second-order differential equations (2.1.15), with four unknowns, L_x, L_y, L_z, q. To solve the problem, one decomposes the constraint force $\mathbf{L}$ into two vector components, $\mathbf{L}_n$ and $\mathbf{L}_t$ (Fig. 2.2). The component $\mathbf{L}_n$ lies in the plane normal to the curve (C) at the point P, while the component $\mathbf{L}_t$ is tangent to the curve and points in the direction of motion of the particle.

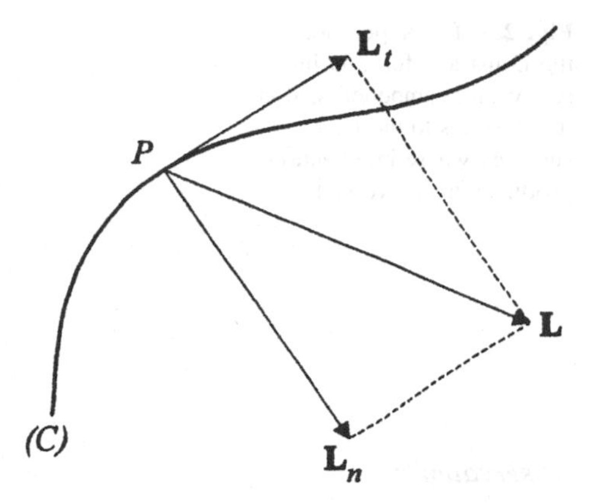

Fig. 2.2 Decomposition of the constraint force **L** into two vector components, one normal and one tangent to the trajectory.

The component $\mathbf{L}_n$ is called the *normal reaction* and $\mathbf{L}_t$ – the *force of friction*. If $\mathbf{L}_t = 0$, the particle moves *without friction* and the curve is *perfectly smooth* or *ideal*. If $\mathbf{L}_n = 0$, the force **L** is tangent to a *perfectly rough* curve.

Assuming $\mathbf{L}_t = 0$, since **v** is always directed along the tangent to the trajectory (which in our case coincides with the constraint), we can write

$$\mathbf{v} \cdot \mathbf{L} = \dot{x}L_x + \dot{y}L_y + \dot{z}L_z = 0. \tag{2.1.17}$$

We are now in possession of four equations (2.1.15) and (2.1.17) for the unknowns L_x, L_y, L_z, q. Therefore, we are able to determine both $q = q(t)$, i.e. the motion of the particle on the curve, and the components of the constraint force.

Second, let us suppose that the fixed curve is given in the implicit form

$$f_1(x, y, z) = 0, \quad f_2(x, y, z) = 0. \tag{2.1.18}$$

In this case, the differential equations (2.1.15), together with the frictionlessness condition (2.1.17) and the constraint equations (2.1.18) form a system of six equations for six unknowns: x, y, z, L_x, L_y, L_z.

The problem can be solved somewhat differently by decomposing the force **L** into two vector components, along the normals to the two surfaces whose intersection produces the curve (C) (Fig. 2.3). Then we may write

$$\mathbf{L} \equiv \mathbf{L}_n = \lambda \,\mathrm{grad}\, f_1 + \mu \,\mathrm{grad}\, f_2, \tag{2.1.19}$$

where λ and μ are two scalar multipliers. The equation of motion is then

$$m\ddot{x}_i = F_i + \lambda \frac{\partial f_1}{\partial x_i} + \mu \frac{\partial f_2}{\partial x_i} \quad (i = 1, 2, 3). \tag{2.1.20}$$

Thus, we are left with a system of five equations (2.1.18) and (2.1.20) for the unknowns x, y, z, λ, μ. In this way, both the motion of the particle and the constraint force are determined.

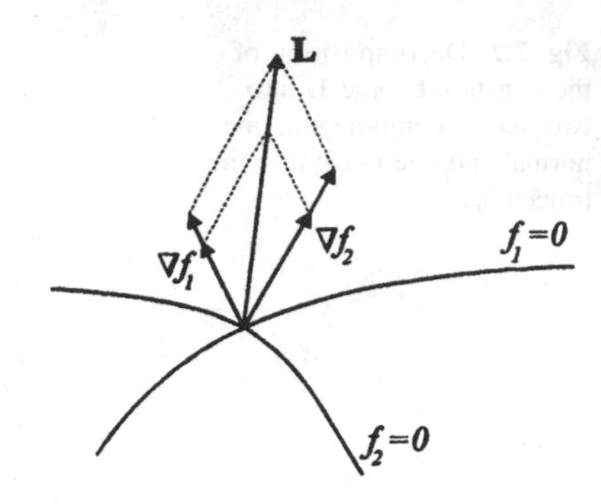

Fig. 2.3 Decomposition of the constraint force **L** into two vector components, along the normals to the two surfaces whose intersection produces the curve (*C*).

Observations:

(a) If the component $\mathbf{L}_t$ is non-zero but known, it can be included in **F**:

$$m\ddot{\mathbf{r}} = \mathbf{F} + \mathbf{L}_t + \mathbf{L}_n = \mathbf{F}' + \mathbf{L}_n \tag{2.1.21}$$

and one then follows the usual procedure.

(b) An alternative form of the equations of motion for a stationary curve is obtained by projecting (2.1.13) on the axes of a natural system of coordinates (see Appendix B):

$$m\dot{v} = F_\tau, \quad \frac{mv^2}{\rho} = F_\nu + L_\nu, \quad 0 = F_\beta + L_\beta, \tag{2.1.22}$$

where the index τ shows the tangent to the curve, ν – the principal normal and β – the bi-normal.

(2) *Motion on a surface*. Here the procedure is similar, though a little more complicated. Following the same order as in the previous case, let us first suppose that the surface is given in the parametric form

$$x_i = x_i(q^1, q^2) \quad (i = 1, 2, 3). \tag{2.1.23}$$

Since

$$\dot{\mathbf{r}} = \sum_{\alpha=1}^{2} \frac{\partial \mathbf{r}}{\partial q^\alpha} \dot{q}^\alpha, \quad \ddot{\mathbf{r}} = \sum_{\alpha,\beta=1}^{2} \frac{\partial^2 \mathbf{r}}{\partial q^\alpha \partial q^\beta} \dot{q}^\alpha \dot{q}^\beta + \sum_{\alpha=1}^{2} \frac{\partial \mathbf{r}}{\partial q^\alpha} \ddot{q}^\alpha,$$

the equation of motion (2.1.13) becomes

$$m\left(\sum_{\alpha,\beta=1}^{2} \frac{\partial^2 \mathbf{r}}{\partial q^\alpha \partial q^\beta} \dot{q}^\alpha \dot{q}^\beta + \sum_{\alpha=1}^{2} \frac{\partial \mathbf{r}}{\partial q^\alpha} \ddot{q}^\alpha \right) = \mathbf{F} + \mathbf{L} \tag{2.1.24}$$

or, in components,

$$m\left(\sum_{\alpha,\beta=1}^{2} \frac{\partial^2 x_i}{\partial q^\alpha \partial q^\beta} \dot{q}^\alpha \dot{q}^\beta + \sum_{\alpha=1}^{2} \frac{\partial x_i}{\partial q^\alpha} \ddot{q}^\alpha \right) = F_i + L_i \quad (i = 1, 2, 3). \tag{2.1.25}$$

There are five unknowns occurring in (2.1.25): q^1, q^2, L_x, L_y, L_z. To solve this problem, one decomposes **L** into two components, $\mathbf{L}_t$ and $\mathbf{L}_n$, the first being tangent to the surface and showing the direction of motion, and the second along the normal to the surface. If $\mathbf{L}_t = 0$, then $\mathbf{L} = \mathbf{L}_n$. To express this property, we observe that **L** is normal to two parametric curves on the surface, i.e.

$$L_i \frac{\partial x_i}{\partial q^\alpha} = 0 \quad (\alpha = 1, 2). \tag{2.1.26}$$

We therefore have five equations, (2.1.25) and (2.1.26), for the unknowns q^1, q^2, L_x, L_y, L_z.

If the surface is perfectly smooth, the constraint force **L** can be eliminated by multiplying (2.1.24) by $\frac{\partial \mathbf{r}}{\partial q^\gamma}$ $(\gamma = 1, 2)$. Then, in view of (2.1.26), we have:

$$m\left(\frac{\partial \mathbf{r}}{\partial q^\gamma} \cdot \sum_{\alpha,\beta=1}^{2} \frac{\partial^2 \mathbf{r}}{\partial q^\alpha \partial q^\beta} \dot{q}^\alpha \dot{q}^\beta + \frac{\partial \mathbf{r}}{\partial q^\gamma} \cdot \sum_{\alpha=1}^{2} \frac{\partial \mathbf{r}}{\partial q^\alpha} \ddot{q}^\alpha\right) = Q_\gamma \quad (\gamma = 1, 2), \tag{2.1.27}$$

where

$$Q_\gamma = \mathbf{F} \cdot \frac{\partial \mathbf{r}}{\partial q^\gamma} \quad (\gamma = 1, 2) \tag{2.1.28}$$

are the covariant components of the applied force **F** along the tangents of any two parametric lines of the surface.

The solution of (2.1.27) is

$$q^\alpha = q^\alpha(t, C_1, \ldots, C_4) \quad (\alpha = 1, 2) \tag{2.1.29}$$

and, if the initial conditions are known, the finite equations of motion can be determined. The solution (2.1.29) is then introduced into (2.1.23) and thus we obtain the motion of the particle in the real (physical) space.

Equation (2.1.27) can be set in a more condensed form by using the *metric tensor* $g_{\alpha\beta}$ (see (2.6.35)), which is defined by

$$g_{\alpha\beta} = \frac{\partial \mathbf{r}}{\partial q^\alpha} \cdot \frac{\partial \mathbf{r}}{\partial q^\beta} \quad (\alpha, \beta = 1, 2). \tag{2.1.30}$$

We have:

$$\frac{\partial g_{\alpha\beta}}{\partial q^\gamma} = \frac{\partial^2 \mathbf{r}}{\partial q^\alpha \partial q^\gamma} \cdot \frac{\partial \mathbf{r}}{\partial q^\beta} + \frac{\partial \mathbf{r}}{\partial q^\alpha} \cdot \frac{\partial^2 \mathbf{r}}{\partial q^\beta q^\gamma}$$

and, making a cyclic permutation of the indices α, β, γ and then combining the three obtained relations, we get

$$\frac{\partial \mathbf{r}}{\partial q^\gamma} \cdot \frac{\partial^2 \mathbf{r}}{\partial q^\alpha \partial q^\beta} = \Gamma_{\alpha\beta,\gamma} \quad (\alpha, \beta, \gamma = 1, 2), \tag{2.1.31}$$

where the quantities

$$\Gamma_{\alpha\beta,\gamma} = \frac{1}{2}\left(\frac{\partial g_{\beta\gamma}}{\partial q^\alpha} + \frac{\partial g_{\gamma\alpha}}{\partial q^\beta} - \frac{\partial g_{\alpha\beta}}{\partial q^\gamma}\right) \tag{2.1.32}$$

are called the *Christoffel symbols of the first kind*. By virtue of (2.1.31), Eq. (2.1.27) can be put in the form

$$m\left(\sum_{\alpha=1}^{2} g_{\gamma\alpha}\ddot{q}^\alpha + \sum_{\alpha,\beta=1}^{2} \Gamma_{\alpha\beta,\gamma}\dot{q}^\alpha\dot{q}^\beta\right) = Q_\gamma \quad (\gamma = 1, 2). \tag{2.1.33}$$

If we multiply this equation by $g^{\gamma\sigma}$ and perform the summation over γ, we find

$$m\left(\ddot{q}^\sigma + \sum_{\alpha,\beta=1}^{2} \Gamma^\sigma_{\alpha\beta}\dot{q}^\alpha\dot{q}^\beta\right) = Q^\sigma \quad (\sigma = 1, 2), \tag{2.1.34}$$

where

$$\Gamma^\sigma_{\alpha\beta} = \sum_{\gamma=1}^{2} g^{\gamma\sigma}\Gamma_{\alpha\beta,\gamma} \tag{2.1.35}$$

are the *Christoffel symbols of the second kind* and

$$Q^\sigma = \sum_{\gamma=1}^{2} g^{\sigma\gamma} Q_\gamma \tag{2.1.36}$$

are the contravariant components of the quantities (2.1.28).

If no applied force **F** acts on the particle, the kinetic energy theorem implies that the particle moves on the surface (2.1.23) with constant speed. In this case, the acceleration vector **a** is oriented along the principal normal to the trajectory which, for $\mathbf{L}_t = 0$, coincides with the normal to the surface. The equations of equilibrium of the particle, written in a *geodetic form*, are then

$$\ddot{q}^\sigma + \sum_{\alpha,\beta=1}^{2} \Gamma^\sigma_{\alpha\beta}\dot{q}^\alpha\dot{q}^\beta = 0 \quad (\sigma = 1, 2). \tag{2.1.37}$$

As an example, let us take $q^1 = \theta$, $q^2 = \varphi$. Then the metric (see Appendix B)

$$ds^2 = r^2 d\theta^2 + r^2 \sin^2\theta\, d\varphi^2 \quad (r = \text{const.})$$

yields

$$g_{11} = r^2, \quad g_{22} = r^2 \sin^2\theta, \quad g_{12} = 0$$

and, together with the condition

$$\sum_{\alpha=1}^{2} g_{\alpha\beta} g^{\alpha\gamma} = \delta_{\beta}^{\gamma},$$

we find

$$g^{11} = \frac{1}{r^2}, \quad g^{22} = \frac{1}{r^2 \sin^2\theta}, \quad g^{12} = 0.$$

We are now able to calculate the Christoffel symbols of the second kind, $\Gamma_{\alpha\beta}^{\gamma}$. The only non-zero symbols are

$$\Gamma_{22}^{1} = -\sin\theta\cos\varphi, \quad \Gamma_{12}^{2} = \Gamma_{21}^{2} = \cot\theta$$

and Eqs. (2.1.37) read

$$\begin{aligned} \ddot{\theta} - \sin\theta\cos\varphi\dot{\varphi}^2 &= 0, \\ \ddot{\varphi} + \cot\theta\dot{\theta}\dot{\varphi} &= 0. \end{aligned}$$

We easily recognize the components of the acceleration vector (1.1.17) along the parametric lines θ, φ, taken for $r =$ const.

If, as a second example, we choose $q^1 = x, q^2 = y$, we have $g_{\alpha\beta} = \delta_{\alpha\beta}$, which leads to the equation of motion

$$\ddot{x} = 0, \ \ddot{y} = 0, \tag{2.1.38}$$

as expected.

If the fixed, ideal surface is given under the implicit form

$$f(x, y, z) = 0, \tag{2.1.39}$$

the components of the fundamental equation of motion are

$$m\ddot{x}_i = F_i + \lambda\frac{\partial f}{\partial x_i} \quad (i = 1, 2, 3). \tag{2.1.40}$$

The three equations (2.1.40), together with the equation of constraint (2.1.39), form a system of four equations in the unknowns x, y, z, λ. Both the motion and the constraint forces can then be determined.

2.1.1.2 Static Equilibrium of a Particle

(1) ***Free particle***. A point mass m is in *equilibrium* relative to a certain frame if the resultant of the forces acting on it is zero

$$m\ddot{\mathbf{r}} = \mathbf{F} = 0. \tag{2.1.41}$$

The equilibrium positions are determined by solving the system of three equations with three unknowns

$$m\ddot{x}_i = F_i = 0. \tag{2.1.42}$$

If, in particular, the solution of the system (2.1.42) is unique, we have only one position of equilibrium.

Assume that the particle is subject to a conservative force field

$$\mathbf{F} = -\operatorname{grad} V = \operatorname{grad} U, \tag{2.1.43}$$

where we denote $U(\mathbf{r}) = -V(\mathbf{r})$. We also assume that $U(\mathbf{r})$ is a function of class C^2. The last two relations give

$$\frac{\partial U}{\partial x_i} = 0 \quad (i = 1, 2, 3). \tag{2.1.44}$$

Consequently, in order for the position P_0 of the particle to be a position of equilibrium, it is necessary to have

$$\left(\frac{\partial U}{\partial x_i}\right)_{P_0} = 0 \quad (i = 1, 2, 3), \tag{2.1.45}$$

meaning that in P_0 the function $U(\mathbf{r})$ has either an extremum, or an inflection point.

A position P_0 of the particle P is a position of *stable equilibrium* if, setting the particle in a position P_1 close to P_0, and giving it a sufficiently small initial velocity $\mathbf{v}_0$, the trajectory of the particle remains in an infinitely small sphere. In other words, the displacement of the particle from the equilibrium position is infinitely small. More rigorously, for any $\varepsilon > 0$ there correspond the functions $\eta_1(\varepsilon) > 0, \eta_2(\varepsilon) > 0$, such that, if $|\overrightarrow{P_0P_1}| < \eta_1(\varepsilon)$ and $|\mathbf{v}_0| < \eta_2(\varepsilon)$, then $|\overrightarrow{P_0P}| < \varepsilon$ for any t.

The position P_0 is a position of maximum (or minimum) for $U(P)$, if there is a vicinity Q_{P_0} of P_0 in which $U(P) \leq U(P_0)$ (or $U(P) \geq U(P_0)$), for any $P \in Q_{P_0}$. When these conditions are fulfilled without the "equal" sign, we have a *strict maximum (minimum)*.

Using these definitions, we shall now demonstrate the *Lagrange–Dirichlet theorem: If in the position P_0 the function U has a strict maximum, then P_0 is a position of stable equilibrium.*

The proof begins with the observation that, since $U(P_0) = \text{max.}$, then $V(P_0) = \text{min.}$ But, as we know, the origin of the potential energy V can be arbitrarily chosen, so that we can take $V(P_0) = 0$. Consequently, there exists a vicinity Q_{P_0} of P_0 (except for P_0) for which $V(P) > 0$. Let L be the maximum value of $V(P)$ on the boundary of the domain Q_{P_0} and let us choose a vicinity $Q'_{P_0} \subset Q_{P_0}$ of P_0, so that $V(P) < \frac{L}{2}$ for any $P \in Q'_{P_0}$. Suppose that at the initial time the particle is in $P_1 \in Q'_{P_0}$ and has the velocity $\mathbf{v}_1$, chosen so as to have $\frac{1}{2}m|\mathbf{v}_1|^2 < \frac{L}{2}$. Applying the kinetic energy theorem (1.3.21), we have:

$$\frac{1}{2}m|\mathbf{v}|^2 - \frac{1}{2}m|\mathbf{v}_1|^2 = U(P) - U(P_1)$$

or, since the system is conservative,

$$\frac{1}{2}m|\mathbf{v}|^2 + V(P) = \frac{1}{2}m|\mathbf{v}_1|^2 + V(P_1) = \text{const.}$$

Thus, we may write

$$\frac{1}{2}m|\mathbf{v}|^2 + V(P) < \frac{L}{2} + \frac{L}{2} = L.$$

The quantities $T(P) = \frac{1}{2}m|\mathbf{v}|^2$ and $V(P)$ are positive, therefore $T(P) < L$, $V(P) < L$, showing that the velocity of the particle in position P cannot be greater than a certain value. Consequently the particle, starting from the position P_1, will never touch the boundary of the domain Q_{P_0}. Recalling the definition of stable equilibrium, the proof is completed.

(2) ***Particle subject to constraints***

(a) *Equilibrium on a surface*. Assuming again a perfectly smooth (ideal) surface, in order for a particle to be in equilibrium it must obey the equation

$$m\ddot{\mathbf{r}} = \mathbf{F} + \mathbf{L}_n = 0. \tag{2.1.46}$$

In other words, the particle is in equilibrium relative to the surface if the resultant of the applied forces is directed along the normal to the surface. If the surface is given in the parametric form (2.1.23), the equations of equilibrium are:

$$F_x + \lambda\frac{\partial(y,z)}{\partial(q^1,q^2)} = 0, \quad F_y + \lambda\frac{\partial(z,x)}{\partial(q^1,q^2)} = 0, \quad F_z + \lambda\frac{\partial(x,y)}{\partial(q^1,q^2)} = 0, \tag{2.1.47}$$

where the functional determinants $\frac{\partial(y,z)}{\partial(q^1,q^2)}$, $\frac{\partial(z,x)}{\partial(q^1,q^2)}$, $\frac{\partial(x,y)}{\partial(q^1,q^2)}$ are the direction parameters of the normal to the surface. The determinant $\frac{\partial(y,z)}{\partial(q^1,q^2)}$, for example, is calculated by

$$\frac{\partial(y,z)}{\partial(q^1,q^2)} = \begin{vmatrix} \frac{\partial y}{\partial q^1} & \frac{\partial y}{\partial q^2} \\ \frac{\partial z}{\partial q^1} & \frac{\partial z}{\partial q^2} \end{vmatrix}.$$

If the surface is given under the implicit form (2.1.39), the equations of equilibrium are

$$F_i + \lambda\frac{\partial f}{\partial x_i} = 0 \quad (i = 1,2,3). \tag{2.1.48}$$

(b) *Equilibrium on a curve*. Following the same procedure, we first consider the case where the curve is given under the parametric form as in (2.1.14). Let x', y', z' be the direction parameters of the tangent to the curve in the point where the particle is. Then, the constraint force $\mathbf{L}$ is normal to the curve if

$$x_i' L_i = 0. \tag{2.1.49}$$

This condition is identically satisfied by the choice:

$$L_x = \lambda z' - \mu y', \quad L_y = \mu x' - \nu z', \quad L_z = \nu y' - \lambda x', \tag{2.1.50}$$

where λ, μ, ν are three arbitrary parameters.

Finally, if the curve is expressed by its implicit equations (2.1.18), in view of (2.1.20), the equilibrium condition reads:

$$F_i + \lambda \frac{\partial f_1}{\partial x_i} + \mu \frac{\partial f_2}{\partial x_i} = 0 \quad (i = 1, 2, 3). \tag{2.1.51}$$

Example. Let us find the equilibrium position of a heavy particle, sliding without friction on a fixed circle of radius R, situated in a vertical plane. The circle can be conceived as given by the intersection of a sphere of radius R and a plane passing through its centre. Choosing the origin of the coordinate system in the centre of the sphere and the x-axis along the descendent vertical, the equations of the circle are

$$f_1(x, y, z) \equiv x^2 + y^2 + z^2 - R^2 = 0, \quad f_2(x, y, z) \equiv z = 0. \tag{2.1.52}$$

The equilibrium positions are obtained by eliminating λ and μ from (2.1.51). Multiplying this equation by $\epsilon_{ijk} \frac{\partial f_1}{\partial x_j} \frac{\partial f_2}{\partial x_k}$, we have:

$$\begin{vmatrix} F_x & F_y & F_z \\ \frac{\partial f_1}{\partial x} & \frac{\partial f_1}{\partial y} & \frac{\partial f_1}{\partial z} \\ \frac{\partial f_2}{\partial x} & \frac{\partial f_2}{\partial y} & \frac{\partial f_2}{\partial z} \end{vmatrix} = 0. \tag{2.1.53}$$

But $F_x = G = mg$, $F_y = F_z = 0$, therefore (2.1.53) gives $y = 0$. These results, when introduced into the first equation of (2.1.52), produce the following two conditions of equilibrium:

$$x = \pm R. \tag{2.1.54}$$

On the other hand, projecting Eq. (2.1.43) on axes, we have:

$$U = mgx + \text{const}. \tag{2.1.55}$$

One observes that U has a maximum for $x = R$ and a minimum for $x = -R$. According to the Lagrange–Dirichlet theorem, the position $P_0(x = +R, y = 0)$ is a position of *stable* equilibrium, while $P_1(x = -R, y = 0)$ is a position of *unstable* equilibrium.

2.1.2 Many-Particle Systems

Let $P_1,\ldots,P_N$ be a system of N particles. At any moment t, the radius-vectors of the particles $\mathbf{r}_1,\ldots,\mathbf{r}_N$ and their velocities $\dot{\mathbf{r}}_1,\ldots,\dot{\mathbf{r}}_N$ can take arbitrary values. A relation of the form

$$f(\mathbf{r}_1,\ldots,\mathbf{r}_N,\dot{\mathbf{r}}_1,\ldots,\dot{\mathbf{r}}_N,t)=0 \tag{2.1.56}$$

is a *constraint* which restricts the motion of the particles. The criteria of classification of constraints for many-particle systems are similar to those encountered in the case of a single particle. For example, the relations

$$f_k(\mathbf{r}_1,\ldots,\mathbf{r}_N,t)=0 \quad (k=\overline{1,s};\ s\le 3N) \tag{2.1.57}$$

express s *bilateral, rheonomous, geometric constraints.* They are also *holonomic* constraints. The number of constraints cannot exceed $3N$; in the case $s=3N$, the N vectors $\mathbf{r}_1,\ldots,\mathbf{r}_N$ would be completely determined by the constraints.

As for a single particle, we consider only those differential constraints which are linear in the velocities:

$$\sum_{i=1}^{N}\mathbf{g}_i^k(\mathbf{r}_1,\ldots,\mathbf{r}_N,t)\cdot\dot{\mathbf{r}}_i+g_0^k(\mathbf{r}_1,\ldots,\mathbf{r}_N,t)=0 \quad (k=\overline{1,s}). \tag{2.1.58}$$

It is seen that (2.1.57) can be written in a form similar to (2.1.58). Indeed, taking its total time derivative, we arrive at

$$\sum_{i=1}^{N}(\mathrm{grad}_i f_k)\cdot\dot{\mathbf{r}}_i+\frac{\partial f_k}{\partial t}=0 \quad (k=\overline{1,s}). \tag{2.1.59}$$

The constraints (2.1.58) can be integrable (holonomic) or non-integrable (non-holonomic). The non-integrable constraints are also called *Pfaffian.*

The fundamental equation of motion, written for the particle P_i of mass m_i of the system, is

$$m_i\ddot{\mathbf{r}}_i=\mathbf{F}_i+\mathbf{L}_i \quad (i=\overline{1,N}), \tag{2.1.60}$$

where $\mathbf{L}_i$ is the resultant of the constraint forces acting on the particle.

Assuming that the holonomic constraints (2.1.57) are ideal, we can generalize the relation (2.1.40) by multiplying (2.1.57) by $\lambda_k(t)$, performing the summation over k, and introducing the result into (2.1.60):

$$m_i\ddot{\mathbf{r}}_i=\mathbf{F}_i+\sum_{k=1}^{s}\lambda_k\mathrm{grad}_i f_k \quad (i=\overline{1,N}). \tag{2.1.61}$$

Equations (2.1.61), together with the constraints (2.1.57), represent $3N+s$ equations in the unknowns $\mathbf{r}_1,\ldots,\mathbf{r}_N$ ($3N$ coordinates) and $\lambda_1,\ldots,\lambda_s$. Equations (2.1.61) are called the *Lagrange equations of the first kind.* They are due to the

Italian-French mathematician *Joseph-Louis Lagrange*, as are many other results, concepts and formalisms which we shall encounter further in this book.

Observation: The problems involving static equilibrium of mechanical systems of particles are discussed in a way similar to that used for a single particle. Notice, nevertheless, that special care must be taken in the case of interacting particles.

2.2 Elementary Displacements

To determine the equilibrium conditions of a system of N particles subject to constraints using the method developed in the previous section, one must separately study the equilibrium of each particle, taking into account that the constraint forces are *a priori* unknown. If the number of particles is large, we have many equations with many unknowns. In this case the aforementioned procedure becomes complicated.

We shall now give a more general and more useful method for solving both dynamic and static problems of mechanics. The main difference from the already known formalism is that the effect of constraints is expressed not by constraint forces, but rather by *elementary displacements* associated with these forces.

Assume that our system is subject to s holonomic, scleronomous constraints

$$f_k(\mathbf{r}_1, \ldots, \mathbf{r}_N) = 0 \quad (k = \overline{1, s}). \tag{2.2.1}$$

Being under the action of applied forces, the particles perform certain displacements which must be *consistent* with the constraints. Let $d\mathbf{r}_i$ be the infinitesimal displacement of the particle P_i during the time interval dt, subject to the applied forces and the initial conditions, and consistent with the constraints. Such a displacement takes place *effectively*, during the time interval dt, being *unique*. It is a *real displacement*. But, if we only fix the position of the particle at time t, we can have an infinite number of velocities $\dot{\mathbf{r}}_1, \ldots, \dot{\mathbf{r}}_N$, consistent with the constraints (2.2.1). The displacements performed by particles under these conditions are called *possible*. The real displacements belong to the multitude of possible displacements, being the subset that satisfies both the equations of motion and the initial conditions.

Now, let us consider a system of displacements $\delta\mathbf{r}_i$ $(i = \overline{1, N})$ that obey only one condition: they are consistent with the constraints. These purely geometric displacements are *synchronic*, i.e. they are taken at an instant t $(\delta t = 0)$. These are usually called *virtual displacements*.

By differentiating (2.2.1), we get

$$\sum_{i=1}^{N} (\mathrm{grad}_i f_k) \cdot d\mathbf{r}_i = 0 \quad (k = \overline{1, s}), \tag{2.2.2}$$

meaning that all the real (or possible) displacements $d\mathbf{r}_i$ lie in the planes tangent to the surfaces $f_1 = 0, \ldots, f_s = 0$. Using the definition of virtual displacements $\delta\mathbf{r}_i$, we infer also:

$$\sum_{i=1}^{N} (\text{grad}_i f_k) \cdot \delta \mathbf{r}_i = 0 \quad (k = \overline{1, s}), \tag{2.2.3}$$

showing that any virtual displacement can become a possible one.

Passing now to the rheonomous constraints, given, for example, by Eq. (2.1.56), we realize that the possible displacements $d\mathbf{r}_i$ $(i = \overline{1, N})$ must obey the relation

$$\sum_{i=1}^{N} (\text{grad}_i f_k) \cdot d\mathbf{r}_i + \frac{\partial f_k}{\partial t} dt = 0 \quad (k = \overline{1, s}), \tag{2.2.4}$$

while the virtual displacements $\delta \mathbf{r}_i$ satisfy an equation similar to Eq. (2.2.3). Writing (2.2.4) for two sets of possible displacements $d\mathbf{r}_i'$ and $d\mathbf{r}_i''$, and then subtracting the obtained relations, we arrive precisely at Eq. (2.2.3), where

$$\delta \mathbf{r}_i = d\mathbf{r}_i' - d\mathbf{r}_i''. \tag{2.2.5}$$

Therefore, *any virtual displacement can be considered as the difference between two possible displacements*. For example, consider a spherical balloon with a fixed centre, taken in the process of inflation (Fig. 2.4). At the moments $t_1 < t_2 < t_3 \ldots$, the radii of the balloon will be $R_1 < R_2 < R_3 \ldots$. An ant moving on the balloon, and being at time t_1 in the position P, could be at time $t_2 > t_1$ in any position P', P'', P''' etc., on the sphere of radius $R_2 > R_1$. The displacements $d\mathbf{r}'$, $d\mathbf{r}''$, $d\mathbf{r}'''$, etc. are *possible* displacements. Depending on the initial conditions (the ant is considered a mechanical system), only one of these displacements is *real*. Any *virtual* displacement $\delta \mathbf{r}$ at time t_2 lies in the plane tangent to the sphere of radius R_2, and obeys the rule (2.2.5). The virtual displacements are *atemporal*, in our example being any displacement on the balloon surface, taken at an instant, while R is fixed.

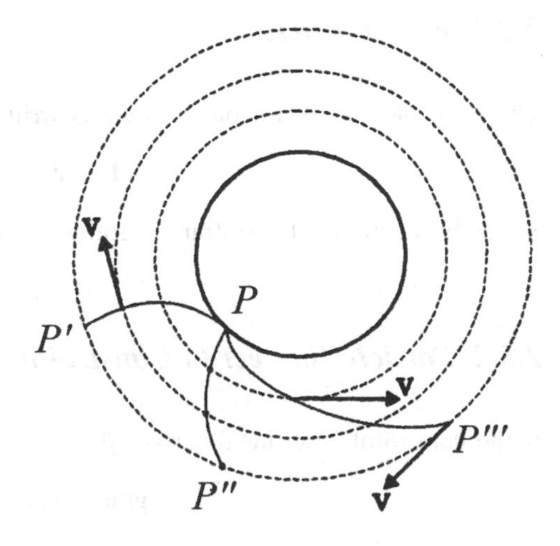

Fig. 2.4 Intuitive examples of real, possible and virtual displacements.

2.3 Principle of Virtual Work

A method providing a very efficient way of eliminating the constraint forces appearing in a mechanical problem is the *principle of virtual work*. Assume again that a system of N particles $P_1, \ldots, P_N$ is in static equilibrium and subject to ideal constraints. If $\delta \mathbf{r}_i$ is a virtual displacement of the particle P_i, consistent with the constraints, then, by definition,

$$\delta W = \mathbf{F}_i \cdot \delta \mathbf{r}_i \tag{2.3.1}$$

is the *virtual work* of the force $\mathbf{F}_i$ relative to the displacement $\delta \mathbf{r}_i$. In the case of static equilibrium ($\ddot{\mathbf{r}}_i = 0$), multiplying (2.1.60) by $\delta \mathbf{r}_i$ and summing for all the particles of the system, yields:

$$\sum_{i=1}^{N} \mathbf{F}_i \cdot \delta \mathbf{r}_i = 0, \tag{2.3.2}$$

where we have used the property of ideal constraints

$$\sum_{i=1}^{N} \mathbf{L}_i \cdot \delta \mathbf{r}_i = 0. \tag{2.3.3}$$

Relation (2.3.2) expresses the *principle of virtual work: The necessary and sufficient condition for static equilibrium of a scleronomous system subject to ideal constraints is that the virtual work of the applied forces, for virtual displacements consistent with the constraints, be zero.* If the particles were free, the displacements $\delta \mathbf{r}_i$ would be arbitrary.

Let us now show that from the principle of virtual work all the conditions of equilibrium discussed in Sect. 2.1 can be derived.

2.3.1 *Free Particle*

The principle (2.3.2) for one particle is written as

$$\mathbf{F} \cdot \delta \mathbf{r} = 0. \tag{2.3.4}$$

Since $\delta \mathbf{r}$ is completely arbitrary, it follows that $\mathbf{F} = 0$, in agreement with (2.1.41).

2.3.2 *Particle Subject to Constraints*

If the constraint is an ideal surface $f(x, y, z) = 0$, then the condition

$$\text{grad } f \cdot \delta \mathbf{r} = 0 \tag{2.3.5}$$

expresses the fact that the particle lies on the surface. Multiplying (2.3.5) by some scalar λ and adding the result to (2.3.4), we fall back on the relation (2.1.48). In the case of an ideal curve, the condition (2.3.4) must be completed with

$$\operatorname{grad} f_1 \cdot \delta\mathbf{r} = 0, \quad \operatorname{grad} f_2 \cdot \delta\mathbf{r} = 0, \tag{2.3.6}$$

leading together to (2.1.51).

2.3.3 System of Free Particles

For arbitrary virtual displacements $\delta\mathbf{r}_i$, we have

$$\sum_{i=1}^{N} \mathbf{F}_i \cdot \delta\mathbf{r}_i = 0, \tag{2.3.7}$$

yielding the conditions of equilibrium

$$\mathbf{F}_i = 0 \quad (i = \overline{1, N}), \tag{2.3.8}$$

which are also obtained from (2.1.60) for $\ddot{\mathbf{r}}_i = 0$, $\mathbf{L}_i = 0$. Notice that, by using the principle of virtual work, the N relations (2.3.8) are replaced by a single relation (2.3.7).

2.3.4 System of Particles Subject to Constraints

Assuming that the constraints are given by (2.1.57), we may write:

$$\sum_{i=1}^{N} \left(\frac{\partial f_k}{\partial x_i} \delta x_i + \frac{\partial f_k}{\partial y_i} \delta y_i + \frac{\partial f_k}{\partial z_i} \delta z_i \right) = 0 \quad (k = \overline{1, s}). \tag{2.3.9}$$

From (2.3.2) we have also

$$\sum_{i=1}^{N} (X_i \delta x_i + Y_i \delta y_i + Z_i \delta z_i) = 0, \tag{2.3.10}$$

where X_i, Y_i, Z_i are the components of the force $\mathbf{F}_i$. The displacements $\delta x_i, \delta y_i, \delta z_i$ are not arbitrary anymore, but they must obey the s relations (2.3.9).

The $(s + 1)$ equations (2.3.9) and (2.3.10) can be written as a single relation by using the method of *Lagrange multipliers*. Let us amplify each of the equations (2.3.9) by λ_k, then perform the summation over the index k and add the result to (2.3.10). We obtain:

$$\sum_{i=1}^{N}\left[\left(X_i+\sum_{k=1}^{s}\lambda_k\frac{\partial f_k}{\partial x_i}\right)\delta x_i+\left(Y_i+\sum_{k=1}^{s}\lambda_k\frac{\partial f_k}{\partial y_i}\right)\delta y_i+\left(Z_i+\sum_{k=1}^{s}\lambda_k\frac{\partial f_k}{\partial z_i}\right)\delta z_i\right]$$
$$=0. \tag{2.3.11}$$

This relation must be satisfied by *any* $\delta x_i, \delta y_i, \delta z_i$ $(i=\overline{1,N})$. Since these variations must obey the s linear homogeneous equations (2.3.10), it follows that $3N-s$ of these displacements can be taken as being independent. Then, in (2.3.11) are determined λ_k so that the parentheses which multiply the s dependent displacements are zero, leading to a number of s equations. Next, we make vanish the parentheses multiplying the $3N-s$ independent displacements, and get more $3N-s$ equations. Finally, we are left with $3N$ equations:

$$X_i+\sum_{k=1}^{s}\lambda_k\frac{\partial f_k}{\partial x_i}=0,\quad Y_i+\sum_{k=1}^{s}\lambda_k\frac{\partial f_k}{\partial y_i}=0,\quad Z_i+\sum_{k=1}^{s}\lambda_k\frac{\partial f_k}{\partial z_i}=0\quad (i=\overline{1,N}). \tag{2.3.12}$$

The $3N$ equations (2.3.12), together with the s equations (2.1.57), form a system of $3N+s$ equations for $3N+s$ unknowns: the equilibrium coordinates x_i, y_i, z_i $(i=\overline{1,N})$ and the multipliers $\lambda_1,\ldots,\lambda_s$.

Observation: The principle of virtual work applies to the study of the equilibrium conditions of a *rigid body* as well. Anticipating, we shall use (4.3.10) to write the velocity $\mathbf{v}_i$ of a particle P_i of the rigid body, relative to a fixed frame $Oxyz$:

$$\mathbf{v}_i=\mathbf{v}_0+\boldsymbol{\omega}\times\mathbf{r}_i'. \tag{2.3.13}$$

Here, $\mathbf{v}_0$ is the velocity of some particle O' of the body and $\mathbf{r}_i'$ the radius-vector of P_i relative to O'. If $\delta\mathbf{r}_i'$ is a virtual displacement of P_i, consistent with the rigidity constraints, we can write:

$$\mathbf{v}_i=\frac{\delta\mathbf{r}_i}{\delta t},\quad \mathbf{v}_0=\frac{\delta\mathbf{r}_0}{\delta t} \tag{2.3.14}$$

and (2.3.13) becomes

$$\delta\mathbf{r}_i=\delta\mathbf{r}_0+(\boldsymbol{\omega}\times\mathbf{r}_i')\delta t. \tag{2.3.15}$$

The principle of virtual work (2.3.2) reads then:

$$\sum_{i=1}^{N}\mathbf{F}_i\cdot\delta\mathbf{r}_i=\delta\mathbf{r}_0\cdot\sum_{i=1}^{N}\mathbf{F}_i+\delta t\,\boldsymbol{\omega}\cdot\sum_{i=1}^{N}\mathbf{r}_i'\times\mathbf{F}_i=0, \tag{2.3.16}$$

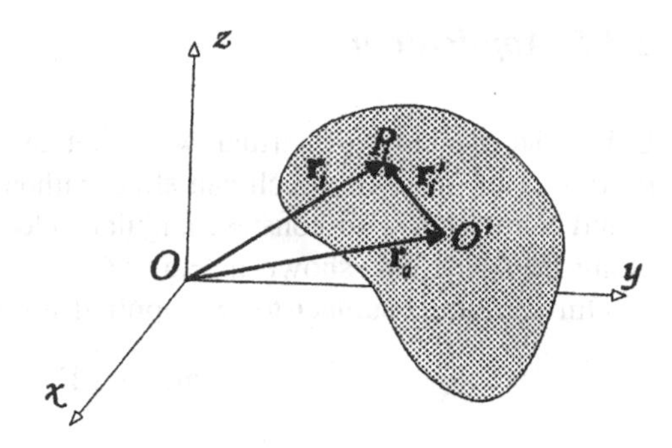

Fig. 2.5 Choice of the systems of coordinates to find the equilibrium conditions of a free rigid body (2.3.20).

being true for completely arbitrary variations $\delta\mathbf{r}_0$ and $\boldsymbol{\omega}\delta t$. Therefore, we obtain:

$$\sum_{i=1}^{N}\mathbf{F}_i = \mathbf{F} = 0, \quad \sum \mathbf{r}'_i \times \mathbf{F}_i = \mathcal{M}' = 0, \tag{2.3.17}$$

where the torque $\mathcal{M}'$ is taken relative to O'. But $\mathbf{r}'_i = \mathbf{r}_i - \mathbf{r}_0$ (see Fig. 2.5), so that

$$\sum_{i=1}^{N}\mathbf{r}'_i \times \mathbf{F}_i = \sum_{i=1}^{N}\mathbf{r}_i \times \mathbf{F}_i - \mathbf{r}_0 \times \sum_{i=1}^{N}\mathbf{F}_i = 0 \tag{2.3.18}$$

and, using (2.3.17),

$$\mathcal{M} = \sum_{i=1}^{N}\mathbf{r}_i \times \mathbf{F}_i = 0. \tag{2.3.19}$$

Since O is arbitrary, we conclude that $\mathcal{M}$ can be taken with respect to *any* point. Therefore, the equilibrium conditions of a free rigid body are:

$$\mathbf{F} = 0, \quad \mathcal{M} = 0. \tag{2.3.20}$$

Note that any point of a free rigid body has two independent virtual vector displacements, $\delta\mathbf{r}_0$ and $\boldsymbol{\omega}\delta t$, equivalent to six components. Consequently, a free rigid body possesses six degrees of freedom.

The equilibrium conditions for a rigid body subject to constraints are obtained in a similar way. For instance, if the body has a fixed point, say O', the reaction force of the point O' can be considered as an applied force, and so the body can be regarded as being free. Hence, in view of (2.3.17),

$$\mathbf{L} + \sum_{i=1}^{N}\mathbf{F}_i = 0, \quad \sum_{i=1}^{N}\mathbf{r}_i \times \mathbf{F}_i = 0. \tag{2.3.21}$$

Since O' is fixed relative to O, we have $\delta\mathbf{r}_0 = 0$, and so

$$\delta\mathbf{r}_i = (\boldsymbol{\omega} \times \mathbf{r}'_i)\delta t, \tag{2.3.22}$$

i.e. a rigid body with a fixed point has three degrees of freedom.

2.3.5 Application

Using the principle of virtual work, let us find the equilibrium positions of a particle A of mass m, which can slide without friction on an ellipse of semi-axes a and b, rotating with constant angular velocity ω about its minor axis, directed along the vertical as shown in Fig. 2.6.

Our particle is subject to two applied forces:

$$\text{force of gravity: } \mathbf{F}_g = m\mathbf{g} = -mg\mathbf{j}, \tag{2.3.23}$$

$$\text{centrifugal force: } \mathbf{F}_{cf} = m\omega^2\mathbf{r} = m\omega^2 x\mathbf{i}, \tag{2.3.24}$$

and a constraint force, due to the restriction of moving on the ellipse,

$$\frac{x^2}{a^2} + \frac{y^2}{b^2} = 1. \tag{2.3.25}$$

The principle of virtual work (2.3.2) yields:

$$\mathbf{F}\cdot\delta\mathbf{r} = m\omega^2 x\delta x - mg\delta y = 0. \tag{2.3.26}$$

On the other hand, by differentiating (2.3.25), we have:

$$\frac{x\delta x}{a^2} + \frac{y\delta y}{b^2} = 0. \tag{2.3.27}$$

Eliminating δy between the last two equations, we obtain

$$x\left(\frac{1}{a^2} + \frac{\omega^2 y}{b^2 g}\right) = 0.$$

This means that either

$$\text{(a)} \quad x = 0, \quad \frac{1}{a^2} + \frac{\omega^2 y}{b^2 g} \neq 0,$$

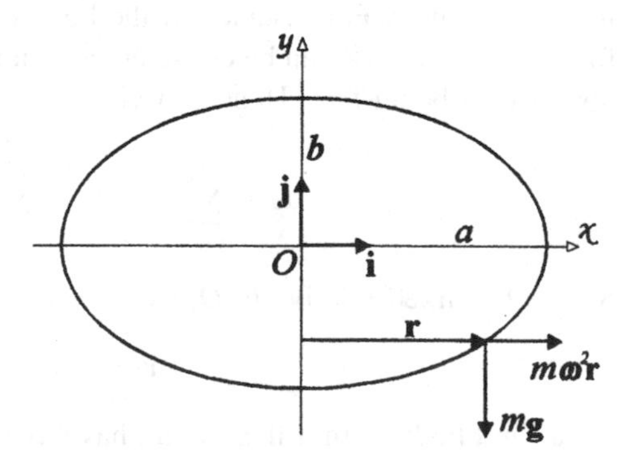

Fig. 2.6 A particle sliding without friction on a rotating ellipse.

or

$$\text{(b)} \quad x \neq 0, \quad \frac{1}{a^2} + \frac{\omega^2 y}{b^2 g} = 0.$$

Consequently, these two cases lead to the following possible equilibrium conditions:

$$\text{(a)} \quad x = 0, \quad y = \pm b, \tag{2.3.28}$$

$$\text{(b)} \quad x = \pm a\sqrt{1 - \frac{b^2 g^2}{a^4 \omega^4}}, \quad y = -\frac{b^2 g}{a^2 \omega^2}. \tag{2.3.29}$$

Obviously, if $\omega \to \infty$, we have $x \to \pm a$, $y \to 0$.

2.4 Generalized Coordinates

Consider again a system of N particles $P_1, \ldots, P_N$, of radius-vectors $\mathbf{r}_1, \ldots, \mathbf{r}_N$ relative to a Cartesian orthogonal frame $Oxyz$, subject to s holonomic independent constraints

$$f_k(\mathbf{r}_1, \ldots, \mathbf{r}_N, t) = 0 \quad (k = \overline{1, s}). \tag{2.4.1}$$

Due to the existence of the constraints, the $3N$ coordinates of particles are not independent, therefore the number of independent coordinates will be

$$3N - s = n, \tag{2.4.2}$$

meaning that our system has $3N - s = n$ degrees of freedom. For instance, a system of two particles, at a fixed distance one from the other, has $6 - 1 = 5$ degrees of freedom.

If the number of particles is large, the presence of constraints makes the determination of the coordinates x_i, y_i, z_i a difficult task. We shall attach to the n degrees of freedom a number of n independent variables $q_1, \ldots, q_n$, called *generalized coordinates* or *Lagrangian variables*. The $3N$ Cartesian coordinates $\mathbf{r}_i$ are then expressed in terms of $q_1, \ldots, q_n$ by

$$\mathbf{r}_i = \mathbf{r}_i(q_1, \ldots, q_n, t) \equiv \mathbf{r}_i(q, t) \quad (i = \overline{1, N}). \tag{2.4.3}$$

The generalized coordinates q_j $(j = \overline{1, n})$ satisfy the following properties:

(a) Any independent variation of $q_1, \ldots, q_n$ yields

$$f_k[\mathbf{r}_1(q, t), \ldots, \mathbf{r}_N(q, t), t] \equiv 0. \tag{2.4.4}$$

(b) Any $\mathbf{r}_1, \ldots, \mathbf{r}_N$, consistent with the constraints (2.4.1), can be obtained from (2.4.3).

(c) There also exists the inverse transformation of (2.4.3), namely

$$q_j = q_j(\mathbf{r}_1,\ldots,\mathbf{r}_N,t) \quad (j=\overline{1,n}), \tag{2.4.5}$$

for $\mathbf{r}_1,\ldots,\mathbf{r}_N$ satisfying (2.4.1).

Similarly to the Cartesian coordinates, the generalized coordinates are assumed to be continuous functions of time, at least twice differentiable. On the other hand, in contrast to the Cartesian coordinates, the generalized coordinates do not necessarily have the dimension of length. We can choose as Lagrangian coordinates any suitable assembly of geometrical objects, such as: segments of straight lines, arcs, angles, surfaces, components of angular velocities, etc.

The choice of generalized coordinates is somewhat arbitrary. It is always possible to find a point transformation

$$q_j \to q'_j = q'_j(q_1,\ldots,q_n,t) \quad (j=\overline{1,n}), \tag{2.4.6}$$

such that $q'_1,\ldots,q'_n$ are a new set of Lagrangian variables.

If the system is not subject to constraints, we can choose as generalized coordinates the $3N$ Cartesian coordinates of the particles, but there are also other possible choices. For instance, the position of a free particle can be defined either by its Cartesian coordinates x, y, z, its spherical coordinates r,θ,φ, or its cylindrical coordinates ρ,φ,z, etc.

Example. A particle P is constrained to remain on the moving sphere

$$(x-at)^2+(y-bt)^2+(z-ct)^2=R^2. \tag{2.4.7}$$

Since $n=2$, we can choose $q_1=\theta$, $q_2=\varphi$. At time t, the centre of the sphere is at the point $(at,\ bt,\ ct)$, therefore we can write

$$x=at+R\sin\theta\cos\varphi,\quad y=bt+R\sin\theta\sin\varphi,\quad z=ct+R\cos\theta, \tag{2.4.8}$$

representing the transition from Cartesian to spherical coordinates.

2.4.1 Configuration Space

The set of radius-vectors $\mathbf{r}_1,\ldots,\mathbf{r}_N$ define the so-called *configuration* of the system of particles, in the real space. If we choose $q_1,\ldots,q_n$ as coordinates of a n-dimensional space R_n, then to each set of values of the variables $q_1,\ldots,q_n$ will correspond a *representative point* in this space, known as the *configuration space*. In other words, any configuration of a mechanical system can be represented by a single point in the configuration space R_n. Note that the configuration space does not generally have an intuitive meaning, as does the Euclidean space used in Newtonian mechanics; but, as we shall prove, the abstract notions of *generalized coordinates* and *configuration space* are very useful not only in mechanics, but in other physical disciplines as well.

As the mechanical system changes its configuration with time, the configuration point traces a curve in configuration space, called *generalized trajectory*. This is by no means any of the real trajectories of particles, but describes the motion of the *whole* system. The generalized trajectory can be conceived as a succession of representative points, each of them corresponding to a certain configuration of the system. To know the *law of motion* in the configuration space means to know

$$q_j = q_j(t) \quad (j = \overline{1, n}). \tag{2.4.9}$$

These are also the *parametric equations* of the generalized trajectory. Once (2.4.9) are known, by means of (2.4.3), the motion of the particles in real space can also be determined.

2.4.2 Generalized Forces

In view of (2.4.3), a real infinitesimal displacement $d\mathbf{r}_i$ of particle P_i, during the time interval dt, is

$$d\mathbf{r}_i = \sum_{j=1}^{n} \frac{\partial \mathbf{r}_i}{\partial q_j} dq_j + \frac{\partial \mathbf{r}_i}{\partial t} dt \quad (i = \overline{1, N}), \tag{2.4.10}$$

while a virtual displacement $\delta\mathbf{r}_i$ satisfies the relation

$$\delta\mathbf{r}_i = \sum_{j=1}^{n} \frac{\partial \mathbf{r}_i}{\partial q_j} \delta q_j \quad (i = \overline{1, N}). \tag{2.4.11}$$

The displacements dq_j and δq_j in the configuration space are similar to the displacements $d\mathbf{r}_i$ and $\delta\mathbf{r}_i$ defined in the real space. Thus, by dq_j we mean *real* (or *possible*) displacements of the representative point during time dt, while δq_j are *virtual displacements*, taken at $t = \text{const.}$ (i.e. $\delta t = 0$). If $q_1, \ldots, q_n$ are independent, $\delta q_1, \ldots, \delta q_n$ are also independent and can be considered as a set of n completely arbitrary displacements at an instant.

Let us now write the virtual work δW, done by applied forces $\mathbf{F}_1, \ldots, \mathbf{F}_N$ on the particles, in terms of virtual displacements in the configuration space. In view of (2.4.3), we have:

$$\delta W = \sum_{i=1}^{N} \mathbf{F}_i \cdot \delta\mathbf{r}_i = \sum_{j=1}^{n} \left(\sum_{i=1}^{N} \mathbf{F}_i \cdot \frac{\partial \mathbf{r}_i}{\partial q_j} \right) \delta q_j.$$

If we define the *generalized forces* by

$$Q_j = \sum_{i=1}^{N} \mathbf{F}_i \cdot \frac{\partial \mathbf{r}_i}{\partial q_j} \quad (j = \overline{1, n}), \tag{2.4.12}$$

the work can be written as

$$\delta W = \sum_{i=1}^{N} \mathbf{F}_i \cdot \delta \mathbf{r}_i = \sum_{j=1}^{n} Q_j \delta q_j. \tag{2.4.13}$$

Since, in general, the forces $\mathbf{F}_i$ are functions of the form

$$\mathbf{F}_i = \mathbf{F}_i(\mathbf{r}_1, \ldots, \mathbf{r}_N,\ \dot{\mathbf{r}}_1, \ldots, \dot{\mathbf{r}}_N,\ t) \quad (i = \overline{1, N}), \tag{2.4.14}$$

we conclude that the generalized forces Q_j have the following functional dependence:

$$Q_j = Q_j(q_1, \ldots, q_n,\ \dot{q}_1, \ldots, \dot{q}_n,\ t) \equiv Q_j(q, \dot{q}, t) \quad (j = \overline{1, n}). \tag{2.4.15}$$

The quantities

$$\dot{q}_j = \frac{dq_j}{dt}$$

are called *generalized velocities* and are related to the real velocities $\mathbf{v}_1, \ldots, \mathbf{v}_N$ by

$$\dot{\mathbf{r}}_i = \mathbf{v}_i = \sum_{j=1}^{n} \frac{\partial \mathbf{r}_i}{\partial q_j} \dot{q}_j + \frac{\partial \mathbf{r}_i}{\partial t} \quad (i = \overline{1, N}). \tag{2.4.16}$$

The physical meaning of the generalized forces Q_j emerges from the significance of their associated generalized coordinates. For example, if the transition from Cartesian coordinates x, y, z to orthogonal curvilinear coordinates q_1, q_2, q_3 is defined by

$$x_i = x_i(q_1, q_2, q_3) \quad (i = \overline{1, N}), \tag{2.4.17}$$

then $\frac{\partial \mathbf{r}}{\partial q_k}$ is a vector tangent to the curve $q_k = \textit{variable}$, while $Q_k = \mathbf{F} \cdot \frac{\partial \mathbf{r}}{\partial q_k}$ is the component of $\mathbf{F}$ on this direction. In particular, the choice $q_1 = r,\ q_2 = \theta,\ q_3 = \varphi$ yields (see Appendix B):

$$\begin{aligned} Q_1 &= Q_r = \mathbf{F} \cdot \mathbf{u}_r = F_r, \\ Q_2 &= Q_\theta = \mathbf{F} \cdot (r\mathbf{u}_\theta) = r\,F_\theta, \\ Q_3 &= Q_\varphi = \mathbf{F} \cdot (r \sin\theta\ \mathbf{u}_\varphi) = r \sin\theta\ F_\varphi. \end{aligned} \tag{2.4.18}$$

The generalized forces do not generally have the dimension of force, but the product $[qQ]$ has always the dimension of work.

If the forces $\mathbf{F}_i$ $(i = \overline{1, N})$ derive from a potential (see (1.3.22)):

$$\mathbf{F}_i = -\mathrm{grad}_i V \quad (i = \overline{1, N}), \tag{2.4.19}$$

then the generalized forces Q_j obey a similar equation:

$$Q_j = -\sum_{i=1}^{N} \frac{\partial V}{\partial \mathbf{r}_i} \cdot \frac{\partial \mathbf{r}_i}{\partial q_j} = -\frac{\partial V}{\partial q_j} \quad (j = \overline{1, n}), \tag{2.4.20}$$

where $V = V(q_1, \ldots, q_n, t)$ is the potential in terms of the new variables.

From (2.4.13) follows that a position of the representative point in the configuration space, at time t, is a position of equilibrium, if

$$\sum_{j=1}^{n} Q_j \, \delta q_j = 0, \tag{2.4.21}$$

which expresses the *principle of virtual work in* R_n. If the virtual displacements δq_j are arbitrary and independent, it results in

$$Q_j = 0 \quad (j = \overline{1,n}), \tag{2.4.22}$$

meaning that: a certain position of a system of particles, subject to holonomic constraints, is a position of equilibrium, if all the generalized forces corresponding to that position are zero.

2.4.3 Kinetic Energy in Generalized Coordinates

It is most useful in the development of our formalism to express the kinetic energy T of the system in terms of the generalized coordinates $q_1, \ldots, q_n$, and the generalized velocities $\dot{q}_1, \ldots, \dot{q}_n$. In view of (2.4.5), we have:

$$T = \frac{1}{2} \sum_{i=1}^{N} m_i \sum_{j=1}^{n} \left(\frac{\partial \mathbf{r}_i}{\partial q_j} \dot{q}_j + \frac{\partial \mathbf{r}_i}{\partial t} \right) \cdot \sum_{k=1}^{n} \left(\frac{\partial \mathbf{r}_i}{\partial q_k} \dot{q}_k + \frac{\partial \mathbf{r}_i}{\partial t} \right).$$

Set

$$a = \frac{1}{2} \sum_{i=1}^{N} m_i \left| \frac{\partial \mathbf{r}_i}{\partial t} \right|^2, \quad a_j = \sum_{i=1}^{N} m_i \frac{\partial \mathbf{r}_i}{\partial q_j} \cdot \frac{\partial \mathbf{r}_i}{\partial t}, \quad a_{jk} = \frac{1}{2} \sum_{i=1}^{N} \frac{\partial \mathbf{r}_i}{\partial q_j} \cdot \frac{\partial \mathbf{r}_i}{\partial q_k}, \tag{2.4.23}$$

where a, a_j, a_{jk} are continuous and differentiable functions of $q_1, \ldots, q_n, t$. Thus,

$$T = a + \sum_{j=1}^{n} a_j \dot{q}_j + \sum_{j=1}^{n} \sum_{k=1}^{n} a_{jk} \dot{q}_j \dot{q}_k = T_0 + T_1 + T_2, \tag{2.4.24}$$

where the meaning of T_0, T_1, T_2 is obvious.

If the constraints are scleronomous, the terms T_0 and T_1 in (2.4.24) vanish and the kinetic energy $T = T_2$ becomes a homogeneous quadratic form of the generalized velocities $\dot{q}_j$:

$$T = \frac{1}{2} \sum_{j=1}^{n} \sum_{k=1}^{n} a_{jk} \dot{q}_j \dot{q}_k. \tag{2.4.25}$$

Keeping in mind the definition of a_{jk}, we see that the quadratic form T_2 is positively defined, $T_2 \geq 0$ (the equality sign is valid only if all $\dot{q}_1, \ldots, \dot{q}_n$ are zero). For example, the kinetic energy of a particle of mass m in spherical coordinates is

$$T = \frac{1}{2} m (\dot{r}^2 + r^2 \dot{\theta}^2 + r^2 \sin^2 \theta \, \dot{\varphi}^2).$$

We can therefore conclude that, in general, the kinetic energy has the following functional dependence:

$$T = T(q, \dot{q}, t). \tag{2.4.26}$$

2.5 Differential and Integral Principles in Analytical Mechanics

As Newtonian mechanics is based on the well-known principles of inertia, of force and of reciprocal interactions, so another formulation of mechanics is constructed on some fundamental axioms, called *principles of analytical mechanics*. These postulates serve to deduce the differential equations of motion in the configuration space. The principles of analytical mechanics are more general then those of Newtonian mechanics; they allow not only to obtain the results of Newtonian mechanics, but also to approach a large variety of non-mechanical problems. As a matter of fact, the methods provided by analytical mechanics play an important role in other physical disciplines, such as: theory of elasticity, quantum field theory, electrodynamics, theory of relativity, etc.

The principles of analytical mechanics can by grouped in two categories:

(a) *Differential principles*, which give us information about the state of a system, at different times, and take care of the behaviour of the system under *infinitesimal variations* of general coordinates and velocities in the configuration space. In general, the differential equations of motion (in both real and configuration spaces) can be considered as mathematical forms of certain differential principles. Such a principle is, for example, *D'Alembert's principle*.

(b) *Integral principles*, which consider the motion of a system during a *finite* time interval. These principles operate with *global* variations in configuration space. In this category fall *variational principles*, that use the methods of variational calculus, for global displacements along the generalized trajectories. The *Hamilton* and the *Maupertuis principles* belong to this category.

The distinction between these two groups is not absolute. As we shall see later on, there is an intimate relation between all the principles of analytical mechanics.

2.5.1 D'Alembert's Principle

Consider again a system of N particles $P_1, \ldots, P_N$, subject to applied and (holonomic) constraint forces. The Newtonian equation of motion of the particle P_i is (see (2.1.60)):

$$m_i \ddot{\mathbf{r}}_i = \mathbf{F}_i + \mathbf{L}_i \quad (i = \overline{1, N}). \tag{2.5.1}$$

Let us denote

$$\mathbf{J}_i = -m_i \ddot{\mathbf{r}}_i \quad (i = \overline{1, N}) \tag{2.5.2}$$

and call it the *inertial force* acting on particle P_i. Then

$$\mathbf{F}_i + \mathbf{L}_i + \mathbf{J}_i = 0 \quad (i = \overline{1, N}). \tag{2.5.3}$$

This vector equation expresses one of the forms of D'Alembert's principle: *there is an equilibrium, at any moment, between the applied, the constraint and the inertial forces acting on a particle*. This is the initial form of the principle, discovered by *Jean-Baptiste le Rond D'Alembert.*

In the case of ideal constraints, D'Alembert's principle can be written in an alternative form, which is very useful in some applications. To find this expression, we multiply (2.5.3) by the virtual displacement $\delta\mathbf{r}_i$, and then we take the sum over all particles of the system. Since the virtual work associated with the ideal constraints is zero, we arrive at

$$\sum_{i=1}^{N} (\mathbf{J}_i + \mathbf{F}_i) \cdot \delta\mathbf{r}_i = 0 \tag{2.5.4}$$

or, in a slightly different form,

$$\sum_{i=1}^{N} (\mathbf{F}_i - m_i \ddot{\mathbf{r}}_i) \cdot \delta\mathbf{r}_i = 0, \tag{2.5.5}$$

meaning that: *The sum of the virtual works of applied and inertial forces, acting on a system subject to ideal constraints, is zero*. This form of D'Alembert's principle is most useful, because it does not contain the constraint forces anymore. It was given by *Lagrange* and is used to deduce the differential equations of motion in configuration space.

2.5.2 Lagrange Equations for Holonomic Systems

We are now prepared to derive the differential equations of motion of a system of N particles, subject to ideal and independent constraints, in terms of generalized coordinates $q_1, \ldots, q_n$. To this end, we shall express both the variations $\delta\mathbf{r}_i$ and the derivatives $\ddot{\mathbf{r}}_i$ occurring in (2.5.5) in the configuration space. The virtual displacements $\delta\mathbf{r}_i$ can be written as (see (2.4.11)):

$$\delta\mathbf{r}_i = \sum_{j=1}^{n} \frac{\partial \mathbf{r}_i}{\partial q_j} \delta q_j \quad (i = \overline{1, N}), \tag{2.5.6}$$

hence

$$\begin{aligned} \sum_{i=1}^{N} m_i \ddot{\mathbf{r}}_i \cdot \delta\mathbf{r}_i &= \sum_{j=1}^{n} \left(\sum_{i=1}^{N} m_i \ddot{\mathbf{r}}_i \cdot \frac{\partial \mathbf{r}_i}{\partial q_j} \right) \delta q_j \\ &= \sum_{j=1}^{n} \left[\frac{d}{dt} \left(\sum_{i=1}^{N} m_i \dot{\mathbf{r}}_i \cdot \frac{\partial \mathbf{r}_i}{\partial q_j} \right) - \sum_{i=1}^{N} m_i \dot{\mathbf{r}}_i \cdot \frac{d}{dt} \left(\frac{\partial \mathbf{r}_i}{\partial q_j} \right) \right] \delta q_j. \end{aligned} \tag{2.5.7}$$

But

$$\begin{aligned} \frac{d}{dt} \left(\frac{\partial \mathbf{r}_i}{\partial q_j} \right) &= \frac{\partial}{\partial t} \left(\frac{\partial \mathbf{r}_i}{\partial q_j} \right) + \sum_{k=1}^{n} \frac{\partial}{\partial q_k} \left(\frac{\partial \mathbf{r}_i}{\partial q_j} \right) \dot{q}_k = \frac{\partial}{\partial q_j} \left(\frac{\partial \mathbf{r}_i}{\partial t} + \sum_{k=1}^{n} \frac{\partial \mathbf{r}_i}{\partial q_k} \dot{q}_k \right) \\ &= \frac{\partial \dot{\mathbf{r}}_i}{\partial q_j}. \end{aligned} \tag{2.5.8}$$

On the other hand, (2.4.16) yields

$$\frac{\partial \dot{\mathbf{r}}_i}{\partial \dot{q}_j} = \frac{\partial \mathbf{r}_i}{\partial q_j}. \tag{2.5.9}$$

The substitution of (2.5.8) and (2.5.9) into (2.5.7) gives

$$\sum_{i=1}^{N} m_i \ddot{\mathbf{r}}_i \cdot \delta\mathbf{r}_i = \sum_{j=1}^{n} \left[\frac{d}{dt} \left(\sum_{i=1}^{N} m_i \dot{\mathbf{r}}_i \cdot \frac{\partial \dot{\mathbf{r}}_i}{\partial \dot{q}_j} \right) - \sum_{i=1}^{N} m_i \dot{\mathbf{r}}_i \cdot \frac{\partial \dot{\mathbf{r}}_i}{\partial q_j} \right] \delta q_j. \tag{2.5.10}$$

Recalling that

$$T = \frac{1}{2} \sum_{i=1}^{N} m_i |\dot{\mathbf{r}}_i|^2$$

is the kinetic energy of the system of particles, it is easy to observe that (2.5.10) becomes

$$\sum_{i=1}^{N} m_i \ddot{\mathbf{r}}_i \cdot \delta\mathbf{r}_i = \sum_{j=1}^{n} \left[\frac{d}{dt} \left(\frac{\partial T}{\partial \dot{q}_j} \right) - \frac{\partial T}{\partial q_j} \right] \delta q_j. \tag{2.5.11}$$

The last step is now to introduce (2.4.13) and (2.5.11) into the expression for D'Alembert's principle (2.5.5). The result is:

$$\sum_{j=1}^{n}\left[\frac{d}{dt}\left(\frac{\partial T}{\partial \dot{q}_j}\right)-\frac{\partial T}{\partial q_j}-Q_j\right]\delta q_j=0. \tag{2.5.12}$$

Since the constraints are independent, the virtual displacements δq_j are completely arbitrary. Therefore (2.5.12) holds true only if all the square brackets are zero, i.e.

$$\frac{d}{dt}\left(\frac{\partial T}{\partial \dot{q}_j}\right)-\frac{\partial T}{\partial q_j}=Q_j \quad (j=\overline{1,n}), \tag{2.5.13}$$

which are called *Lagrange equations of the second kind.* From now on, we shall use these equations under the shorter name of *Lagrange equations.* They represent a system of n second-order differential equations in the variables q_j. The general integral of (2.5.13),

$$q_j=q_j(t,C_1,\ldots,C_{2n}) \quad (j=\overline{1,n}), \tag{2.5.14}$$

expresses the *law of motion* in the configuration space R_n. The $2n$ arbitrary constants $C_1,\ldots,C_{2n}$ are determined by $2n$ initial conditions: at time $t=t_0$, we must know both the generalized coordinates and the generalized velocities,

$$q_j^0=q_j(t_0,C_1,\ldots,C_{2n}), \quad \dot{q}_j^0=\dot{q}_j(t_0,C_1,\ldots,C_{2n}). \tag{2.5.15}$$

Once the motion in configuration space is determined, the solution (2.5.14) is introduced into (2.4.3), giving the motion in real space.

If, in particular, there are no constraints acting on the particles, we can choose as generalized coordinates the Cartesian coordinates, thus falling on Newton's second law discussed in Chap. 1.

Assume now that the applied forces $\mathbf{F}_i$ are *potential.* Then, according to (2.4.20), the generalized forces Q_j are also potentials and we obtain:

$$\frac{d}{dt}\left(\frac{\partial T}{\partial \dot{q}_j}\right)-\frac{\partial T}{\partial q_j}-\frac{\partial V}{\partial q_j}=0 \quad (j=\overline{1,n}),$$

where $V=V(q,t)$. Introducing the *Lagrangian function* or, simply, the *Lagrangian* L by

$$L(q,\dot{q},t)=T(q,\dot{q},t)-V(q,t), \tag{2.5.16}$$

we finally arrive at

$$\frac{d}{dt}\left(\frac{\partial L}{\partial \dot{q}_j}\right)-\frac{\partial L}{\partial q_j}=0 \quad (j=\overline{1,n}). \tag{2.5.17}$$

These equations are remarkably useful for several reasons. First, as we have already mentioned, they do not contain constraint forces. Second, all the information regarding the behaviour of the system is contained in a single scalar function, the Lagrangian. These equations are widely applied in many branches of physics, as we shall show in our further development of this formalism.

To solve a problem using the Lagrangian technique, one should proceed as follows:

1. Identify the n degrees of freedom of the system and choose suitable generalized coordinates q_j;
2. Construct either the functions T, Q_j, or the Lagrangian L;
3. Impose initial conditions;
4. Integrate the Lagrange equations and then, if necessary, determine the trajectories of the particles;
5. Obtain the constraint forces by means of (2.1.60):

$$\mathbf{L}_i = m_i \ddot{\mathbf{r}}_i - \mathbf{F}_i \quad (i = \overline{1, N}). \tag{2.5.18}$$

In particular, if there is no applied force acting on the particles, the Lagrange equations determine the geodesics of the configuration space R_n. The already known form of the equation of geodesics (see (2.1.37)) is obtained recalling that, for scleronomous constraints, the kinetic energy can be written as (see (2.4.25)):

$$T = \frac{1}{2} \sum_{j=1}^{n} \sum_{k=1}^{n} a_{jk}(q) \dot{q}_j \dot{q}_k. \tag{2.5.19}$$

The Lagrange equations (2.5.13) then yield:

$$\sum_{k=1}^{n} a_{jk} \ddot{q}_k + \sum_{k=1}^{n} \sum_{l=1}^{n} \frac{\partial a_{jk}}{\partial q_l} \dot{q}_k \dot{q}_l - \frac{1}{2} \sum_{k=1}^{n} \sum_{l=1}^{n} \frac{\partial a_{kl}}{\partial q_j} \dot{q}_k \dot{q}_l = Q_j \quad (j = \overline{1, n}).$$

Introducing the Christoffel symbols of the first kind,

$$\Gamma_{kl,j} = \frac{1}{2} \left(\frac{\partial a_{jk}}{\partial q_l} + \frac{\partial a_{lj}}{\partial q_k} - \frac{\partial a_{kl}}{\partial q_j} \right), \tag{2.5.20}$$

we have:

$$\sum_{k=1}^{n} a_{jk} \ddot{q}_k + \sum_{k=1}^{n} \sum_{l=1}^{n} \Gamma_{kl,j} \dot{q}_k \dot{q}_l = Q_j. \tag{2.5.21}$$

If $Q_j = 0$, we arrive at the geodetic form of the equilibrium equations in configuration space R_n (see (2.1.37)).

2.5.3 Velocity-Dependent Potential

Let us show that the Lagrange equations (2.5.17) keep their form in the case of a *generalized* or *velocity-dependent* potential $V(q, \dot{q}, t)$, linear in $\dot{q}_j$, if Q_j can be taken as

$$Q_j = \frac{d}{dt}\left(\frac{\partial V}{\partial \dot{q}_j}\right) - \frac{\partial V}{\partial q_j} \quad (j = \overline{1,n}). \tag{2.5.22}$$

Consider the potential

$$V(q, \dot{q}, t) = \sum_{j=1}^{n} \alpha_j \dot{q}_j + V_0 = V_1 + V_0, \tag{2.5.23}$$

where $\alpha_j \; (j = \overline{1,n})$ and V_0 are functions of q_j and t, and add the quantity

$$\frac{\partial V}{\partial q_j} - \frac{d}{dt}\left(\frac{\partial V}{\partial \dot{q}_j}\right)$$

to both sides of (2.5.13). Then it is obvious that, if (2.5.23) is true, we arrive at the Lagrange equations (2.5.17), where

$$L(q, \dot{q}, t) = T(q, \dot{q}, t) - V(q, \dot{q}, t). \tag{2.5.24}$$

A classic example of generalized potential is offered by the motion of an electrically charged particle in an external electromagnetic field. It is well known that the electromagnetic force acting on a particle of mass m and charge e, moving with velocity $\mathbf{v}$ in the field $\mathbf{E}$, $\mathbf{B}$, is:

$$\mathbf{F} = e(\mathbf{E} + \mathbf{v} \times \mathbf{B}). \tag{2.5.25}$$

The fields $\mathbf{E}$ and $\mathbf{B}$ are usually given in terms of the electromagnetic potentials $\mathbf{A}(\mathbf{r}, t)$ and $\phi(\mathbf{r}, t)$ as

$$\mathbf{E} = -\nabla\phi - \frac{\partial \mathbf{A}}{\partial t}, \quad \mathbf{B} = \nabla \times \mathbf{A}. \tag{2.5.26}$$

Since the particle is free, it has three degrees of freedom. We choose $q_i = x_i, \; \dot{q}_i = \dot{x}_i = v_i \; (i = 1, 2, 3)$. Recalling that q_i and $\dot{q}_i$ are independent with respect to each other, we can write (see Appendix B):

$$\nabla(\mathbf{v} \cdot \mathbf{A}) = \mathbf{v} \times (\nabla \times \mathbf{A}) + (\mathbf{v} \cdot \nabla)\mathbf{A}. \tag{2.5.27}$$

We also have:

$$\frac{d\mathbf{A}}{dt} = \frac{\partial \mathbf{A}}{\partial t} + (\mathbf{v} \cdot \nabla)\mathbf{A}. \tag{2.5.28}$$

Using (2.5.26)–(2.5.28), we get from (2.5.25):

$$F_i = e\left\{-\frac{\partial \phi}{\partial x_i} - \frac{\partial A_i}{\partial t} + [\mathbf{v} \times (\nabla \times \mathbf{A})]_i\right\}$$
$$= e\left\{-\frac{\partial}{\partial x_i}(\phi - \mathbf{v}\cdot\mathbf{A}) - \frac{d}{dt}\left[\frac{\partial}{\partial v_i}(\mathbf{v}\cdot\mathbf{A})\right]\right\}.$$

If we define the velocity-dependent potential by

$$V = e(\phi - \mathbf{v}\cdot\mathbf{A}), \tag{2.5.29}$$

which is of the form (2.5.23), we see that F_i can indeed be derived from (2.5.29). Therefore, the Lagrangian of our system is

$$L = \frac{1}{2}m|\mathbf{v}|^2 - e\phi + e\mathbf{v}\cdot\mathbf{A}. \tag{2.5.30}$$

In this example we started from the equation of motion (2.5.25) and arrived at the Lagrangian (2.5.30), but usually the problem is posed in an inverse way: given the Lagrangian, we are supposed to find the differential equations of motion.

Observations:

(a) Systems admitting a simple or a generalized potential are called *natural*. In view of the definition (2.5.16), we can write:

$$L = L_0 + L_1 + L_2, \tag{2.5.31}$$

where

$$L_0 = b, \quad L_1 = \sum_{j=1}^{n} b_j\dot{q}_j, \quad L_2 = \sum_{j=1}^{n}\sum_{k=1}^{n} b_{jk}\dot{q}_j\dot{q}_k.$$

The coefficients b, b_j, b_{jk} are functions of $q_1, \ldots, q_n, t$. Taking into account (2.4.24), in the case of a simple potential $V(q, t)$,

$$L_0 = T_0 - V, \quad L_1 = T_1, \quad L_2 = T_2, \tag{2.5.32}$$

while for a generalized potential $V(q, \dot{q}, t)$ (see (2.5.22))

$$L_0 = T_0 - V, \quad L_1 = T_1 - V_1, \quad L_2 = T_2. \tag{2.5.32'}$$

(b) Conservative forces represent a particular case of potential forces, therefore the Lagrange equations are used in the form (2.5.17), observing that now the function V is the *potential energy* of the system.

(c) If the Lagrangian L does not depend on one of the generalized coordinates $q_1, \ldots, q_n$, say q_k (k fixed), the Lagrange equations (2.5.17) yield:

$$\frac{\partial L}{\partial \dot{q}_k} = \text{const.} \tag{2.5.33}$$

Such a generalized coordinate is called *cyclic* or *ignorable*, and (2.5.33) is a first integral of (2.5.17).

(d) Let us add to the Lagrangian a term which is the total time derivative of some function $F(q, t)$:

$$L' = L(q, \dot{q}, t) + \frac{d}{dt} F(q, t). \tag{2.5.34}$$

Introducing (2.5.34) into the Lagrange equations (2.5.17), the terms containing F give zero, and we obtain the same system of equations for L'. This simple exercise is left to the reader. In conclusion, the terms having the form of a total time derivative can be omitted from a Lagrangian. In other words, two Lagrangian functions which differ from one another by terms being total time derivatives give the same description of the motion and therefore the two Lagrangians L and L' are *equivalent*.

A Heavy Particle Moving on a Spherical Surface

Let us find the differential equations of motion of a particle of mass m, moving without friction under the influence of gravity on a fixed spherical surface of radius l (a spherical pendulum) (Fig. 2.7).

This system has two degrees of freedom. Using spherical coordinates and choosing $q_1 = \theta$, $q_2 = \varphi$, we have:

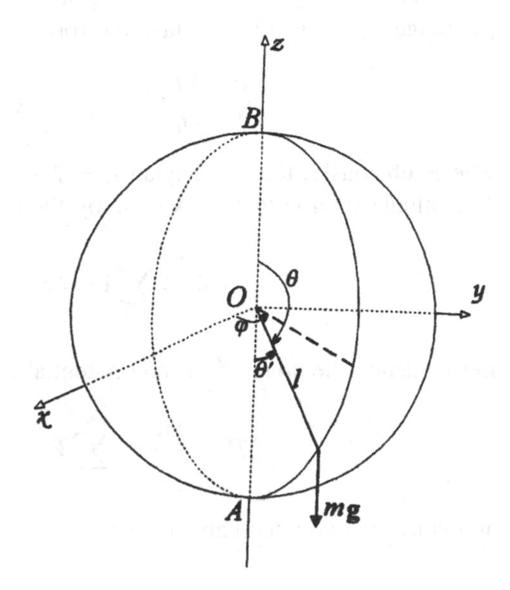

Fig. 2.7 A particle moving without friction on a fixed sphere (spherical pendulum).

$$T = \frac{1}{2} ml^2(\dot{\theta}^2 + \sin^2\theta\dot{\varphi}^2), \quad V = mgl\cos\theta,$$

hence

$$L = \frac{1}{2} ml^2(\dot{\theta}^2 + \sin^2\theta\,\dot{\varphi}^2) - mgl\cos\theta. \tag{2.5.35}$$

Performing the calculations in (2.5.17), we obtain the equations of motion:

$$\ddot{\theta} - \sin\theta\cos\theta\dot{\varphi}^2 - \frac{g}{l}\sin\theta = 0, \tag{2.5.36}$$

$$\ddot{\varphi} + 2\cot\theta\dot{\theta}\dot{\varphi} = 0. \tag{2.5.37}$$

These two second-order differential equations are non-linear. If $\varphi = \text{const.}$, i.e. if the motion is performed on a vertical circle of radius l, we are left with a single equation:

$$\ddot{\theta} - \frac{g}{l}\sin\theta = \ddot{\theta}' + \frac{g}{l}\sin\theta' = 0, \tag{2.5.38}$$

which is the differential equation of a *plane* (or *simple*) pendulum. The problem of the mathematical pendulum will be thoroughly discussed in Chap. 3.

2.5.4 Non-potential Forces

Assume that on a system of particles act both potential and non-potential forces. If we denote by $\tilde{Q}_j\,(j = \overline{1,n})$ the generalized non-potential forces, then the Lagrange equations (2.5.17) take the form

$$\frac{d}{dt}\left(\frac{\partial L}{\partial \dot{q}_j}\right) - \frac{\partial L}{\partial q_j} = \tilde{Q}_j \quad (j = \overline{1,n}), \tag{2.5.39}$$

where, obviously, the Lagrangian $L = T - V$ includes only the potential forces. The infinitesimal virtual work done by the non-potential forces is

$$\delta\tilde{W} = \sum_{i=1}^{N} \tilde{\mathbf{F}}_i \cdot \delta\mathbf{r}_i = \sum_{j=1}^{n} \tilde{Q}_j \delta q_j. \tag{2.5.40}$$

Let us define the *power* $\tilde{\mathcal{P}}$ of non-potential forces:

$$\tilde{\mathcal{P}} = \frac{\delta\tilde{W}}{\delta t} = \sum_{i=1}^{N} \tilde{\mathbf{F}}_i \cdot \mathbf{v}_i = \sum_{j=1}^{n} \tilde{Q}_j \dot{q}_j, \tag{2.5.41}$$

and consider two remarkable cases:

(1) The non-potential forces of negative power ($\tilde{\mathcal{P}}<0$) are called *dissipative forces*. Such a force is, for example, the friction force. If this can be written as

$$\mathbf{F}_i^f = -k\mathbf{v}_i \quad (i = \overline{1,N},\ k > 0), \tag{2.5.42}$$

then there exists a scalar function $\mathcal{T}$,

$$\mathcal{T} = \frac{1}{2}k\sum_{i=1}^{N}|\mathbf{v}_i|^2, \tag{2.5.43}$$

so that

$$\mathbf{F}_i^f = -\frac{\partial \mathcal{T}}{\partial \mathbf{v}_i} = -\nabla_{\mathbf{v}_i}\mathcal{T} \quad (i = \overline{1,N}), \tag{2.5.44}$$

where $\nabla_{\mathbf{v}_i}$ stands for the partial derivative with respect to $\mathbf{v}_i$. The function $\mathcal{T}$ is called the *Rayleigh dissipation function*. It is obvious that Rayleigh's function for a scleronomous system is quadratic and homogeneous in the generalized velocities $\dot{q}_j$:

$$\mathcal{T} = \frac{1}{2}\sum_{j=1}^{n}\sum_{k=1}^{n} C_{jk}\dot{q}_j\dot{q}_k. \tag{2.5.45}$$

The physical significance of $\mathcal{T}$ is found by writing the power of the friction forces:

$$\tilde{\mathcal{P}} = \sum_{i=1}^{N}\mathbf{F}_i^f\cdot\mathbf{v}_i = -k\sum_{i=1}^{N}|\mathbf{v}_i|^2 = -2\mathcal{T}, \tag{2.5.46}$$

i.e. the function $\mathcal{T}$ is equal to half of the power dissipated by friction.

The generalized forces $\tilde{Q}_j^f$, associated with the friction forces $\mathbf{F}_i^f$, are:

$$\tilde{Q}_j^f = \sum_{i=1}^{N}\mathbf{F}_i^f\cdot\frac{\partial \mathbf{r}_i}{\partial q_j} = -\frac{\partial \mathcal{T}}{\partial \dot{q}_j} \quad (j = \overline{1,n}). \tag{2.5.47}$$

Therefore, in our case, the Lagrange equations (2.5.39) become

$$\frac{d}{dt}\left(\frac{\partial L}{\partial \dot{q}_j}\right) - \frac{\partial L}{\partial q_j} - \frac{\partial \mathcal{T}}{\partial \dot{q}_j} = 0 \quad (j = \overline{1,n}). \tag{2.5.48}$$

(2) If the power of non-potential forces is zero,

$$\sum_{i=1}^{N}\tilde{\mathbf{F}}_i\cdot\mathbf{v}_i = \sum_{j=1}^{n}\tilde{Q}_j\dot{q}_j = 0, \tag{2.5.49}$$

we deal with *gyroscopic forces*. Remark that, in order for (2.5.49) to be valid, $\tilde{\mathbf{F}}_i$ must be written as a cross product of two vectors, where one of them is $\mathbf{v}_i$ or, equivalently, $\tilde{Q}_j$ must have the form

$$\tilde{Q}_j = \sum_{k=1}^{n} h_{jk}\dot{q}_k \quad (j = \overline{1,n}), \tag{2.5.50}$$

the coefficients h_{jk} being antisymmetric:

$$h_{jk} = -h_{kj}. \tag{2.5.51}$$

As two examples of gyroscopic forces, we give the *Lorentz force* acting on a particle of charge e:

$$\mathbf{F}_L = e(\mathbf{v} \times \mathbf{B}), \tag{2.5.52}$$

and the *Coriolis force* (see (4.3.7)):

$$\mathbf{F}_i = -2m_i(\boldsymbol{\omega} \times \mathbf{v}_i), \tag{2.5.53}$$

where $\mathbf{v}_i$ is the relative velocity of the particle m_i and $\boldsymbol{\omega}$ is the instantaneous vector of rotation.

The definition (2.5.49) shows that the instantaneous rate of the work done by a scleronomous system subject to gyroscopic forces is zero:

$$\frac{d}{dt}(\delta\tilde{W}) = \sum_{j=1}^{n}\sum_{k=1}^{n} h_{jk}\dot{q}_j\dot{q}_k = 0, \tag{2.5.54}$$

and therefore there exists the energy first integral.

In the next two chapters we shall give special attention to the application of this formalism on concrete models of dissipative and gyroscopic systems.

Observation: The Lagrange equations for non-holonomic systems are derived and applied in Chap. 4 (see (4.6.52)).

2.6 Elements of Calculus of Variations

Hamilton's principle (see Sect. 2.7), which is one of the most important principles of theoretical physics, belongs to the category of *variational principles*. For a better understanding of the formalism implied by the use of this principle, let us briefly review some elements of variational calculus.

The calculus of variations deals with the study of extremum values of functions depending on a curve, or on another function, rather than a real number. For the beginning, let us consider a function *f*(*x*) of class at least C^2 (i.e. continuous, together with its second partial derivatives), and expand it according to Taylor's formula, about a fixed value x_0:

$$f(x) = f(x_0) + \frac{x - x_0}{1!} f'(x_0) + \frac{(x - x_0)^2}{2!} f''(x_0) + \cdots \tag{2.6.1}$$

The quantity

$$\delta f = (x - x_0) f'(x_0) = f'(x_0) \delta x$$

is called the *first variation* of f at the point x_0. The *necessary* and *sufficient* condition that the function f has a *stationary value* at x_0 is that $\delta f = 0$, for any arbitrary variation δx. This yields:

$$f'(x_0) = \left(\frac{\partial f}{\partial x} \right)_{x = x_0} \equiv \left(\frac{\partial f}{\partial x} \right)_0 = 0, \tag{2.6.2}$$

which reminds us of the condition of static equilibrium (2.1.45). Going further, we can define the *second variation* of f as

$$\delta^2 f = \frac{1}{2} (x - x_0)^2 f''(x_0). \tag{2.6.3}$$

If $f''(x_0) \geq 0$, we have a local minimum at x_0, while if $f''(x_0) \leq 0$, x_0 is a local maximum.

Assume now that f is of the form $f(x_1, \ldots, x_n)$. Then its first variation at $(x_1^0, \ldots, x_n^0)$ is

$$\delta f = \sum_{j=1}^{n} \left(\frac{\partial f}{\partial x_j} \right)_0 \delta x_j, \quad \delta x_j = x_j - x_j^0, \tag{2.6.4}$$

while the condition that f has a *stationary* value at $(x_1^0, \ldots, x_n^0)$, for independent and arbitrary δx_j, reads:

$$\left(\frac{\partial f}{\partial x_j} \right)_0 = 0 \quad (j = \overline{1, n}). \tag{2.6.5}$$

If the variables $x_1, \ldots, x_n$ must obey s independent constraint equations

$$g_k(x_1, \ldots, x_n) = 0 \quad (k = \overline{1, s}), \tag{2.6.6}$$

where $g_1, \ldots, g_s$ are functions of class C^2, the variations δx_j are no longer independent, but must satisfy the system of s equations

$$\delta g_k = \sum_{j=1}^{n} \left(\frac{\partial g_k}{\partial x_j} \right)_0 \delta x_j = 0 \quad (k = \overline{1, s}). \tag{2.6.7}$$

To find the stationary conditions in the presence of constraints, one multiplies (2.6.7) by some arbitrary Lagrange multipliers λ_k $(k = \overline{1, s})$ and add the result to (2.6.4). Thus, we have:

$$\sum_{j=1}^{n}\left[\left(\frac{\partial f}{\partial x_j}\right)_0 + \sum_{k=1}^{s}\lambda_k\left(\frac{\partial g_k}{\partial x_j}\right)_0\right]\delta x_j = 0. \tag{2.6.8}$$

Here the variations δx_j are independent, therefore the stationarity condition reads:

$$\left(\frac{\partial f}{\partial x_j}\right)_0 + \sum_{k=1}^{s}\lambda_k\left(\frac{\partial g_k}{\partial x_j}\right)_0 = 0 \quad (j = \overline{1, n}). \tag{2.6.9}$$

Let us now consider the definite integral

$$I[y(x)] = \int_{x_1}^{x_2} f(x, y, y')dx, \tag{2.6.10}$$

where $y = y(x)$ is a curve in the xy-plane and $y' = \frac{dy}{dx}$. The function $f(x, y, y')$ is of class C^2 in each of its arguments. The integral (2.6.10) is a *functional* of $y(x)$, giving the correspondence between the function f and the number I, associated to the curve $y = y(x)$.

One of the central problems of the calculus of variations is to find the curve $y = y(x)$ for which the associated integral (2.6.10) is an extremum in the given interval $x_1 \leq x \leq x_2$.

Denote by (C) the path $y = y(x)$ that makes the integral (2.6.10) an extremum and consider a neighbouring curve (C^*), given by

$$y^*(x) = y(x) + \epsilon\eta(x), \tag{2.6.11}$$

where ϵ is a small parameter independent of x, while $\eta(x)$ is a function of class C^1 which satisfies the conditions

$$\eta(x_1) = \eta(x_2) = 0. \tag{2.6.12}$$

Therefore, the two paths (C) and (C^*) have the same initial and final points, $P_1(x_1, y_1)$ and $P_2(x_2, y_2)$ (Fig. 2.8). Varying the parameter ε, we obtain a family of curves $C_1^*, C_2^*, \ldots$, all of them passing through P_1 and P_2. The functional associated to the curve (C^*) is

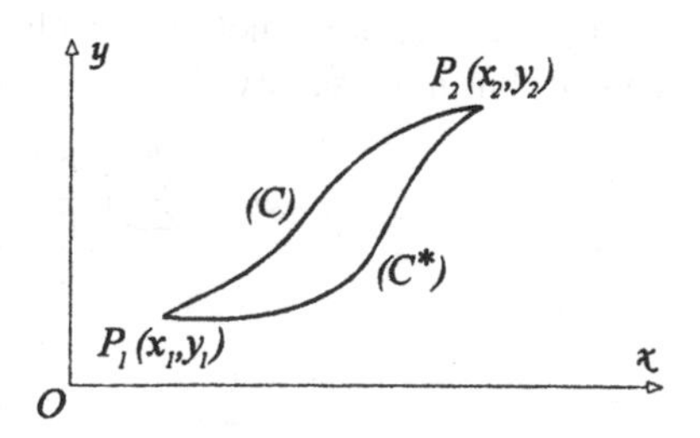

Fig. 2.8 The two paths (C) and (C^*), having the same initial and final points.

$$I[y^*(x)] = \int_{x_1}^{x_2} f(x, y^*, y^{*\prime})dx. \tag{2.6.13}$$

Since

$$f(x, y^*, y^{*\prime}) = f(x, y + \epsilon\eta, y' + \epsilon\eta') = f(x, y, y') + \epsilon\eta\frac{\partial f}{\partial y} + \epsilon\eta'\frac{\partial f}{\partial y'} + \cdots,$$

the first variation of the integral I is

$$\delta I = \epsilon\int_{x_1}^{x_2}\left(\eta\frac{\partial f}{\partial y} + \eta'\frac{\partial f}{\partial y'}\right)dx, \tag{2.6.14}$$

or, upon integration by parts of the second term and using (2.6.12):

$$\delta I = \epsilon\int_{x_1}^{x_2}\eta\left[\frac{\partial f}{\partial y} - \frac{d}{dx}\left(\frac{\partial f}{\partial y'}\right)\right]dx. \tag{2.6.15}$$

Recalling that $\eta(x)$ is arbitrary, except for the condition (2.6.12), the necessary and sufficient condition for a stationary value of I is

$$\frac{\partial f}{\partial y} - \frac{d}{dx}\left(\frac{\partial f}{\partial y'}\right) = 0. \tag{2.6.16}$$

Consequently, among all curves passing through the fixed points $P_1(x_1, y_1)$ and $P_2(x_2, y_2)$, the curve which makes the integral I stationary satisfies Eq. (2.6.16).

These considerations can be generalized for functionals of the type

$$I[y_1(x), \ldots, y_n(x)] = \int_{x_1}^{x_2} f(x, y_1, \ldots, y_n, y_1', \ldots, y_n')dx, \tag{2.6.17}$$

where $y_i = y_i(x)$ $(i = \overline{1, n})$ are n functions of class C^2 and $y_i' = \frac{dy_i}{dx}$ $(i = \overline{1, n})$. Let $x, y_1, \ldots, y_n$ be the coordinates of a point in a $(n + 1)$-dimensional space Q_{n+1} and let $P_1(x^1, y_k^1)$, $P_2(x^2, y_k^2)$ be two fixed points in Q_{n+1}. Then, if

$$y_i = y_i(x) \quad (i = \overline{1, n}) \tag{2.6.18}$$

are the equations of the curve which makes (2.6.17) an extremum, and

$$y_i^* = y_i(x) + \epsilon\eta_i(x) \tag{2.6.19}$$

are the equations of a neighbouring curve passing through the same initial and final points, then, following a similar procedure, we obtain the condition which $y_i(x)$ has to obey so that the integral (2.6.17) be an extremum, in the form

$$\frac{\partial f}{\partial y_i} - \frac{d}{dx}\left(\frac{\partial f}{\partial y_i'}\right) = 0 \quad (i = \overline{1,n}). \tag{2.6.20}$$

These equations were first obtained by *Leonhard Euler* in 1744 and later used by *Lagrange* in mechanics. They are usually called the *Euler–Lagrange equations.*

Before going any further, we shall apply this formalism to some classical problems of variational calculus.

2.6.1 Shortest Distance Between Two Points in a Plane

Our aim is to minimize the integral

$$I = \int_{x_1}^{x_2} ds = \int_{x_1}^{x_2} \sqrt{1 + y'^2}dx, \tag{2.6.21}$$

where $s = \sqrt{x^2 + y^2}$ is the arc length in the xy-plane. Comparing (2.6.21) with (2.6.10), we get $f = \sqrt{1 + y'^2}$, and the Euler–Lagrange equation (2.6.16) yields:

$$\frac{\partial f}{\partial y'} = \frac{y'}{\sqrt{1 + y'^2}}, \quad \frac{\partial f}{\partial y} = 0,$$

hence

$$\frac{y'}{\sqrt{1 + y'^2}} = C, \quad y' = \frac{C}{\sqrt{1 - C^2}} = C_1,$$

leading by integration to the equation of a *straight line*, $y = C_1 x + C_2$. The constants C_1 and C_2 are determined by the condition that the curve must pass through the points $P_1(x_1, y_1)$, $P_2(x_2, y_2)$. Note that our solution produces an *extremum* for (2.6.21) and we cannot know the nature of this extremum at the beginning. But the investigation of the problem, together with our common sense, tells us that the extremum is a *minimum.*

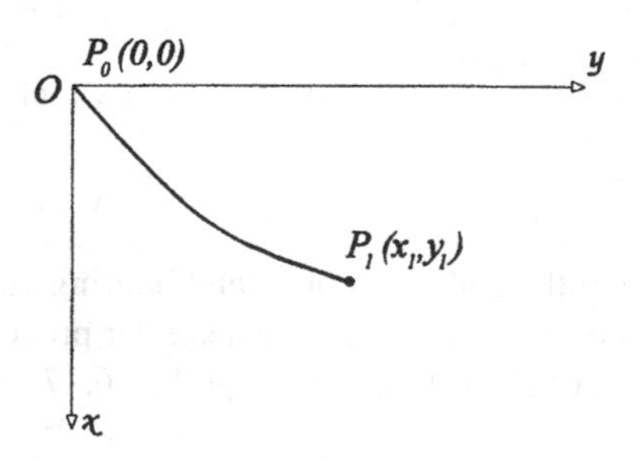

Fig. 2.9 Choice of coordinates for the brachistochrone problem. The points $P_0(0, 0)$ and $P_1(x_1, y_1)$ are fixed.

2.6.2 Brachistochrone Problem

Among all curves lying in a vertical plane and passing through two fixed points, find the one for which a heavy particle would slide down the curve without friction in minimum (extremum) time.

This problem was formulated in 1696 by the Swiss mathematician *Johann Bernoulli*, being the problem which lead to the calculus of variations. The word *brachistochrone* derives from the Greek *brachistos* (shortest) and *chronos* (time).

Choosing the coordinates as in Fig. 2.9, with the fixed points $P_0(0, 0)$ and $P_1(x_1, y_1)$, the time of descent from P_0 to P_1, on any curve, can be written as

$$t = \int_{P_0}^{P_1} \frac{ds}{v}, \tag{2.6.22}$$

where v is the speed of the particle along the curve. Using the kinetic energy theorem (1.3.21),

$$mgx = \frac{1}{2}mv^2, \tag{2.6.23}$$

and the relation $ds = \sqrt{1 + y'^2}dx$ for the path element, we obtain:

$$t = \frac{1}{\sqrt{2g}} \int_0^{x_1} \sqrt{\frac{1 + y'^2}{x}} dx. \tag{2.6.24}$$

Next, we apply the Euler–Lagrange equation (2.6.16), where $f = \sqrt{\frac{1+y'^2}{x}}$. Performing simple calculations, we have:

$$\frac{\partial f}{\partial y} = 0, \quad \frac{\partial f}{\partial y'} = \frac{y'}{\sqrt{x(1 + y'^2)}} = \frac{1}{\sqrt{2a}}, \tag{2.6.25}$$

where a is a constant. Separating the variables and integrating, we arrive at

$$y = \int_0^x \sqrt{\frac{x}{2a - x}} dx. \tag{2.6.26}$$

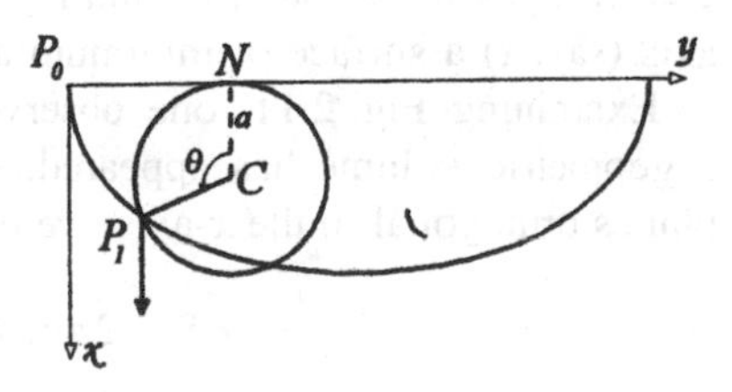

Fig. 2.10 Generation of the cycloid $x = a(1 - \cos\theta)$.

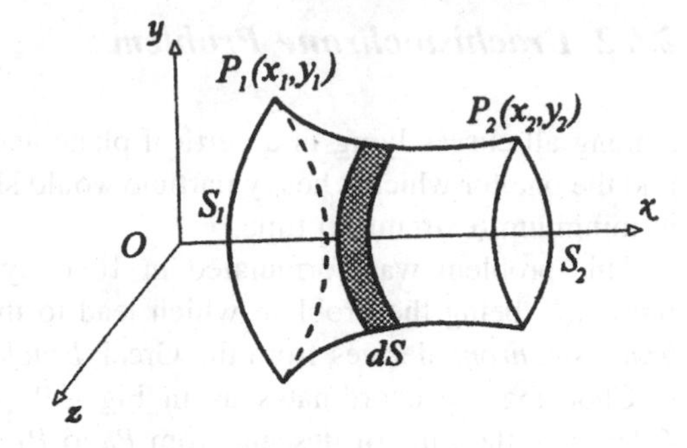

Fig. 2.11 Surface of revolution of minimum area. The points $P_1(x_1, y_1)$ and $P_2(x_2, y_2)$ are fixed.

To perform the integration, one makes the change of variable

$$\frac{x}{2a - x} = u^2, \tag{2.6.27}$$

hence

$$y = -2a \int_0^u u\, d\left(\frac{1}{1+u^2}\right) = -2a\left(\left[\frac{u}{1+u^2}\right]_0^u - [\arctan u]_0^u\right).$$

A new substitution

$$u = \tan\frac{\theta}{2} \tag{2.6.28}$$

yields finally

$$y = a(\theta - \sin\theta), \tag{2.6.29}$$

while x is found from (2.6.27) and (2.6.28):

$$x = a(1 - \cos\theta). \tag{2.6.30}$$

Equations (2.6.29) and (2.6.30) are the parametric equations of a cycloid, having the y-axis as basis and the concavity upwards. The constant a is the radius of the circle that generates the cycloid (Fig. 2.10). In fact, we have shown that the path of the cycloid insures a stationary value of t, but it is obvious that the extremum must be a minimum.

2.6.3 Surface of Revolution of Minimum Area

Let $P_1(x_1, y_1)$ and $P_2(x_2, y_2)$ be two fixed points in the xy-plane. Find the curve $y = y(x)$ passing through P_1 and P_2 which would generate by revolution about an axis (say, x) a surface of minimum area.

Examining Fig. 2.11, one observes that upon a revolution about the x-axis, a geometric volume has appeared, with fixed basis areas S_1 and S_2. With two planes orthogonal to the x-axis, we delimit an elementary cylinder, of lateral area

$$dS = 2\pi y\, ds = 2\pi y\sqrt{1 + y'^2}dx.$$

The area generated by the curve passing through P_1 and P_2 is

$$S = 2\pi \int_{x_1}^{x_2} y\sqrt{1+y'^2}dx. \tag{2.6.31}$$

To make (2.6.31) a minimum (a maximum would not make any sense), the integrand $f = y\sqrt{1+y'^2}$ must satisfy the Euler–Lagrange equation (2.6.16). The easiest way to get the result is to observe that, since f does not explicitly depend on x, Eq. (2.6.16) admits the first integral

$$y'\frac{\partial f}{\partial y'} - f = \text{const.} \tag{2.6.32}$$

Since

$$\frac{\partial f}{\partial y'} = \frac{yy'}{\sqrt{1+y'^2}},$$

from (2.6.32) we obtain:

$$\frac{yy'^2}{\sqrt{1+y'^2}} - y\sqrt{1+y'^2} = C_1,$$

and, separating the variables and integrating,

$$x = C_1 \int \frac{dy}{\sqrt{y^2 - C_1^2}} + C_2 = C_1 \operatorname{arccosh}\frac{y}{C_1} + C_2,$$

yielding finally:

$$y = C_1 \cosh\frac{x - C_2}{C_1}, \tag{2.6.33}$$

which is the equation of a *catenary* (from the Latin *catena*, meaning *chain*). This is the shape, for instance, of a uniform, flexible heavy chain under gravity, when it is held fix at two points. The constants C_1 and C_2 are determined from the boundary conditions.

2.6.4 Geodesics

A *geodesic* is defined as the shortest distance between two points in a given space. We have already encountered this notion earlier in this chapter, but only in two particular cases. We wish now to give a general theory of geodesics, useful not only in classical mechanics, but also in the general theory of relativity.

First, we must define the *metric tensor*. Let E_m be an Euclidean space with the Cartesian coordinates $y_1, \ldots, y_m$, and write the line element in the form

$$ds^2 = \sum_{j=1}^{m} dy_j dy_j. \tag{2.6.34}$$

Let now $\mathcal{R}_n$ $(n<m)$ be a n-dimensional manifold in E_m and let $x^1, \ldots, x^n$ be the coordinates of a point in $\mathcal{R}_n$. Since $y_j = y_j(x^1, \ldots, x^n)$, we have:

$$ds^2 = \sum_{j=1}^{m} \sum_{i,k=1}^{n} \frac{\partial y_j}{\partial x^i} \frac{\partial y_j}{\partial x^k} dx^i dx^k$$

and, with the notation

$$g_{ik}(x^1, \ldots, x^n) = g_{ki} = \sum_{j=1}^{m} \frac{\partial y_j}{\partial x^i} \frac{\partial y_j}{\partial x^k} \tag{2.6.35}$$

for the *metric tensor*, we arrive at the following form of the *metric* (squared line element):

$$ds^2 = \sum_{i,k=1}^{n} g_{ik} dx^i dx^k. \tag{2.6.36}$$

If $g_{ik} = \delta_{ik}$, i.e. if the manifold $\mathcal{R}_n$ is Euclidean, we fall back on the metric (2.6.34).

If the metric (2.6.36) is invariant under a general coordinate transformation

$$x'^i = x'^i(x^1, \ldots, x^n) \quad (i = \overline{1,n}), \tag{2.6.37}$$

the manifold $\mathcal{R}_n$ is called *Riemannian.* In our case, the metric tensor g_{ik} is associated with transition from Cartesian to general coordinates, but the transformation can be performed between two manifolds of the same dimension. We can also write

$$ds^2 = \sum_{i=1}^{n} dx_i dx^i, \tag{2.6.38}$$

hence

$$dx_i = \sum_{k=1}^{n} g_{ik} dx^k \quad (i = \overline{1,n}). \tag{2.6.39}$$

This can be considered as a system of n linear algebraic equations in the unknown quantities $dx^1, \ldots, dx^n$. Solving the system by Cramer's rule, we get:

$$dx^k = \sum_{i=1}^{n} g^{ki} dx_i \quad (k = \overline{1,n}), \tag{2.6.40}$$

where

$$g^{ki} = g^{ik} = \frac{G_{ki}}{g} \tag{2.6.41}$$

are the contravariant components of the metric tensor, while G_{ki} is the algebraic complement of the element g_{ki} in the determinant

$$g = \det(g_{ki}). \tag{2.6.42}$$

Since

$$dx_i = \sum_{k=1}^{n} g_{ik} dx^k = \sum_{k,l=1}^{n} g_{ik} g^{kl} dx_l,$$

we must have

$$\sum_{k=1}^{n} g_{ik} g^{kl} = g_i^l = \delta_i^l \quad (i, l = \overline{1, n}). \tag{2.6.43}$$

These elements of tensor calculus are useful in the derivation of the differential equation of geodesics.

Let $x^i (i = \overline{1, n})$ be the coordinates of a particle in $\mathcal{R}_n$ and let

$$x^i = x^i(s) \quad (i = \overline{1, n}) \tag{2.6.44}$$

be the parametric equations of a curve passing through the given points 1 and 2. The arc length between the two points is

$$L = \int_1^2 ds = \int_1^2 \sqrt{\sum_{i,k=1}^{n} g_{ik} \dot{x}^i \dot{x}^k}\, ds, \quad \dot{x}^i = \frac{dx^i}{ds}. \tag{2.6.45}$$

In order for the curve (2.6.44) to be a geodesic, the function

$$f(x, \dot{x}, s) = \sqrt{\sum_{i,k=1}^{n} g_{ik} \dot{x}^i \dot{x}^k} \tag{2.6.46}$$

must satisfy the Euler–Lagrange equations (2.6.20):

$$\frac{d}{ds}\left(\frac{\partial f}{\partial \dot{x}^m}\right) - \frac{\partial f}{\partial x^m} = 0 \quad (m = \overline{1, n}). \tag{2.6.47}$$

Evaluating the derivatives, we have successively:

$$\frac{\partial f}{\partial \dot{x}^m} = \sum_{k=1}^{n} g_{mk}\dot{x}^k,$$
$$\frac{d}{ds}\left(\frac{\partial f}{\partial \dot{x}^m}\right) = \sum_{k=1}^{n} g_{mk}\ddot{x}^k + \sum_{i,k=1}^{n} \frac{\partial g_{mk}}{\partial x^i}\dot{x}^i\dot{x}^k,$$
$$\frac{\partial f}{\partial x^m} = \frac{1}{2}\sum_{i,k=1}^{n} \frac{\partial g_{ik}}{\partial x^m}\dot{x}^i\dot{x}^k,$$

and (2.6.47) yields:

$$\sum_{k=1}^{n} g_{mk}\ddot{x}^k + \frac{1}{2}\sum_{i,k=1}^{n}\left(\frac{\partial g_{mk}}{\partial x^i} + \frac{\partial g_{im}}{\partial x^k} - \frac{\partial g_{ik}}{\partial x^m}\right)\dot{x}^i\dot{x}^k = 0.$$

If we denote

$$\Gamma_{ik,m} = \frac{1}{2}\left(\frac{\partial g_{mk}}{\partial x^i} + \frac{\partial g_{im}}{\partial x^k} - \frac{\partial g_{ik}}{\partial x^m}\right), \tag{2.6.48}$$

the last equation becomes

$$\sum_{k=1}^{n} g_{mk}\ddot{x}^k + \sum_{i,k=1}^{n} \Gamma_{ik,m}\dot{x}^i\dot{x}^k = 0. \tag{2.6.49}$$

The quantities (2.6.48) are the *Christoffel symbols of the first kind*. Multiplying (2.6.49) by g^{ml}, then summing over m and using (2.6.43), we finally arrive at the differential equation of *geodesics* in $\mathcal{R}_n$:

$$\ddot{x}^l + \sum_{i,k=1}^{n} \Gamma^l_{ik}\dot{x}^i\dot{x}^k = 0 \quad (l = \overline{1,n}), \tag{2.6.50}$$

where

$$\Gamma^l_{ik} = \sum_{m=1}^{n} g^{ml}\Gamma_{ik,m} \tag{2.6.51}$$

are the *Christoffel symbols of the second kind*. It can be shown that the Christoffel symbols are not tensors (except for linear transformations).

If we take four dimensions, then (2.6.50) are the equations of geodesics in the Riemannian manifold $\mathcal{R}_4$, used in the relativistic theory of gravitation. Here, Γ^l_{ik} determine the *intensity* of the field, while the components of the metric tensor g_{ik} play the role of *potentials* of the gravitational field.

Observation: If we denote

$$\phi = \frac{1}{2}\sum_{i,k=1}^{n} g_{ik}\dot{x}^i\dot{x}^k, \tag{2.6.52}$$

then the equations

$$\frac{d}{ds}\left(\frac{\partial \phi}{\partial \dot{x}^m}\right) - \frac{\partial \phi}{\partial x^m} = 0 \quad (m = \overline{1, n}) \tag{2.6.53}$$

yield the same result (2.6.50). Consequently, the following two variational equations:

$$\delta \int_1^2 \sqrt{\sum_{i,k=1}^{n} g_{ik}\dot{x}^i \dot{x}^k ds} = 0$$

and

$$\delta \int_1^2 \sum_{i,k=1}^{n} g_{ik}\dot{x}^i \dot{x}^k \, ds = 0, \tag{2.6.54}$$

are *equivalent*.

Geodesics of a Sphere

Let us find the geodesics of a sphere of constant radius $R = 1$. The sphere can be imagined as a two-dimensional Riemannian manifold embedded in the three-dimensional Euclidean space E_3. The arc element on the sphere of unit radius is

$$ds^2 = d\theta^2 + \sin^2\theta \, d\varphi^2, \tag{2.6.55}$$

and thus our variational principle can be put in the form

$$\delta \int ds = \delta \int \frac{ds^2}{ds^2} ds = \delta \int (\dot{\theta}^2 + \sin^2\theta \dot{\varphi}^2) \, ds = 0, \tag{2.6.56}$$

where $\dot{\theta} = \frac{d\theta}{ds}$ and $\dot{\varphi} = \frac{d\varphi}{ds}$. Obviously, in our case

$$f = \dot{\theta}^2 + \sin^2\theta \dot{\varphi}^2 = 1. \tag{2.6.57}$$

The geodesic for the variable φ is obtained by using the Euler–Lagrange equations (2.6.20). Performing elementary derivatives, we arrive at:

$$\ddot{\varphi} + 2\cot\theta \, \dot{\theta}\dot{\varphi} = 0. \tag{2.6.58}$$

To obtain the *explicit* equation of the geodesic $\varphi = \varphi(\theta)$, we must eliminate the parameter s between the last two equations. First, we observe that (2.6.58) can be written as

$$d\dot{\varphi} + 2\dot{\varphi}\cot\theta \, d\theta = 0,$$

giving by integration

$$\dot{\varphi} = \frac{C}{\sin^2\theta}, \tag{2.6.59}$$

where C is a constant. Then, we can write:

$$\dot{\theta} = \frac{d\theta}{ds} = \frac{d\theta}{d\varphi}\dot{\varphi}$$

and, in view of (2.6.57) and (2.6.59),

$$\sqrt{1 - \frac{C^2}{\sin^2\theta}} = \frac{d\theta}{d\varphi}\frac{C}{\sin^2\theta}.$$

Separating the variables, we have:

$$d\varphi = \frac{C}{\sin\theta\sqrt{\sin^2\theta - C^2}}\,d\theta,$$

therefore

$$\cos(\varphi - \varphi_0) = \frac{C}{\sqrt{1 - C^2}}\cot\theta, \tag{2.6.60}$$

where φ_0 is a constant of integration. This is the equation of a plane through the origin of the coordinate system, which is also the centre of the sphere. Being at the intersection of this plane with the sphere, the geodesics of our problem are *great circles*. To make our result more obvious, let us write (2.6.60) in Cartesian coordinates. Using the well-known formula for $\cos(\varphi - \varphi_0)$ and the relations of transformation

$$x = \sin\theta\cos\varphi, \quad y = \sin\theta\sin\varphi, \quad z = \cos\theta,$$

we find the equation of our plane in the *normal* form:

$$\sqrt{1 - C^2}\,(x\cos\varphi_0 + y\sin\varphi_0) - Cz = 0. \tag{2.6.61}$$

The constants φ_0 and C are determined by the choice of the fixed points.

2.7 Hamilton's Principle

The purpose of the previous section was to prepare the reader with regard to the characteristics of the variational principles of mechanics. For a better understanding of the importance and usefulness of these principles, we shall begin our study in real, physical space.

Let us consider a system of N particles, subject to ideal holonomic constraints of the form (2.1.57), and suppose we know the *real* motion of the particles during the time interval (t_1, t_2), i.e. we know the functions

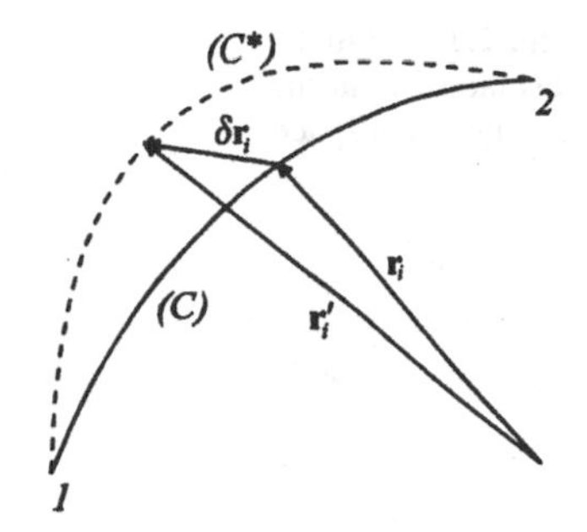

Fig. 2.12 Virtual variation of the radius-vector of a particle from one point of the real trajectory (*C*) to the corresponding point of the varied path (C^*).

$$\mathbf{r}_i = \mathbf{r}_i(t), \quad t_1 \le t \le t_2 \quad (i = \overline{1,N}). \tag{2.7.1}$$

Let us also consider another law of motion, given by

$$\mathbf{r}_i^* = \mathbf{r}_i^*(t), \tag{2.7.2}$$

which is consistent with the constraints (2.1.57), but does not obey the equations of motion, expressing, as we already know, a *virtual* motion of the system. We also assume that

$$\mathbf{r}_i(t_\alpha) = \mathbf{r}_i^*(t_\alpha) \quad (\alpha = 1, 2,\ i = \overline{1,N}), \tag{2.7.3}$$

meaning that both the real and the virtual trajectories pass, at times t_1 and t_2, through the same fixed points of the real three-dimensional space. Hence, the virtual displacements

$$\delta\mathbf{r}_i = \mathbf{r}_i(t) - \mathbf{r}_i^*(t) \quad (i = \overline{1,N}) \tag{2.7.4}$$

represent the variation of the radius-vector of the particle P_i from one point of the real trajectory (*C*) to the corresponding point of the varied path (C^*) (Fig. 2.12). It follows that

$$\delta\mathbf{r}_i(t_1) = \delta\mathbf{r}_i(t_2) = 0 \quad (i = \overline{1,N}), \tag{2.7.5}$$

as well as

$$\frac{d}{dt}(\delta\mathbf{r}_i) = \frac{d}{dt}(\mathbf{r}_i - \mathbf{r}_i^*) = \mathbf{v}_i - \mathbf{v}_i^* = \delta\mathbf{v}_i \quad (i = \overline{1,N}). \tag{2.7.6}$$

Let us now direct our attention to the kinetic energy T^*, associated with the virtual motion. Supposing the trajectory (C^*) is infinitely close to (*C*), we may write:

$$T^* \simeq \frac{1}{2}\sum_{i=1}^{N} m_i(|\dot{\mathbf{r}}_i|^2 - 2\dot{\mathbf{r}}_i\cdot\delta\dot{\mathbf{r}}_i) = T - \sum_{i=1}^{N} m_i\dot{\mathbf{r}}_i\cdot\delta\dot{\mathbf{r}}_i. \tag{2.7.7}$$

Using this result, we shall make some transformation in D'Alembert's principle (2.5.5), as follows:

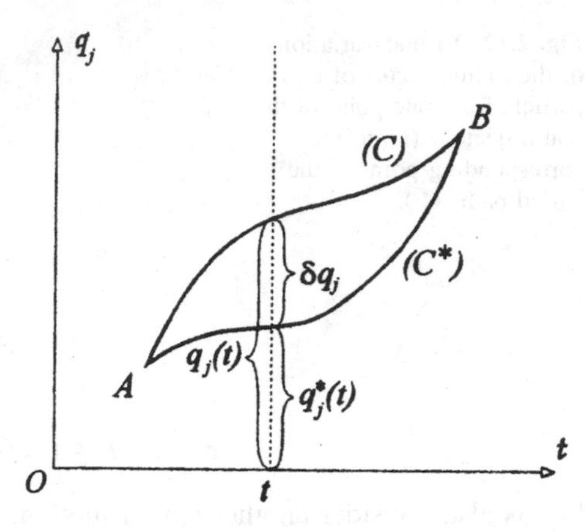

Fig. 2.13 Virtual displacement in the configuration space.

$$\sum_{i=1}^{N} m_i\ddot{\mathbf{r}}_i \cdot \delta\mathbf{r}_i = \frac{d}{dt}\left(\sum_{i=1}^{N} m_i\dot{\mathbf{r}}_i \cdot \delta\mathbf{r}_i\right) - \sum_{i=1}^{N} m_i\dot{\mathbf{r}}_i \cdot \delta\dot{\mathbf{r}}_i = \frac{d}{dt}\left(\sum_{i=1}^{N} m_i\dot{\mathbf{r}}_i \cdot \delta\mathbf{r}_i\right) - \delta T, \tag{2.7.8}$$

where $\delta T = T - T^*$ is given by (2.7.7). In view of (2.4.13), we can write:

$$\frac{d}{dt}\left(\sum_{i=1}^{N} m_i\dot{\mathbf{r}}_i \cdot \delta\mathbf{r}_i\right) = \delta(T + W). \tag{2.7.9}$$

Integrating with respect to time between the fixed limits t_1 and t_2 and using (2.7.5), we obtain:

$$\int \delta(T + W)dt = 0. \tag{2.7.10}$$

In the case of potential forces, the virtual potential V^* can be calculated in terms of the real potential V in a similar way:

$$\begin{aligned} V^* &= V(\mathbf{r}_1^*, \ldots, \mathbf{r}_N^*, t) = V(\mathbf{r}_1 - \delta\mathbf{r}_1, \ldots, \mathbf{r}_N - \delta\mathbf{r}_N, t) \\ &\simeq V(\mathbf{r}_1, \ldots, \mathbf{r}_N, t) - \sum_{i=1}^{N} (\mathrm{grad}_i V) \cdot \delta\mathbf{r}_i, \end{aligned} \tag{2.7.11}$$

therefore the virtual work done by the system is:

$$\delta W = \sum_{i=1}^{N} \mathbf{F}_i \cdot \delta\mathbf{r}_i = V^* - V = -\delta V, \tag{2.7.12}$$

and relation (2.7.10), with notation (2.5.16), becomes:

$$\int_{t_1}^{t_2} \delta L\, dt = \delta \int_{t_1}^{t_2} L\, dt = 0. \tag{2.7.13}$$

This relation expresses *Hamilton's principle*. Since $\mathbf{r}_i$ and $\mathbf{r}_i^*$ are not independent variables, but must satisfy the same constraint relations, this principle is usually used in the configuration space. Let $q_j = q_j(t)$ and $q_j^* = q_j^*(t)$ $(j = \overline{1,n})$ be the parametric equations of generalized trajectories corresponding to the real and (one of the) virtual motions, respectively, and define the virtual displacements δq_j by

$$\delta q_j(t) = q_j(t) - q_j^*(t) \quad (j = \overline{1,n}). \tag{2.7.14}$$

According to the condition (2.7.5), the generalized trajectories corresponding to the real and virtual displacements pass through the same points in configuration space, i.e.

$$\delta q_j(t_1) = \delta q_j(t_2) = 0 \quad (j = \overline{1,n}). \tag{2.7.15}$$

Except for (2.7.15), the virtual variations δq_j are *independent* and, as pointed out in Fig. 2.13, they are orthogonal to the t-axis. Therefore, we can write (2.7.13) as

$$\delta \int_{t_1}^{t_2} L(q, \dot{q}, t)\, dt = 0. \tag{2.7.16}$$

The integral

$$S = \int_{t_1}^{t_2} L(q, \dot{q}, t)\, dt \tag{2.7.17}$$

is called the *action integral*. We are now able to formulate Hamilton's principle: *Out of all the possible generalized paths passing through two fixed points, corresponding to the times t_1 and t_2, the real motion is performed on that path for which the action is stationary*. Hamilton's principle is also called the *principle of stationary action*. Since, in general, the stationary extremum is a minimum, sometimes it is named the *principle of least action*.

The principle was published in 1834 by *William Rowan Hamilton*. Its discovery played an important role in the development of various aspects of theoretical physics and we shall support this statement by many examples.

2.7.1 Euler–Lagrange Equations for the Action Integral

Let us prove that the Lagrange differential equations of motion, for both potential and non-potential forces, can be derived from Hamilton's principle. Taking the first (virtual) variation of the action, we have:

$$\delta S = \int_{t_1}^{t_2} \sum_{j=1}^{n} \left(\frac{\partial L}{\partial q_j} \delta q_j + \frac{\partial L}{\partial \dot{q}_j} \delta \dot{q}_j \right) dt,$$

or, if we integrate by parts the second term,

$$\delta S = \left[\sum_{j=1}^{n} \frac{\partial L}{\partial \dot{q}_j} \delta q_j \right]_{t_1}^{t_2} - \int_{t_1}^{t_2} \sum_{j=1}^{n} \left[\frac{d}{dt}\left(\frac{\partial L}{\partial \dot{q}_j} \right) - \frac{\partial L}{\partial q_j} \right] \delta q_j \, dt.$$

Using (2.7.15) and the arbitrariness of the variations δq_j, we arrive at the Lagrange equations (2.5.17).

In a similar way, starting from (2.7.10), we can obtain the Lagrange equations for non-potential systems (2.5.13). Indeed, we may write

$$\int_{t_1}^{t_2} \sum_{j=1}^{n} \left(\frac{\partial T}{\partial q_j} \delta q_j + \frac{\partial T}{\partial \dot{q}_j} \delta \dot{q}_j + Q_j \delta q_j \right) dt = 0$$

and, after an integration by parts,

$$\left[\sum_{j=1}^{n} \frac{\partial T}{\partial \dot{q}_j} \delta q_j \right]_{t_1}^{t_2} + \int_{t_1}^{t_2} \left[\frac{\partial T}{\partial q_j} - \frac{d}{dt}\left(\frac{\partial T}{\partial \dot{q}_j} \right) + Q_j \right] \delta q_j \, dt = 0.$$

Since $\delta q_j(t_1) = \delta q_j(t_2) = 0$, for independent and arbitrary δq_j we arrive at (2.5.13), as expected.

The condition (2.7.10), sometimes called *generalized Hamilton's principle*, can also be used to derive the Lagrange equations in case of existence of a velocity-dependent potential $V(q, \dot{q}, t)$. To show this, we must prove that there exists a function $L = T - V(q, \dot{q}, t)$, such that (2.7.10) is equivalent with (2.7.16) if the condition (2.5.23) is satisfied. Indeed, we have:

$$\int_{t_1}^{t_2} (\delta T + \delta W) \, dt = \int_{t_1}^{t_2} \left\{ \delta T + \sum_{j=1}^{n} \left[\frac{d}{dt}\left(\frac{\partial V}{\partial \dot{q}_j} \right) - \frac{\partial V}{\partial q_j} \right] \delta q_j \right\} dt = 0$$

and, after integration by parts,

$$\int_{t_1}^{t_2} \left[\delta T - \sum_{j=1}^{n} \left(\frac{\partial V}{\partial q_j} \delta q_j + \frac{\partial V}{\partial \dot{q}_j} \delta \dot{q}_j \right) \right] dt = \int_{t_1}^{t_2} \delta(T - V) \, dt = \delta \int_{t_1}^{t_2} L \, dt = 0,$$

which completes the proof.

Finally, let us show that Hamilton's principle (2.7.16) is a *variational principle*. Comparing (2.6.10) with (2.7.17), it appears obvious that the action S is a *functional* of $q_1, \ldots, q_n$, while the correspondence

$$x \to t, \quad y_i(x) \to q_j(t), \quad f(y, y', x) \to L(q, \dot{q}, t),$$
$$I(y_1, \ldots, y_n) \to S(q_1, \ldots, q_n)$$

shows that the Euler–Lagrange equations (2.6.20) are formally identical with the Lagrange equations for natural systems (2.5.17).

Before closing this section, we wish to emphasize the importance of the variational principles, Hamilton's principle being one of them. (Another variational principle will be discussed in Sect. 2.9; its understanding needs some extra background, which by then will be given.) Compared with several other formulations of mechanics, this principle offers certain advantages. First, since we deal with quantities defined with respect to any frame, the principle does not depend on the choice of coordinates. Second, once a single scalar function, the Lagrangian, is known, one can obtain both the differential equations of motion and the associated laws of conservation, in a direct and simple way. Third, the variational principles can be used for a unitary description of some other systems, like fields. This extension is possible because the Lagrangian has the dimension of energy and this quantity can be defined for any type of motion, while not all interactions can be described by forces. As we shall see later on, the fundamental equations of electrodynamics (Maxwell's equations), of the theory of linear elasticity (Lamé's equations), of quantum mechanics (Schrödinger equation), etc. can be derived from Hamilton's principle.

2.7.2 *Criteria for the Construction of Lagrangians*

There are several criteria which must be obeyed in constructing the Lagrangian function used in our formalism. They are:

(a) *Superposition principle*. If the physical system consists of two (or more) interacting particles, the Lagrangian is composed of three groups of terms: (i) The Lagrangians of each particle, when the others are absent; (ii) Terms expressing the interaction between particles; (iii) Terms describing the interaction between the system and the exterior fields (if there are any).
(b) *Invariance principle*. The action must be invariant with respect to the appropriate group of transformations (e.g. Galilei group in Newtonian mechanics, Lorentz group in relativistic mechanics).
(c) *Correspondence principle*. The Lagrangian must be constructed in such a way, that all results of Newtonian mechanics be obtained by Hamilton's principle.
(d) *Principle of physical symmetry*. The choice of the generalized coordinates must provide a Lagrangian function not only simple, but also useful, i.e. suitable to the symmetry properties of the system.

2.8 Symmetry Properties and Conservation Theorems

We already know that the motion of a mechanical system can be determined, in principle, by integrating the Lagrange equations (2.5.17). We say 'in principle', because there are circumstances when this operation is neither useful nor even possible. Nevertheless, there exist some cases when it is possible to obtain information about our system without a full integration of the equations of motion. This is done by using the *first integrals.*

2.8.1 First Integrals as Constants of Motion

Consider a system of N particles with n degrees of freedom, subject to holonomic constraints, and assume that we found a relation of the type:

$$f(q_1,\ldots,q_n,\dot{q}_1,\ldots,\dot{q}_n,t) = C(\text{const.}), \tag{2.8.1}$$

which is identically satisfied by *any* solution of the Lagrange equations and for *any* initial conditions. Then (2.8.1) is called a *first integral* of (2.5.17) or a *constant of motion* (see Chap. 1, Sect. 1.2)

Suppose we know h distinct first integrals

$$f_s(q,\dot{q},t) = C_s \quad (s = \overline{1,h}). \tag{2.8.2}$$

Then any function

$$F(f_1,\ldots,f_h) = \text{const.} \tag{2.8.3}$$

is also a constant of motion, but not independent of (2.8.2). Since the general integral of the Lagrange equations (2.5.17),

$$q_j = q_j(t, C_1,\ldots,C_{2n}) \quad (j = \overline{1,n}), \tag{2.8.4}$$

depends on $2n$ arbitrary independent constants, it follows that the maximum number of *distinct* first integrals is $2n$. The constants $C_1,\ldots,C_{2n}$ are determined from the initial conditions:

$$q_j^0 = q_j(t_0, C_1,\ldots,C_{2n}), \quad \dot{q}_j^0 = \dot{q}_j(t_0, C_1,\ldots,C_{2n}) \quad (j = \overline{1,n}). \tag{2.8.5}$$

The integration of the Lagrange equations is considerably facilitated by the application of first integrals, because:

(i) Finding a solution of a *first-order* differential equation is an easier task;
(ii) A first integral offers information on the physical nature of the system, as well as its symmetry properties;
(iii) In some cases, first integrals express the conservation of fundamental physical quantities, such as linear and angular momenta, energy, etc.

Consequently, finding the first integrals (if there are any) is a necessary step in solving a problem by the Lagrangian technique.

As we have already mentioned in Sect. 2.5, if q_k is a cyclic coordinate, then the quantity $\frac{\partial L}{\partial \dot{q}_k}$ is a constant of motion. Let us define the quantities

$$p_j = \frac{\partial L}{\partial \dot{q}_j} \quad (j = \overline{1,n}) \tag{2.8.6}$$

and call them the *generalized momenta* associated (or conjugated) to the generalized coordinates q_j. The dimensions of p_j are given by those of q_j: if q is a distance, then p is a linear momentum; if q is an angle, then p is an angular momentum, etc. But in any case, we must have:

$$[p_j \dot{q}_j] = [\text{ENERGY}] = ML^2T^{-2}. \tag{2.8.7}$$

Introducing (2.8.6) into the Lagrange equations (2.5.17), we obtain:

$$\dot{p}_j = \frac{\partial L}{\partial q_j} \quad (j = \overline{1,n}). \tag{2.8.8}$$

If the coordinate q_k is cyclic, then (2.8.8) yields

$$p_k = \text{const.}, \tag{2.8.9}$$

expressing the *conservation of the generalized momentum associated with a cyclic coordinate.*

Equation (2.8.9) gives either the conservation of linear momentum, or that of angular momentum. It is valid not only in mechanics, but also for other physical systems. For example, since the coordinate x_k does not appear in the Lagrangian (2.5.30) describing the behaviour of a charged particle in an external electromagnetic field, the conjugated momentum is conserved:

$$p_k = mv_k + eA_k = \text{const.} \tag{2.8.10}$$

Therefore, in solving a concrete problem we should follow the rule: look for cyclic coordinates, each of them being associated with a first integral. Then, if there are not any, search for another set of generalized coordinates q'_j $(j = \overline{1,n})$, of which at least one being cyclic.

A useful example is offered by a particle moving in a central field. When expressed in Cartesian coordinates, the Lagrangian

$$L = \frac{1}{2}m(\dot{x}^2 + \dot{y}^2) - V(x, y)$$

does not display any cyclic coordinate, but if it is written in terms of polar coordinates,

$$L = \frac{1}{2}m(\dot{r}^2 + r^2\dot{\varphi}^2) - V(r),$$

it shows the ignorable coordinate φ, leading to the first integral

$$p_\varphi = mr^2\dot{\varphi} = \text{const.} \tag{2.8.11}$$

2.8.2 Symmetry Transformations

As we have mentioned, the first integrals are related with the conservation of the fundamental physical quantities. This fact emerges from the intrinsic properties of the space–time continuum and expresses the connection between different types of motion and the conservation of associated quantities.

The study of a large amount of experimental data has led to the conclusion that the real, physical space is *homogeneous* (there are no privileged reference frames) and *isotropic* (there are no privileged directions), while time passes *uniformly* (there are no privileged moments of time). The homogeneity of space results in the fact that the properties of an isolated mechanical system do not change if all particles of the system perform infinitesimal translations with the same velocity **v**, while the isotropy yields the conservation of the properties of the system if all particles execute infinitesimal rotations of the same angle, about the same direction. Finally, the uniformity of time shows that the origin of the time interval can be arbitrarily chosen, meaning that the properties of the system remain unchanged at an infinitesimal displacement of the time origin.

An important role in the study of physical systems is played by those transformations which leave the form of the differential equations of motion unchanged. These are called *symmetry transformations*. For example, we can cite space–time transformations, gauge transformations, etc. In the first category are included the *space displacements* (translations, rotations) and *time transformations*. The *gauge transformations* appear when one (or more) physical quantity is not completely determined by its equation of definition, and we shall familiarize the readers with them later on in this book.

2.8.3 Noether's Theorem

As we mentioned earlier in this section, the cyclic coordinates lead to constants of motion, expressing the symmetry of the Lagrangian with respect to certain space–time transformations. But not all constants of motion come from evident symmetry properties of the system, or have a simple form. That is why there appears the necessity of giving a general method to obtain the first integrals. Such formalism was provided in 1918 by the German Jewish mathematician *Emmy Noether*.[1]

Let us consider a physical system, described by the generalized coordinates q_j and velocities $\dot{q}_j$, and assume that the Lagrangian $L(q, \dot{q}, t)$ of the system is known. A transformation of coordinates and time

$$q'_j = q'_j(q_1, \ldots, q_n, t), \quad t' = t'(q_1, \ldots, q_n, t) \tag{2.8.12}$$

[1] Noether, E.: Invariante Variationsprobleme. Nachr. Kgl. Ges. Wiss. Göttingen **2**, 235 (1918).

is a *symmetry transformation* if it leaves Hamilton's principle invariant or, in view of (2.5.34), if

$$L\left(q', \frac{dq'}{dt'}, t'\right) dt' = \left[L(q, \dot{q}, t) + \frac{dF(q,t)}{dt}\right] dt. \quad (2.8.13)$$

Let us now specify the transformation (2.8.12), namely that it is an *infinitesimal* transformation of the form:

$$q'_j = q_j + \epsilon \eta_j(q,t) \quad (j = \overline{1,n}), \quad t' = t + \epsilon \tau(q,t), \quad (2.8.14)$$

where η_j and τ are arbitrary functions, while the parameter ϵ is small enough as to keep only terms linear in it. Since

$$\frac{dt'}{dt} = 1 + \epsilon \frac{d\tau}{dt}, \quad \frac{dt}{dt'} \simeq 1 - \epsilon \frac{d\tau}{dt}, \quad (2.8.15)$$

we have:

$$\frac{dq'}{dt'} = \frac{dq'}{dt}\frac{dt}{dt'} \simeq \left(\dot{q}_j + \epsilon \frac{d\eta_j}{dt}\right)\left(1 - \epsilon \frac{d\tau}{dt}\right) \simeq \dot{q}_j + \epsilon\left(\frac{d\eta_j}{dt} - \dot{q}_j \frac{d\tau}{dt}\right). \quad (2.8.16)$$

Introducing (2.8.15) and (2.8.16) into (2.8.13), we arrive at:

$$L\left[q + \epsilon\eta, \dot{q} + \epsilon\left(\frac{d\eta}{dt} - \dot{q}\frac{d\tau}{dt}\right), t + \epsilon\tau\right]\left(1 + \epsilon\frac{d\tau}{dt}\right) = L(q, \dot{q}, t) + \epsilon\frac{d\phi}{dt},$$

where we took $F(q,t) = \epsilon\phi(q,t)$ because, obviously, F must be infinitesimal and linear in ϵ. Using Taylor's formula for series expansion in the l.h.s. and keeping only terms linear in ϵ, after some reduction and rearranging of terms we are left with

$$\epsilon\left[L\frac{d\tau}{dt} + \sum_{j=1}^{n} \eta_j \frac{\partial L}{\partial q_j} + \sum_{j=1}^{n}\left(\frac{d\eta_j}{dt} - \dot{q}_j\frac{d\tau}{dt}\right)\frac{\partial L}{\partial \dot{q}_j} + \tau\frac{\partial L}{\partial t} - \frac{d\phi}{dt}\right] = 0. \quad (2.8.17)$$

The transformation (2.8.14) is a symmetry transformation if, for a given L, there exist some functions $\eta_j\,(j = \overline{1,n})$ and τ, so that the l.h.s. of (2.8.17) is a total time derivative of a function of q_j, t. The first integrals of motion are obtained from the condition of stationary action S, on any path where the equations of motion (2.5.17) are satisfied. Before applying the action principle, we should mention that the variations $\delta q_j = q'_j - q_j\,(j = \overline{1,n})$ and $\delta t = t' - t$ differ from those previously used by the fact that they perform a transition between two possible trajectories. Therefore, the first variation of the action S reads:

$$\delta S = \epsilon \int_{t_1}^{t_2} \left[\sum_{j=1}^{n} \eta_j \frac{\partial L}{\partial q_j} + \sum_{j=1}^{n}\left(\frac{d\eta_j}{dt} - \dot{q}_j\frac{d\tau}{dt}\right)\frac{\partial L}{\partial \dot{q}_j} + L\frac{d\tau}{dt} + \tau\frac{\partial L}{\partial t} - \frac{d\phi}{dt}\right] dt,$$

or, recalling that L is a function of $q_j, \dot{q}_j, t$:

$$\delta S = \epsilon \int_{t_1}^{t_2} \left\{ \frac{d}{dt}\left[\sum_{j=1}^{n} \eta_j \frac{\partial L}{\partial \dot{q}_j} - \tau\left(\sum_{j=1}^{n} \dot{q}_j \frac{\partial L}{\partial \dot{q}_j} - L \right) - \phi \right] + \sum_{j=1}^{n} (\tau \dot{q}_j - \eta_j)\left[\frac{d}{dt}\left(\frac{\partial L}{\partial \dot{q}_j} \right) - \frac{\partial L}{\partial q_j} \right] \right\} dt. \tag{2.8.18}$$

The invariance of Hamilton's principle under the symmetry transformation (2.8.14) means $\delta S = 0$ for any time interval within which the Lagrange equations are valid. This implies

$$\sum_{j=1}^{n} \eta_j \frac{\partial L}{\partial \dot{q}_j} - \tau\left(\sum_{j=1}^{n} \frac{\partial L}{\partial \dot{q}_j}\dot{q}_j - L \right) - \phi = C, \tag{2.8.19}$$

where C is a constant. This equation expresses *Noether's theorem* for discrete systems of particles: *To any continuous symmetry transformation* (2.8.14), *one can associate a first integral* (2.8.19). Noether's theorem can also be written for infinitesimal quantities $\epsilon\eta_j$, $\epsilon\tau$, $\epsilon\phi$, ϵC, which is useful in some applications.

We shall now consider some particular cases, which will show the connection between Noether's theorem and the general theorems of mechanics discussed in Chap. 1.

1. Let us consider an isolated system of N particles ($V^{(e)} = 0$) with n degrees of freedom and assume that the particle P_i performs an infinitesimal space displacement of the form

$$\mathbf{r}'_i = \mathbf{r}_i + \delta\mathbf{r}_i \quad (i = \overline{1,N}), \quad t' = t. \tag{2.8.20}$$

Since

$$\frac{\partial L}{\partial \dot{q}_j} = \frac{\partial T}{\partial \dot{q}_j} = \sum_{i=1}^{N} m_i \dot{\mathbf{r}}_i \cdot \frac{\partial \dot{\mathbf{r}}_i}{\partial \dot{q}_j} = \sum_{i=1}^{N} m_i \dot{\mathbf{r}}_i \cdot \frac{\partial \mathbf{r}_i}{\partial q_j},$$

we obtain:

$$\epsilon \sum_{j=1}^{n} \eta_j \frac{\partial L}{\partial \dot{q}_j} = \epsilon \sum_{i=1}^{N} \left(\sum_{j=1}^{n} m_i \dot{\mathbf{r}}_i \cdot \frac{\partial \mathbf{r}_i}{\partial q_j} \eta_j \right) = \sum_{i=1}^{N} m_i \dot{\mathbf{r}}_i \cdot \delta\mathbf{r}_i. \tag{2.8.21}$$

Suppose now that our space displacement is a *translation*, i.e. all particles of the system perform a straight motion in the same direction with the same velocity. Then we have

$$\delta\mathbf{r}_i \equiv \delta\mathbf{r} = \mathbf{n}\,\delta r, \tag{2.8.22}$$

$\mathbf{n}$ being the unit vector along the direction of translation. Taking $\phi = 0$ in (2.8.19), we get:

$$\delta r\,\mathbf{n}\cdot\sum_{i=1}^{N} m_i\dot{\mathbf{r}}_i = \text{const.}, \tag{2.8.23}$$

showing the conservation of total linear momentum in the direction of translation (Chap. 1, Sect. 1.3)

2. If the infinitesimal space transformation (2.8.20) is a *rotation* of all the particles, about the same axis and in the same direction:

$$\delta\mathbf{r}_i = \delta\boldsymbol{\theta}\times\mathbf{r}_i = \delta\theta\mathbf{s}\times\mathbf{r}_i, \tag{2.8.24}$$

where $\mathbf{s}$ is the unit vector along the axis of rotation and $\delta\theta$ is the constant angle of rotation, then:

$$\epsilon\sum_{j=1}^{n}\eta_j\frac{\partial L}{\partial\dot{q}_j} = \sum_{i=1}^{N} m_i\dot{\mathbf{r}}_i\cdot(\delta\boldsymbol{\theta}\times\mathbf{r}_i) = \delta\theta\mathbf{s}\cdot\sum_{i=1}^{N} m_i\mathbf{r}_i\times\dot{\mathbf{r}}_i$$

and, according to Noether's theorem (2.8.19),

$$\delta\theta\,\mathbf{s}\cdot\sum_{i=1}^{N} m_i\mathbf{r}_i\times\dot{\mathbf{r}}_i = \text{const.}, \tag{2.8.25}$$

which is nothing else but the conservation of the total angular momentum (Chap. 1, Sect. 1.3)

3. A special type of space transformation is that associated with the Newtonian mechanics principle which states that two inertial frames are equivalent in describing the motion of a mechanical system. The transition from one frame to another is given by an *infinitesimal Galilean transformation*:

$$\mathbf{r}'_i = \mathbf{r}_i + (\delta\mathbf{v}_0)t, \tag{2.8.26}$$

where the infinitesimal constant vector $\delta\mathbf{v}_0$ is the relative velocity of the frames. Indeed, taking the time derivative of (2.8.26), we have:

$$\mathbf{v}'_i = \mathbf{v}_i + \delta\mathbf{v}_0. \tag{2.8.27}$$

Choosing

$$\epsilon\phi = -\delta\mathbf{v}_0\cdot\sum_{i=1}^{N} m_i\mathbf{r}_i, \quad \epsilon C = -\left(\sum_{i=1}^{N} m_i\right)\mathbf{r}_G^0\cdot\delta\mathbf{v}_0 \tag{2.8.28}$$

in (2.8.19) and using (2.8.26), we obtain:

$$\delta\mathbf{v}_0\cdot\left(\frac{\sum_{i=1}^{N} m_i\mathbf{r}_i}{\sum_{i=1}^{N} m_i} - \mathbf{r}_G^0 - \frac{\sum_{i=1}^{N} m_i\mathbf{v}_i}{\sum_{i=1}^{N} m_i}t\right) = 0,$$

or

$$\mathbf{n}\cdot(\mathbf{r}_G - \mathbf{r}_G^0 - \mathbf{v}_G t) = 0, \tag{2.8.29}$$

which is the centre of mass theorem for isolated systems (Chap. 1, Sect. 1.3). Here, $\mathbf{n}$ is the unit vector of $\delta \mathbf{v}_0$, while the meaning of $\mathbf{r}_G$ and $\mathbf{v}_G$ is obvious.

4. Let us now consider a pure time transformation and take

$$\tau = 1, \quad \delta q_j = 0, \quad \phi = 0$$

in (2.8.19). Hence:

$$\sum_{j=1}^{n} \frac{\partial L}{\partial \dot{q}_j} \dot{q}_j - L = C. \tag{2.8.30}$$

To understand the physical significance of this equation of conservation, we shall first make some comments on the function

$$H = \sum_{j=1}^{n} p_j \dot{q}_j - L, \tag{2.8.31}$$

where p_j is given by (2.8.6). This function is called the *Hamiltonian* of the system. Recalling Euler's theorem for homogeneous functions:

$$\sum_{i=1}^{n} \frac{\partial f}{\partial x_i} x_i = m f, \tag{2.8.32}$$

where $f(x_1, \ldots, x_n)$ is a homogeneous function of grade m, and using (2.5.31), we may write:

$$\begin{aligned} H &= \sum_{j=1}^{n} \frac{\partial L_1}{\partial \dot{q}_j} \dot{q}_j + \sum_{j=1}^{n} \frac{\partial L_2}{\partial \dot{q}_j} \dot{q}_j - L = L_1 + 2L_2 - (L_0 + L_1 + L_2) = L_2 - L_0 \\ &= T_2 - T_0 + V. \end{aligned} \tag{2.8.33}$$

If the constraints are scleronomous, then $T_0 = 0$, $T = T_2$, and we arrive at

$$H = T + V = \text{const.}, \tag{2.8.34}$$

which shows that *the Hamiltonian of a conservative system represents the total energy*, being a constant of motion.

The function H is of great importance in analytical mechanics. We shall encounter it again in Chap. 5 and discuss there its properties more thoroughly.

2.9 Principle of Least Action

This principle was discovered in 1745 by *Pierre-Louis Moreau de Maupertuis* and it is the first integral principle of mechanics. Its initial formulation was nebulous and it was the merit of Euler, Lagrange and Jacobi that the principle acquired its current form.

In the discussion of Hamilton's principle, we used the notion of virtual displacements consistent with the constraints δq_j, being performed by the representative point in the configuration space, when passing from a point P of the real generalized trajectory (C), to some point P^* of an infinitely close trajectory (C^*) *at the same time t* (synchronic variations):

$$P(q,t) \to P^*(q+\delta q, t). \tag{2.9.1}$$

Since the boundary points were supposed to be fixed, we also had:

$$\delta q_j(t_1) = \delta q_j(t_2) = 0. \tag{2.9.2}$$

Summarizing, we can state that δ is a linear operator that satisfies the following conditions:

(i) δq_j are arbitrary except for the end points, where $\delta q_j = 0$;
(ii) $\delta t = 0$.

Let us now introduce a new operator Δ, including the variation of both space and time variables, defined by:

$$\Delta = \delta + \Delta t \frac{d}{dt}, \tag{2.9.3}$$

with the properties:

(i) Δq_j are arbitrary, except for end points, where $\Delta q_j = 0$;
(ii) Δt is arbitrary.

As we can see, the *asynchronous* variations given by Δ are less restrictive than those produced by δ. Applying Δ to q_j, we have:

$$\Delta q_j = \delta q_j + \dot{q}_j \Delta t, \tag{2.9.4}$$

expressing the correspondence between two points, one on the real and the other on the neighbouring path (Fig. 2.14). In the end points $\delta q_j(t_1) \neq 0$, $\delta q_j(t_2) \neq 0$, but

$$\Delta q_j(t_1) = \Delta q_j(t_2) = 0. \tag{2.9.5}$$

Let us now apply the operator Δ to some function $f(q, \dot{q}, t)$:

$$\begin{aligned} \Delta f = \delta f + \frac{df}{dt}\Delta t &= \sum_{j=1}^{n}\left(\frac{\partial f}{\partial q_j}\delta q_j + \frac{\partial f}{\partial \dot{q}_j}\delta \dot{q}_j\right) + \sum_{j=1}^{n}\left(\frac{\partial f}{\partial q_j}\dot{q}_j + \frac{\partial f}{\partial \dot{q}_j}\ddot{q}_j\right)\Delta t + \frac{\partial f}{\partial t}\Delta t \\ &= \sum_{j=1}^{n}\left(\frac{\partial f}{\partial q_j}\Delta q_j + \frac{\partial f}{\partial \dot{q}_j}\Delta \dot{q}_j\right) + \frac{\partial f}{\partial t}\Delta t, \end{aligned} \tag{2.9.6}$$

which is the usual differential of $f(q, \dot{q}, t)$. Next, we apply Δ to the action integral

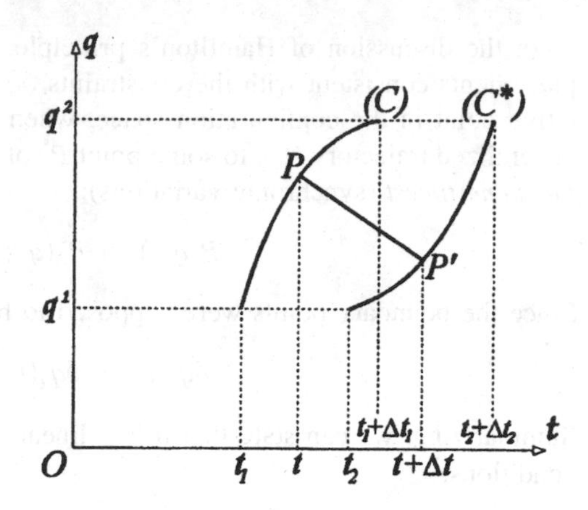

Fig. 2.14 Correspondence between two points, one on the real and the other on a neighbouring path, when both space and time variations are considered.

$$S = \int_{t_1}^{t_2} L(q, \dot{q}, t)\, dt, \tag{2.9.7}$$

where the limits t_1 and t_2 are now variable. Let $A(t)$ be the primitive function of the Lagrangian $L[q(t), \dot{q}(t), t]$. It then follows that

$$\Delta S = \Delta \int_{t_1}^{t_2} L\, dt = \Delta A(t_2) - \Delta A(t_1), \tag{2.9.8}$$

or, in view of (2.9.3),

$$\Delta S = \delta A(t_2) - \delta A(t_1) + \dot{A}(t_2)\Delta t_2 - \dot{A}(t_1)\Delta t_1 = \delta \int_{t_1}^{t_2} L\, dt + [L\Delta t]_{t_1}^{t_2}. \tag{2.9.9}$$

On the other hand, the Lagrange equations (2.5.17) allow us to write

$$\begin{aligned}\delta L &= \sum_{j=1}^{n} \left(\frac{\partial L}{\partial q_j} \delta q_j + \frac{\partial L}{\partial \dot{q}_j} \delta \dot{q}_j \right) = \sum_{j=1}^{n} (\dot{p}_j \delta q_j + p_j \delta \dot{q}_j) \\ &= \frac{d}{dt} \left(\sum_{j=1}^{n} p_j \delta q_j \right) = \frac{d}{dt} \sum_{j=1}^{n} (p_j \Delta q_j - p_j \dot{q}_j \Delta t). \end{aligned} \tag{2.9.10}$$

Therefore, using (2.9.5) and (2.8.31), we obtain:

$$\Delta S = \Delta \int_{t_1}^{t_2} L\, dt = \left[\left(\sum_{j=1}^{n} -p_j \dot{q}_j + L \right) \Delta t \right]_{t_1}^{t_2} = -[H\Delta t]_{t_1}^{t_2}. \tag{2.9.11}$$

If the system is conservative, then

$$H = E = T + V = \text{const.}, \tag{2.9.12}$$

meaning that on any varied path (C^*), the energy has the same value as on the real path (C). Therefore,

$$[H\Delta t]_{t_1}^{t_2} = H(\Delta t_2 - \Delta t_1) = H\Delta \int_{t_1}^{t_2} dt = \Delta \int_{t_1}^{t_2} H\, dt. \tag{2.9.13}$$

Introducing this result into (2.9.11), we arrive at

$$\Delta \int_{t_1}^{t_2} \sum_{j=1}^{n} p_j \dot{q}_j\, dt = 0, \tag{2.9.14}$$

which is one of the forms of the *principle of least action*: *The action taken for a real generalized trajectory is stationary with respect to any neighbouring isoenergetic path.* The quantity

$$W = \int_{t_1}^{t_2} \sum_{j=1}^{n} p_j \dot{q}_j dt \tag{2.9.15}$$

is called *Maupertuisian action.*

The principle of least action can be written in different equivalent forms. For example, recalling that the system is conservative, the kinetic energy T is a quadratic homogeneous form of generalized velocities:

$$\sum_{j=1}^{n} p_j \dot{q}_j = \sum_{j=1}^{n} \frac{\partial T}{\partial \dot{q}_j} \dot{q}_j = 2T$$

and (2.9.14) yields:

$$\Delta \int_{t_1}^{t_2} 2T\, dt = 0. \tag{2.9.16}$$

Another form of this principle was given by *Carl Jacobi.* To write it, let us extract dt from the kinetic energy formula,

$$T = \frac{1}{2} \sum_{i=1}^{N} m_i \left| \frac{d\mathbf{r}_i}{dt} \right|^2 = \frac{1}{2} \sum_{i=1}^{N} m_i \left(\frac{ds_i}{dt} \right)^2,$$

and then introduce it into (2.9.16):

$$\Delta \int_{P_1}^{P_2} \sqrt{2(E - V) \sum_{i=1}^{N} m_i\, ds_i^2} = 0, \tag{2.9.17}$$

where P_1 and P_2 are the positions of the system at the times t_1 and t_2, in the real space. If ds_i^2 is expressed in terms of q_j, then P_1 and P_2 are end points in configuration space. For a single particle, (2.9.17) reduces to

$$\Delta \int_{P_1}^{P_2} \sqrt{2m(E-V)}\, ds = 0. \tag{2.9.18}$$

Let us define the n-dimensional manifold $\mathcal{R}_n$ by the metric

$$d\sigma^2 = 2(E-V) \sum_{i=1}^{N} m_i\, ds_i^2. \tag{2.9.19}$$

Then, as we know,

$$\Delta \int_{P_1}^{P_2} d\sigma = 0$$

defines the *geodesic line* in $\mathcal{R}_n$, between P_1 and P_2. If, in particular, $\mathcal{R}_n$ is the configuration space R_n, then the metric is (see (2.4.24)):

$$d\sigma^2 = 2(E-V) \sum_{j=1}^{n} \sum_{k=1}^{n} a_{jk}\, dq_j dq_k \tag{2.9.20}$$

and the principle of least action finally acquires the form:

$$\Delta \int_{P_1}^{P_2} \sqrt{2(E-V) \sum_{j=1}^{n} \sum_{k=1}^{n} a_{jk}\, dq_j dq_k} = 0. \tag{2.9.21}$$

Equal-Action Wave Front

Let us first determine the connection between the Hamiltonian (S) and Maupertuisian (W) actions. Assuming again that the system is conservative, we have:

$$S = \int_{t_1}^{t_2} L\, dt = \int_{t_1}^{t_2} (T-V)\, dt = \int_{t_1}^{t_2} (2T-E)\, dt = W - E(t_2 - t_1).$$

Taking $t_2 = t$, $t_1 = 0$, we arrive at:

$$S(q,t) = -Et + W(q). \tag{2.9.22}$$

On the other hand, Hamilton's principle (2.7.16) and the definition of the Hamiltonian (2.8.31) yield:

$$\frac{dS}{dt} = \frac{\partial S}{\partial t} + \sum_{j=1}^{n} \frac{\partial S}{\partial q_j} \dot{q}_j = \sum_{j=1}^{n} p_j \dot{q}_j - H, \tag{2.9.23}$$

therefore, in view of (2.9.22),

$$p_j = \frac{\partial S}{\partial q_j} \quad (j = \overline{1,n}). \tag{2.9.24}$$

(This is only a rough deduction of generalized momenta in terms of S, needed in this application; for more details, see Chap. 5).

Let our conservative system be a single particle and choose $q_j = x_j$ $(j = 1,2,3)$. Then

$$p_j = \frac{\partial S}{\partial q_j} = \frac{\partial W}{\partial q_j} = (\text{grad } W)_j \quad (j = 1,2,3) \tag{2.9.25}$$

which, together with the formula of the Hamiltonian,

$$\frac{1}{2m}|\text{grad } W|^2 + V = E,$$

yields

$$|\text{grad } W| = \sqrt{2m(E-V)}. \tag{2.9.26}$$

With our choice of coordinates, the configuration space coincides with the real, three-dimensional space, while the generalized trajectory is just the real path of the particle. Then the equation

$$W(x,y,z) = \text{const.} \tag{2.9.27}$$

represents a family of fixed surfaces, while

$$S(x,y,z,t) = \text{const.} \tag{2.9.28}$$

stands for a family of moving surfaces. For example, if at $t = 0$ the surfaces S_1 and W_1 coincide, after a time interval dt, the surface S_1 has passed from W_1 to $W_1 + E\,dt$ (Fig. 2.15) and so on, similarly to the propagation of a wave front. If ds is the elementary displacement of the wave front in normal direction, we can write

$$dW = |\text{grad } W|ds = E\,dt,$$

which helps to write the phase velocity of the wave front:

$$u = \frac{E}{|\text{grad } W|} = \frac{E}{\sqrt{2m(E-V)}} = \frac{E}{\sqrt{2mT}} = \frac{E}{p}, \tag{2.9.29}$$

where p is the momentum of the particle. In other words, (2.9.29) gives the phase velocity of propagation of the *equal-action wave front.*

To study the nature of these waves, one must find certain characteristic quantities, such as frequency and wavelength. These can be obtained by making an analogy with the propagation of light waves, whose equation is:

$$\Delta\psi - \frac{n^2}{c^2}\frac{\partial^2\psi}{\partial t^2} = 0. \tag{2.9.30}$$

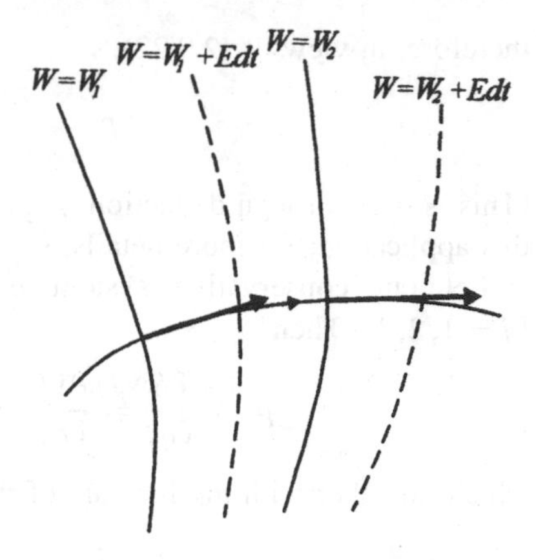

Fig. 2.15 Geometry of fixed ($W = $ const.) and moving ($W + E\,dt = $ const.) surfaces, propagating as a wave front.

Here, n is the index of refraction of the medium. If $n = $ const., the solution of (2.9.30) is

$$\psi = \psi_0(x, y, z)e^{i(\mathbf{k}\cdot\mathbf{r}-\omega t)}, \tag{2.9.31}$$

where $k = \frac{n\omega}{c} = nk_0$ is the wave number. Taking $\mathbf{k}$ in the positive direction of the x-axis, we have:

$$\psi = \psi_0(x, y, z)e^{i(k_0 nx-\omega t)}. \tag{2.9.32}$$

If n is no longer a constant, but its variation is smooth, the solution of (2.9.30) is close to the form

$$\psi = \psi_0(x, y, z)e^{i[k_0 L(x,y,z)-\omega t]}. \tag{2.9.33}$$

The quantity L is called *eikonal*. If $n = $ const., then $L = nx$ (optical path length). Introducing (2.9.33) into (2.9.30), evaluating the derivatives and then separating the real and imaginary parts, we obtain:

$$\psi_0 k_0^2[n^2 - |\text{grad}\, L|^2] + \Delta\psi_0 = 0, \tag{2.9.34}$$

$$\psi_0 \Delta L + 2(\text{grad}\,\psi_0)\cdot(\text{grad}\, L) = 0. \tag{2.9.35}$$

Suppose that the wavelength is small compared to the distance on which the medium displays its non-homogeneity. Then the presence of k_0^2 makes the first term in (2.9.34) much greater than the second, which results in

$$|\text{grad}\, L| = n. \tag{2.9.36}$$

This is the *eikonal* equation, fundamental in geometrical optics. The equal-phase surfaces are given by

$$f(x, y, z, t) = k_0 L(x, y, z) - \omega t = \text{const.} \tag{2.9.37}$$

Comparing (2.9.26) with the eikonal equation (2.9.36), we realize that they are similar: the quantity $\sqrt{2m(E-V)}$ plays the role of the refraction index and the function W that of the eikonal, while the surfaces $S = \text{const.}$ are analogous to the surfaces $f = \text{const.}$ Therefore, we can set:

$$S = \alpha f, \quad W = \alpha k_0 L(x, y, z), \quad E = \alpha\omega, \tag{2.9.38}$$

where α is an arbitrary constant. Hence:

$$n = |\text{grad}\, L| = \frac{1}{\alpha k_0}|\text{grad}\, W| = \frac{c}{\alpha\omega}\sqrt{2m(E - V)}. \tag{2.9.39}$$

This analogy shows that the propagation of equal-action waves and light waves are similar phenomena. In their remarkable papers, *Erwin Schrödinger* and *Louis de Broglie* showed that the relation between wave and geometrical optics is similar to that between quantum and classical mechanics. If we apply the principle of least action in the form (2.9.18) and observe that according to (2.9.39) the integrand is proportional to n, we have:

$$\Delta \int n\, ds = 0, \tag{2.9.40}$$

which is nothing else but the well-known *Fermat principle* of geometrical optics. Concluding our discussion, we can state that: *If applied forces are absent, the trajectory described by a particle of light (photon) is a geodesic (minimal optical path).*

2.10 Problems

1. Determine the covariant and contravariant components of the metric tensor g_{ik} in spherical coordinates r, θ, φ.
2. Study the tensor properties of the Christoffel symbols $\Gamma_{ik,l}$ and Γ^l_{ik}.
3. Determine the shape of the curve traced by a catenary of mass m and length l, whose fixed ends are at the same height. The distance between ends is a.
4. Find the plane closed curve of given perimeter, which encloses the maximum area (isoperimetric problem).
5. Study the motion of a heavy particle, constrained to move without friction on the surface of a cone.

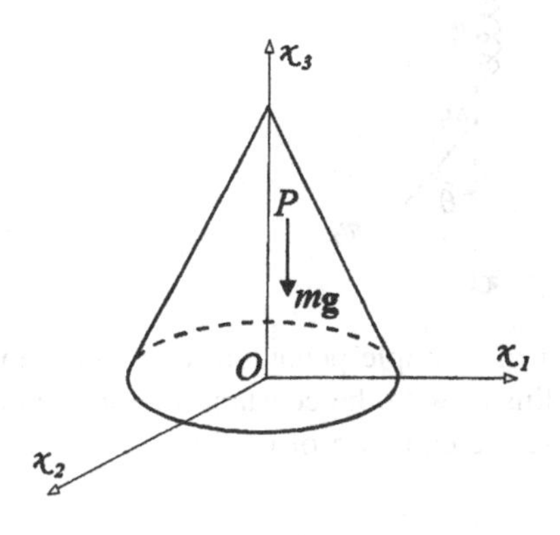

6. A particle of mass m and velocity v_1 passes from a semi-space in which its potential energy U_1 is constant, to a semi-space in which its potential energy U_2 is also constant. Determine the change in the direction of the particle.
7. A particle P of mass m moves without friction on the curve $y = f(x)$ passing through the origin. Assuming that the curve rotates about the vertical axis Oy with constant angular velocity ω_0, find the shape of the curve so that the particle remains at rest with respect to the curve.
8. Construct the Lagrangian for a system of N charged particles interacting via Coulomb law, and placed in an external variable electromagnetic field **E**, **B**.
9. Find the Lagrangian of a double coplanar pendulum and write the differential equations of motion. Linearize these equations for small motions.
10. Two masses m_1 and m_2 are fastened at the ends of an inextensible, flexible rope, running over a massless pulley (Atwood machine). Determine the law of motion and the force of constraint.

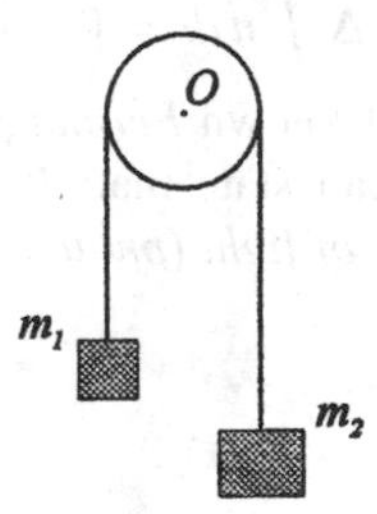

11. Investigate the motion of a plane pendulum of mass m_1 whose point of support of mass m_2 is able to perform one of the following motions:

    (a) A displacement on a horizontal straight line;
    (b) A displacement on a vertical circle with constant angular velocity ω;
    (c) Oscillations along a horizontal line according to the law $a \cos \omega t$;
    (d) Oscillations along a vertical line according to the law $a \sin \omega t$.

12. Determine the equations of motion and the period of small oscillations of the system shown in the figure.

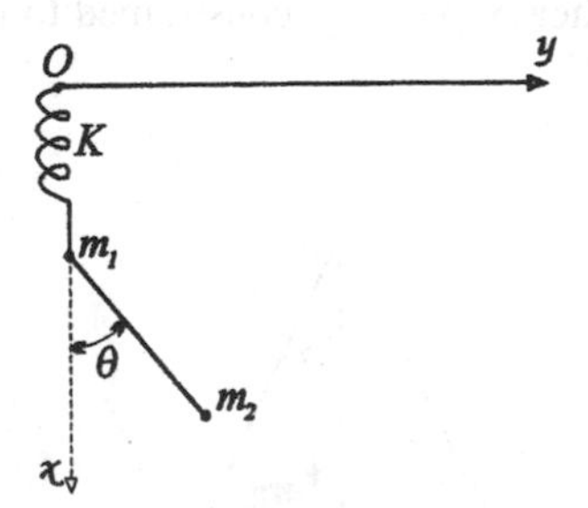

13. The point of support of a simple pendulum of mass m moves uniformly on a vertical circle of radius R, with the constant angular velocity ω. Construct the Lagrangian and write the equation of motion.

14. A system is composed of a particle of mass M and n particles of mass m. Separate the motion of the centre of mass and reduce the problem to the motion of n particles.
15. Discuss the one-dimensional motion of a particle of mass m in the field with potential energy $U(x) = U_0(e^{-2\alpha x} - e^{-\alpha x})$ (the Morse potential). Here, $U_0 > 0$, $\alpha > 0$.
16. Assume that a particle of mass m moves in a field whose potential is either $(a)\, U = U(\rho)$, or $(b)\, U = U(z)$, where ρ, φ, z are cylindric coordinates. Find the first integrals of motion in both cases.
17. A particle moves on a helix of equations $\rho = a$, $z = b\varphi$, where a and b are constants. If the potential energy is

$$V = \frac{1}{2}k(\rho^2 + z^2),$$

where k is another constant, find the law of motion of the particle and the force of constraint.
18. Two particles m_1 and m_2 are connected by a cord passing through a hole in a horizontal table. The mass m_1 moves like a simple pendulum, while the mass m_2 slides without friction on the table. Identify the constraints and write the equations of motion.
19. Show that the transformation

$$x'_i = x_i + \eta_i a_i \sin \omega t, \quad t' = t \quad (i = 1, 2, 3),$$

where $\mathbf{a}(a_1, a_2, a_3)$ is an arbitrary constant vector and $\omega^2 = k/m$, is a symmetry transformation for the Lagrangian

$$L = \frac{m}{2}\dot{\mathbf{r}}^2 - \frac{1}{2}kr^2$$

(space oscillator) and find the first integral of motion associated with this transformation.
20. Using the first integral found in the previous problem, as well as some other first integrals corresponding to the motion described by this Lagrangian, determine the law of motion and the trajectory associated with the following initial conditions:

$$\mathbf{r}(0) = (x_0, 0, 0), \quad \dot{\mathbf{r}}(0) = (0, v_0, 0).$$

Chapter 3
Applications of the Lagrangian Formalism in the Study of Discrete Particle Systems

3.1 Central Force Fields

3.1.1 Two-Body Problem

Consider a system of N bodies, supposed to be particles, and assume that they are subject only to internal forces $\mathbf{F}_{ij}$ (gravitational, electrostatic, etc.). The problem of determining the motion of each body in the presence of the other $N-1$ is known as the *problem of the N bodies*. The difficulty of solving such a problem is obviously dependent on the number of the bodies involved. The simplest case ($N=2$) is found in classical systems, like Sun–Earth, or nucleus–electron, and is called the *two-body problem*.

Let us consider an isolated system of two particles of masses m_1 and m_2 and let $\mathbf{r}_1$, $\mathbf{r}_2$ be their radius-vectors relative to the origin of coordinates. The Lagrangian of the system is

$$L = T(\dot{\mathbf{r}}_1, \dot{\mathbf{r}}_2) - V(\mathbf{r}_1 - \mathbf{r}_2, \dot{\mathbf{r}}_1 - \dot{\mathbf{r}}_2).$$

If we denote by $\mathbf{r}_G$ the radius-vector of the centre of mass, by $\mathbf{r}'_1$, $\mathbf{r}'_2$ the radius vectors of particles relative to G and by $\mathbf{r}$ the difference $\mathbf{r} = \mathbf{r}_2 - \mathbf{r}_1 = \mathbf{r}'_2 - \mathbf{r}'_1$, in view of (1.3.59), we have:

$$m_1\mathbf{r}'_1 + m_2\mathbf{r}'_2 = 0, \quad \mathbf{r}'_\alpha = \mathbf{r}_\alpha - \mathbf{r}_G \quad (\alpha = 1, 2).$$

Hence,

$$\mathbf{r}'_1 = -\frac{m_2}{M}\mathbf{r}, \quad \mathbf{r}'_2 = \frac{m_1}{M}\mathbf{r}, \tag{3.1.1}$$

where $M = m_1 + m_2$. Using König's theorem (1.3.61), the Lagrangian becomes:

$$L = \frac{1}{2}M|\mathbf{v}_g|^2 + \frac{1}{2}\mu|\dot{\mathbf{r}}|^2 - V(\mathbf{r}, \dot{\mathbf{r}}), \tag{3.1.2}$$

M. Chaichian et al., *Mechanics*, DOI: 10.1007/978-3-642-17234-2_3,

with the notation

$$\mu = \frac{m_1 m_2}{m_1 + m_2} = \frac{m_1 m_2}{M} \tag{3.1.3}$$

for the *reduced mass* of the system.

Since the components of $\mathbf{r}_G$ are cyclic variables, we have

$$\mathbf{P}_G = M\mathbf{v}_G = \text{const.},$$

meaning that the centre of mass moves *independently* of the motion of particles, being completely determined by the initial conditions. Consequently, we can choose the origin of coordinates in the centre of mass and the Lagrangian becomes:

$$L = \frac{1}{2}\mu|\dot{\mathbf{r}}|^2 - V(\mathbf{r}, \dot{\mathbf{r}}). \tag{3.1.4}$$

Therefore, the motion of the system of two particles can be studied as a motion of a *fictitious* particle of mass μ and radius-vector $\mathbf{r}$, in an external force field of potential V. If we can write (see (2.5.23))

$$\mathbf{F}(\mathbf{r}, \dot{\mathbf{r}}) = \frac{d}{dt}\left(\frac{\partial V}{\partial \dot{\mathbf{r}}}\right) - \frac{\partial V}{\partial \mathbf{r}},$$

where it is assumed that V depends linearly on the velocity $\dot{\mathbf{r}}$, then the particle with reduced mass μ obeys the equation of motion

$$\mu\ddot{\mathbf{r}} = \mathbf{F}(\mathbf{r}, \dot{\mathbf{r}}). \tag{3.1.5}$$

If one of the particles, say m_2, is much more massive than the other, $m_2 \gg m_1$, then

$$\mu = \frac{m_1}{1 + \frac{m_1}{m_2}} \simeq m_1$$

and the centre of mass almost coincides with the particle m_2. In this case, the problem can be studied as motion of the lighter particle in the field of the more massive one, which is assumed to be fixed.

A particularly interesting case is that in which the interaction potential between particles does not depend on their relative velocity and depends only on the relative distance $r = |\mathbf{r}|$ between the particles, i.e. $V = V(r)$. Then

$$\mathbf{F} = \mathbf{F}(\mathbf{r}) = f(r)\frac{\mathbf{r}}{r}, \tag{3.1.6}$$

with

$$f(r) = -\frac{dV}{dr}. \tag{3.1.7}$$

The field produced by such a force is called *central force field.* If $V(r)$ increases with r in some region, meaning $\frac{dV}{dr} > 0$, or $f(r) < 0$, the force points to the centre of the field, being a force of attraction in that region. In the opposite case, the force is repulsive.

3.1.2 General Properties of Motion in Central Field

Observing that $\mathcal{M} = \mathbf{r} \times \mathbf{F} = 0$, the angular momentum theorem (1.3.10) yields

$$\mathbf{l} = m\mathbf{r} \times \dot{\mathbf{r}} = \mathbf{C}\ \text{(const.)}. \tag{3.1.8}$$

Taking the scalar product of this equation and $\mathbf{r}$, we obtain:

$$\mathbf{r} \cdot \mathbf{C} = xC_x + yC_y + zC_z = 0. \tag{3.1.9}$$

This is the normal form of the equation of a plane passing through the points $P(x, y, z)$ and O. Consequently, a particle subject to a central force describes a *plane trajectory*.

Let the plane of motion be the x_1x_2-plane (we re-label the Cartesian coordinates x, y and z by x_1, x_2 and x_3, respectively). Due to the spherical symmetry of our problem, it is convenient to choose $q_1 = r$, $q_2 = \varphi$ as generalized coordinates. Written in these coordinates, the Lagrangian of the system is

$$L = \frac{1}{2}m(\dot{r}^2 + r^2\dot{\varphi}^2) - V(r). \tag{3.1.10}$$

Since φ is a cyclic coordinate, the associated generalized momentum is a constant of motion:

$$p_\varphi = \frac{\partial L}{\partial \dot{\varphi}} = mr^2\dot{\varphi} = C_1. \tag{3.1.11}$$

The physical significance of (3.1.11) is found by projecting Eq. (3.1.8) on the $x_3 = z$-axis:

$$l_z = m[r\mathbf{u}_r \times (\dot{r}\mathbf{u}_r + r\dot{\varphi}\mathbf{u}_\varphi)] \cdot \mathbf{k} = mr^2\dot{\varphi} = C_z. \tag{3.1.12}$$

Comparing the last two relations, we conclude that $C_1 = C_z = C$, meaning the conservation of angular momentum, in projection on the axis of rotation (z).

Another first integral is obtained from the fact that the system is conservative. Indeed, the Lagrangian (3.1.10) does not explicitly depend on time, therefore the total energy E is constant:

$$E = \frac{1}{2}m(\dot{r}^2 + r^2\dot{\varphi}^2) + V(r). \tag{3.1.13}$$

The elimination of $\dot{\varphi}$ between (3.1.11) and (3.1.13) yields a relation between the total energy of the particle and the variable r:

$$E = \frac{1}{2}m\dot{r}^2 + V_{eff}(r), \tag{3.1.14}$$

where the quantity

$$V_{eff}(r) = V(r) + \frac{C^2}{2mr^2} \tag{3.1.15}$$

is called *effective potential*. It can be considered as giving rise to an *effective force*:

$$f_{eff}(r) = -\frac{dV_{eff}}{dr} = f(r) + \frac{C^2}{mr^3} = f(r) + \frac{mv_\varphi^2}{r}. \tag{3.1.16}$$

The second term on the r.h.s. of (3.1.16) is a centrifugal force and therefore the term $\frac{C^2}{2mr^2}$ in (3.1.15) is called *centrifugal potential*. This term becomes important near the origin.

The existence of the two first integrals (3.1.11) and (3.1.13) makes possible the derivation of the *finite* equations of motion by a straightforward integration. Indeed, (3.1.14) leads to

$$t = \pm\sqrt{\frac{m}{2}}\int \frac{dr}{\sqrt{E - V_{eff}(r)}} + \text{const.}, \tag{3.1.17}$$

which determines $r = r(t)$. Introducing now $r(t)$ into (3.1.11), we obtain the other coordinate, $\varphi = \varphi(t)$:

$$\varphi = \frac{C}{m}\int \frac{dt}{[r(t)]^2} + \text{const.} \tag{3.1.18}$$

The *explicit equation* of the trajectory, $\varphi = \varphi(r)$, is obtained by eliminating the variable t between the last two equations:

$$\varphi = \pm\int \frac{C}{r^2}\frac{dr}{\sqrt{2m(E - V_{eff}(r))}} + \text{const.} \tag{3.1.19}$$

If we denote $\frac{1}{r} = u$, the last equation takes the form:

$$\varphi = \mp\int \frac{du}{\sqrt{\frac{2mE}{C^2} - \frac{2mV}{C^2} - u^2}} + \text{const.} \tag{3.1.20}$$

All constants which appear in our calculations, including C and E, are determined from the initial conditions.

Although we have already found the finite equations of motion and of trajectory as well, there are several good reasons to derive the *differential* equation of trajectory. Thus, observing that

$$\dot{r} = \frac{dr}{d\varphi}\dot{\varphi} = \frac{C}{mr^2}\frac{dr}{d\varphi} = -\frac{C}{m}\frac{d}{d\varphi}\left(\frac{1}{r}\right),$$

we obtain from (3.1.14):

$$E = \frac{C^2}{2mr^2}\left[\frac{1}{r^2}\left(\frac{dr}{d\varphi}\right)^2 + 1\right] + V(r), \qquad (3.1.21)$$

or, in terms of u,

$$E = \frac{C^2}{2m}\left[\left(\frac{du}{d\varphi}\right)^2 + u^2\right] + V\left(\frac{1}{u}\right), \qquad (3.1.22)$$

which is *Binet's equation.* If $V(r)$ is known, this equation can be integrated by separation of variables.

Binet's equation receives an alternative form by using the differential equations of motion. Thus, the Lagrange equation (2.5.17) for the variable r gives:

$$m\ddot{r} - mr\dot{\varphi}^2 = f(r). \qquad (3.1.23)$$

But

$$\ddot{r} = -\frac{C^2}{m^2r^2}\frac{d^2}{d\varphi^2}\left(\frac{1}{r}\right),$$

hence

$$-\frac{C^2}{mr^2}\left[\frac{d^2}{d\varphi^2}\left(\frac{1}{r}\right) + \frac{1}{r}\right] = f(r), \qquad (3.1.24)$$

or

$$-\frac{C^2}{m}u^2\left(\frac{d^2u}{d\varphi^2} + u\right) = f\left(\frac{1}{u}\right). \qquad (3.1.25)$$

Binet's equation can also be used to determine the potential energy $V(r)$ or the force $f(r)$, if the trajectory $r = r(\varphi)$ is known.

3.1.3 Discussion of Trajectories

The formulas obtained in the previous section permit the determination of both the trajectory and the law of motion, if the potential energy $V(r)$ is known. Nevertheless, certain general characteristics of trajectories can be found without knowing the analytical structure of this function.

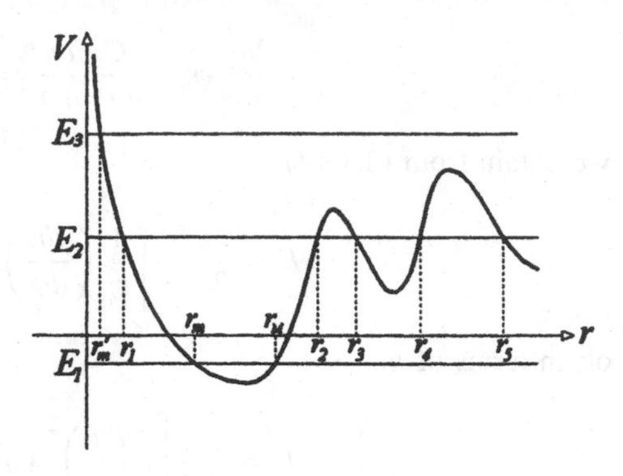

Fig. 3.1 Graphical representation of an arbitrary potential energy V as a function of the distance r.

Since $\dot{r}$, t, φ are real quantities, it follows from (3.1.14), (3.1.17) and (3.1.19) that one must have $V_{eff}(r) \leq E$. This relation determines the domain of variation of r, for given values of E and C. The boundary of this domain is provided by the equality

$$V_{eff}(r) = E. \tag{3.1.26}$$

According to (3.1.14), on the boundary the radial component of velocity vanishes ($\dot{r} = 0$), but $\dot{\varphi} \neq 0$, if $C \neq 0$. Consequently, the condition $\dot{r} = 0$ determines a *turning point* of the trajectory: the function $r(t)$ changes its sense of variation, i.e. $\dot{r}$ changes its sign. According to (3.1.11), $\dot{\varphi}$ does not change its sign, therefore φ is a monotonic function of time. This means that in (3.1.17)–(3.1.19) the limits must be chosen as to correspond to a monotonic interval of variation of r, while the sign must be suitably taken.

There are maximum and minimum values of r among the roots of (3.1.26), say r_M and r_m, respectively. The relation $V_{eff}(r) \leq E$ will therefore determine the domain of variation of r:

$$0 \leq r_m \leq r \leq r_M. \tag{3.1.27}$$

The roots of Eq. (3.1.26) can be determined graphically, as being given by the intersection points of the curve $V = V_{eff}(r)$ with the straight line $V = E$ (Fig. 3.1). As one can see, the number of roots depends on the structure of $V(r)$, as well as on the values of the constants C and E.

3.1.3.1 Bound Orbits

If r_M is finite, we have *bound* trajectories. In this case, assuming $r_m > 0$, the motion takes place within the circular crown (annulus) determined by the concentric circles of radii r_m and r_M (Fig. 3.2). The turning points for which $r = r_m$ are called *pericentres*, while those corresponding to $r = r_M$ are called *apocentres*. If the centre of force is the Earth, they are known as *perigee* and *apogee*, while around the Sun, the trajectory of each planet has a *perihelion* and an *aphelion*.

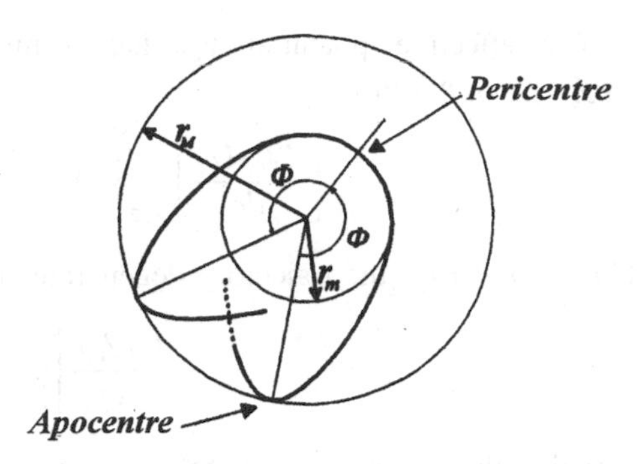

Fig. 3.2 Bound trajectories of a planet. The circles of radii r_m and r_M are concentric.

The trajectory is symmetric about any turning point. Indeed, if we choose the coordinate axes so as to have $\varphi = 0$ in a turning point, then according to (3.1.19) near this point we have two pairs of symmetrical positions, (r, φ) and $(r, -\varphi)$, given by

$$\varphi = \pm \int_{r_0}^{r} \frac{C}{r^2} \frac{dr}{\sqrt{2m(E - V_{eff})}},$$

where r_0 is r_m for a pericentre, and r_M for an apocentre. The angle at centre between a pericentre and the next apocentre is

$$\Phi = \int_{r_m}^{r_M} \frac{C}{r^2} \frac{dr}{\sqrt{2m(E - V_{eff})}}, \tag{3.1.28}$$

and, in view of the aforementioned discussion, the angle between two consecutive pericentres (apocentres) is $\Delta\varphi = 2\Phi$.

A *bound* orbit could be *closed* or *open*. To be closed, an orbit must satisfy the relation:

$$\Delta\varphi = 2\Phi = 2\pi \frac{n'}{n} \quad (n, n' \text{ integer numbers}), \tag{3.1.29}$$

i.e. after n revolutions about the centre, the radius-vector sweeps a multiple n' of 2π radians. If this condition is not fulfilled, the orbit is *open*, which means that the orbit never passes twice through a given point. There exist only two types of central fields for which all bound orbits are also *closed*, namely those characterized by the potentials:

$$V(r) = a\, r^2, \quad a > 0, \tag{3.1.30}$$

$$V(r) = \frac{k}{r}, \quad k < 0. \tag{3.1.31}$$

This statement expresses Bertrand's theorem and will be proved in Sect. 3.1.4.

The effective potential V_{eff} has a minimum in a point of the annulus $r_m \leq r \leq r_M$, given by:

$$\left[\frac{dV_{eff}(r)}{dr}\right]_{r=\bar{r}} = 0 \quad \rightarrow \quad r_m \leq \bar{r} \leq r_M. \tag{3.1.32}$$

If $r_m = r_M = r_c$, the trajectory degenerates in a circle of radius r_c, and we have:

$$\left[\frac{dV_{eff}}{dr}\right]_{r=r_c} = 0, \tag{3.1.33}$$

meaning that r_c corresponds to that value of r for which $V_{eff} = (V_{eff})_{min}$, and is realized when the energy E equals $(V_{eff})_{min}$.

3.1.3.2 Unbound Orbits

If the domain of variation of r is given by $r \geq r_{min}$, the particles may go to infinity. It then follows that:

$$E \geq \lim_{r \to \infty} V_{eff}(r) = \lim_{r \to \infty} V(r) = V_\infty.$$

The limit V_∞ is finite and one usually chooses $V_\infty = 0$, so that the "escape to infinity" condition is

$$E \geq 0.$$

The physical interpretation of this condition is simple: at infinity the interaction between the particle and the centre of force ceases, $V_\infty = 0$, and the particle is left with kinetic energy only: $E = T > 0$. Note that the symmetry of the orbit with respect to the straight line passing through the centre of force and pericentre remains valid. A particle describing an unbound trajectory passes only *once* through pericentre.

3.1.3.3 Falling on the Centre of Force

If $r_m = 0$, the particle passes through (or stops at) the centre of force. Assuming $C \neq 0$ ($C = 0$ would correspond to a motion along a straight line), the centrifugal term in V prevents the particle from falling on the centre, even if the force is attractive. The condition of passing through the centre of force is $V_{eff} \leq E$, which can be written as

$$r^2 V(r) + \frac{C^2}{2m} \leq Er^2.$$

For r to take the value $r_m = 0$, one must have

$$\lim_{r \to \infty} [r^2 V(r)] \leq -\frac{C^2}{2m},$$

i.e. near the origin the potential must decrease at least as

$$-Ar^{-2}, \quad A > \frac{C^2}{2m}, \tag{3.1.34}$$

or

$$-Ar^{-n}, \quad A > 0,\ n > 2. \tag{3.1.35}$$

3.1.4 Bertrand's Theorem

Let us prove that the only central fields having the property that all bound orbits are also closed, are those given by (3.1.30) and (3.1.31) (*Bertrand's theorem*).

The proof is based both on condition (3.1.29) and on the fact that there is a bound and closed orbit for which $V_{eff}(r)$ admits a minimum. Such an orbit is the circular orbit of radius $\bar{r}$ given by condition (3.1.32). In other words, for a given central field, we can have a circular orbit if the angular momentum of the particle has a value that satisfies (3.1.33), while the kinetic energy fulfills the condition (3.1.26),

$$\overline{E} = V_{eff}(\bar{r}) = V(\bar{r}) + \frac{C^2}{2m\bar{r}^2}, \tag{3.1.36}$$

where, according to (3.1.32), $V_{eff}(\bar{r})$ is the value of V_{eff} corresponding to the considered minimum. We also have:

$$V'_{eff}(\bar{r}) = 0, \tag{3.1.37}$$

$$V''_{eff}(\bar{r}) > 0. \tag{3.1.38}$$

Let us now consider an orbit corresponding to an energy E, close to the circular shape, and demand that this orbit is also closed. It is essential that the condition (3.1.38) be fulfilled, because in case of a maximum, the circular orbit would not be stable, the orbit would be unbound, and the problem would become meaningless.

The angle at centre between two consecutive turning points of the almost circular orbit, in view of (3.1.28), is:

$$\Phi = \frac{C}{\sqrt{m}} \int_{u_M}^{u_m} \frac{du}{\sqrt{2E - 2\mathcal{V}(u)}}. \tag{3.1.39}$$

Here, $u_m = \frac{1}{r_m}$, $u_M = \frac{1}{r_M}$, $\mathcal{V}(u) = V_{eff}(\frac{1}{u})$. We also have $E \to E + \epsilon$, with $\frac{\epsilon}{E} \ll 1$. Expanding $\mathcal{V}(u)$ according to Taylor's formula about $\overline{u} = \frac{1}{\overline{r}}$, we have:

$$\mathcal{V}(u) = \mathcal{V}(\overline{u}) + \mathcal{V}'(\overline{u})(u - \overline{u}) + \frac{1}{2}\mathcal{V}''(\overline{u})(u - \overline{u})^2 + \cdots,$$

or, in view of (3.1.36) and (3.1.37),

$$\mathcal{V}(u) = \overline{E} + \frac{1}{2}\mathcal{V}''(\overline{u})(u - \overline{u})^2 + \cdots$$

With this approximation, (3.1.39) becomes:

$$\begin{aligned}\Phi &= \frac{C}{\sqrt{m}} \int_{u_M}^{u_m} [2\epsilon - \mathcal{V}''(\overline{u})(u - \overline{u})^2]^{-\frac{1}{2}} du \\ &= \frac{C}{\sqrt{m}} [\mathcal{V}''(\overline{u})]^{-\frac{1}{2}} \arcsin \left[\sqrt{\frac{\mathcal{V}''(\overline{u})}{2\epsilon}} (u - \overline{u}) \right]_{u_M}^{u_m},\end{aligned}$$

where $u_{m,M} = \overline{u} \pm [\frac{2\epsilon}{\mathcal{V}''(\overline{u})}]^{\frac{1}{2}}$ in the new variables. It then results that

$$\Phi = C[m\mathcal{V}''(\overline{u})]^{-\frac{1}{2}}. \tag{3.1.40}$$

Using the condition (3.1.33), we obtain:

$$\frac{m\mathcal{V}''(\overline{u})}{C^2} = \alpha^2 > 0, \tag{3.1.41}$$

where α is a rational number, so as to have

$$\Phi = \frac{\pi}{\alpha}. \tag{3.1.42}$$

These considerations are valid for almost circular orbits. But it is clear that, once the potential $V(r)$ is found, by varying the pair of quantities (E, C), we can pass from one circular orbit to another by a continuous variation of $\overline{r}$. This means that the discrete parameter α must be the same, for any circular orbit, i.e. for any $\overline{r} = \frac{1}{\overline{u}}$. Under these circumstances, Eq. (3.1.41) serves to determine the potential $V(r)$. In view of (3.1.15), we then have:

$$r\frac{d^2V}{dr^2} = (\alpha^2 - 3)\frac{dV}{dr},$$

with the solution:

$$V(r) = Ar^{\alpha^2-2}. \tag{3.1.43}$$

Since $\alpha = \sqrt{2}$ is not a rational number, we can take $\alpha^2 \neq 2$.

Let us first discuss the case $\alpha^2 > 2$. This leads to $V_{eff} \to \infty$ for $r \to 0$. But $E \geq V_{eff}$, meaning that E also becomes infinite, and therefore $r_m = 0$ is a turning point of infinite energy. In view of (3.1.43), the function $\mathcal{V}(u)$ can be written as follows:

$$\mathcal{V}(u) = \frac{C^2}{2m}u^2 + Au^{2-\alpha^2}. \tag{3.1.44}$$

Introducing (3.1.44) into (3.1.39), and taking into account the fact that now $E = \mathcal{V}(u_m)$, we have:

$$\Phi = \frac{C}{\sqrt{m}} \int_{u_M}^{u_m} \left[\frac{C^2}{m}(u_m^2 - u^2) + 2A(u_m^{2-\alpha^2} - u^{2-\alpha^2}) \right]^{-\frac{1}{2}} du.$$

Making the substitution $u = u_m x$, we obtain:

$$\Phi = \int_{\frac{u_M}{u_m}}^{1} \left[1 - x^2 + \frac{2mA}{C^2} u_m^{2-\alpha^2} (1 - x^{2-\alpha^2}) \right]^{-\frac{1}{2}} dx.$$

In the limit $E \to \infty$, i.e. $u_m \to \infty$, this integral reduces to

$$\lim_{E\to\infty} \Phi = \int_0^1 (1 - x^2)^{-\frac{1}{2}} dx = \frac{\pi}{2}. \tag{3.1.45}$$

Comparing this relation with (3.1.42), we obtain $\alpha = 2$, and (3.1.43) leads to (3.1.30). Here, the constant α must be positive, otherwise we would have only unbound orbits.

Let us finally discuss the case $0 < \alpha^2 < 2$. One can see that the effective potential tends to zero by negative values, when $r \to \infty$. The point $r_M = \infty$ $(u_M = 0)$ is therefore a turning point of zero energy. There exists also another turning point of zero energy, given by $V_{eff}(r_m) = 0$, or

$$\frac{C^2}{2m} + Au_m^{-\alpha^2} = 0. \tag{3.1.46}$$

Relation (3.1.39) yields:

$$\Phi = \frac{C}{\sqrt{m}} \int_0^{u_m} \left(-\frac{C^2}{m}u^2 - 2Au^{2-\alpha^2} \right)^{-\frac{1}{2}} du,$$

or, in view of (3.1.46),

$$\Phi = \int_0^{u_m} (u_m^{\alpha^2} u^{2-\alpha^2} - u^2)^{-\frac{1}{2}} du.$$

The integral is worked out by the substitution $u = u_m x^{\frac{2}{\alpha^2}}$, which gives:

$$\Phi = \frac{2}{\alpha^2} \int_0^1 (1 - x^2)^{-\frac{1}{2}}\, dx = \frac{\pi}{\alpha^2}. \qquad (3.1.47)$$

Comparing (3.1.47) with (3.1.42), we obtain $\alpha = 1$, therefore (3.1.43) yields (3.1.31).

In conclusion, we have shown that only the fields having a potential of the form (3.1.30) or (3.1.31) can produce closed orbits. The proof of the theorem is complete.

3.2 Kepler's Problem

In this section we shall study the central field with potential energy of the type

$$V(r) = \frac{k}{r}. \qquad (3.2.1)$$

In this category fall the gravitational ($k < 0$) and Coulombian ($k<0,\ k > 0$) fields. The structure of orbits corresponding to this potential can be qualitatively analyzed by drawing the graphic of the function

$$V_{eff}(r) = \frac{k}{r} + \frac{C^2}{2mr^2}. \qquad (3.2.2)$$

If the force is one of attraction ($k < 0$) (Fig. 3.3), we distinguish the following types of orbits.

For $\overline{E} = (V_{eff})_{min} = V_{eff}(\overline{r})$, we have a circular orbit of radius

$$\overline{r} = \frac{C^2}{m|k|}, \qquad (3.2.3)$$

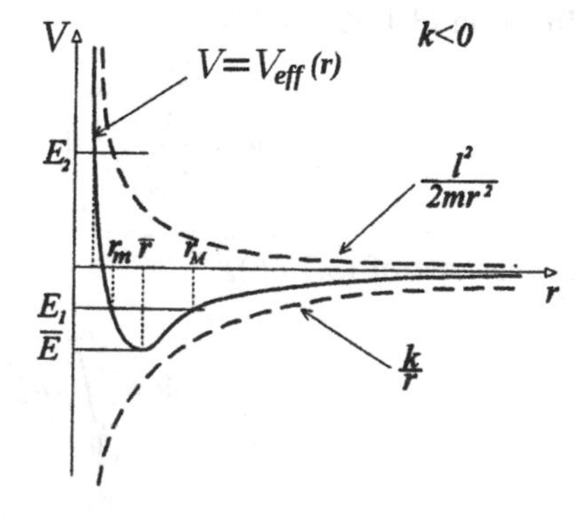

Fig. 3.3 Graphical representation of V as a function of r for an attractive force ($k < 0$).

Fig. 3.4 Graphical representation of V as a function of r for a repulsive force ($k > 0$).

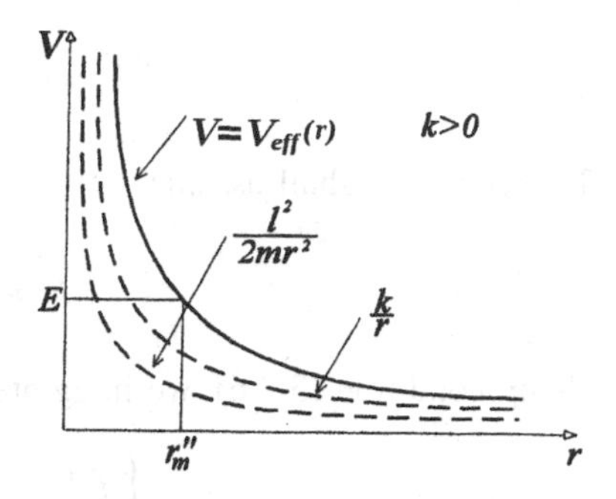

which is obtained from (3.1.32). For the energy E_1, with $(V_{eff})_{min} < E_1 < 0$, we have bound orbits in the annulus defined by r_m and r_M. These curves are also closed, and we shall show that they are ellipses. For another energy $E_2 \geq 0$, the orbits are unbound, the pericentre being given, say, by r'_m.

If the force is repulsive ($k > 0$) (Fig. 3.4), we have only unbound orbits, with the pericentre r''_m, for any $E > 0$.

3.2.1 Determination of Trajectories

Choosing the x_1-axis to pass through the pericentre ($\varphi = 0$ corresponds to $r = r_m$), we obtain from (3.1.19):

$$\varphi = \int_{r_m}^{r} \frac{C^2}{r^2}\left[2m\left(E - \frac{k}{r} - \frac{C^2}{2mr^2}\right)\right]^{-\frac{1}{2}} dr, \tag{3.2.4}$$

or, if we pass to the new variable $u = \frac{1}{r}$ (see (3.1.20)),

$$\varphi = \int_{u}^{u_m} \left[\frac{2m}{C^2}(E - ku) - u^2\right]^{-\frac{1}{2}} du = \int_{u}^{u_m} \left[\frac{2mE}{C^2} + \frac{m^2k^2}{C^4} - \left(u + \frac{mk}{C^2}\right)^2\right]^{-\frac{1}{2}} du. \tag{3.2.5}$$

If $E = \overline{E}(<0)$, which is possible only for $k < 0$, one obtains a circular orbit. Indeed, using (3.1.14) and (3.2.3) to write

$$\dot{r}^2 = -\left(\frac{k}{C} + \frac{C}{mr^2}\right),$$

we get the solution $\dot{r} = 0$, that is

$$r = \overline{r} = -\frac{C^2}{mk} = \frac{C^2}{m|k|}.$$

Therefore, we shall assume

$$E > \overline{E} = -\frac{mk^2}{2C^2}. \tag{3.2.6}$$

Using condition (3.2.6), we integrate (3.2.5) and obtain:

$$\varphi = \arccos\left\{\left(\frac{1}{r} + \frac{mk}{C^2}\right)\left(\frac{2mE}{C^2} + \frac{m^2k^2}{C^4}\right)^{-\frac{1}{2}}\right\},$$

which yields

$$\frac{1}{r} = \frac{1}{p}\left(-\frac{k}{|k|} + \epsilon\cos\varphi\right). \tag{3.2.7}$$

This is the equation of a conic, having one focus in the centre of force O. Here,

$$p = \frac{C^2}{m|k|} \tag{3.2.8}$$

is the parameter of the conic and

$$\epsilon = \left(1 + \frac{2EC^2}{mk^2}\right)^{\frac{1}{2}} \tag{3.2.9}$$

is its eccentricity.

One can also write the equation of the conic in Cartesian coordinates. To do this, we observe that our choice of axes yields

$$\cos\varphi = \frac{x_1}{r} = \frac{x_1}{\sqrt{x_1^2 + x_2^2}},$$

hence Eq. (3.2.7) leads to:

$$x_1^2 + x_2^2 - (\epsilon x_1 - p)^2 = 0. \tag{3.2.10}$$

If $E < 0$, which is possible only if $k < 0$, we have a bound orbit. In this case $\epsilon < 1$, and (3.2.10) yields:

$$(1 - \epsilon^2)x_1^2 + x_2^2 + 2p\epsilon x_1 = p^2,$$

or

$$\frac{\left(x_1 + \frac{p\epsilon}{1-\epsilon^2}\right)^2}{\left(\frac{p}{1-\epsilon^2}\right)^2} + \frac{x_2^2}{\left(\frac{p}{\sqrt{1-\epsilon^2}}\right)^2} = 1, \tag{3.2.11}$$

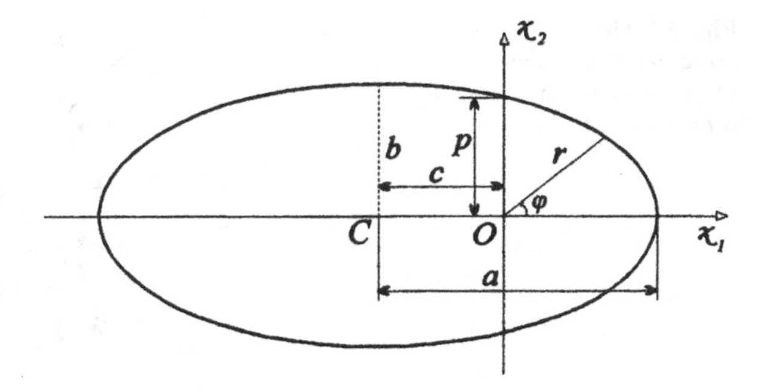

Fig. 3.5 Graphical representation in the Ox_1x_2-plane of a closed elliptic orbit.

which is the equation of an ellipse of semi-axes (see Fig. 3.5)

$$a = \frac{p}{1-\epsilon^2} = \frac{|k|}{2|E|} = \frac{k}{2E}, \tag{3.2.12}$$

$$b = \frac{p}{\sqrt{1-\epsilon^2}} = \frac{C}{\sqrt{2m|E|}}, \tag{3.2.13}$$

while the focal distance (distance from the centre to a focus) is

$$c = \frac{p\epsilon}{1-\epsilon^2} = \epsilon a = \sqrt{a^2 - b^2}. \tag{3.2.14}$$

The distances from the centre to the pericentre and apocentre are, respectively:

$$r_m = a - c = (1-\epsilon)a = \frac{p}{1+\epsilon}, \quad r_M = a + c = (1+\epsilon)a = \frac{p}{1-\epsilon}. \tag{3.2.15}$$

It is interesting to note that the semi-major axis of the ellipse, a, depends only on the mechanical energy of the particle, being equal to $\frac{r_m + r_M}{2}$.

If $E > 0$, the trajectory is unbound, irrespective of the sign of k, while $\epsilon > 1$. Thus, Eq. (3.2.10) reads:

$$(\epsilon^2 - 1)x_1^2 - x_2^2 - 2p\epsilon x_1 = -p^2,$$

or

$$\frac{\left(x_1 - \frac{p\epsilon}{\epsilon^2-1}\right)^2}{\left(\frac{p}{\epsilon^2-1}\right)^2} - \frac{x_2^2}{\left(\frac{p}{\sqrt{\epsilon^2-1}}\right)^2} = 1, \tag{3.2.16}$$

which is the equation of a hyperbola, of semi-axes:

$$a = \frac{p}{\epsilon^2 - 1} = \frac{|k|}{2E}, \tag{3.2.17}$$

$$b = \frac{p}{\sqrt{\epsilon^2 - 1}} = \frac{C}{\sqrt{2mE}} \tag{3.2.18}$$

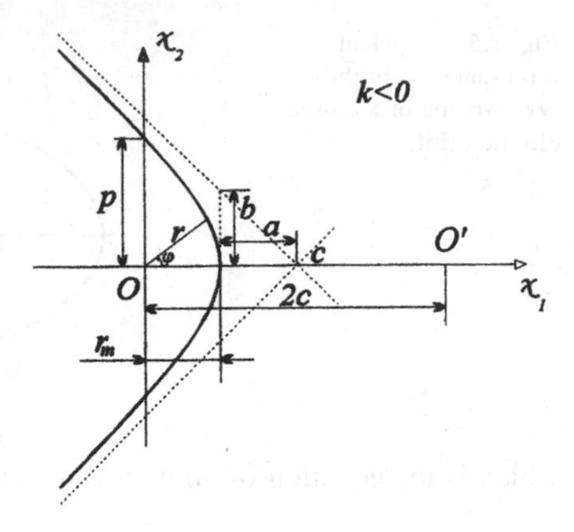

Fig. 3.6 Graphical representation in the Ox_1x_2-plane of an unbound orbit for $k < 0$.

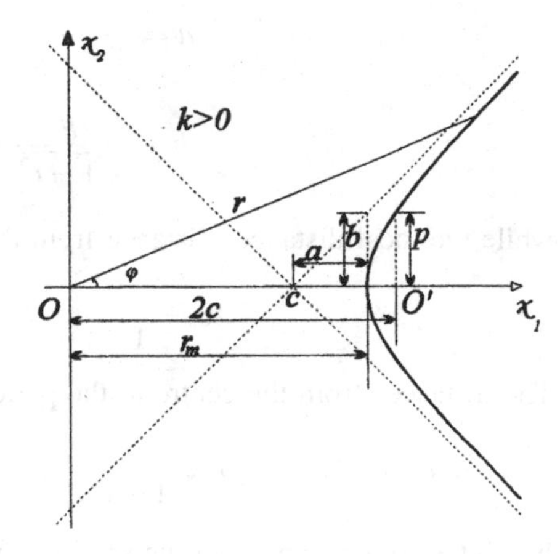

Fig. 3.7 Graphical representation in the Ox_1x_2-plane of an unbound orbit for $k > 0$.

and having the focal distance

$$c = \frac{p\epsilon}{\epsilon^2 - 1} = \epsilon a = \sqrt{a^2 + b^2}. \tag{3.2.19}$$

The angle made by the asymptote with the x_1-axis is given by

$$\frac{b}{a} = \sqrt{\epsilon^2 - 1} = \frac{2CE}{|k|}.$$

As shown in Figs. 3.6 and 3.7, if $k < 0$, the particle moves along the branch of the hyperbola which surrounds the centre of force, while for $k > 0$, it moves on the other branch. The distance to pericentre is calculated by putting $\varphi = 0$ in (3.2.7):

$$r_m = \frac{p}{\epsilon + 1} = a(\epsilon - 1) < c \quad (k < 0), \tag{3.2.20}$$

$$r_m = \frac{p}{\epsilon - 1} = a(\epsilon + 1) > c \quad (k > 0). \tag{3.2.21}$$

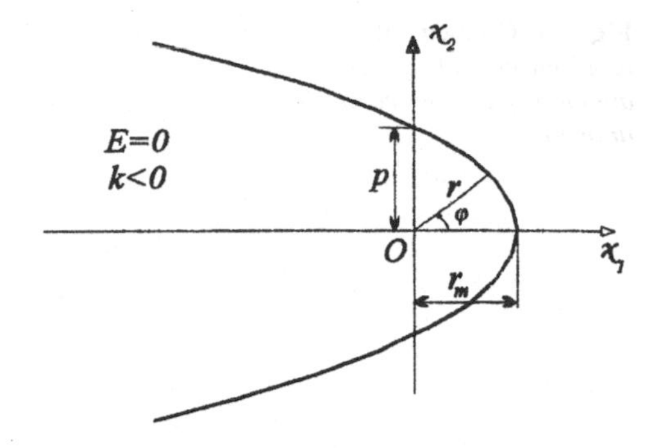

Fig. 3.8 Graphical representation in the Ox_1x_2-plane of an unbound orbit for $k < 0$ and $E = 0$.

It is seen that, during the motion of the particle on the trajectory, the value r_m is reached only once.

If $E = 0$, which is possible only if $k < 0$, we have unbound orbits, the particle being at rest at infinity. Then (3.2.9) yields $\epsilon = 1$ and (3.2.10) reads

$$x_2^2 = -2p\left(x_1 - \frac{p}{2}\right), \tag{3.2.22}$$

which is a parabola. The distance to pericentre is

$$r_m = \frac{p}{2} = \frac{C^2}{2m|k|}, \tag{3.2.23}$$

while the points of intersection of the parabola with the x_2-axis are $x_2 = \pm p$ (Fig. 3.8).

Summarizing, we conclude that the type of orbit is completely determined by the possible values of mechanical energy, which directly determine the eccentricity:

(a) $E < 0 \rightarrow \epsilon < 1$: ellipse;
(b) $E > 0 \rightarrow \epsilon > 1$: hyperbola;
(c) $E = 0 \rightarrow \epsilon = 1$: parabola;
(d) $E = \overline{E} = -\frac{mk^2}{2C^2} \rightarrow \epsilon = 0$: circle.

Cases (a), (c) and (d) can be realized only for attractive forces, while (b) is possible for both attractive and repulsive potentials.

3.2.2 *Law of Motion*

To determine the finite equations of motion, i.e. $r = r(t)$, $\varphi = \varphi(t)$, we assume that at $t = 0$ the particle passes through the centre. Then the integral (3.1.17) yields:

$$t = \int_{r_m}^{r} \left[\frac{2E}{m}r^2 - \frac{2k}{m}r - \frac{C^2}{m^2}\right]^{-\frac{1}{2}} r\,dr. \tag{3.2.24}$$

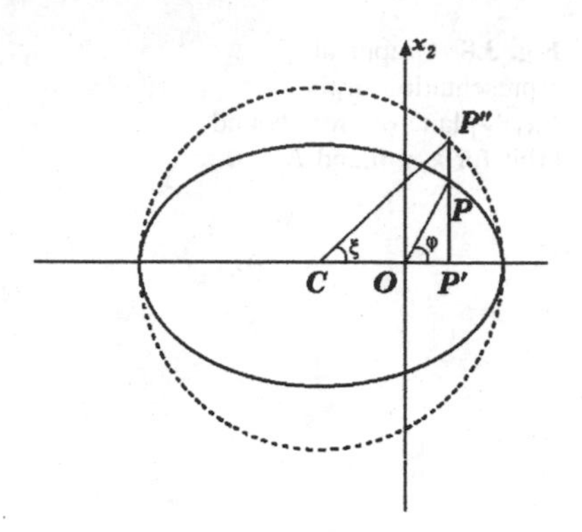

Fig. 3.9 Geometrical representation of the *real anomaly* φ and the *eccentric anomaly* ξ.

We shall calculate this integral for the situations (a), (b) and (c) presented above.

(a) $E < 0$ $(k < 0)$. Relations (3.2.12)–(3.2.14) yield:

$$k = 2aE, \quad C^2 = 2m|E|b^2 = 2m|E|a^2(1-\epsilon^2),$$

and, in view of (3.2.15),

$$t = \sqrt{\frac{m}{2|E|}} \int_{(1-\epsilon)a}^{r} [a^2\epsilon^2 - (a-r)^2]^{-\frac{1}{2}} dr.$$

With the substitution

$$r = a(1 - \epsilon \cos \xi), \tag{3.2.25}$$

the result of integration is:

$$t = \sqrt{\frac{ma^3}{|k|}} (\xi - \epsilon \sin \xi). \tag{3.2.26}$$

The last two relations give r, t in terms of the parameter ξ, while φ can be obtained from (3.2.7). If we put $\left(\frac{|k|}{ma^3}\right)^{\frac{1}{2}} = \nu$ in (3.2.26), we obtain:

$$\xi - \epsilon \sin \xi = \nu t, \tag{3.2.27}$$

known as *Kepler's equation*.

The parameter ξ has a simple geometric interpretation. Assume that the semi-major axis a of the ellipse is taken along the x_1-axis, one of the foci being at O and the centre at C (Fig. 3.9). Consider a point P on the ellipse, corresponding to a value φ of the polar angle, and draw a perpendicular at P on x_1. Let P', P'' be the intersection points of this perpendicular with the x_1-axis and with the circle of radius a and centre at C, respectively. If we denote $\xi = \widehat{P''CO}$, then

$$a \cos \xi = c + r \cos \varphi. \tag{3.2.28}$$

Using now (3.2.7), we arrive at (3.2.25), which means that the parameter ξ identifies with the angle $\widehat{P''CO}$. In astronomy, ξ is known as the *eccentric anomaly*, while φ is called the *real anomaly*.

Relation (3.2.28) shows that a complete revolution of the particle on the ellipse corresponds to a variation of 2π of ξ. The period of revolution, τ, results from (3.2.27):

$$\nu\tau = 2\pi,$$

yielding

$$\frac{\tau^2}{a^3} = \frac{4\pi^2 m}{|k|},$$

which is *Kepler's third law*.
The Cartesian coordinates of the particle are:

$$x_1 = r \cos \varphi = a(\cos \xi - \epsilon), \tag{3.2.29}$$

$$x_2 = \sqrt{r^2 - x_1^2} = a\sqrt{1 - \epsilon^2} \sin \xi.$$

(b) $E > 0$. From the relations (3.2.17)–(3.2.19), we obtain:

$$|k| = 2aE, \quad C^2 = 2mb^2 E = 2ma^2 E(\epsilon^2 - 1),$$

while (3.2.20) and (3.2.21) can be written under a unified form

$$r_m = a\left(\epsilon + \frac{k}{|k|}\right).$$

Thus, the integral (3.2.24) reads:

$$t = \sqrt{\frac{m}{2E}} \int_{r_m}^{r} \left[\left(r - \frac{k}{|k|}a\right)^2 - a^2\epsilon^2\right]^{-\frac{1}{2}} r dr.$$

Making the substitution

$$r = a\left(\epsilon \cosh \xi + \frac{k}{|k|}\right), \tag{3.2.30}$$

the integration is easily performed and we obtain:

$$t = \sqrt{\frac{m}{2E}} a\left(\epsilon \sinh \xi + \frac{k}{|k|}\xi\right),$$

or

$$t = \sqrt{\frac{ma^3}{|k|}}\left(\epsilon \sinh \xi + \frac{k}{|k|}\xi\right). \tag{3.2.31}$$

This time, the parameter ξ can take all values from $-\infty$ to $+\infty$ and does not have a simple geometric interpretation anymore. Passing to Cartesian coordinates, we have:

$$\begin{aligned} x_1 &= a\left(\epsilon + \frac{k}{|k|}\cosh \xi\right), \\ x_2 &= a\sqrt{\epsilon^2 - 1}\sinh \xi. \end{aligned} \tag{3.2.32}$$

(c) $E = 0\,(k<0)$. In view of (3.2.23), the relation (3.2.24) reads:

$$t = \sqrt{\frac{m}{2|k|}}\int_{\frac{p}{2}}^{r}\left(r - \frac{p}{2}\right)^{-\frac{1}{2}} r\,dr,$$

which can be solved easily. The result is:

$$t = \sqrt{\frac{m}{2|k|}}\cdot\left[\frac{2}{3}\left(r - \frac{p}{2}\right)^{\frac{3}{2}} + p\left(r - \frac{p}{2}\right)^{\frac{1}{2}}\right]. \tag{3.2.33}$$

Using now (3.2.7), with $\epsilon = 1$, $k<0, \eta = \tan\frac{\varphi}{2}$, we have:

$$r = \frac{p}{2}(1 + \eta^2). \tag{3.2.34}$$

Introducing (3.2.34) into (3.2.33), we obtain:

$$t = \frac{1}{2}\sqrt{\frac{mp^3}{|k|}}\eta\left(1 + \frac{\eta^2}{3}\right). \tag{3.2.35}$$

Finally, the Cartesian coordinates of the particle are given by:

$$\begin{aligned} x_1 &= \frac{p}{2}(1 - \eta^2), \\ x_2 &= p\eta. \end{aligned} \tag{3.2.36}$$

3.2.3 Runge–Lenz Vector

We wish to show that the field characterized by $V(r) = \frac{k}{r}$ admits a third first integral. In view of (3.1.6) and (3.1.7), Newton's fundamental equation reads:

$$\ddot{\mathbf{r}} = \frac{k}{m}\frac{\mathbf{r}}{r^3}. \tag{3.2.37}$$

Using this formula, let us calculate the cross product $\ddot{\mathbf{r}} \times \mathbf{l}$, where $\mathbf{l}$ is the angular momentum of the particle:

$$\ddot{\mathbf{r}} \times \mathbf{l} = k\frac{\mathbf{r}(\mathbf{r}\cdot\dot{\mathbf{r}})}{r^3} - k\frac{\dot{\mathbf{r}}}{r}.$$

But $\mathbf{r}\cdot\dot{\mathbf{r}} = r\dot{r}$, therefore

$$\ddot{\mathbf{r}} \times \mathbf{l} = -k\mathbf{r}\frac{d}{dt}\left(\frac{1}{r}\right) - k\frac{\dot{\mathbf{r}}}{r} = -\frac{d}{dt}\left(k\frac{\mathbf{r}}{r}\right).$$

On the other hand, since $\dot{\mathbf{l}} = 0$, we may write:

$$\ddot{\mathbf{r}} \times \mathbf{l} = \frac{d}{dt}(\dot{\mathbf{r}} \times \mathbf{l}).$$

The last two relations give

$$\frac{d}{dt}\left(\dot{\mathbf{r}} \times \mathbf{l} + k\frac{\mathbf{r}}{r}\right) = 0,$$

which shows that the vector quantity

$$\mathcal{R} = \dot{\mathbf{r}} \times \mathbf{l} + k\frac{\mathbf{r}}{r} \tag{3.2.38}$$

is a first integral. It is called the *Runge–Lenz vector*.

The magnitude of $\mathcal{R}$ is calculated observing that

$$|\dot{\mathbf{r}} \times \mathbf{l}|^2 = \frac{2l^2}{m}\left(E - \frac{k}{r}\right)$$

and

$$\mathbf{r}\cdot(\dot{\mathbf{r}} \times \mathbf{l}) = \mathbf{l}\cdot(\mathbf{r} \times \dot{\mathbf{r}}) = \frac{1}{m}l^2. \tag{3.2.39}$$

Therefore, recalling that in our case $l = C$ (const.), we have:

$$|\mathcal{R}|^2 = \frac{2EC^2}{m} + k^2,$$

or, if we use the notation (3.2.9),

$$|\mathcal{R}| = |k|\epsilon. \tag{3.2.40}$$

Since $\mathcal{R} \cdot \mathbf{l} = 0$, the vector $\mathcal{R}$ passes through O and lies in the plane of motion. If we direct the x_1-axis along $\mathcal{R}$, the angle between $\mathbf{r}$ and $\mathcal{R}$ is the polar angle φ. Then, in view of (3.2.39), we have:

$$\mathbf{r} \cdot \mathcal{R} = r|\mathcal{R}| \cos \varphi = \frac{C^2}{m} + kr,$$

or, if we use (3.2.40),

$$r(|k|\epsilon \cos \varphi - k) = \frac{C^2}{m}.$$

If we now introduce the notation (3.2.8), we obtain the equation of a conic in the form given by (3.2.7). This shows that the Runge–Lenz vector is oriented from the centre of force to the pericentre, irrespective of the sign of k.

3.2.4 Artificial Satellites of the Earth. Cosmic Velocities

Assuming the Earth as an isotropic sphere of radius R, the potential energy of a body in the gravitational field of the Earth is given by (3.2.1), with the parameter $k = -\gamma\, m m_E$, where m_E is the mass of the Earth and γ – the gravitational constant. If we neglect the rotation of the Earth about its axis, we have:

$$mg = -\gamma \frac{m m_E}{R^2},$$

which yields $k = -mgR^2$, and thus

$$V(r) = -\frac{mgR^2}{r}. \tag{3.2.41}$$

Consider an artificial satellite of mass m, which takes off from the leading rocket at a height h relative to the Earth's surface (point P, Fig. 3.10). The velocity of the satellite with respect to the Earth is $\mathbf{v}_0$, while the angle between $\mathbf{v}_0$ and the vertical in P is α.

In this application we shall use our previous results. In order to have a closed orbit, the following conditions must be fulfilled: $k<0$, $E<0$. Since

$$E = \frac{m v_0^2}{2} - \frac{mgR^2}{R+h}, \tag{3.2.42}$$

the condition $E < 0$ produces:

$$v_0^2 < \frac{2gR^2}{R+h}. \tag{3.2.43}$$

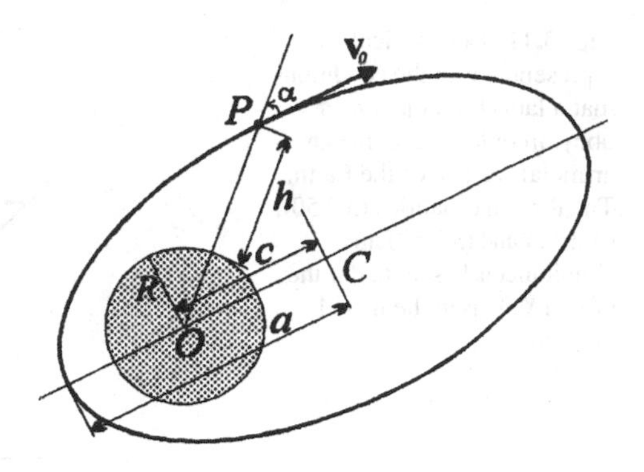

Fig. 3.10 Geometrical representation of the trajectory of an artificial satellite of the Earth.

On the other hand, we must consider the fact that the satellite should not intersect the Earth's surface, therefore according to (3.2.15), $a - c \geq R$. Since $c = \epsilon a$, we have

$$\epsilon \leq 1 - \frac{R}{a}. \tag{3.2.44}$$

This condition is fulfilled if $a \geq R$, which, in view of (3.2.12) and (3.2.42), leads to the inequality

$$\frac{v_0^2}{gR} \geq \frac{R-h}{R+h}, \tag{3.2.45}$$

or, using (3.2.9),

$$\frac{C^2}{m^2 g R^2} \geq 2\left(R - \frac{|E|}{mg}\right).$$

From (3.2.42), observing that $l = mv_0(R+h)\sin\alpha$, we get:

$$\frac{v_0^2}{gR}\left[\frac{h(2R+h)}{R^2}\sin^2\alpha - \cos^2\alpha\right] \geq \frac{2h}{R+h}. \tag{3.2.46}$$

In order to solve simultaneously the inequalities (3.2.43), (3.2.45) and (3.2.46), we introduce the notations:

$$v_I \equiv \sqrt{gR} \simeq 7,906\,\text{m/s}, \tag{3.2.47}$$

$$X \equiv \frac{v_0}{v_I}\cos\alpha, \quad Y \equiv \frac{v_0}{v_I}\sin\alpha, \tag{3.2.48a, b}$$

$$n = \frac{h}{R}, \tag{3.2.49}$$

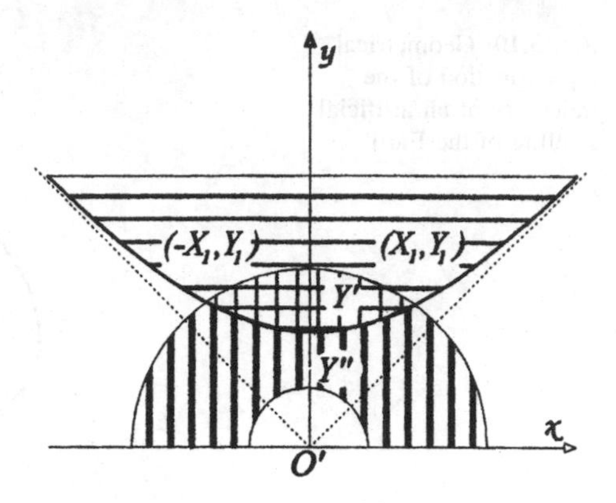

Fig. 3.11 Geometrical representation of the conditions that a launched object must obey, in order to become an artificial satellite of the Earth. The three inequalities (3.2.50), (3.2.51) and (3.2.52) are simultaneously satisfied if the point (X, Y) is in the netted domain.

leading to the new form of the three inequalities:

$$X^2 + Y^2 < \frac{2}{n+1}, \tag{3.2.50}$$

$$X^2 + Y^2 \geq \frac{1-n}{1+n}, \tag{3.2.51}$$

$$-X^2 + n(n+2)Y^2 \geq \frac{2n}{n+1}. \tag{3.2.52}$$

Consider the $O'xy$-plane, where to each point is attached a pair of variables (X, Y). Then all three inequalities are simultaneously satisfied if the point (X, Y) is inside both the annulus determined by the circles of radii $(\frac{1-n}{1+n})^{\frac{1}{2}}$ and $(\frac{2}{1+n})^{\frac{1}{2}}$ (the domain marked by vertical lines in Fig. 3.11) and the hyperbola

$$-\frac{X^2}{\frac{2n}{n+1}} + \frac{Y^2}{\frac{2}{(n+1)(n+2)}} = 1, \tag{3.2.53}$$

(the region indicated by horizontal lines). Consequently, the inequalities are simultaneously satisfied by the points in the netted domain.

Let Y', Y'' be the points of intersection of the hyperbola (3.2.53) and of the circle of radius $(\frac{1-n}{1+n})^{\frac{1}{2}}$ with the $O'y$-axis. Then:

$$Y' = \left[\frac{2}{(n+1)(n+2)}\right]^{\frac{1}{2}}, \quad Y'' = \left(\frac{1-n}{1+n}\right)^{\frac{1}{2}}.$$

Since $Y' > Y''$, the condition (3.2.51) (or (3.2.45)) is *always* satisfied. The hyperbola always intersects the circle of radius $(\frac{2}{n+1})^{\frac{1}{2}}$ in the points (X_1, Y_1), $(-X_1, Y_1)$, with

$$X_1 = \frac{\sqrt{2n}}{n+1}, \quad Y_1 = \frac{\sqrt{2}}{n+1}, \tag{3.2.54}$$

meaning that the netted domain exists always, i.e. for any h (any n) it is possible to choose v_0 and α in such a way that the launched object become an artificial satellite of the Earth.

Assume that the launching takes place near the Earth's surface. In this case, since $h \ll R$, we can take $n \approx 0$ in (3.2.52) and obtain $X \approx 0$, or $\alpha \simeq \frac{\pi}{2}$, meaning that the launching has to be done parallel to the Earth's surface. Taking $X \approx 0$, $n \approx 0$ in (3.2.50) and (3.2.51), we obtain $1 \le Y \le \sqrt{2}$, or

$$v_I \le v_0 \le v_{II},$$

where $v_{II} = \sqrt{2} v_I \approx 11,180\,\mathrm{m/s}$. The velocities v_I and v_{II} are called the *first* and, respectively, the *second cosmic velocity*. It is obvious that v_I and v_{II} represent the minimum and the maximum speeds of launching a body from the Earth's surface, in order to become a satellite. If $v_0 > v_{II}$, the object either leaves the gravitational field, or falls back on the Earth. From (3.2.54) we can also obtain the possible interval of variation of the angle α, for given h and v_0:

$$\frac{\pi}{2} - \alpha_0 \le \alpha \le \frac{\pi}{2} + \alpha_0,$$

where $\alpha_0 = \arctan \sqrt{n}$.

3.3 Classical Theory of Collisions Between Particles

3.3.1 Collisions Between Two Particles

Consider two particles in motion, which initially are far away from each other, so that each of them can be considered free. If the distance between particles becomes small enough, they begin to interact and, if certain conditions are fulfilled, they *collide*. The collision is called *elastic* if it does not produce any change in the internal state of the particles, and *inelastic*, if such a change happens. In the case that the particles remain glued together after the interaction, the collision is termed *completely inelastic*.

Our system of two particles of masses m_1 and m_2 is supposed to be isolated, therefore the total momentum is conserved:

$$m_1 \mathbf{v}_1 + m_2 \mathbf{v}_2 = m_1 \mathbf{v}'_1 + m_2 \mathbf{v}'_2, \tag{3.3.1a}$$

or

$$\mathbf{p}_1 + \mathbf{p}_2 = \mathbf{p}'_1 + \mathbf{p}'_2, \tag{3.3.1b}$$

where unprimed and primed letters stand for velocities and momenta before and after collision, respectively. Introducing the *relative* velocities:

$$\mathbf{v} = \mathbf{v}_2 - \mathbf{v}_1, \quad \mathbf{v}' = \mathbf{v}'_2 - \mathbf{v}'_1, \tag{3.3.2}$$

and the velocity of the centre of mass:

$$\mathbf{V} = \frac{m_1\mathbf{v}_1 + m_2\mathbf{v}_2}{M} = \frac{\mathbf{p}_1 + \mathbf{p}_2}{M}, \quad M = m_1 + m_2, \tag{3.3.3}$$

we can write the velocities of the particles before and after collision as

$$\mathbf{v}_1 = \mathbf{V} - \frac{m_2}{M}\mathbf{v}, \quad \mathbf{v}_2 = \mathbf{V} + \frac{m_1}{M}\mathbf{v}, \tag{3.3.4}$$

$$\mathbf{v'}_1 = \mathbf{V} - \frac{m_2}{M}\mathbf{v}', \quad \mathbf{v'}_2 = \mathbf{V} + \frac{m_1}{M}\mathbf{v}'. \tag{3.3.5}$$

If the collision is elastic, the field of interaction between particles is conservative, while the interaction potential $V_{int}(r)$ must vanish at infinity:

$$\lim_{r\to\infty} V_{int}(r) = 0.$$

In this case, the total kinetic energy is conserved:

$$T = \frac{1}{2}MV^2 + \frac{1}{2}\mu v^2 = \frac{1}{2}MV^2 + \frac{1}{2}\mu v'^2, \quad \mu = \frac{m_1 m_2}{m_1 + m_2},$$

which leads to the conservation of the magnitude of the relative velocity:

$$|\mathbf{v}| = |\mathbf{v}'|. \tag{3.3.6}$$

Using the fact that the centre of mass moves uniformly in a straight line, let us consider an inertial reference system with respect to which the centre of mass is at rest. Such a frame is called the *centre of mass system* (CMS). We choose the origin of CMS in the centre of mass. Denoting by a star each vector defined with respect to CMS, in this frame we have

$$\mathbf{p}_1^* = m_1\mathbf{v}_1^* = m_1(\mathbf{v}_1 - \mathbf{V}) = -\mu\mathbf{v}, \quad \mathbf{p}_2^* = m_2\mathbf{v}_2^* = m_2(\mathbf{v}_2 - \mathbf{V}) = \mu\mathbf{v},$$

which yields:

$$\mathbf{p}_1^* + \mathbf{p}_2^* = \mathbf{p'}_1^* + \mathbf{p'}_2^* = 0. \tag{3.3.7}$$

This relation can be taken as definition of the centre of mass. In view of (3.3.6), we remark that

$$|\mathbf{p}_1^*| = |\mathbf{p}_2^*| = |\mathbf{p'}_1^*| = |\mathbf{p'}_2^*|. \tag{3.3.8}$$

The relation between momenta relative to CMS is represented graphically in Fig. 3.12. Before collision, the momenta of the two particles, equal in magnitude but opposite in direction, lie on a straight line. This property survives after collision, but the momenta are directed along a different straight line. The angle θ_{CM} between these two lines is called *scattering angle*.

Usually, one of the particles is considered to be the 'target', being at rest relative to the experimental device. Such a reference system is termed *laboratory*

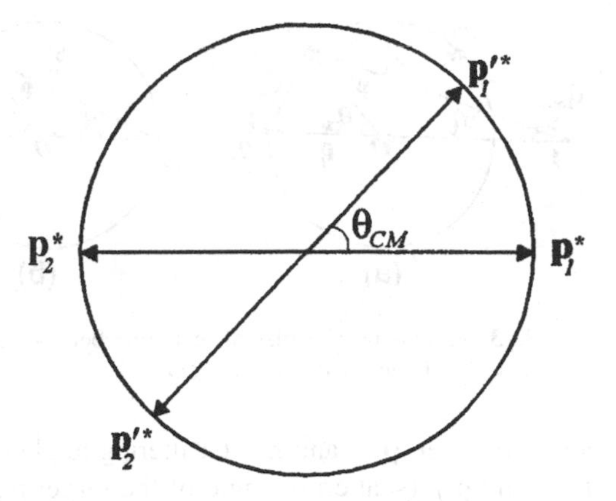

Fig. 3.12 Relation between momenta relative to the centre of mass reference system in a collision of two particles.

system (LS). If we choose particle 2 as being at rest, then LS is defined by the condition

$$\mathbf{v}_{2L} = 0.$$

In the following, to simplify the notation, we shall use the letter L to specify the vectors defined relative to LS, and remove the star in case of the vectors defined with respect to CMS. The velocity of CMS relative to LS is then

$$\mathbf{V} = \frac{m_1 \mathbf{v}_{1L}}{M}. \tag{3.3.9}$$

Let us find the relations of transition between the two systems. Since $\mathbf{v}_2 = \mathbf{v}_{2L} - \mathbf{V} = -\mathbf{V}$, it follows that:

$$\mathbf{V} = -\mathbf{v}_2 = -\frac{\mathbf{p}_2}{m_2} = \frac{\mathbf{p}_1}{m_1}, \tag{3.3.10}$$

hence

$$\mathbf{p}_{1L} = m_1 \mathbf{v}_{1L} = M\mathbf{V} = \frac{M}{m_2}\mathbf{p}_1 = \left(1 + \frac{m_1}{m_2}\right)\mathbf{p}_1. \tag{3.3.11}$$

After collision, the particle 1 has the momentum:

$$\mathbf{p}'_{1L} = m_1 \mathbf{v}'_{1L} = m_1(\mathbf{v}'_1 + \mathbf{V}) = \mathbf{p}'_1 + \frac{m_1}{m_2}\mathbf{p}_1, \tag{3.3.12}$$

while the momentum of particle 2 (the 'target') is:

$$\mathbf{p}'_{2L} = \mathbf{p}_{1L} - \mathbf{p}'_{1L} = \mathbf{p}_1 - \mathbf{p}'_1. \tag{3.3.13}$$

Relations (3.3.11)–(3.3.13) have a simple geometric interpretation. We can distinguish three cases: $\frac{m_1}{m_2} > 1, =1, <1$. In Fig. 3.13 the vector $\vec{OB}$ stands for $\mathbf{p}_1$, $\vec{OC}$ stands for $\mathbf{p}'_1$ and $\theta_{CM} = \widehat{BOC}$. The vector $\vec{AO}$ (collinear with $\vec{OB}$) is $\frac{m_1}{m_2}\mathbf{p}_1$, therefore $\vec{AB}$ is $\mathbf{p}_{1L}$ given by (3.3.11). From (3.3.12), $\vec{AC} = \vec{OC} + \vec{AO}$ is $\mathbf{p}'_{1L}$, while, in agreement with (3.3.13), $\vec{CB} = \vec{OB} - \vec{OC}$ is $\mathbf{p}'_{2L}$. If we denote by θ_L the

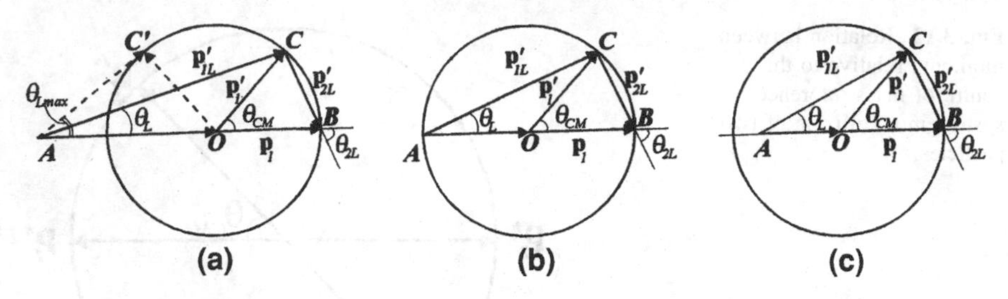

Fig. 3.13 Various possibilities of scattering between two particles, in terms of the ratio m_1/m_2: (a) $m_1 > m_2$, (b) $m_1 = m_2$, (c) $m_1 < m_2$.

angle between $\mathbf{p}'_{1L}$ and $\mathbf{p}_{1L}$ (scattering angle in LS) and by θ_{2L} the angle between $\mathbf{p}'_{2L}$ and $\mathbf{p}_{1L}$ (scattering angle of the target particle), in all three cases we have

$$\theta_{2L} = \frac{1}{2}(\pi - \theta_{CM}). \tag{3.3.14}$$

Let us now find the relation between the scattering angles θ_L and θ_{CM}. Examining Fig. 3.13, we can write:

$$\tan\theta_L = \frac{|\vec{AC}|\sin\theta_L}{|\vec{AC}|\cos\theta_L} = \frac{|\vec{OC}|\sin\theta_{CM}}{|\vec{AO}| + |\vec{OC}|\cos\theta_{CM}}.$$

But $|\vec{OC}| = |\mathbf{p}'_1| = |\mathbf{p}_1|$ and $|\vec{AO}| = \frac{m_1}{m_2}|\mathbf{p}_1|$, hence:

$$\tan\theta_L = \frac{m_2\sin\theta_{CM}}{m_1 + m_2\cos\theta_{CM}}. \tag{3.3.15}$$

We shall now analyze various cases, depending on the value of the ratio $\frac{m_1}{m_2}$. If $m_1 > m_2$, the point A is outside the circle of radius $|\mathbf{p}_1|$ (Fig. 3.13a). In this case, θ_L takes values in the interval $[0, \theta_{Lmax}]$, where θ_{Lmax} is the angle $\widehat{C'AO}$, while θ_{CM} takes values in the interval $[0,\pi]$. Note that

$$\sin\theta_{Lmax} = \frac{|\mathbf{p}'_1|}{\frac{m_1}{m_2}|\mathbf{p}_1|} = \frac{m_2}{m_1}, \tag{3.3.16}$$

which implies $\theta_{Lmax} \le \frac{\pi}{2}$.

If $m_1 = m_2$ (Fig. 3.13b), the point A is on the circle of radius $|\mathbf{p}_1|$, therefore $\theta_L = \frac{\theta_{CM}}{2}$, i.e. $\theta_{CM} \in \left[0, \frac{\pi}{2}\right]$. In view of (3.3.14), we obtain

$$\theta_L + \theta_{2L} = \frac{\pi}{2},$$

which means that, after collision, the two particles move in LS along orthogonal directions.

Finally, if $m_1 < m_2$ (Fig. 3.13c), the point A is inside the circle, and the intervals of variation of θ_L and θ_{CM} are the same.

We realize that by means of the equations of conservation of momentum and energy, we can determine only the magnitude of momenta in CMS. To find the scattering angle (either θ_{CM} or θ_L), we must know both the law of interaction between particles and their mutual position. This means that we have to find the law of motion *during* collision and we shall discuss this matter in the next section. Nevertheless, in some particular cases we can obtain some information from the aforementioned formalism. For example, as we have shown, if $m_1 > m_2$, the angle θ_L is limited by θ_{Lmax} (see (3.3.16)), which means that, if $m_1 \gg m_2$, we have $\theta_{Lmax} \approx 0$, leading to $\theta_L \approx 0$, i.e., the particle does not change its initial direction after collision. If $m_1 < m_2$, θ_L, as well as θ_{CM}, can take the value π. For $\theta_L = \theta_{CM} = \pi$, Fig. 3.13c shows that $\mathbf{p}'_1 = -\mathbf{p}_1$, and the relations (3.3.12), (3.3.13) yield:

$$\mathbf{p}'_{1L} = \mathbf{p}_1\left(\frac{m_1}{m_2} - 1\right), \quad \mathbf{p}'_{2L} = 2\mathbf{p}_1.$$

The kinetic energy of the 'target' after collision, measured in LS, is

$$T'_{2L} = \frac{1}{2m_2}|\mathbf{p}'_{2L}|^2 = \frac{1}{m_2}|\mathbf{p}_1|^2(1 - \cos\theta_{CM})$$

and, for $\theta_{CM} = \pi$, in view of (3.3.11),

$$(T'_{2L})_{max} = \frac{2}{m_2}|\mathbf{p}_1|^2 \frac{2m_2}{m^2}|\mathbf{p}_{1L}|^2 = \frac{4\mu}{M}T_{1L}. \tag{3.3.17}$$

Therefore, in case of the 'backwards scattering' ($\theta_L = \theta_{CM} = \pi$), the energy obtained by the target particle after collision is always smaller than the initial kinetic energy of the incident particle, which is equal to the total initial energy. One can verify that this is the largest energy of the target particle after collision (if θ_L varies from 0 to 2π). If $m_1 \ll m_2$, we have $\theta_L \approx \theta_{CM}$, but this time θ_L is not limited (like in case $m_1 \gg m_2$), because the target particle remains at rest in LS, while the motion of the particle 1 depends on the law of interaction and the initial conditions.

3.3.2 Effective Scattering Cross Section

In physical applications we have to deal with the scattering of a large number of particles by a centre of force (Fig. 3.14), rather than with the deviation of a single particle. The *incident beam* is formed by independent, free particles, having the same mass m_1 and the same velocity $\mathbf{v}_1$. Coming close to the centre of force, also called *the centre of scattering*, all particles are subject to the same law of interaction given by the potential $V = V(r)$. After scattering, the particles are considered free again, but having different velocities and moving in different directions. We shall analyze here only elastic scattering. Assuming that at infinity the non-interaction condition is given by $\lim_{r\to\infty} V(r) = 0$ and recalling that $T_\infty = E \geq 0$,

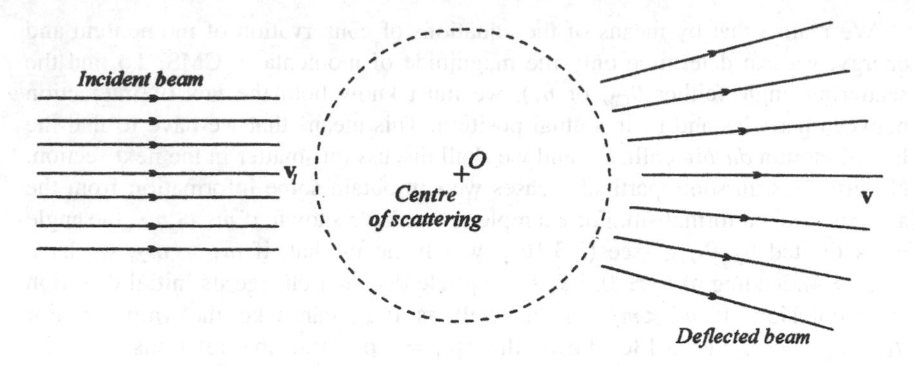

Fig. 3.14 Schematic representation of a scattering.

this means that in the process of collision the mechanical energy of the particles is positive ($E > 0$).

To describe the scattering, let us take the origin of coordinates O at the centre and the x_3-axis parallel to $\mathbf{v}_1$. We call *impact parameter* b of a particle of the beam the distance between the centre of scattering O and the initial direction of motion of the incident particle (Fig. 3.14). Since the central force is conservative, the energy of each particle remains constant:

$$E = T_1 = \frac{m_1 v_1^2}{2}. \tag{3.3.18}$$

On the other hand, the plane of the trajectory of each particle is determined by its conserving angular momentum. Since $l = m_1 v r_\perp$, where $r_\perp$ is the projection of $\mathbf{r}$ on $\mathbf{v}$, with $r_\perp = b$ at infinity, we can write the angular momentum in terms of the impact parameter:

$$l = m_1 v_1 b. \tag{3.3.19}$$

The trajectories of the particles are symmetrical relative to a line drawn from the centre O to the pericentre P (Fig. 3.15). The *scattering angle* θ is the angle made by the two asymptotes to the orbit, being the same for all particles having an identical b.

In the theory of particle scattering, an important quantity is the *effective scattering cross section* $d\sigma$, defined as the ratio

$$d\sigma = \frac{dN}{N_0},$$

where dN is the number of particles scattered per unit time within the solid angle $d\Omega$ and N_0 is the number of particles passing in unit time and in normal direction a unit area of the beam cross section. Observing that the scattering angle is a monotonically decreasing function of b, the particles having the impact parameter in the interval $(b, b + db)$ will be deflected at angles in the interval $(\theta, \theta + d\theta)$.

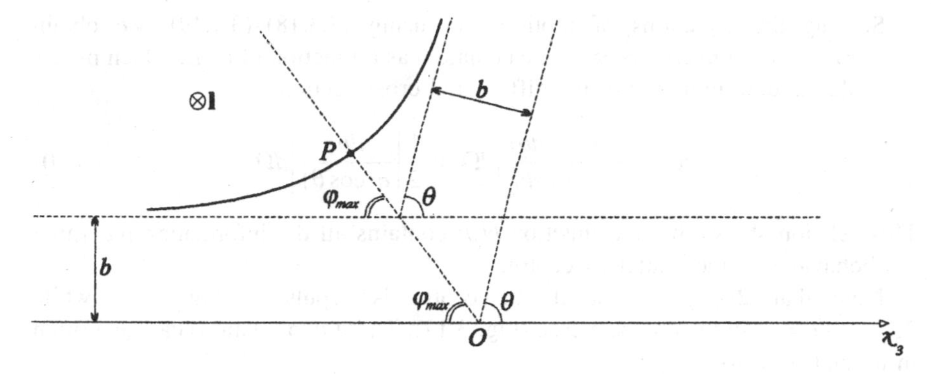

Fig. 3.15 Trajectory of a scattered particle in a repulsive potential V, and the scattering angle θ.

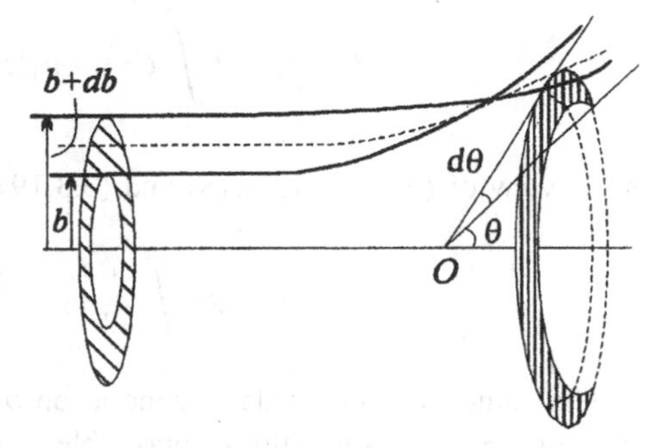

Fig. 3.16 Geometrical representation of scattering, in terms of the impact parameter b, for a repulsive centre.

These particles pass in the normal direction through the area $b\,db\,d\Phi$, where Φ is the polar angle of rotation about the x_3-axis. Thus,

$$dN = N_0 b\,db\,d\Phi.$$

Due to the axial symmetry, none of the quantities involved depend on Φ, therefore we may consider particles passing through the annulus formed by the circles of radii b, $b + db$, and which, after scattering, lie in a conic shell (Fig. 3.16), i.e.

$$dN = 2\pi N_0 b\,db.$$

Denoting by $d\Omega = 2\pi \sin\theta d\theta$ the solid angle defined by the directions $(\theta,\ \theta + d\theta)$, we have:

$$dN = N_0 \frac{b}{\sin\theta}\left|\frac{db}{d\theta}\right| d\Omega,$$

where the choice of the absolute value of the derivative $\frac{db}{d\theta}$ is determined by the fact that it is usually negative, while dN is positive by definition.

Solving the equations of motion and using (3.3.18)–(3.3.19), we obtain $\theta = \theta(E, b)$, or conversely, we can calculate b as a function of θ, E, which makes possible the determination of the differential cross section:

$$d\sigma = \frac{1}{2\sin\theta}\left|\frac{db^2}{d\theta}\right| d\Omega = \frac{1}{2}\left|\frac{db^2}{d(\cos\theta)}\right| d\Omega. \tag{3.3.20}$$

This relation shows that the function $b(\theta)$ contains all the information regarding the behaviour of the scattering centre.

Note that $2\varphi_{max} + \theta = \pi$ if the centre is repulsive (Fig. 3.15), while $2\varphi_{max} - \theta = \pi$ in the opposite case (Fig. 3.17). These two relations can be written in a condensed form:

$$\theta = |\pi - 2\varphi_{max}|.$$

Using (3.1.19), we have:

$$\theta = \left|\pi - 2\int_{r_m}^{\infty} Cr^2\{2m_1[E - V_{eff}(r)]\}^{-\frac{1}{2}} dr\right|,$$

or, in view of (3.1.12), (3.3.18) and (3.3.19):

$$\theta = \left|\pi - 2b\int_{r_m}^{\infty} \frac{1}{r^2}\left[1 - \frac{V(r)}{E} - \frac{b^2}{r^2}\right]^{-\frac{1}{2}} dr\right|. \tag{3.3.21}$$

The value of r_m depends in general on b, and this gives rise to a complicated relation $\theta = \theta(b)$, sometimes impossible to invert.

Our theory is based on a model involving the motion of a beam of particles in an external central field. But we can also conceive this process as a number of collisions between pairs of particle, one particle of each pair being the 'target'. Each of these systems, in view of the theory developed in Sect. 3.1, can be considered as a 'particle in a central field', which means that all our formulas remain valid in CMS, if instead of the mass m_1 of the incident particle we put the reduced mass μ of the pair.

Recall that the differential cross section $d\sigma$, being defined as a ratio of numbers of particles, is independent on the reference frame, therefore it does not change when passing from CMS to LS. But the solid angle changes, so that the cross section per unit of solid angle is:

$$d\sigma = \left(\frac{d\sigma}{d\Omega}\right) d\Omega = \left(\frac{d\sigma}{d\Omega}\right)_L d\Omega_L.$$

Since the axial symmetry is manifest also in LS, we have

$$d\Omega_L = 2\pi \sin\theta_L d\theta_L = -2\pi d(\cos\theta_L),$$

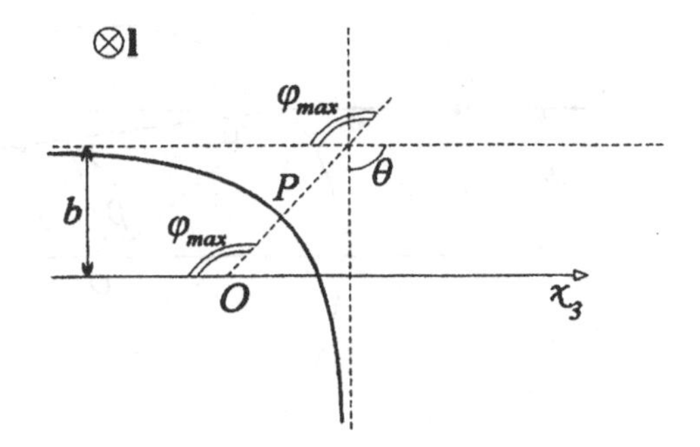

Fig. 3.17 Geometrical representation of scattering, in terms of the impact parameter b, for an attractive centre.

which can also be written as

$$d\Omega_L = 2\pi \sin\theta d\theta \left| \frac{d(\cos\theta_L)}{d(\cos\theta)} \right| = \left| \frac{d(\cos\theta_L)}{d(\cos\theta)} \right| d\Omega.$$

For the incident particles, in view of (3.3.15), we have:

$$\frac{d\sigma}{d\Omega} = \frac{1 + \frac{m_1}{m_2}\cos\theta}{\left[1 + 2\frac{m_1}{m_2}\cos\theta + \frac{m_1^2}{m_2^2}\right]^{\frac{3}{2}}} \left(\frac{d\sigma}{d\Omega}\right)_{1L}, \tag{3.3.22}$$

where m_2 is the mass of the target particles and $(d\sigma/d\Omega)_{1L}$ is the differential cross section of the incident particles. If θ_{2L} is the angle between the directions of the incident particles and the target after collision, using (3.3.14) we obtain:

$$\frac{d\sigma}{d\Omega} = \frac{1}{4|\sin\frac{\theta}{2}|} \left(\frac{d\sigma}{d\Omega}\right)_{2L} = \frac{1}{4|\cos\theta_{2L}|} \left(\frac{d\sigma}{d\Omega}\right)_{2L}, \tag{3.3.23}$$

where $(d\sigma/d\Omega)_{2L}$ is the differential cross section of the target particles.

3.3.3 Scattering on a Spherical Potential Well

By 'spherical potential well' we mean a central field with the potential

$$V(r) = 0, \quad r > R,$$
$$V(r) = -V_0, \quad r \le R,$$

where V_0 is a constant. On each portion of its trajectory, the particle is free. Its velocities outside and inside the well are $\mathbf{v}_1$ and $\mathbf{v}$, respectively, both being constant. Since the field is central and conservative, both energy and angular momentum are conserved:

$$E = \frac{m_1 v_1^2}{2} = \frac{m_1 v^2}{2} - V_0 > 0, \quad l = m_1 v_1 b = m_1 v r_m,$$

where $r_m = \overline{OP}$ (Fig. 3.18). From these relations we get

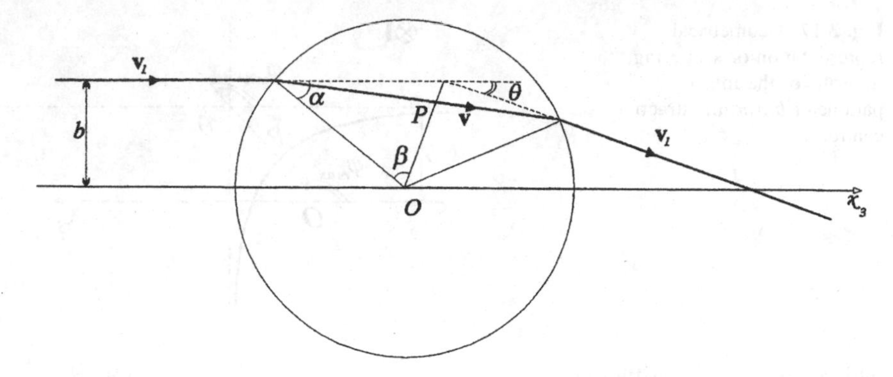

Fig. 3.18 Scattering on a spherical potential well.

$$b = \frac{v}{v_1} r_m = n r_m, \tag{3.3.24}$$

with

$$n \equiv \frac{v}{v_1} = \sqrt{1 + \frac{V_0}{E}} > 1, \tag{3.3.25}$$

known as the *refraction index* of the potential well. Using Fig. 3.18, we find

$$r_m = R \cos \beta = R \cos\left(\frac{\pi}{2} + \frac{\theta}{2} - \alpha\right),$$

or, after some trigonometric calculations,

$$r_m = -\sqrt{R^2 - b^2} \sin \frac{\theta}{2} + b \cos \frac{\theta}{2}.$$

Introducing this relation into (3.3.24), we obtain:

$$b^2 = \frac{n^2 \sin^2 \frac{\theta}{2}}{n^2 - 2n \cos \frac{\theta}{2} + 1} R^2 \tag{3.3.26}$$

and formula (3.3.20) yields:

$$\frac{d\sigma}{d\Omega} = \frac{(n \cos \frac{\theta}{2} - 1)(\cos \frac{\theta}{2} - n)}{\cos \frac{\theta}{2} (n^2 - 2n \cos \frac{\theta}{2} + 1)^2} \frac{n^2 R^2}{4}. \tag{3.3.27}$$

The impact parameter b varies from 0 to R (for $b > R$, the particles pass without deviation), while the scattering angle takes values from 0 to θ_{max}, given by

$$\theta_{max} = 2 \arccos \frac{1}{n}.$$

Integrating over the solid angle $2\theta_{max}$, which contains all scattered particles, we arrive at the total effective cross section:

$$\sigma = 2\pi \int_0^{\theta_{max}} \frac{d\sigma}{d\Omega} \sin\theta \, d\theta = 2\pi n^2 R^2 \int_{\frac{1}{n}}^{1} \frac{(nz-1)(n-z)}{(n^2 - 2nz + 1)^2} dz.$$

The result of integration is πR^2, as expected.

3.3.4 Rutherford's Formula

The scattering of electrically charged particles in the Coulomb field of the targets is a process which served as a model in the well-known experiment of *Ernest Rutherford*, leading to the discovery of the atomic nucleus. Since the particles involved in this type of interaction are elementary, a correct treatment of the problem can be performed only by using the quantum mechanical formalism. Nevertheless, some useful results are deduced within the classical approach.

Let us consider the scattering of a beam of particles of mass m_1 and charge q_1 on the target formed by particles of mass m_2 and charge q_2. Assuming that the potential is of the type (3.2.1), with

$$k = \frac{q_1 q_2}{4\pi\epsilon_0}, \tag{3.3.28}$$

we shall determine the scattering angle θ in CMS. The field being Coulombian, the relation between φ_{max} and θ can be written as

$$\varphi_{max} = \frac{\pi}{2} - \frac{k}{|k|}\frac{\theta}{2},$$

thus accounting for both the attraction and repulsion forces. Since $\varphi_{max} = \lim_{r\to\infty} \varphi(r)$, Eq. (3.2.7) yields:

$$\sin\frac{\theta}{2} = \frac{1}{\epsilon}.$$

On the other hand, setting $l = \mu v_0 b$ and $E + \frac{\mu v_0^2}{2}$ in (3.2.9), we have:

$$\epsilon^2 = 1 + \left(\frac{\mu v_0^2 b}{k}\right)^2,$$

hence

$$b^2 = \frac{k^2}{\mu^2 v_0^4} \cot^2\frac{\theta}{2}.$$

Introducing this relation into (3.3.20), we arrive at *Rutherford's formula*:

$$\frac{d\sigma}{d\Omega} = \left(\frac{k}{2\mu v_0^2}\right)^2 \frac{1}{\sin^4 \frac{\theta}{2}}. \tag{3.3.29}$$

The effective cross section for the incident particles in LS, $\left(\frac{d\sigma}{d\Omega}\right)_{1L}$, is obtained by substituting θ by θ_L in (3.3.22) and (3.3.29), and by using (3.3.15), which leads to a very complicated relation. Much simpler is the expression for the effective cross section of the 'targets'. Since $2\theta_{2L} = \pi - \theta$, we get from (3.3.23) and (3.3.29):

$$\left(\frac{d\sigma}{d\Omega}\right)_{2L} = \left(\frac{k}{\mu v_0^2}\right)^2 |\cos\theta_{2L}|^{-3}. \tag{3.3.30}$$

A simple, but approximate formula can also be found for the cross section of incident particles if $m_1 \ll m_2$. In this case, $\mu \approx m_1$, $\theta_L \approx \theta$, therefore

$$\left(\frac{d\sigma}{d\Omega}\right)_{1L} \simeq \left(\frac{k}{4E_1}\right)^2 \left(\sin\frac{\theta_{1L}}{2}\right)^{-4}, \tag{3.3.31}$$

where $E_1 = m_1 \frac{v_0^2}{2} \approx \mu \frac{v_0^2}{2}$ is the energy of the incident particles.

In the case $m_1 = m_2 = 2\mu$, since $\theta_{1L} = \frac{\theta}{2} = \theta_L$, $\theta_{2L} = \frac{\pi}{2} - \frac{\theta}{2} = \frac{\pi}{2} - \theta_L$, both cross sections $\left(\frac{d\sigma}{d\Omega}\right)_{1L}$ and $\left(\frac{d\sigma}{d\Omega}\right)_{2L}$ are easily calculated. In this case, after scattering we cannot make any more distinction between incident and target particles. Then, we can define the differential cross section as

$$d\sigma = \left[\left(\frac{d\sigma}{d\Omega}\right)_{1L} d\Omega_{1L} + \left(\frac{d\sigma}{d\Omega}\right)_{2L} d\Omega_{2L}\right]_{\theta_{1L}=\theta_L,\, \theta_{2L}=\frac{\pi}{2}-\theta_L},$$

hence

$$\left(\frac{d\sigma}{d\Omega}\right)_L = \left(\frac{k}{E_1}\right)^2 \left(\frac{1}{\sin^4\theta_L} + \frac{1}{\cos^4\theta_L}\right) \cos\theta_L.$$

The Coulomb field is characterized by an infinite total effective cross section:

$$\sigma = \int_{4\pi} \frac{d\sigma}{d\Omega} d\Omega,$$

as a result of its infinite radius of action.

3.4 Periodical Motion of a Particle Under the Influence of Gravity

A particle constrained to move on a curve or a surface, subject to the force of gravity and performing periodical motions about a fixed point, is a *mathematical pendulum.* We shall discuss three classic models: simple, cycloidal and spherical pendulums.

3.4.1 Simple Pendulum

The *simple* or *plane* pendulum is a system formed by a heavy particle, moving without friction on a vertical circle. Let R be the radius of the circle, the motion being performed in the x_1x_2-plane, with $\overrightarrow{Ox_1} \parallel \mathbf{G}$ (Fig. 3.19). The (holonomic) constraints are:

$$x_1^2 + x_2^2 = R^2, \quad x_3 = 0,$$

therefore the system has one degree of freedom. Such a model is realized by a particle P of mass m, fastened at one end to a massless rigid rod, the other end being suspended in a fixed point O. If the rod is replaced by a non-extensible but flexible wire, the constraint becomes non-holonomic: $x_1^2 + x_2^2 \leq R^2$.

The most convenient choice for the generalized coordinate is the angle θ between OP and the x_1-axis. Since $x_1 = r\cos\theta$, $x_2 = r\sin\theta$, the kinetic and potential energies are:

$$T = \frac{m}{2}R^2\dot{\theta}^2, \quad V = -mgR\cos\theta$$

and the Lagrangian reads:

$$L = \frac{m}{2}r^2\dot{\theta}^2 + mgR\cos\theta. \tag{3.4.1}$$

The differential equation of motion is obtained by using the Lagrange equations (2.5.17):

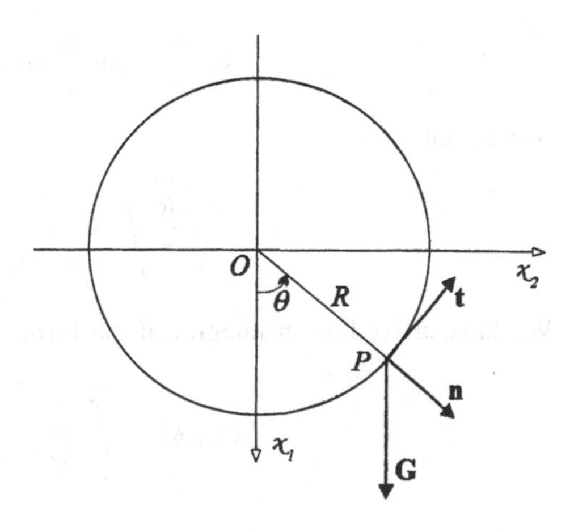

Fig. 3.19 Simple (plane) pendulum. The system has one degree of freedom.

$$\ddot{\theta} + \frac{g}{R} \sin\theta = 0. \tag{3.4.2}$$

This is a non-linear differential equation, which yields $\theta = \theta(t)$. In the following, we shall determine the exact solution of (3.4.2). To this end, we use the first integral of motion expressing the conservation of mechanical energy:

$$E = \frac{m}{2} R^2 \dot{\theta}^2 - mgR\cos\theta = \text{const.}$$

The constant is determined by the initial conditions: $\theta(0) = \theta_0$, $\dot{\theta}(0) = 0$. Then, $E = -mgR\cos\theta_0$ and we obtain:

$$\dot{\theta} = \pm\sqrt{\frac{2g}{R}(\cos\theta - \cos\theta_0)}. \tag{3.4.3}$$

Since $\dot{\theta}$ is a real quantity, we must have $|\theta| \leq \theta_0$. This means that at the beginning of the motion θ decreases, i.e. $\dot{\theta} < 0$ (which corresponds to the minus sign in (3.4.3)), varying from zero at $\theta\,(0) = \theta_0$ to its minimum value at $\theta = 0$, then increases to zero, while θ decreases from zero to $-\theta_0$. Next, the motion repeats itself, but in the opposite sense: the angle θ increases from $-\theta_0$ to 0, while $\dot{\theta} > 0$ (the plus sign in (3.4.3)) and increases from 0 to its maximum value, etc. The turning points of the trajectory are, therefore, $\theta = \pm\theta_0$. The motion repeats itself periodically, with the angular amplitude θ_0. It is obvious that in each of the intervals $(\theta_0, 0)$, $(0, -\theta_0)$, $(-\theta_0, 0)$, $(0, \theta_0)$ the motion is similar, which means that the period of the pendulum is

$$\tau = 4\int_{\theta_0}^{0} \frac{1}{-\sqrt{\frac{2g}{R}(\cos\theta - \cos\theta_0)}} d\theta.$$

Using the substitution (note that φ does not have the meaning of an extra degree of freedom besides θ!)

$$\sin\frac{\theta}{2} = \sin\frac{\theta_0}{2}\sin\varphi = k\sin\varphi$$

one obtains:

$$\tau = 4\sqrt{\frac{R}{g}} \int_0^{\frac{\pi}{2}} \frac{1}{\sqrt{1 - \sin^2\frac{\theta_0}{2}\sin^2\varphi}} d\varphi. \tag{3.4.4}$$

We have arrived at an integral of the form

$$F(\varphi_0, k) = \int_0^{\varphi_0} \frac{1}{\sqrt{1 - k^2\sin^2\varphi}} d\varphi, \tag{3.4.5}$$

which is an *elliptic integral of the first kind*. The function φ is the *amplitude* and k is the *modulus* of the integral. The integral $K(k) = F(\frac{\pi}{2}, k)$ is termed *complete elliptic integral of the first kind*. Tables of elliptic integrals can be found in special mathematical publications. Using our notation, we can write the period of motion as

$$\tau = 4\sqrt{\frac{R}{g}}\, F\left(\frac{\pi}{2}, k\right) = 4\sqrt{\frac{R}{g}}\, K(k). \tag{3.4.6}$$

Excluding the singular value $\theta_0 = \pi$, we may take $|k| < 1$. Then we can perform a series expansion of the integrand in (3.4.5):

$$\begin{aligned}(1 - k^2 \sin^2 \varphi)^{-\frac{1}{2}} &= 1 + \frac{1}{2}k^2 \sin^2 \varphi + \frac{1 \cdot 3}{2 \cdot 4}k^4 \sin^4 \varphi + \cdots \\ &= \sum_{n=0}^{\infty} \frac{1 \cdot 3 \cdot 5 \ldots (2n-1)}{2 \cdot 4 \cdot 6 \ldots 2n} k^{2n} \sin^{2n} \varphi.\end{aligned} \tag{3.4.7}$$

This series is absolutely and uniformly convergent for $|k| < 1$ in the interval $(0, \pi)$, which means that we can integrate it term by term. To do this, we shall deduce an auxiliary formula, first obtained by the English mathematician *John Wallis*. Let us denote

$$A_{2n} = \int_0^{\frac{\pi}{2}} \sin^{2n} \varphi d\varphi, \tag{3.4.8}$$

where n is an integer. We can write

$$A_{2n} = \int_0^{\frac{\pi}{2}} \sin^{2n-2} \varphi(1 - \cos^2 \varphi)d\varphi = A_{2n-2} - \int_0^{\frac{\pi}{2}} \sin^{2n-2} \varphi \cos^2 \varphi d\varphi,$$

or, if we integrate by parts,

$$A_{2n} = A_{2n-2} - \frac{1}{2n-1} \int_0^{\frac{\pi}{2}} \sin^{2n} \varphi d\varphi.$$

In view of (3.4.8), we can write the following recurrence relation:

$$2n\, A_{2n} = (2n-1)A_{2n-2}. \tag{3.4.9}$$

Giving to n all values from 0 to n and then taking the product of the obtained relations, we find

$$A_{2n} = \frac{1 \cdot 3 \cdot 5 \ldots (2n-1)}{2 \cdot 4 \cdot 6 \ldots 2n} A_0,$$

or

$$\int_0^{\frac{\pi}{2}} \sin^{2n} \varphi d\varphi = \frac{1 \cdot 3 \cdot 5 \ldots (2n-1)}{2 \cdot 4 \cdot 6 \ldots 2n} \frac{\pi}{2} = \frac{(2n-1)!!}{2^n n!} \frac{\pi}{2}, \tag{3.4.10}$$

which is *Wallis' formula*. Using this result in (3.4.6), we obtain finally the period of the pendulum as

$$\tau = 2\pi \sqrt{\frac{R}{g}} \sum_{n=0}^{\infty} \left[\frac{(2n-1)!!}{2^n n!} \right]^2 \sin^{2n} \frac{\theta_0}{2}. \tag{3.4.11}$$

This result is *exact* (we have not made any approximation). If $\theta_0 < \frac{\pi}{2}$, we can expand in series $\sin \frac{\theta_0}{2}$ and obtain

$$\tau = 2\pi \sqrt{\frac{R}{g}} \left(1 + \frac{1}{16} \theta_0^2 + \frac{11}{3072} \theta_0^4 + \cdots \right). \tag{3.4.12}$$

For small amplitudes ($\theta_0 < 4°$), the terms containing θ_0^2, θ_0^4, etc. are negligible with respect to 1, and we arrive at the well-known approximative formula

$$\tau = 2\pi \sqrt{\frac{R}{g}},$$

which says that the small oscillations of pendulum are *tautochronous* (from the Greek *tauto* (the same) and *chronos* (time)). This property is used in the construction of astronomical clocks ($\theta_0 = 1°30'$).

Let us now calculate the force of constraint. The tangent and normal projections of the equation $\mathbf{L} = m\ddot{\mathbf{r}} - \mathbf{G}$ are:

$$L_t = mR\ddot{\theta} + mg \sin \theta, \quad L_n = -mR\dot{\theta}^2 - mg \cos \theta.$$

In agreement with (3.4.2), we have $L_t = 0$, as expected. Using (3.4.3), we find

$$L_n = mg(2\cos \theta_0 - 3\cos \theta). \tag{3.4.13}$$

If $\theta_0 < \frac{\pi}{2}$, then for any θ the vector $\mathbf{L}$ points towards the centre O. In this case, the constraint remains holonomic if the rigid rod is replaced by an inextensible, but flexible wire, whose initial length is R.

3.4.2 Cycloidal Pendulum

A heavy particle moving without friction on a vertical cycloid with the concavity upwards is a *cycloidal pendulum* (Fig. 3.20). The parametric equations of a cycloid are (see (2.6.29), (2.6.30)):

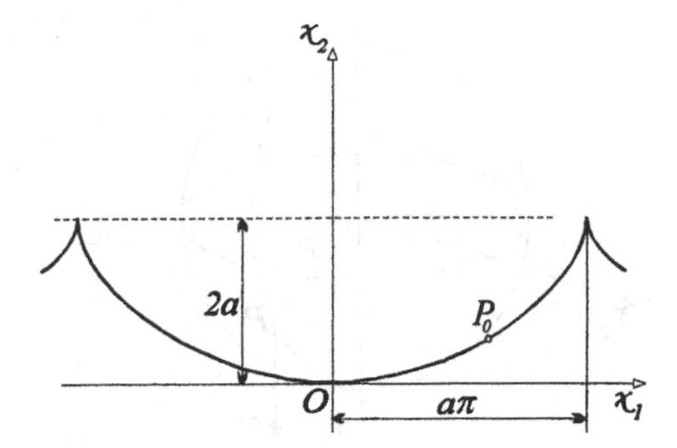

Fig. 3.20 Cycloidal pendulum. The system has one degree of freedom.

$$x_1 = a(\theta + \sin\theta), \quad x_2 = a(1 - \cos\theta), \quad x_3 = 0 \tag{3.4.14}$$

and we have again a system with one degree of freedom. Taking θ as generalized coordinate, the Lagrangian reads:

$$L = ma^2(1 + \cos\theta)\dot{\theta}^2 + mga\cos\theta, \tag{3.4.15}$$

leading to the equation of motion:

$$2(1 + \cos\theta)\ddot{\theta} - \sin\theta\dot{\theta}^2 + \frac{g}{a}\sin\theta = 0,$$

or

$$\frac{d^2}{dt^2}\left(\sin\frac{\theta}{2}\right) + \frac{g}{4a}\sin\frac{\theta}{2} = 0.$$

Taking $\theta(0) = \theta_0$ and $\dot{\theta}(0) = 0$ as initial conditions, the last equation yields the solution

$$\sin\frac{\theta}{2} = \sin\frac{\theta_0}{2}\cos\left(\sqrt{\frac{g}{4a}}\,t\right). \tag{3.4.16}$$

Consequently, the motion is periodical, with the period $4\pi\sqrt{\frac{a}{g}}$. The same result is obtained using the energy first integral.

The time needed for the particle to move from P_0 to $O\,(\theta = 0)$ is

$$t_1 = \pi\sqrt{\frac{a}{g}}, \tag{3.4.17}$$

i.e. the periodical motion on a cycloid is tautochronous, for *any* angle θ.

3.4.3 Spherical Pendulum

By *spherical pendulum* we mean a heavy particle moving without friction on a fixed sphere. Such a system can be obtained by suppressing the constraint which

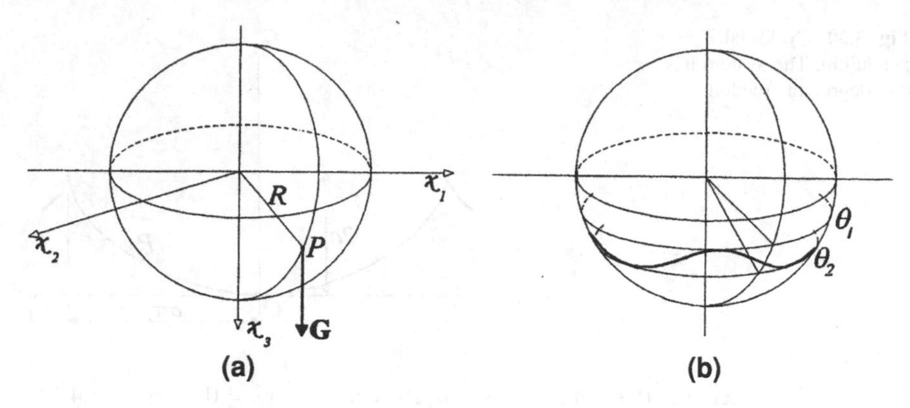

Fig. 3.21 Spherical pendulum. The system has two degrees of freedom.

determines the simple pendulum to move in a plane. Choosing the coordinate system $Ox_1x_2x_3$, with the origin at the centre of the sphere, and $\vec{Ox_3} \parallel \mathbf{G}$ (Fig. 3.21a), we have a single constraint,

$$x_1^2 + x_2^2 + x_3^2 = R^2, \tag{3.4.18}$$

which means that the system possesses two degrees of freedom. Taking the polar and azimuthal angles θ, φ as generalized coordinates, the Lagrangian reads:

$$L = \frac{m}{2} R^2(\dot{\theta}^2 + \sin^2\theta\dot{\varphi}^2) + mgR\cos\theta. \tag{3.4.19}$$

As we already know, since φ is cyclic, $p_\varphi = l_3$ is a first integral of motion:

$$p_\varphi = mR^2 \sin^2\theta\dot{\varphi} = l_3 = \text{const.} \tag{3.4.20}$$

Another first integral is furnished by the conservation of the total energy:

$$E = \frac{mR^2}{2}\dot{\theta}^2 + \frac{l_3^2}{2mR^2\sin^2\theta} - mgR\cos\theta. \tag{3.4.21}$$

Defining the effective potential

$$V_{eff}(\theta) = \frac{l_3^2}{2mR^2\sin^2\theta} - mgR\cos\theta, \tag{3.4.22}$$

the relation (3.4.21) yields:

$$\dot{\theta}^2 = \frac{2}{mR^2}[E - V_{eff}(\theta)]. \tag{3.4.23}$$

During its motion, the particle will reach only those values of θ which correspond to $E \geq V_{eff}(\theta)$. The limit values of θ can be obtained as the intersection points of the curve $V = V_{eff}(\theta)$ with the straight line $V = E$, in the plane of variables (θ, V).

Since $V_{eff}|_{\theta=0} = V_{eff}|_{\theta=\pi} = \infty$, the pendulum never passes through the two poles of the sphere. These points are equilibrium positions of the particle: stable equilibrium for $\theta = 0$ and unstable for $\theta = \pi$. If $\dot{\varphi} = 0$, then $l_3 = 0$, which yields $\varphi =$ const., and we arrive at the particular case of the simple pendulum. Therefore, in the following we shall assume $\dot{\varphi}(0) \neq 0$, $l_3 \neq 0$.

Introducing the new variable $u = \cos\theta$ and using the last two equations, we obtain

$$\dot{u}^2 = P(u), \tag{3.4.24}$$

where $P(u)$ is a polynomial of third degree in u:

$$P(u) = \left[\frac{2E}{mR^2} + \frac{2g}{R}u\right](1-u^2) - \left(\frac{l_3}{mR^2}\right)^2. \tag{3.4.25}$$

The motion is possible for those values of u, for which $P(u) \geq 0$. The law of motion for the variable θ is obtained from (3.4.24) as an implicit function:

$$t = \pm \int_{u_0}^{u} [P(u)]^{-\frac{1}{2}} du,$$

where the sign is chosen as to produce a monotonically increasing time. The trajectory is found by using (3.4.20):

$$\varphi - \varphi_0 = \frac{l_3}{mR^2} \int_{u_0}^{u} \frac{du}{\pm(1-u^2)\sqrt{P(u)}}.$$

To analyze the roots of the polynomial $P(u)$, let us first observe that we must have $P(u_0) > 0$, where $u_0 = \cos\theta(0) \in (-1, 1)$. Since $P(\pm 1) = -\left(\frac{l_3}{mR^2}\right)^2 < 0$, it results that $P(u)$ has two roots in the interval $(-1, +1)$, say u_1 and u_2 $(u_1 < u_2)$. Observing that $\lim_{u\to\pm\infty} P(u) = \pm\infty$, there is a third real root $u_3 \in (-\infty, -1)$, which does not have a physical meaning.

Since $u_0 \in [u_1, u_2]$, during the motion we must have $u \in [u_1, u_2]$, meaning that the trajectory is between the parallel circles θ_1, θ_2, with $\theta_1 > \theta_2$ (see Fig. 3.21b). In view of (3.4.25), we can write:

$$u_1 u_2 + u_1 u_3 + u_2 u_3 = -1,$$

or

$$u_3(u_1 + u_2) = -1 - u_1 u_2.$$

Since $|u_1 u_2| < 1$, while $u_3 < 0$, we must have

$$u_1 + u_2 > 0.$$

This means that at least $u_2 > 0$, i.e. $\theta_2 < \frac{\pi}{2}$, which shows that the trajectory of the pendulum cannot be situated *only* in the upper hemisphere (Fig. 3.21b).

We conclude that θ is a periodical function, with values between θ_1 and θ_2, corresponding to $\dot{\theta} = 0$. If the initial conditions are such that $E = (V_{eff})_{min}$, then $P(u)$ has a double root $u_1 = u_2$ in the interval $(-1, +1)$. In this case, the condition $P(u) \geq 0$ yields $P(u) = 0$, which is the constant value u_1 for u and, since $u_0 \in [u_1, u_2]$, it follows that $u = u_0$ all the time. Therefore the particle describes a horizontal circular trajectory, defined by $\theta = \theta_0$. During this motion, the radius vector OP generates a cone of angle $2\theta_0$, for which reason the system is called a *conic pendulum*. If the energy E is not much different from $(V_{eff})_{min}$, the trajectory is close to the circular shape. This shows that the trajectory corresponding to the horizontal circle $\theta = \theta_0$, with $\left[\frac{dV_{eff}}{d\theta}\right]_{\theta=\theta_0} = 0$, is stable.

3.5 Motion of a Particle Subject to an Elastic Force

In this section we shall discuss the motion of a particle under the influence of the attractive central force field

$$\mathbf{F} = -k\mathbf{r}, \tag{3.5.1}$$

corresponding to the potential energy (see (3.1.30))

$$V(\mathbf{r}) = \frac{1}{2}kr^2, \tag{3.5.2}$$

where k is positive.

3.5.1 Harmonic Linear Oscillator

A *harmonic linear oscillator* is a particle of mass m, constrained to move along a straight line, subject to an elastic force, periodically passing through a fixed point of the straight line. Taking this point as the origin and choosing as generalized coordinate the oriented distance x from O to the particle, we have:

$$F = -kx, \quad V = \frac{1}{2}kx^2. \tag{3.5.3}$$

Writing the Lagrangian

$$L = \frac{1}{2}m\dot{x}^2 - \frac{1}{2}kx^2 \tag{3.5.4}$$

and using the Lagrange equations (2.5.17), we arrive at the equation of motion:

$$\ddot{x} + \omega^2 x = 0, \tag{3.5.5}$$

where

$$\omega = \sqrt{\frac{k}{m}}. \tag{3.5.6}$$

The two linearly independent solutions of Eq. (3.5.5) can be written in the condensed form

$$x(t) = a\cos(\omega t + \varphi), \tag{3.5.7}$$

showing that the particle performs harmonic oscillations about the equilibrium position O. The deviation x from the equilibrium position O is called *elongation*, a is the *amplitude* of oscillation, while ω is the *circular* (or *angular*) *frequency* of the periodical motion. The argument of the cosine is the *phase* of the motion, φ being the *initial phase*. The constants a and φ are determined from the initial conditions: assuming $x(0) = x_0$, $\dot{x}(0) = v_0$, we have:

$$a = \sqrt{x_0^2 + \frac{v_0^2}{\omega^2}}, \quad \tan\varphi = -\frac{v_0}{\omega x_0}.$$

Sometimes it is more convenient to use the complex solution of (3.5.5),

$$\tilde{x}(t) = Ae^{\pm i(\omega t + \varphi)},$$

instead of the real solution (3.5.7). Here, the constant A is complex and φ is real.

Our model is conservative, consequently we have the energy first integral:

$$E = \frac{1}{2}m\dot{x}^2 + \frac{1}{2}kx^2 = \text{const.}$$

The linear harmonic oscillator is a very simple model, but it can be applied in the study of various physical systems. The reader is acquainted with the theory of damped and forced oscillations from the courses on general physics. Here, we shall discuss only some applications, especially useful in physical chemistry.

3.5.2 Space Oscillator

A *space oscillator* is a mechanical system with three degrees of freedom, formed by a particle subject to an elastic central force (3.5.1). As we already know, the trajectory lies in a plane. Assuming the origin of coordinates at the centre of force and $Ox_3 \parallel \mathbf{l}$, the motion will take place in the plane Ox_1x_2. Therefore, it is convenient to choose r, φ as generalized coordinates.

In order to determine the trajectory, we shall use (3.1.19), where the effective potential energy V_{eff} is:

$$V_{eff} = \frac{k}{2}r^2 + \frac{l^2}{2mr^2} = \frac{m\omega^2}{2}r^2 + \frac{l^2}{2mr^2}.$$

If we take the x_1-axis to pass through the pericentre, then

$$\varphi = -\frac{1}{2}\int_{\frac{1}{r_m^2}}^{\frac{1}{r^2}} \left[\frac{m^2}{l^4}(E^2 - l^2\omega^2) - \left(u^2 - \frac{mE}{l^2}\right)^2\right]^{-\frac{1}{2}} du^2, \tag{3.5.8}$$

where r_m is the solution of the equation $E - V_{eff} = 0$, i.e.

$$r_m^2 = \frac{1}{m\omega^2}\left[E - \sqrt{E^2 - l^2\omega^2}\right]. \tag{3.5.9}$$

Since $E^2 - l^2\omega^2 \geq 0$, the quantity r_m^2 is always positive. It is easy to verify that $(V_{eff})_{min} = l\omega$.

Working out the integral (3.5.8), we obtain:

$$2\varphi + \pi = \arccos\frac{\frac{1}{r^2} - \frac{mE}{l^2}}{\frac{m}{l^2}\sqrt{E^2 - l^2\omega^2}},$$

or

$$\frac{1}{r^2} = \frac{mE}{l^2}\left(1 - \sqrt{1 - \frac{l^2\omega^2}{E^2}\cos^2\varphi}\right), \tag{3.5.10}$$

which represents an ellipse of semi-axes

$$a = \left(\frac{E + \sqrt{E^2 - l^2\omega^2}}{m\omega^2}\right)^{\frac{1}{2}}, \quad b = \left(\frac{E - \sqrt{E^2 - l^2\omega^2}}{m\omega^2}\right)^{\frac{1}{2}}. \tag{3.5.11}$$

The same result is obtained in Cartesian coordinates, using the Lagrangian

$$L = \frac{m}{2}(\dot{x}_1^2 + \dot{x}_2^2) - \frac{m\omega^2}{2}(x_1^2 + x_2^2).$$

Then the Lagrange equations lead to two independent equations

$$\ddot{x}_1 + \omega^2 x_1 = 0, \quad \ddot{x}_2 + \omega^2 x_2 = 0, \tag{3.5.12}$$

which have solutions of the form (3.5.7).

3.5.3 Non-linear Oscillations

In all cases previously analyzed we had to deal, from the mathematical point of view, with second-order linear differential equations, whose general form is:

$$\ddot{x} + \alpha(t)\dot{x} + \beta(t)x = \gamma(t). \tag{3.5.13}$$

The systems described by such equations perform *linear motions*. But there are cases when the behaviour of one-dimensional oscillating systems is described by equations of the type

$$\ddot{x} + f(x, \dot{x}, t) = 0, \tag{3.5.14}$$

where $f(x, \dot{x}, t)$ is a *non-linear* function of the variables $x, \dot{x}, t$, i.e. it contains terms like: x^2, $x\dot{x}$, xt, etc. The oscillations performed by such systems are called *non-linear*.

If the displacements from the equilibrium positions are not small enough, the formulas (3.5.1) and (3.5.2) are not valid anymore, because they do not approximate well enough the physical reality. In this case, the Lagrangian must contain terms of higher order, producing non-linear differential equations of the type (3.5.14). As a matter of fact, most physical systems are described by non-linear differential equations and the linear approximation appears as a particular case, corresponding to small oscillations.

There are no general methods of integration of non-linear differential equations, but there exist special techniques, like the method of successive approximations, or the expansion in Fourier series, etc. These problems are studied by *non-linear mechanics*.

As an example, let us consider the free motion of a plane pendulum. Here, for small angles, the non-linear equation (3.4.2) becomes:

$$\ddot{\theta} + \frac{g}{R}\theta = 0,$$

similar to (3.5.5), where we assumed $\sin\theta \approx \theta$. If we take into consideration the second term in the series expansion of $\sin\theta$, the corresponding equation is:

$$\ddot{\theta} + \frac{g}{R}\theta - \frac{g}{6R}\theta^3 = 0, \tag{3.5.15}$$

which is a non-linear equation with constant coefficients. It is not our purpose to find a solution of this equation, but, at least, we can find the Lagrangian leading to (3.5.15). Indeed, expanding in series the potential energy about $\theta = 0$, we have:

$$V(\theta) = -mgR\cos\theta \simeq -mgR\left(1 - \frac{\theta^2}{2} + \frac{\theta^4}{24}\right).$$

The Lagrangian of the system is:

$$L = \frac{1}{2} m R^2 \dot{\theta}^2 + mgR\left(1 - \frac{\theta^2}{2} + \frac{\theta^4}{24}\right). \quad (3.5.16)$$

It is easy to prove that the Lagrange equations, for the Lagrangian (3.5.16), lead to (3.5.15). The problem is solved in a similar way if the motion takes place in a resistant medium.

We conclude that in most physical problems, the higher we go with the approximation, the closer to the reality becomes the result, implying a more complicated mathematical formalism.

3.6 Small Oscillations About a Position of Stable Equilibrium

In the case of systems with many degrees of freedom, the Lagrange equations are very complicated, appearing as non-linear coupled equations, and their exact integration is not possible. In the following, we shall give a method of approximating the solution, with many applications in different branches of physics.

Consider a conservative system of N particles, subject to (at most) scleronomous constraints. A configuration $\{\mathbf{r}_i^0,\ i = \overline{1,N}\}$ is an *equilibrium configuration* at the time $t_0 = 0$, if at any $t \geq 0$ are satisfied the conditions:

$$\mathbf{r}_i(t) = \mathbf{r}_i(0) = \mathbf{r}_i^0, \quad \dot{\mathbf{r}}_i(t) = \dot{\mathbf{r}}_i(0) = 0 \quad (i = \overline{1,N}).$$

In our case, these conditions are fulfilled if

$$(\mathbf{F}_i + \mathbf{L}_i)_{\mathbf{r}_1^0,\ldots,\mathbf{r}_N^0} = 0 \quad (i = \overline{1,N}). \quad (3.6.1)$$

Let us now pass to the generalized coordinates $q_1, .., q_n$. Denoting by $q^0 : \{q_j^0,\ j = \overline{1,n}\}$ the coordinates corresponding to the equilibrium position and using (2.1.61), (2.4.12), the generalized forces $Q_j^0 = Q_j(q^0)$ associated to this position are

$$Q_j^0 = \left[\sum_{i=1}^{N} \mathbf{F}_i \cdot \frac{\partial \mathbf{r}_i}{\partial q_j}\right]_{q=q^0} = -\left[\sum_{i=1}^{N}\sum_{k=1}^{s} \lambda_k [\mathrm{grad}_i f_k(\mathbf{r}_1, .., \mathbf{r}_N)] \cdot \frac{\partial \mathbf{r}_i}{\partial q_j}\right]_{q=q^0}.$$

But

$$\sum_{i=1}^{N} [\mathrm{grad}_i f_k(\mathbf{r}_1, .., \mathbf{r}_N)] \cdot \frac{\partial \mathbf{r}_i}{\partial q_j} = \frac{\partial f_k}{\partial q_j} = 0 \quad (k = \overline{1,s},\ j = \overline{1,n}),$$

therefore $Q_j(q^0) = 0$. Since the system is supposed to be conservative, there exists a function $V(q)$ with the property

$$Q_j(q^0) = -\left[\frac{\partial V}{\partial q_j}\right]_{q=q^0} = 0 \quad (j = \overline{1,n}). \quad (3.6.2)$$

In other words, the equilibrium configuration corresponds to a minimum of the potential energy. According to the Lagrange–Dirichlet theorem (see Chap. 2), the equilibrium is stable if this extremum is a strict (absolute) minimum. Indeed, in case of a maximum, say $V(q^0) = E_0$, if we furnish to the system an energy a little larger than E_0, then the bigger the distance from the equilibrium position, the bigger is the difference $E - V = T$, which means that the equilibrium is unstable.

Let us introduce as new generalized coordinates

$$\xi_j = q_j - q_j^0 \quad (j = \overline{1,n}), \tag{3.6.3}$$

to express the small deviations from the equilibrium position. These new coordinates are considered small, to allow us to retain only terms quadratic in ξ in any calculation. Expanding $V(q)$ by Taylor's formula, we have:

$$V(q) = V_0 + \sum_{j=1}^{n} \left(\frac{\partial V}{\partial q_j}\right)_0 (q_j - q_j^0) + \frac{1}{2} \sum_{j=1}^{n} \sum_{k=1}^{n} \left(\frac{\partial^2 V}{\partial q_j \partial q_k}\right)_0 (q_j - q_j^0)(q_k - q_k^0) + \cdots, \tag{3.6.4}$$

where the subscript 'zero' means that the derivatives are calculated at the equilibrium configuration. The first term in (3.6.4) is an additive constant which can be taken zero, while the second term vanishes by virtue of (3.6.2). Denoting

$$V_{jk} = \left(\frac{\partial^2 V}{\partial q_j \partial q_k}\right)_0 = V_{kj}, \tag{3.6.5}$$

we obtain the first approximation for the potential energy as

$$V = \frac{1}{2} \sum_{j=1}^{n} \sum_{k=1}^{n} V_{jk} \xi_j \xi_k. \tag{3.6.6}$$

If the configuration $\{q_j,\ j = \overline{1,n}\}$ is of stable equilibrium, then according to the Lagrange–Dirichlet theorem, the form (3.6.6) is positive definite. From (2.4.25), the kinetic energy is:

$$T = \frac{1}{2} \sum_{j=1}^{n} \sum_{k=1}^{n} a_{jk}(q) \dot{q}_j \dot{q}_k = \frac{1}{2} \sum_{j=1}^{n} \sum_{k=1}^{n} a_{jk}(q) \dot{\xi}_j \dot{\xi}_k. \tag{3.6.7}$$

But

$$a_{jk}(q) = (a_{jk})_0 + \sum_{l=1}^{n} \left(\frac{\partial a_{jk}}{\partial q_l}\right)_0 (q_l - q_l^0) + \cdots, \tag{3.6.8}$$

therefore, using our convention,

$$a_{jk}(q) \simeq (a_{jk})_0 = a_{jk} = a_{kj}, \tag{3.6.9}$$

and the kinetic energy reads:

$$T = \frac{1}{2} \sum_{j=1}^{n} \sum_{k=1}^{n} a_{jk}\, \dot{\xi}_j\, \dot{\xi}_k. \tag{3.6.10}$$

3.6.1 Equations of Motion. Normal Coordinates

The Lagrangian of the system is

$$L = \frac{1}{2} \sum_{j=1}^{n} \sum_{k=1}^{n} (a_{jk}\, \dot{\xi}_j\, \dot{\xi}_k - V_{jk}\, \xi_j\, \xi_k) \tag{3.6.11}$$

and the Lagrange equations yield

$$\sum_{k=1}^{n} (a_{jk}\, \ddot{\xi}_k + V_{jk}\, \xi_k) = 0 \quad (j = \overline{1,n}). \tag{3.6.12}$$

This is a system of n coupled linear differential equations. Let us show that, if the variables are suitably chosen, the solution of the problem reduces to the integration of n equations of the form (3.5.5). To this end, we shall look for solutions of the form

$$Ae^{i\omega t},$$

the coordinate ξ_j being the real part of this complex quantity. Introducing this solution into (3.6.12), we obtain the linear, homogeneous, algebraic system of equations

$$\sum_{k=1}^{n} (V_{jk} - \omega^2 a_{jk}) A_k = 0 \quad (j = \overline{1,n}). \tag{3.6.13}$$

This system has non-trivial solutions for the amplitudes A_k only if

$$\det (V_{jk} - \omega^2 a_{jk}) = 0, \tag{3.6.14}$$

which is an equation of degree n in ω^2, called the *characteristic equation.* Its roots are real and positive (see Appendix A). This property follows immediately if one multiplies (3.6.13) by the complex conjugated A_j^* and sums over j, which yields

$$\omega^2 = \frac{\sum_{j=1}^{n} \sum_{k=1}^{n} V_{jk}\, A_j\, A_k^*}{\sum_{j=1}^{n} \sum_{k=1}^{n} a_{jk}\, A_j\, A_k^*},$$

where both the numerator and denominator are real and positive. In general, we have n distinct solutions ω_s^2 $(s = \overline{1,n})$; if two or more solutions are equal, we have a *degeneracy*. The corresponding frequencies are called *normal* frequencies. To each normal frequency corresponds a solution of Eq. (3.6.12); thus, the general solution is:

$$\xi_j = \sum_{s=1}^{n} c_j^s Re\{|A_s|e^{i(\omega_s t+\varphi_s)}\} = \sum_{s=1}^{n} Re\{a_j^s e^{i(\omega_s t+\varphi_s)}\}, \tag{3.6.15}$$

where the constants c_j^s, a_j^s are real. Considering the particular complex solutions

$$\tilde{\xi}_j = a_j^s e^{i(\omega_s t+\varphi_s)},$$

the characteristic system of equations becomes:

$$\omega_s^2 \sum_{k=1}^{n} a_{jk} a_k^s = \sum_{k=1}^{n} V_{jk} a_k^s \quad (\text{no summation over } s;\ s, j = \overline{1,n}). \tag{3.6.16}$$

Since these equations are homogeneous, only some of the amplitudes a_j^s are determined. In the non-degenerate case, the matrix $\hat{V} - \omega^2 \hat{a}$ is of order $n - 1$, which means that, for each s, we can express $n - 1$ amplitudes as functions of an arbitrary one, say a_n^s. All together, we have n undetermined amplitudes and n initial phases, φ_s, i.e. $2n$ constants which are determined from the initial conditions. The situation is similar to that encountered in the degenerate case. Suppose, for example, that $\omega_1 = \omega_2 = \cdots = \omega_m$, i.e. ω_1 is a multiple root of order m, the rest of the roots being distinct. In this case, the rank of the matrix $\hat{V} - \omega_s^2 \hat{a}$ is $(n - m)$, and so m amplitudes $A_j^{(1)}$ remain undetermined. For each $s = m + 1$, $m + 2, \ldots, n$ there remains an undetermined amplitude, which means $(n - m)$ amplitudes. The total number of arbitrary amplitudes is, again, n, and together with the initial phases, there are $2n$ arbitrary constants.

The choice of generalized coordinates is, to a great extent, arbitrary, which allows us to represent the system by a system of harmonic oscillators, each of them being associated with a degree of freedom. In other words, with each normal frequency one associates a generalized coordinate, which varies periodically with time, with the respective frequency. These coordinates are called *normal.* To make an appropriate choice for normal coordinates, let us first write a few useful relations.

Multiplying (3.6.16) by a_j^r $(r \neq s;\ r = \overline{1,n})$ and performing summation over j, we obtain:

$$\omega_s^2 \sum_{j=1}^{n} \sum_{k=1}^{n} a_{jk} a_j^r a_k^s = \sum_{j=1}^{n} \sum_{k=1}^{n} V_{jk} a_j^r a_k^s.$$

Interchanging now r and s in the last relation, then subtracting the result from the original expression, we have:

$$(\omega_s^2 - \omega_r^2)\sum_{j=1}^{n}\sum_{k=1}^{n} a_{jk}\, a_j^r\, a_j^s = 0. \tag{3.6.17}$$

In the non-degenerated case, for $r \neq s$ we have $\omega_r \neq \omega_s$, therefore the double sum in (3.6.17) must be zero. If $r = s$, the double sum cannot be zero (this would mean that all the amplitudes a_j^s would be zero), so we can choose the amplitudes such that

$$\sum_{j=1}^{n}\sum_{k=1}^{n} a_{jk}\, a_j^s\, a_k^s = 1 \quad (\text{no summation over}\, s;\ s = \overline{1,n}). \tag{3.6.18}$$

These relations represent n supplementary conditions which yield the n amplitudes, remained undetermined after solving (3.6.16). All these conditions can be written in the compact form

$$\sum_{j=1}^{n}\sum_{k=1}^{n} a_{jk}\, a_j^r\, a_k^s = \delta_{rs} \quad (r, s = \overline{1,n}). \tag{3.6.19}$$

We shall now define the *normal coordinates* as the following linear combinations of ξ_j:

$$\eta_s = Re\sum_{j=1}^{n}\sum_{k=1}^{n} a_{jk}\, a_k^s\, \xi_j \quad (s = \overline{1,n}). \tag{3.6.20}$$

In view of (3.6.15) and (3.6.18), we obtain:

$$\eta_s = Re\{e^{i(\omega_s t + \varphi_s)}\} = \cos(\omega_s t + \varphi_s) \quad (s = \overline{1,n}), \tag{3.6.21}$$

which, for any s, satisfies equations of the form:

$$\ddot{\eta}_s + \omega_s^2 \eta_s = 0 \quad (s = \overline{1,n}). \tag{3.6.22}$$

Using (3.6.5) and (3.6.12), we can write the coordinates ξ_j as functions of the normal coordinates:

$$\xi_j = a_j^s \eta_s \quad (j = \overline{1,n}). \tag{3.6.23}$$

Conversely, the definition (3.6.21) of normal coordinates can be obtained by inverting the linear system (3.6.23).

Let us show, finally, that the use of the linear transformations (3.6.23) in the kinetic energy T, the potential energy V and the Lagrangian L, turns these functions into diagonal forms (sums of squared quantities). Indeed,

$$T = \frac{1}{2}\sum_{j=1}^{n}\sum_{k=1}^{n} a_{jk}(a_j^r\, \dot{\eta}_r)(a_k^s\, \dot{\eta}_s),$$

which, in view of (3.6.19), becomes:

$$T = \frac{1}{2}\sum_{s=1}^{n} \dot{\eta}_s^2. \tag{3.6.24}$$

Similarly, from (3.6.6) and (3.6.16), using (3.6.23), we arrive at

$$V = \frac{1}{2}\sum_{s=1}^{n} \omega_s^2 \eta_s^2, \tag{3.6.25}$$

therefore

$$L = \frac{1}{2}\sum_{s=1}^{n} (\dot{\eta}_s^2 - \omega_s^2 \eta_s^2) \tag{3.6.26}$$

is the Lagrangian of n independent harmonic oscillators, of unit mass and normal frequencies.

To summarize, the importance of this formalism is that once the functions V, T, L are determined, we are able to find a linear transformation leading to a set of new generalized parameters – the normal coordinates. Written in terms of these coordinates, the Lagrangian becomes diagonal and provides Lagrange equations of the type (3.6.22), which are easy to integrate.

Observation: The condition (3.6.18) is not indispensable. If it is not imposed, then each parenthesis in the expression for the Lagrangian (3.6.26) will be amplified by a constant, which obviously does not modify the equations of motion.

3.6.2 Small Oscillations of Molecules

The formalism developed in the previous section can be applied in the analysis of small oscillations of coupled pendulums or elastic rods, but more interesting and useful is the study of molecule vibrations. For example, the investigation of small oscillations in a one-dimensional crystal provides important information concerning the thermodynamical properties of a solid body.

Assuming the molecule to be formed of N atoms, we should first note that not all the motions corresponding to the $3N$ degrees of freedom have the meaning of oscillations about the equilibrium configuration. Indeed, there can be a translation and/or a rotation of the whole molecule, each of these motions possessing three degrees of freedom, meaning that only $3N - 6$ degrees of freedom are left for oscillations. A special situation is encountered in a linear molecule: in the state of equilibrium, all atoms are disposed along a straight line. Here, we have only two degrees of freedom for rotation, therefore we are left with $3N - 5$ degrees of freedom for oscillation.

If we are interested only in molecule oscillations, it is convenient to eliminate the motions of rotation and translation. To remove the rotation, we shall demand that the centre of mass be at rest, which means that the study is done in CMS. Denoting by $\mathbf{r}_i^0$ $(i = \overline{1,N})$ the radius-vectors of the atoms at equilibrium and by ξ_i $(i = \overline{1,N})$ the radius-vector of the atom i relative to its equilibrium position, in view of (1.3.54), we have:

$$\sum_{i=1}^{N} m_i \mathbf{r}_i^0 = \sum_{i=1}^{N} m_i (\mathbf{r}_i^0 + \xi_i),$$

or

$$\sum_{i=1}^{N} m_i \xi_i = 0. \tag{3.6.27}$$

To eliminate the rotation, we impose the condition that the total angular momentum vanish:

$$0 = \sum_{i=1}^{N} m_i (\mathbf{r}_i^0 + \xi_i) \times \dot{\xi}_i \simeq \sum_{i=1}^{N} m_i (\mathbf{r}_i^0 \times \dot{\xi}_i),$$

where the small term $\sum_{i=1}^{N} m_i \xi_i \times \dot{\xi}_i$ has been neglected. Integrating with respect to time the last relation, we obtain

$$\sum_{i=1}^{N} m_i \mathbf{r}_i^0 \times \xi_i = 0. \tag{3.6.28}$$

Here, the integration constant was taken to be zero, because all ξ_i vanish at equilibrium.

Let us discuss, as an example, the symmetrical linear triatomic molecule, which happens to be the situation in the molecule CO_2. There are two degrees of freedom, i.e. two normal oscillations along the molecule ($N - 1$, in general). For non-longitudinal displacements, we have two degrees of freedom ($2N - 4$, in general), but only one normal oscillation ($N - 2$), because due to symmetry reasons, the oscillations taking place in two planes mutually orthogonal and passing through the molecule must be identical.

We shall deal first with the longitudinal motion. Assuming the x-axis along the molecule (Fig. 3.22a), let x_1, x_2 be the coordinates of the two atoms of mass m, while x_3 defines the position of the centre of mass M. If x_i^0 $(i = 1, 2, 3)$ are the coordinates of the stable equilibrium position, to study the small oscillations it is convenient to define as generalized coordinates

$$\xi_i = x_i - x_i^0 \quad (i = 1, 2, 3).$$

The kinetic energy is then

$$T = \frac{1}{2}[m(\dot{\xi}_1^2 + \dot{\xi}_2^2) + M\dot{\xi}_3^2], \tag{3.6.29}$$

while the potential energy, considering only the interaction of the atoms of mass m with the central atom of mass M, is of the form:

$$V = V(x_2 - x_3, x_3 - x_1).$$

In the approximation of small oscillations, we have

$$V = \frac{1}{2}[a(\xi_2 - \xi_3)^2 + 2b(\xi_2 - \xi_3)(\xi_3 - \xi_1) + c(\xi_3 - \xi_1)^2],$$

where

$$a = \left[\frac{\partial^2 V(u,v)}{\partial u^2}\right]_{u_0,v_0}, \quad b = \left[\frac{\partial^2 V(u,v)}{\partial u \partial v}\right]_{u_0,v_0}, \quad c = \left[\frac{\partial^2 V(u,v)}{\partial v^2}\right]_{u_0,v_0},$$

with $u_0 = x_2^0 - x_3^0$, $v_0 = x_3^0 - x_1^0$. The molecule is symmetrical, therefore the interchange of the atoms 1 and 2 does not modify its structure. Consequently, under the interchange, the potential energy V remains the same, which leads to $c = a$, hence:

$$V = \frac{1}{2}\{a[(\xi_2 - \xi_3)^2 + (\xi_3 - \xi_1)^2] + 2b(\xi_2 - \xi_3)(\xi_3 - \xi_1)\}. \tag{3.6.30}$$

Then the condition (3.6.27), which eliminates the translation, becomes

$$\xi_3 = -\frac{m}{M}(\xi_1 + \xi_2),$$

and serves to eliminate ξ_3 from (3.6.29) and (3.6.30):

$$\begin{aligned} T &= \frac{1}{2}[m(\dot{\xi}_1^2 + \dot{\xi}_2^2) + \frac{m^2}{M}(\dot{\xi}_1 + \dot{\xi}_2)^2], \\ V &= \frac{1}{2}\left\{\frac{a}{M^2}[(M^2 + 2m^2 + 2mM)(\xi_1^2 + \xi_2^2) + 4m(m+M)\xi_1\xi_2 \right. \\ &\quad \left. - \frac{2b}{M^2}[(M^2 + 2m^2 + 2mM)\xi_1\xi_2 + m(m+M)(\xi_1^2 + \xi_2^2)]\right\}. \end{aligned}$$

It is easily seen that these two expressions can be diagonalized by the change of variable $\xi_1 = A\eta_a + B\eta_s$, $\xi_2 = A\eta_a - B\eta_s$. Imposing that the kinetic energy has the form (3.6.24) in the new coordinates, we obtain ξ_1, ξ_2 in terms of the normal coordinates η_a, η_s:

$$\xi_1 = \frac{1}{\sqrt{2m}}\left[\eta_s + \sqrt{\frac{M}{2m+M}}\eta_a\right], \quad \xi_2 = \frac{1}{\sqrt{2m}}\left[-\eta_s + \sqrt{\frac{M}{2m+M}}\eta_a\right].$$

The kinetic and potential energies will be thus given by

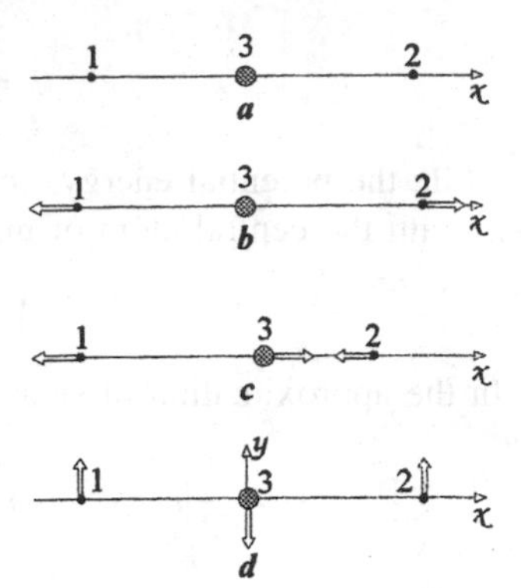

Fig. 3.22 Various possibilities of small oscillations of molecules.

$$T=\frac{1}{2}(\dot{\eta}_s^2+\dot{\eta}_a^2) \tag{3.6.31}$$

and

$$V=\frac{1}{2}\left[\frac{a+b}{m}\eta_s^2+\frac{(a-b)(2m+M)}{mM}\eta_a^2\right].$$

The normal frequencies are then:

$$\omega_s=\left(\frac{a+b}{m}\right)^{\frac{1}{2}}, \tag{3.6.32}$$

$$\omega_a=\left[\frac{(a-b)(2m+M)}{mM}\right]^{\frac{1}{2}}. \tag{3.6.33}$$

If $\eta_a=0$, we have $\xi_2=-\xi_1$, $\xi_3=0$, meaning that ω_s corresponds to a symmetrical oscillation (Fig. 3.22b). On the other hand, if we take $\eta_s=0$, we arrive at $\xi_2=\xi_1$, $\xi_3=-\frac{2m}{M}\xi_1$ (Fig. 3.22c). There can occur an accidental degeneracy, for $ma=(m+M)b$, in which case both the symmetrical and antisymmetrical oscillations have the same frequency.

In order to investigate the transverse displacements, we shall also consider the oscillations perpendicular to the x-axis, in the xy-plane. Since in the equilibrium configuration the atoms are on the x-axis, the generalized coordinates suitable to our investigation are y_1, y_2, y_3. The conditions (3.6.27) and (3.6.28) then yield

$$y_3=-\frac{2m}{M}y_1,\quad y_2=y_1,$$

hence

$$T=\frac{1}{2}\frac{2m(2m+M)}{M}\dot{y}_1^2, \tag{3.6.34}$$

$$V = \frac{1}{2}ky_1^2. \tag{3.6.35}$$

Introducing the normal coordinate

$$\eta_y = \sqrt{\frac{2m(2m+M)}{M}}\,y_1,$$

we obtain:

$$T = \frac{1}{2}\dot{\eta}_y^2, \tag{3.6.36}$$

$$V = \frac{1}{2}\left[\frac{Mk}{2m(2m+M)}\right]^{\frac{1}{2}}\eta_y^2, \tag{3.6.37}$$

corresponding to the normal frequency

$$\omega_y = \left[\frac{Mk}{2m(2m+M)}\right]^{\frac{1}{2}} \tag{3.6.38}$$

and to the transverse oscillation shown in Fig. 3.22d. This oscillation, as we already know, is degenerated, because a transverse oscillation, having the same frequency $\omega_z = \omega_y$, will also appear in the xz-plane.

3.7 Analogy Between Mechanical and Electric Systems

3.7.1 Kirchhoff's Rule Relative to the Loops of an Electric Circuit

The study of alternating current circuits displayed an analogy between these systems and mechanical systems of particles. This analogy was identified in the 19th century by *James Clerk Maxwell* and allows the application of the Lagrangian formalism in the study of electric circuits. In the following, we shall consider the case of alternating current circuits in steady state.

Consider a circuit composed of many branches, each of them having loops and junction points, and assume that there are resistors, capacitors, coils and electromotive forces on each branch of every loop. Let $I_k = \dot{q}_k$ be an *arbitrary* current, having the same magnitude in each point of the loop k. By q we denote the electric charge. If ϕ_i is the magnetic flux generated by the current I_k, which passes through the neighbouring loop i, then the quantities ϕ_i and I_k are related by

$$\varphi_i = M_{ik}I_k = M_{ik}\,\dot{q}_k \quad \text{(no summation)}, \tag{3.7.1}$$

where M_{ik} is the *mutual inductance* between the loops i and k. For $i = k, M_{ii} = L_i$ is the *self-inductance* of the loop i. If the circuit is composed of n loops, one must perform summation over k:

$$\varphi_i = \sum_{k=1}^{n} M_{ik} I_k = \sum_{k=1}^{n} M_{ik}\, \dot{q}_k. \tag{3.7.2}$$

The magnetic energy of the circuit is:

$$W_{mag} = \frac{1}{2} \sum_{i=1}^{n} \sum_{k=1}^{n} M_{ik}\, \dot{q}_i\, \dot{q}_k. \tag{3.7.3}$$

In a similar way, we can find the electric energy of the circuit. Let V_i be the electric potential of the conductor i and q_k the electric charge distributed on conductor k and producing the potential V_i. The relation between V_i and q_k is

$$V_i = S_{ik} q_k \quad \text{(no summation)}, \tag{3.7.4}$$

where the coefficient S_{ik} is called *elastance* and represents the reciprocal of the *influence coefficients* C_{ik}: $S_{ik} = C_{ik}^{-1}$. For $i = k, C_{ii} = C_i$ are called *capacitance coefficients*. If the potential V_i is produced by all n conductors of the circuit, then

$$V_i = \sum_{k=1}^{n} S_{ik}\, q_k. \tag{3.7.5}$$

The electric energy of the circuit is then:

$$W_{el} = \frac{1}{2} \sum_{i=1}^{n} \sum_{k=1}^{n} S_{ik}\, q_i\, q_k. \tag{3.7.6}$$

According to Ohm's law, for the loop i we can write:

$$U_i = \sum_{k=1}^{n} R_{ik} I_k = \sum_{k=1}^{n} R_{ik}\, \dot{q}_k, \tag{3.7.7}$$

where R_{ik} are some constant coefficients, with the significance of electrical resistance, common for both loops i and k. For $i = k$, $R_{ii} = R_i$ is the *self-resistance* of the loop i.

We also assume that the loop i has voltage generators, the total electromotive force produced by them being $\mathcal{E}_i(t)$.

The analysis of the relations (3.7.1)–(3.7.7) suggests the following correspondence between electric and mechanical quantities:

Generalized coordinate $q(t)$	$\to$	Electric charge $q(t)$
Generalized velocity $\dot{q}(t)$	$\to$	Electric current $I(t)$
External periodical force F	$\to$	Electromotive force $\mathcal{E}(t)$
Mass m	$\to$	Inductance L
Elastic constant k	$\to$	Elastance S
Damping force constant r	$\to$	Electric resistance R
Kinetic energy T	$\to$	Magnetic energy W_{mag}
Potential energy V	$\to$	Electric energy W_{el}

According to this analogy, our electric circuit can be conceived as an oscillating system with n degrees of freedom, subject to two potential forces and a non-potential, dissipative force. The potential energy of the system is:

$$V = \frac{1}{2}\sum_{i=1}^{n}\sum_{k=1}^{n} S_{ik}q_i q_k - \sum_{i=1}^{n} q_i\mathcal{E}_i, \tag{3.7.8}$$

while the Rayleigh dissipation function (see (2.5.45)) is:

$$\mathcal{T} = \frac{1}{2}\sum_{i=1}^{n}\sum_{k=1}^{n} R_{ik}\dot{q}_i\dot{q}_k. \tag{3.7.9}$$

Using the Lagrange equations in the form (2.5.48), with the Lagrangian

$$L = \frac{1}{2}\sum_{i=1}^{n}\sum_{k=1}^{n}(M_{ik}\dot{q}_i\dot{q}_k - S_{ik}q_i q_k) + \sum_{i=1}^{n} q_i\mathcal{E}_i(t), \tag{3.7.10}$$

we arrive at

$$\sum_{k=1}^{n}(M_{ik}\ddot{q}_k + R_{ik}\dot{q}_k + S_{ik}q_k) = \mathcal{E}_i(t) \quad (i = \overline{1,n}), \tag{3.7.11}$$

which is nothing else but *Kirchhoff's rule relative to the loop i* of the circuit. An equivalent form of this rule is:

$$\sum_{k=1}^{n}\left(M_{ik}\frac{dI_k}{dt} + R_{ik}I_k + S_{ik}\int I_k dt\right) = \mathcal{E}_i(t) \quad (i = \overline{1,n}). \tag{3.7.12}$$

Each model of oscillating mechanical system has a correspondent model among the alternating current circuits. Let us discuss two simple examples.

3.7.1.1 LC Series Circuit. Free Oscillations

Consider the circuit shown in Fig. 3.23, where C is the capacitance of the capacitor and L is the inductance of the coil. If the capacitor is charged by some method and then discharged through the coil, the instantaneous electric charge $q(t)$ on the capacitor, according to Kirchhoff's rule (3.7.11), is the solution of the equation

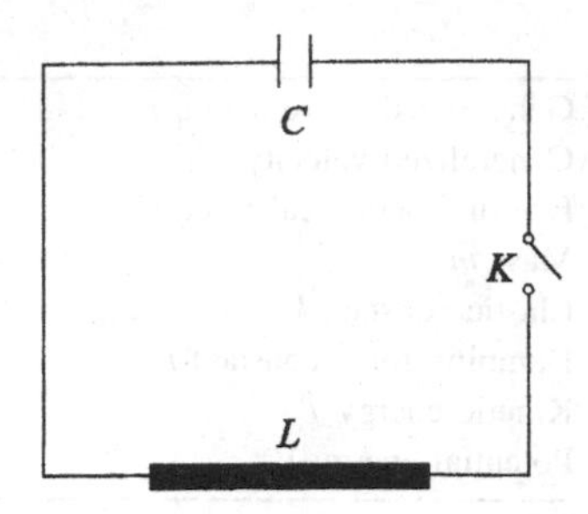

Fig. 3.23 LC series circuit. Free oscillations.

$$\ddot{q} + \omega_0^2 q = 0, \tag{3.7.13}$$

where we made the notation

$$\omega_0^2 = \frac{1}{LC}. \tag{3.7.14}$$

The solution of (3.7.13) is

$$q = A\cos\omega_0 t + B\sin\omega_0 t. \tag{3.7.15}$$

The constants A and B are determined from the initial conditions: $q(0) = q_0$, $I(0) = I_0 = \dot{q}(0) = 0$, and the solution reads:

$$q = q_0 \cos\omega_0 t. \tag{3.7.16}$$

We conclude that the discharge of the capacitor through the coil is a periodical phenomenon, with the period $T = 2\pi\sqrt{LC}$ (Thomson's formula).

3.7.1.2 RLC Series Circuit. Forced Oscillations

Assume the circuit shown in Fig. 3.24, where $U = U_0 \sin\omega t$, and let us determine the instantaneous value of the current passing through the circuit. In view of Kirchhoff's rule (3.7.12), we have:

$$L\frac{dI}{dt} + RI + \frac{1}{C}\int I\,dt = U_0 \sin\omega t,$$

or, by taking the derivative with respect to time,

$$\frac{d^2I}{dt^2} + \frac{R}{L}\frac{dI}{dt} + \frac{1}{LC}I = \frac{U_0\omega}{L}\cos\omega t. \tag{3.7.17}$$

This equation is similar to that obtained in the study of a mechanical oscillator, subject to both friction and external (periodical) forces. The steady-state part of the solution is

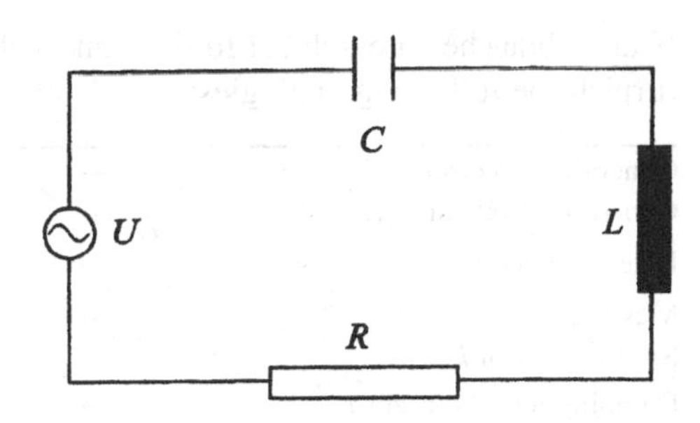

Fig. 3.24 RLC series circuit. Forced oscillations.

$$I = \frac{U_0}{\pm\sqrt{(X_L - X_C)^2 + R^2}} \cos(\omega t + \beta), \tag{3.7.18}$$

where $X_L = \omega L$ is the *reactive inductance*, $X_C = (\omega C)^{-1}$ is the *reactive capacitance* and

$$\beta = \arctan \frac{R}{X_L - X_C}.$$

Since

$$\cos(\omega t + \beta) = -\sin\left(\omega t + \beta - \frac{\pi}{2}\right) = -\sin\left(\omega t - \arctan \frac{X_L - X_C}{R}\right),$$

by choosing the minus sign in (3.7.18), we obtain the solution

$$I = \frac{U_0}{|Z|} \sin(\omega t - \varphi), \tag{3.7.19}$$

where

$$|Z| = \sqrt{(X_L - X_C)^2 + R^2} \tag{3.7.20}$$

is the magnitude of the *impedance* of the circuit and

$$\varphi = \arctan \frac{X_L - X_C}{R} \tag{3.7.21}$$

is the *phase angle* between U and I. In particular, for a given R, the current is maximum for $X_L = X_C$ (resonance), and we arrive again at Thomson's formula (3.7.14), as expected.

3.7.2 Kirchhoff's Rule Relative to the Junction Points of an Electric Circuit

Let us consider a junction point of an electric circuit, as the intersection point of n branches, and let I_i be the current which enters the junction point on the branch i.

If all n branches are subject to the same voltage $U(t)$, then the experimental data furnish the following analogies:

Generalized coordinate $q(t)$	$\rightarrow$	Electric voltage $U(t)$
Generalized velocity $\dot{q}(t)$	$\rightarrow$	$\frac{dU(t)}{dt}$
External force F	$\rightarrow$	$\frac{dI(t)}{dt}$
Mass m	$\rightarrow$	Capacitance C
Spring constant k	$\rightarrow$	Reciprocal of inductance, $\mathcal{L} = \frac{1}{L}$
Damping force constant r	$\rightarrow$	Conductance λ
Kinetic energy T	$\rightarrow$	$\frac{1}{2}\sum_{i=1}^n \sum_{k=1}^n C_{ik}\dot{U}_i\dot{U}_k$
Potential energy V	$\rightarrow$	$\frac{1}{2}\sum_{i=1}^n \sum_{k=1}^n \mathcal{L}_{ik}U_iU_k$
Rayleigh function $\mathcal{T}$	$\rightarrow$	$\frac{1}{2}\sum_{i=1}^n \sum_{k=1}^n \lambda_{ik}\dot{U}_i\dot{U}_k.$

The Lagrangian of the system is then

$$L = \frac{1}{2}\sum_{i=1}^{n}\sum_{k=1}^{n}(C_{ik}\dot{U}_i\dot{U}_k - \mathcal{L}_{ik}U_iU_k) + \sum_{i=1}^{n} U_i\frac{dI_i}{dt}, \tag{3.7.22}$$

and, using the Lagrange equations (2.5.48), we arrive at:

$$\sum_{k=1}^{n}(C_{ik}\ddot{U}_k + \lambda_{ik}\dot{U}_k + \mathcal{L}_{ik}U_k) = \frac{dI_i}{dt} \quad (i = \overline{1,n}). \tag{3.7.23}$$

Integrating with respect to t, we finally find

$$\sum_{k=1}^{n}\left(C_{ik}\dot{U}_k + \lambda_{ik}U_k + \mathcal{L}_{ik}\int U_k dt\right) = I_i(t) \quad (i = \overline{1,n}), \tag{3.7.24}$$

which is *Kirchhoff's rule relative to the junction point i* of the circuit. Here is an example.

3.7.2.1 RLC Parallel Circuit. Forced Oscillations

From Fig. 3.25 we notice that all component elements R, L, C of the circuit are subject to the same voltage $U(t)$. Using Kirchhoff's rule (3.7.23), we can write

$$C\ddot{U} + \frac{1}{R}\dot{U} + \frac{1}{L}U = I_0\omega\cos\omega t,$$

where we assumed that the time variation law of I is: $I = I_0 \sin\omega t$. We can rewrite the latter formula as

$$\ddot{U} + \frac{1}{RC}\dot{U} + \frac{1}{LC}U = \frac{I_0\omega}{C}\cos\omega t.$$

Fig. 3.25 RLC parallel circuit. Forced oscillations.

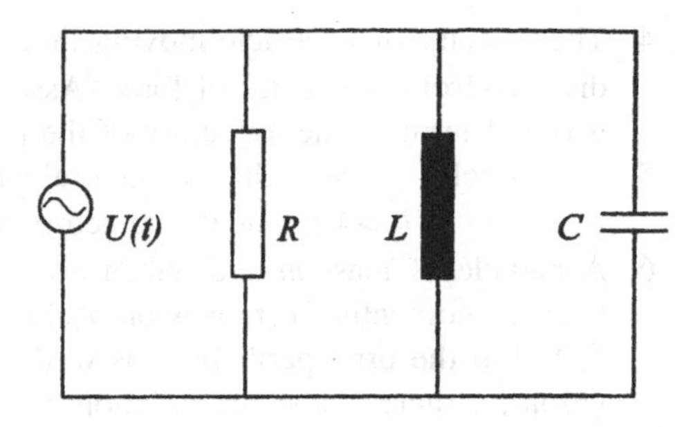

The steady-state part of the solution of this equation is

$$U = \frac{I_0}{|Z|}\sin(\omega t - \varphi), \tag{3.7.25}$$

where

$$|Z| = \sqrt{\left(\frac{1}{\omega L} - \omega C\right)^2 + \frac{1}{R^2}}, \tag{3.7.26}$$

and

$$\varphi = \arctan R\left(\frac{1}{L\omega} - \omega C\right). \tag{3.7.27}$$

If $X_L > X_C$, the current I *leads* the voltage U in phase by φ, while if $X_L < X_C$, the current *lags* the voltage in phase by φ.

The reader is advised to choose some other examples of electric circuits and obtain by means of the Lagrangian formalism all the main formulas encountered in the general course of electricity and magnetism.

3.8 Problems

1. Determine the explicit equation of the trajectory (3.2.7), as well as the eccentricity (3.2.9) for the choice of the potential energy (3.1.31), by using Binet's formula (3.1.24) and the initial conditions: $r(0) = r_0, \varphi(0) = \varphi_0$, $\dot{r}(0) = \dot{r}_0, \dot{\varphi}(0) = \dot{\varphi}_0$.
2. Show that there is no central field with a straight line as the trajectory.
3. A particle P of mass m is at the distance r_0 from the centre of force C in a field of potential energy $U(r) = \frac{1}{3}kr^3$. Its velocity $\mathbf{v}_0$ makes an angle $\alpha = \pm\frac{\pi}{2}$ with respect to the straight line PC. Find the magnitude v_0 of the velocity for which the trajectory is a circle.

4. The velocity of a particle moving in a central field is $v = \frac{a}{r^n}$, where r is the distance from the centre of force. Assuming that the angular momentum l is given, determine the trajectory of the particle and the law of force.
5. A particle moves in a central field whose potential energy is $U(r) = -\frac{k}{r^2}\,(k > 0)$. Find the trajectory of the particle if its total energy is zero.
6. A particle of mass m situated in a central field $U(r)$, having energy E and angular momentum l, moves on a closed orbit. Determine the displacement $\delta(\Delta\varphi)$ of the orbit perihelion, as well as the variation $\delta\tau$ of the radial oscillations, if there is a small variation $\delta U(r)$ in the field potential.
7. Show that the transformation

$$\mathbf{r}' = \mathbf{r} + \epsilon[\mathbf{v} \times (\mathbf{r} \times \dot{\mathbf{r}}) + \mathbf{r} \times (\mathbf{v} \times \dot{\mathbf{r}})], \quad t' = t,$$

where $\mathbf{v}$ is any fixed vector, is a symmetry transformation of the Lagrangian

$$L = \frac{1}{2} m |\dot{\mathbf{r}}|^2 - \frac{k}{r},$$

and find the corresponding first integral of the motion.
8. Show that the trajectory of a spherical pendulum performing small oscillations about its position of stable equilibrium is an ellipse.
9. Consider a linear, homogeneous and neutral medium (gas), and let N be the number of electrons per unit volume, each atom having an electron of mass m and charge $-e$, the electron being elastically connected with its nucleus. Assume that each electron is subject to: (a) an elastic force $\mathbf{F}_e = -m\omega_0^2\mathbf{r}$; (b) a damping force $\mathbf{F}_d = -m\gamma\dot{\mathbf{r}}$; (c) a Lorentz force $\mathbf{F}_L = -e(\mathbf{E} + \mathbf{v} \times \mathbf{B}) \simeq -e\mathbf{E}$. If in the medium propagates an electromagnetic plane wave, given by

$$\mathbf{E} = \mathbf{E}_0 e^{i(kx - \omega t)}, \quad \mathbf{B} = \mathbf{B}_0 e^{i(kx - \omega t)},$$

using the model of the damped oscillator with an electric driving force, find the dispersion equation $n = n(\omega)$ of this physical model.
10. A particle of mass m moving in the vertical direction is subject to gravitational force and friction force, the latter being proportional to the speed of particle. Find the law of motion of the particle and determine the approximate solution containing terms up to t^3.
11. The point of support of mass m of a plane pendulum of mass M is able to move on a straight horizontal line, which lies in the vertical plane of motion. Find the finite equation of motion.

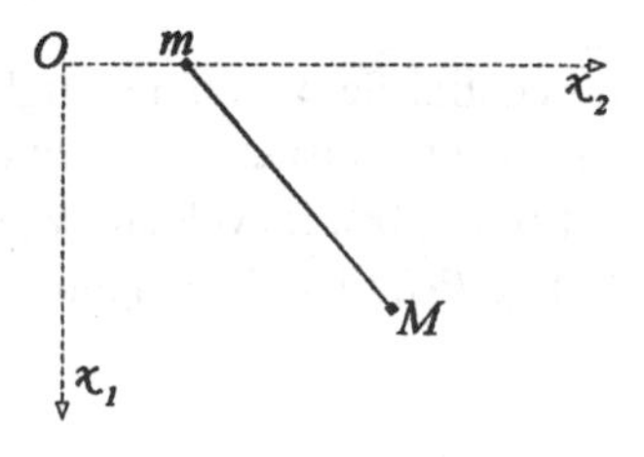

12. Determine the effective cross section of scattering of particles from a sphere of radius R. It is assumed that $V_{(r<R)} = \infty$, $V_{(r>R)} = 0$.
13. Find the small oscillations of a coplanar double pendulum.
14. Determine the effective cross section of the spherical 'gap' of potential

$$U = \begin{cases} -U_0\,(U_0 > 0), & 0 \leq r \leq R \\ 0, & r > 0 \end{cases}$$

15. Show that the cross section of scattering produced by a central force $f = k/r^3\ (k > 0)$ is

$$\sigma(\theta) = \frac{\pi^2 k}{2E} \frac{(\pi - \theta)}{\theta^2 (2\pi - \theta)^2 \sin\theta},$$

where E is the energy of a particle.
16. Two fixed points A and B act on a particle P of mass m with the elastic forces $\mathbf{f}_1$ and $\mathbf{f}_2$. (The elastic constant k is the same for both forces.) Determine the initial conditions under which the trajectory is a circle passing through A and B.
17. Determine the self oscillations of a linear chain of N identical particles, coupled by identical strings of elastic constant k.
18. A pendulum is formed by a bead P of mass m situated at one end of a massless rigid rod of length l, the other end being suspended in a point M of mass m. The point M can slide without friction along a horizontal rod, having two springs of elastic constant k, as shown in the figure. The distance AB is $2l_0$, where l_0 is the length of one spring at rest. Determine the frequencies of the small oscillations of the system.

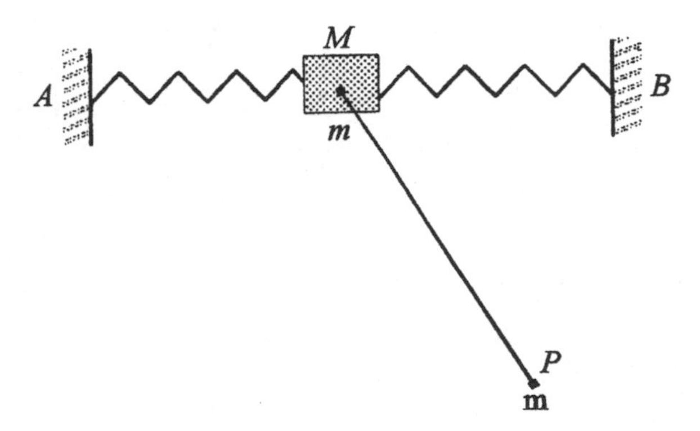

19. Investigate the oscillations of the system corresponding to the Lagrangian

$$L = \frac{1}{2}(\dot{x}_1^2 + \dot{x}_2^2) - \frac{1}{2}(\omega_1^2 x_1^2 + \omega_2^2 x_2^2) + \kappa x_1 x_2,$$

where ω_1, ω_2 and κ are constants.

20. The radial electric field between two homogeneous coaxial cylinders of radii R_1 and $R_2(>R_1)$, maintained at potentials Φ_1 and Φ_2, is

$$E = \frac{1}{r}\frac{\Phi_1 - \Phi_2}{\log \frac{R_2}{R_1}}.$$

Find the differential equation of the trajectory of a charged particle introduced between these electrodes. It is assumed that the initial velocity lies in a plane orthogonal to the common axis of the cylinders.

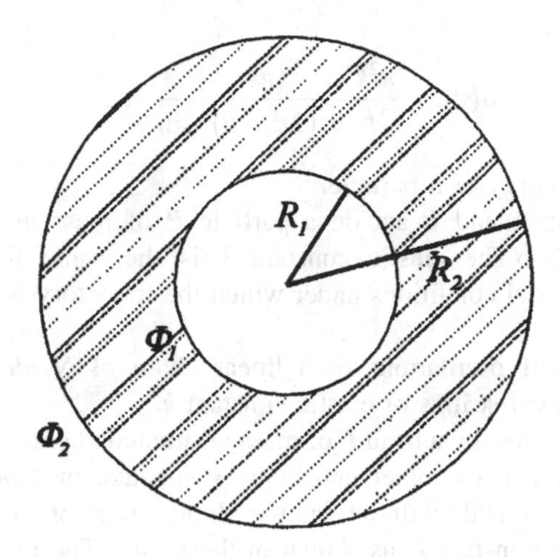

Chapter 4
Rigid Body Mechanics

4.1 General Considerations

By *rigid* or *non-deformable* body we mean a continuous or discrete system of particles, with the property that the distance between any two particles does not change during the motion. Under normal conditions, within certain pressure and temperature limits, bodies made out of metal, glass, stone, etc. can be considered rigid.

Let P_i and P_k be any two particles of the system and $\mathbf{r}_i$, $\mathbf{r}_k$ their radius-vectors relative to the origin of the frame $S(Oxyz)$ (Fig. 4.1). The rigidity condition is then expressed as

$$|\mathbf{r}_i - \mathbf{r}_k| = |\mathbf{r}_{ik}| = \text{const.} \tag{4.1.1}$$

The number of degrees of freedom of a rigid body is six. To prove this, we observe that the position of the body relative to O is fixed by the position of any three non-collinear particles. If these particles were free, then their positions would be determined by nine independent parameters. But, since the coordinates of the particles are related by three *constraints* of the form (4.1.1), the number of independent parameters reduces to six. It is obvious that the number of degrees of freedom of a rigid body subject to external constraints (e.g. a body with a fixed point, or having a fixed axis, etc.) is smaller.

As a convenient choice of the generalized coordinates associated with the six degrees of freedom of a rigid body, one usually takes a certain point O' of the body and defines the Cartesian orthogonal frame $S'(O'x'y'z')$, invariably related to the body (body coordinates) (Fig. 4.2). The position of the body is then determined by three coordinates of O' (translation coordinates) and three more coordinates that define the motion of the body about O', i.e. the orientation of the axes $O'x', O'y', O'z'$, relative to Ox, Oy, Oz (rotation coordinates). Let us analyze the motion of rotation (see also Appendix A).

If $\mathbf{r}$ is the radius-vector of some point P of the rigid body relative to O and if we assume that $O \equiv O'$, since $\mathbf{r}(x, y, z) \equiv \mathbf{r}(x', y', z')$, we can write

M. Chaichian et al., *Mechanics*, DOI: 10.1007/978-3-642-17234-2_4,

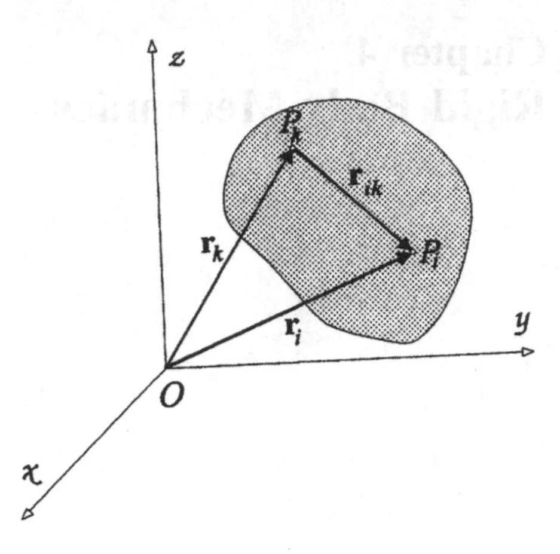

Fig. 4.1 A rigid body in a system of coordinates.

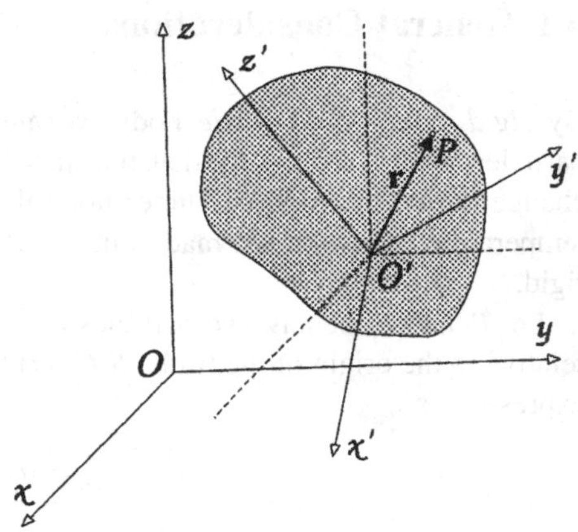

Fig. 4.2 Translation and rotation degrees of freedom of a rigid body.

$$x_i'\mathbf{u}_i' = x_i\mathbf{u}_i \quad (i = 1, 2, 3), \tag{4.1.2}$$

where $\mathbf{u}_i$ and $\mathbf{u}_i'$ $(i = 1, 2, 3)$ are orthonormal vectors in S and S', and the relation (1.1.12) and the summation convention have been used. Taking the scalar product of (4.1.2) with $\mathbf{u}_k'$ and observing that $x_i'\delta_{ik} = x_k'$, we have:

$$x_k' = a_{ki}x_i \quad (i, k = 1, 2, 3), \tag{4.1.3}$$

where $a_{ki} = \mathbf{u}_k' \cdot \mathbf{u}_i$ is the cosine of the angle between the axes $O'x_k'$ and Ox_i.

Since the distance between any two particles of the rigid body must be invariant $(x_i'x_i' = x_ix_i)$, we deduce that the coefficients a_{ik} satisfy the orthogonality condition

$$a_{ik}a_{il} = \delta_{kl} \quad (i, k, l = 1, 2, 3). \tag{4.1.4}$$

The linear transformation (4.1.3), in which the coefficients a_{ik} obey the condition (4.1.4), is called an *orthogonal transformation*. Since the nine direction cosines a_{ik} are related by the six relations (4.1.4), there are three independent coefficients a_{ik}. They can be chosen as generalized coordinates.

The coefficients a_{ik} can be regarded as elements of the transformation matrix $A = (a_{ik})$:

$$A = \begin{pmatrix} a_{11} & a_{12} & a_{13} \\ a_{21} & a_{22} & a_{23} \\ a_{31} & a_{32} & a_{33} \end{pmatrix}.$$

The matrix A is the operator carrying out the transition $x \to x'$. If, in particular, the trajectories described by the particles of the rigid body are parallel to a steady plane, for example Oxy, then (4.1.3) will give a rotation about an axis passing through O and orthogonal to the plane. In this case, the four direction cosines a_{ik} $(i, k = 1, 2)$ will be subject to the three orthogonality conditions (4.1.4), and thus only a single independent parameter remains. Let this be the angle α between $O'x'$ and Ox. The transformation matrix is then

$$A = \begin{pmatrix} \cos\alpha & \sin\alpha & 0 \\ -\sin\alpha & \cos\alpha & 0 \\ 0 & 0 & 1 \end{pmatrix}. \tag{4.1.5}$$

The relation (4.1.3) can also be written in matrix form:

$$\mathbf{x}' = A\mathbf{x}, \tag{4.1.6}$$

where $\mathbf{x}'$ and $\mathbf{x}$ are one-column matrices. For more details, see Appendix A.

4.2 Distribution of Velocities and Accelerations in a Rigid Body

To study the motion of a free rigid body, it is necessary to know the velocity and acceleration fields associated to its six degrees of freedom. In this respect, we shall analyze the general case of the so-called *relative motion* of the particle which, in particular, will lead us to the distribution of the velocities and accelerations of the particles of a rigid body. The influence of the inertia forces on the motion of the bodies at the surface of the Earth will be also considered in this chapter.

Let us consider the motion of a particle P with respect to two reference frames, S $(Oxyz)$ and $S'(O'x'y'z')$. The frame S is supposed to be fixed (inertial) and S' is non-inertial relative to S (Fig. 4.3). As an example, we can consider the motion of a car on the Earth's surface, the Earth being, in its turn, in motion with respect to the Sun.

In order to distinguish the motion of the particle P relative to the two frames, we shall call *absolute* its motion with respect to S and *relative* – the one with respect to S'. If the particle P is fixed with respect to S', then the motion of S' relative to S is named *transport motion.* Such a motion of the particle is called *composite motion.* In the Universe, *any* motion can be studied as a composite motion.

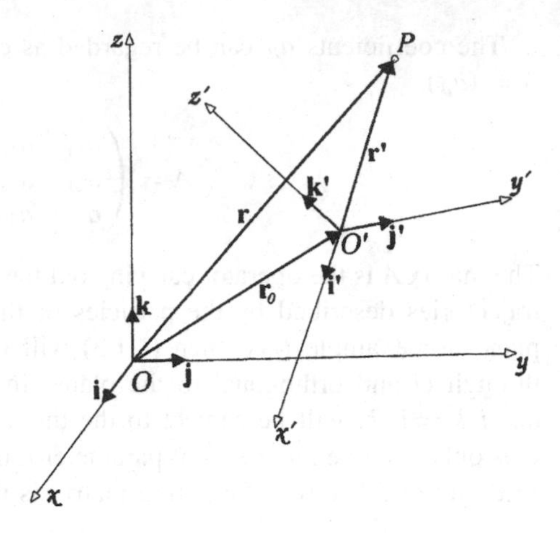

Fig. 4.3 Choice of the coordinate systems in order to obtain the distribution of velocities and accelerations in a rigid body.

From Fig. 4.3, we have

$$\mathbf{r} = \mathbf{r}_0 + \mathbf{r}'. \tag{4.2.1}$$

In view of the principle of absolute simultaneity (see Chap. 1), the time derivative of (4.2.1) gives:

$$\mathbf{v} = \dot{\mathbf{r}} = \dot{\mathbf{r}}_0 + \dot{\mathbf{r}}', \tag{4.2.2}$$

where $\mathbf{v}$ is the velocity of the particle relative to S, called *absolute velocity*. Using the summation convention, we can write $\mathbf{r}' = x'_k \mathbf{u}'_k$ $(k = 1,2,3)$. Thus, keeping in mind that $\mathbf{u}'_k$ vary in time, we have:

$$\dot{\mathbf{r}}' = \dot{x}'_k \mathbf{u}'_k + x'_k \dot{\mathbf{u}}'_k = \mathbf{v}_r + x'_k \dot{\mathbf{u}}'_k, \tag{4.2.3}$$

where $\mathbf{v}_r = \dot{x}'_k \mathbf{u}'_k$ is the *relative velocity*, i.e. the velocity of P with respect to S'.

To understand the significance of the term $x'_k \dot{\mathbf{u}}'_k$ in (4.2.3), let ω'_k $(k = 1,2,3)$ be the components of the vectors $\dot{\mathbf{u}}'_k$ in the orthonormal basis $\mathbf{u}'_k$, namely

$$\dot{\mathbf{u}}'_k = \omega'_{ks} \mathbf{u}'_s. \tag{4.2.4}$$

The basis $\mathbf{u}'_k$ being orthonormal, by taking the time derivative of the orthogonality condition $\mathbf{u}'_k \cdot \mathbf{u}'_s = \delta_{ks}$ and using (4.2.4), we find:

$$\omega'_{ks} + \omega'_{sk} = 0, \tag{4.2.5}$$

which means that the coefficients ω'_{ks} are the components of a second-order antisymmetric tensor. Let $\boldsymbol{\omega}$ be the axial vector associated with this tensor, and having the components (see Appendix A):

$$\omega'_{ks} = \epsilon_{ksi} \omega'_i \quad (i,k,s = 1,2,3). \tag{4.2.6}$$

Substituting (4.2.6) into (4.2.4), we get:

$$\dot{\mathbf{u}}'_k = \omega'_i(\epsilon_{iks}\,\mathbf{u}'_s) = (\omega'_i\,\mathbf{u}'_i) \times \mathbf{u}'_k,$$

i.e.

$$\dot{\mathbf{u}}'_k = \boldsymbol{\omega} \times \mathbf{u}'_k \quad (k = 1, 2, 3), \tag{4.2.7}$$

called *Poisson's formula*. Using (4.2.2), (4.2.3) and (4.2.7), we find the absolute velocity of the particle:

$$\mathbf{v} = \mathbf{v}_0 + \mathbf{v}_r + \boldsymbol{\omega} \times \mathbf{r}', \tag{4.2.8}$$

where $\mathbf{v}_0 = \dot{\mathbf{r}}_0$ is the velocity of the origin O'.

The *absolute acceleration* is found by taking the time derivative of (4.2.8):

$$\mathbf{a} = \dot{\mathbf{v}}_0 + \ddot{x}'_k\,\mathbf{u}'_k + \dot{x}'_k\,\dot{\mathbf{u}}'_k + \dot{\boldsymbol{\omega}} \times \mathbf{r}' + \boldsymbol{\omega} \times \dot{\mathbf{r}}'$$

or, using (4.2.3) and (4.2.7),

$$\mathbf{a} = \mathbf{a}_0 + \mathbf{a}_r + \dot{\boldsymbol{\omega}} \times \mathbf{r}' + \boldsymbol{\omega} \times (\boldsymbol{\omega} \times \mathbf{r}') + 2\boldsymbol{\omega} \times \mathbf{v}_r. \tag{4.2.9}$$

Here, $\mathbf{a}_0 = \dot{\mathbf{v}}_0$ is the acceleration of O' and $\mathbf{a}_r = \ddot{x}'_k\,\mathbf{u}'_k$ – the acceleration of the point P with respect to O', called *relative acceleration*. The term $\boldsymbol{\omega} \times (\boldsymbol{\omega} \times \mathbf{r}')$ is named *centripetal acceleration* and the term $2\boldsymbol{\omega} \times \mathbf{v}_r$ is the *Coriolis acceleration*.

Suppose now that P is a certain point of a rigid body R, invariably related to the frame S'. In other words, the rigid body *identifies* with the frame S'. In this case, it is obvious that $\mathbf{v}_r = 0$, $\mathbf{a}_r = 0$, while the formulas (4.2.8) and (4.2.9) become:

$$\mathbf{v}_{tr} = \mathbf{v}_0 + \boldsymbol{\omega} \times \mathbf{r}', \tag{4.2.10}$$

$$\mathbf{a}_{tr} = \mathbf{a}_0 + \dot{\boldsymbol{\omega}} \times \mathbf{r}' + \boldsymbol{\omega} \times (\boldsymbol{\omega} \times \mathbf{r}'), \tag{4.2.11}$$

where $\mathbf{v}_{tr}$ and $\mathbf{a}_{tr}$ denote the *transport velocity* and *transport acceleration*, respectively. We may also write:

$$\mathbf{v} = \mathbf{v}_r + \mathbf{v}_{tr}, \tag{4.2.12}$$

$$\mathbf{a} = \mathbf{a}_r + \mathbf{a}_{tr} + \mathbf{a}_c. \tag{4.2.13}$$

If the origins O and O' of the two reference frames coincide (i.e. there is no translation motion), then $\mathbf{r}_0 = 0$, $\mathbf{v}_0 = 0$, $\mathbf{a}_0 = 0$, and we are left with

$$\mathbf{v} = \boldsymbol{\omega} \times \mathbf{r}, \tag{4.2.14}$$

$$\mathbf{a} = \dot{\boldsymbol{\omega}} \times \mathbf{r} + \boldsymbol{\omega} \times (\boldsymbol{\omega} \times \mathbf{r}). \tag{4.2.15}$$

Physical Significance of the Vector $\boldsymbol{\omega}$

We know so far that $\boldsymbol{\omega}$ is an axial vector, associated with the antisymmetric tensor ω'_{sk}, its components having been introduced as coefficients of the linear expansion (4.2.4). In order to find the physical significance of this vector, let us consider a

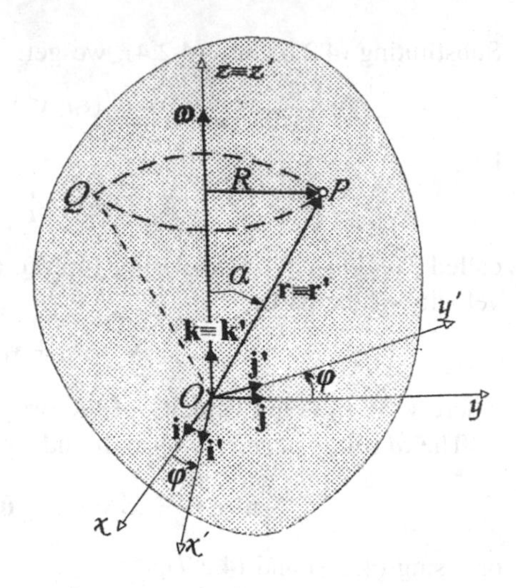

Fig. 4.4 Physical significance of the vector ω: it is directed along the axis of rotation and its magnitude is equal to the angular velocity $\dot{\varphi}$.

rigid body moving around a fixed axis. It is obvious that such a system possesses one degree of freedom. There is no loss of generality if we take $O \equiv O'$ and $Oz \equiv Oz'$ as the fixed axis (Fig. 4.4).

With this choice, (4.2.7) leads to $\dot{\mathbf{k}} = \boldsymbol{\omega} \times \mathbf{k} = 0$, which says that $\boldsymbol{\omega}$ is collinear to the axis of rotation: $\boldsymbol{\omega} = \omega\mathbf{k}$. On the other hand, (4.2.14) leads to the conclusion that the velocity $\mathbf{v}$ of any point P of the body is orthogonal to the plane defined by vectors $\boldsymbol{\omega}$ and $\mathbf{r}$ and has the magnitude

$$|\mathbf{v}| = |\boldsymbol{\omega} \times \mathbf{r}| = \omega r \sin\alpha = \omega R. \tag{4.2.16}$$

But, in view of (1.1.18),

$$|\mathbf{v}| = \dot{\varphi} R. \tag{4.2.17}$$

The last two relations give then $\omega = \dot{\varphi}$, i.e. the vector $\boldsymbol{\omega}(t)$ is directed along the axis of rotation and its magnitude is equal to the angular velocity $\dot{\varphi}$. It is called *instantaneous vector of rotation.*

4.3 Inertial Forces

Let us again consider the two reference frames S and S', S being inertial and S' non-inertial relative to S. As we know, the fundamental equation of motion written for a particle of mass m is

$$m\mathbf{a} = \mathbf{F}. \tag{4.3.1}$$

By using the Lagrangian formalism developed in the previous chapters, we wish now to find the form of the equation of motion of the same particle, relative to the non-inertial frame S'. To this end, we shall use the fact that the Lagrange equations do not change their form when the reference frame changes, provided that the

Lagrangian in the new frame is suitably chosen. Supposing that a conservative force field acts on the particle, the Lagrangian in the frame S is given by

$$L = \frac{1}{2} m|\mathbf{v}|^2 - V(\mathbf{r}), \tag{4.3.2}$$

where $\mathbf{v}$ is the velocity of the particle with respect to S and $V(\mathbf{r})$ is its potential energy.

In order to write the equation of motion of the particle in S', it is necessary to express the Lagrangian L as a function of the coordinates x'_i $(i = 1, 2, 3)$ of the point and the components $\dot{x}'_i$ of its velocity relative to S'. Using (4.2.8), we obtain:

$$\begin{aligned} L = \frac{1}{2} m[&|\mathbf{v}_r|^2 + |\boldsymbol{\omega} \times \mathbf{r}'|^2 + 2\mathbf{v}_r \cdot (\boldsymbol{\omega} \times \mathbf{r}') + |\mathbf{v}_0|^2 + 2\mathbf{v}_0 \cdot \mathbf{v}_r \\ &+ 2\mathbf{v}_0 \cdot (\boldsymbol{\omega} \times \mathbf{r}')] - V(\mathbf{r}'). \end{aligned} \tag{4.3.3}$$

The last three terms in the square bracket can be written as follows:

$$\begin{aligned} |\mathbf{v}_0|^2 + 2\mathbf{v}_0.\mathbf{v}_r + 2\mathbf{v}_0 \cdot (\boldsymbol{\omega} \times \mathbf{r}') &= 2\mathbf{v}_0 \cdot \mathbf{v} - |\mathbf{v}_0|^2 = \mathbf{v}_0 \cdot \frac{d}{dt}(2\mathbf{r} - \mathbf{r}_0) = \mathbf{v}_0 \cdot \frac{d}{dt}(\mathbf{r}_0 + 2\mathbf{r}') \\ &= \frac{d}{dt}[\mathbf{v}_0 \cdot (2\mathbf{r}' + \mathbf{r}_0)] - 2\mathbf{a}_0 \cdot \mathbf{r}' - \mathbf{a}_0 \cdot \mathbf{r}_0 \cdot \end{aligned}$$

Since the Lagrangian L is defined up to a term which is the total derivative with respect to time of any scalar function of generalized coordinates q_j and time t, we may omit the total time derivative in the last relation. In its turn, the term $\mathbf{a}_0 \cdot \mathbf{r}_0$ is a function of time only, i.e. it can be written as a total time derivative, meaning that it also can be omitted. Thus, we can finally write the Lagrangian in the form:

$$L = \frac{1}{2} m\,|\mathbf{v}_r|^2 + \frac{1}{2} m|\boldsymbol{\omega} \times \mathbf{r}'|^2 + m\mathbf{v}_r \cdot (\boldsymbol{\omega} \times \mathbf{r}') - m\mathbf{a}_0 \cdot \mathbf{r}' - V(\mathbf{r}'). \tag{4.3.4}$$

In order to obtain the equation of motion of the particle relative to the frame S', it is convenient to write the Lagrangian as

$$\begin{aligned} L = &\frac{1}{2} m\dot{x}'_i\dot{x}'_i + \frac{1}{2} m(\omega'_k\omega'_k)(x'_ix'_i) - \frac{1}{2} m(x'_i\omega'_i)(x'_k\omega'_k) + m\epsilon_{ijk}\,\dot{x}'_i\,\omega'_j\,x'_k \\ &- ma'_{0i}x'_i - V(x'_k), \end{aligned} \tag{4.3.5}$$

where $x'_i, \dot{x}'_i, \omega'_i, a'_{0i}$ are the components of $\mathbf{r}', \mathbf{v}_r, \boldsymbol{\omega}, \mathbf{a}_0$ in the frame S'. Then we have, successively:

$$\begin{aligned} \frac{\partial L}{\partial \dot{x}'_s} &= m\dot{x}'_s + m\epsilon_{sjk}\,\omega'_j x'_k, \\ \frac{d}{dt}\left(\frac{\partial L}{\partial \dot{x}'_s}\right) &= m\ddot{x}'_s + m\epsilon_{sjk}\,\dot{\omega}'_j\,x'_k + m\epsilon_{sjk}\,\omega'_j\,\dot{x}'_k, \\ \frac{\partial L}{\partial x'_s} &= m(\omega'_k\omega'_k)x'_s - m(x'_i\omega'_i)\omega'_s + m\epsilon_{sij}\,\dot{x}'_i\,\omega'_j - ma'_{0s} - \frac{\partial V}{\partial x'_s} \cdot \end{aligned}$$

Next, we use the Lagrange equations (2.5.17) and take x'_s as generalized coordinates:

$$\frac{d}{dt}\left(\frac{\partial L}{\partial \dot{x}'_s}\right) - \frac{\partial L}{\partial x'_s} = 0 \quad (s = 1, 2, 3),$$

arriving at the second-order differential equation

$$m\ddot{x}'_s + m(\dot{\boldsymbol{\omega}} \times \mathbf{r}')_s + ma'_{0s} + m[\boldsymbol{\omega} \times (\boldsymbol{\omega} \times \mathbf{r}')]_s + 2m(\boldsymbol{\omega} \times \mathbf{v}_r)_s + \frac{\partial V}{\partial x'_s} = 0,$$

which is the x'_s-component of the vector equation

$$m\mathbf{a}_r = \mathbf{F} - m\mathbf{a}_0 - m\dot{\boldsymbol{\omega}} \times \mathbf{r}' - m\boldsymbol{\omega} \times (\boldsymbol{\omega} \times \mathbf{r}') - 2m\boldsymbol{\omega} \times \mathbf{v}_r. \tag{4.3.6}$$

Taking into account (4.2.11) and introducing the notations

$$\mathbf{F}_{tr} = -m\mathbf{a}_{tr}, \quad \mathbf{F}_c = -m\mathbf{a}_c, \tag{4.3.7}$$

the equation of motion of the particle with respect to the non-inertial frame S' reads:

$$m\mathbf{a}_r = \mathbf{F} + \mathbf{F}_{tr} + \mathbf{F}_c. \tag{4.3.8}$$

It then follows that in the frame S' Newton's fundamental equation does not keep its form, as we expected. There are two more forces, in addition to the applied force $\mathbf{F}$. The force $\mathbf{F}$ is customarily called *real* or *actual*, while the forces $\mathbf{F}_{tr}$ and $\mathbf{F}_c$ are known as *inertial*, or *apparent*, or, still, *complementary*. They have a fictitious character, in the sense that they cannot give rise to, transmit, or maintain the motion. They occur only *during* the motion and *because* of the motion of the non-inertial frame S'.

One observes that if $\mathbf{a}_0 = 0$, $\boldsymbol{\omega} = 0$, i.e. if S' is, in its turn, an inertial frame, then Eq. (4.3.8) acquires the form (4.3.1), as it has to happen by virtue of the principle of classical relativity.

4.3.1 Action of the Coriolis Force on the Motion of Bodies at the Surface of the Earth

Let the origin O of the frame S be at the centre of the Earth and have its axes fixed (e.g. pointing towards three fixed stars). Rigorously speaking, such a frame is non-inertial, since the Earth performs a non-uniform motion around the Sun. However, for a short time interval, the trajectory of the point O can be considered as being straight and its motion uniform. Under these assumptions, the frame S can be considered inertial.

Next, we shall take the origin O' of the non-inertial frame S' at a fixed point on the surface of the Earth, its axes being chosen as follows: $O'z'$ along the ascending vertical, $O'y'$ tangent to the parallel going through O' and pointing West, and $O'x'$ tangent to the meridian through O' and pointing North (Fig. 4.5).

Fig. 4.5 Action of the Coriolis force on the motion of the bodies at the surface of the Earth.

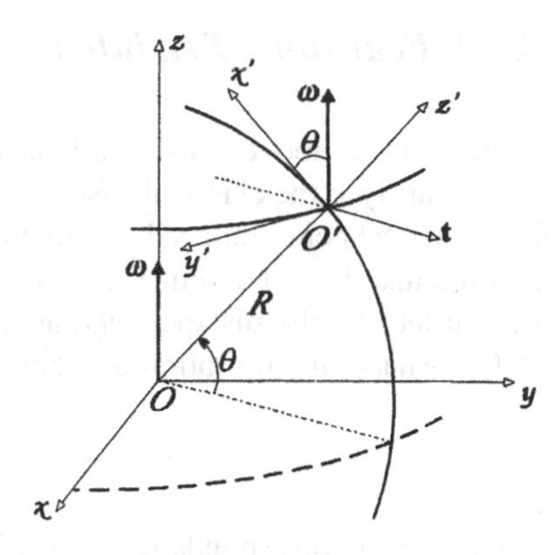

In order to study the motion of a body (considered as a particle) at the surface of the Earth, we shall use the Lagrangian (4.3.4). Observing that the point O' describes a circle of radius $R\cos\theta$, we may write:

$$\begin{aligned}\mathbf{a}_0 &= -R\omega^2\mathbf{t}\cos\theta \\ &= -R\omega^2(\mathbf{k}'\cos\theta - \mathbf{i}'\sin\theta)\cos\theta,\end{aligned} \tag{4.3.9}$$

where $\mathbf{t}$ is the unit vector of the radius of the parallel circle passing through O' and pointing outwards. Since $\boldsymbol{\omega}$ is small in magnitude ($\omega \simeq 7.27 \times 10^{-5}\,\mathrm{s}^{-1}$), we may neglect the terms in ω^2 which occur in (4.3.4). On the other hand, since we deal with quantities determined in S' only, we shall temporarily omit the index 'prime'. (This convention is valid only in this application).

The only applied force acting on the particle is its gravity, so that the Lagrangian is:

$$L = \frac{1}{2}m|\mathbf{v}_r|^2 + m\mathbf{v}_r \cdot (\boldsymbol{\omega} \times \mathbf{r}) + m\mathbf{g} \cdot \mathbf{r}, \tag{4.3.10}$$

where we took into account that the components of $\mathbf{g}$ are $(0, 0, -g)$, which leads to the relation $-mgz = m\mathbf{g} \cdot \mathbf{r}$.

If we choose x_i and v_i as generalized coordinates and generalized velocities, respectively, and recall that $\boldsymbol{\omega}$ is a constant vector, the Lagrange equations (see Chap. 2) lead to the differential equations of motion of the particle:

$$m\ddot{x}_i = mg_i - 2m\epsilon_{ijk}\,\omega_j\,\dot{x}_k,$$

equivalent to the vector equation

$$m\ddot{\mathbf{r}} = m\mathbf{g} - 2m\boldsymbol{\omega} \times \mathbf{v}. \tag{4.3.11}$$

This equation tells us that the motion of a body at the surface of the Earth is affected by the Coriolis force, orthogonal to both the pole axis and the direction of motion of the body.

4.3.2 Foucault's Pendulum

In view of the above considerations and using the Lagrangian formalism, we wish now to analyze the classical case of the *Foucault pendulum*. Roughly speaking, this is a spherical pendulum, subject to both the gravitational and Coriolis forces. Let us assume that the pendulum P is suspended in the point Q, situated above the point P, and let l be the distance PQ, as measured on the vertical line passing through P. The equations of motion are then obtained from the Lagrangian (see (4.3.10))

$$L = \frac{1}{2}m|\mathbf{v}|^2 + m\boldsymbol{\omega}\cdot(\mathbf{r}\times\mathbf{v}) - mgz, \tag{4.3.12}$$

where $z = 0$ corresponds to the vertical position of the pendulum. If the oscillations are small enough, the coordinate z can be deduced with the help of the constraint $x^2 + y^2 + (z-l)^2 = l^2$. Since $z^2 \ll x^2 + y^2$, we can write

$$z = \frac{1}{2l}(x^2 + y^2). \tag{4.3.13}$$

In the same way, taking into account that $|\boldsymbol{\omega}|$ is small, we neglect the terms in which products of the type $\omega_i z$ or $\omega_i \dot{z}$ $(i = 1, 2, 3)$ occur in the mixed product $\boldsymbol{\omega}\cdot(\mathbf{r}\times\mathbf{v})$. Then

$$\boldsymbol{\omega}\cdot(\mathbf{r}\times\mathbf{v}) = \omega_z(x\dot{y} - y\dot{x})$$

and the Lagrangian (4.3.12) becomes:

$$L = \frac{1}{2}m(\dot{x}^2 + \dot{y}^2) - \frac{mg}{2l}(x^2 + y^2) + m\omega_z(x\dot{y} - y\dot{x}). \tag{4.3.14}$$

Instead of the Cartesian coordinates x, y, it is more convenient to use the polar coordinates r, φ of the projection P' of the point P on the $O'xy$ plane. Since $x = r\cos\varphi$, $y = r\sin\varphi$, we find:

$$L = \frac{1}{2}m(\dot{r}^2 + r^2\dot{\varphi}^2) - \frac{mg}{2l}r^2 + m\omega_z r^2\dot{\varphi}. \tag{4.3.15}$$

This form is particularly convenient, because it displays the cyclic coordinate φ, associated with the first integral

$$p_\varphi = \frac{\partial L}{\partial\dot{\varphi}} = mr^2\dot{\varphi} + mr^2\omega_z = C,$$

or

$$r^2\dot{\varphi} + r^2\omega_z = C_1, \tag{4.3.16}$$

where C and $C_1 = \frac{C}{m}$ are constants.

The physical system being conservative, we also have the energy first integral

$$E = \frac{1}{2}m(\dot{r}^2 + r^2\dot{\varphi}^2) + \frac{mg}{2l}r^2 = h_1(\text{const.}),$$

or

$$\dot{r}^2 + r^2\dot{\varphi}^2 = h - \frac{g}{l}r^2 \qquad \left(h = \frac{2h_1}{m}\right). \tag{4.3.17}$$

In order to understand the physical significance of (4.3.16) and (4.3.17), let us make a clockwise rotation of the $O'xyz$ frame about the $O'z$ axis, of angle $\omega_z t$, arriving at the frame $O'x_1y_1z$. (Remember that we have omitted the index 'prime' of the coordinates in this application). In the frame $O'x_1y_1$ the polar coordinates of P' are r_1, α, where $\alpha = \varphi + \omega_z t$ (Fig. 4.6). The new Cartesian coordinates x_1, y_1 are then found via the matrix relation:

$$\begin{pmatrix} x_1 \\ y_1 \end{pmatrix} = \begin{pmatrix} \cos\omega_z t & \sin\omega_z t \\ -\sin\omega_z t & \cos\omega_z t \end{pmatrix} \begin{pmatrix} x \\ y \end{pmatrix}. \tag{4.3.18}$$

Therefore, $x_1 = x\cos\omega_z t + y\sin\omega_z t$, $y_1 = -x\sin\omega_z t + y\cos\omega_z t$, leading to $r_1^2 = r^2$. Since in the two first integrals (4.3.16) and (4.3.17) occurs only the second power of r, we shall keep the former notation r. Then the first integral (4.3.16) becomes:

$$r^2\dot{\alpha} = C_1, \tag{4.3.19}$$

which is the projection on the $O'z$-axis of the areas theorem. Introducing $\varphi = \alpha - \omega_z t$ into (4.3.17) and neglecting the terms in ω^2, one finds:

$$\dot{r}^2 + \frac{C_1^2}{r^2} + \frac{g}{l}r^2 = h_2, \quad h_2 = h + 2C_1\omega_z. \tag{4.3.20}$$

Recalling the results obtained in the study of the central force problem, we observe that the trajectory of the spherical pendulum is an ellipse, its centre being

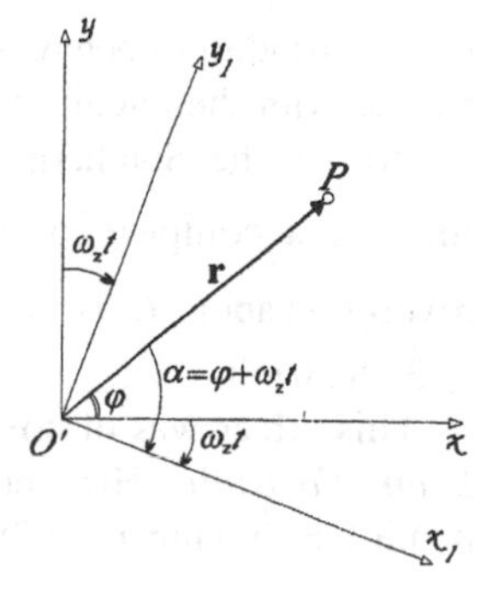

Fig. 4.6 Rotation of the coordinate axes of the Foucault pendulum.

at O' and its axes uniformly rotating clockwise about the point O', in the $O'xy$-plane, with the angular velocity $\omega_z = \omega \sin\theta$. If we choose the initial conditions as:

$$r(0) = r_0, \quad \varphi(0) = \varphi_0,$$
$$\dot{r}(0) = 0, \quad \dot{\varphi}(0) = 0,$$

i.e. the pendulum is pulled out of its equilibrium position and then allowed to move freely, we have $|\mathbf{v}_0| = r_0\omega_z$, meaning that its initial velocity $\mathbf{v}_0$ (relative to O') is orthogonal to $\mathbf{r}_0$. It follows from (4.3.19) that $C_1 = r_0^2\omega_z$, which leads to $h_2 = r_0^2(\omega_0^2 + \omega_z^2)$, where $\omega_0^2 = \frac{g}{l}$. We have:

$$\dot{r}^2 = \omega_0^2(r_0^2 - r^2) + r_0^2\omega_z^2 - \frac{C_1^2}{r^2}. \tag{4.3.21}$$

The turning points of the trajectory are given by the roots of the equation $\dot{r}^2 = 0$, i.e.

$$r^4 - \left(1 + \frac{\omega_z^2}{\omega_0^2}\right) r_0^2 r^2 + \frac{\omega_z^2}{\omega_0^2} r_0^4 = 0. \tag{4.3.22}$$

The solutions to the biquadratic equation (4.3.22) are: $r_I = r_0$, $r_{II} = r_0\frac{\omega_z}{\omega_0}$, meaning that the ends of the ellipse axes describe two circles of diameters $2a$ and $2b$, where $a = r_0$ is the major semi-axis and $b = r_0\frac{\omega_z}{\omega_0}$ is the minor semi-axis. The trajectory of the particle is then situated between the two circles. One observes that a depends on the initial conditions only, while b varies with the latitude of the location of the experiment on Earth. Since ω is small, the ratio

$$\frac{b}{a} = \frac{\omega_z}{\omega_0} = \omega\sqrt{\frac{l}{g}} \cos\theta$$

is also very small, meaning that the ellipse is very flattened and may practically be identified with a straight line. In particular, if we take $\dot{\varphi} = 0$, then $\dot{\alpha} = \omega_z$, and (4.3.20) gives:

$$\ddot{r} + \omega_1^2 r = 0 \quad (\omega_1^2 = \omega_0^2 + \omega_z^2), \tag{4.3.23}$$

i.e. the pendulum behaves like a harmonic oscillator. The oscillation plane $O'x_1z$ rotates with the angular velocity ω_z about $O'z$, in the direction East-South-West-North. (In the Southern hemisphere, the direction of rotation is opposite.) The time of a complete oscillation is $\tau_1 = \frac{2\pi}{\omega_1} = 2\pi\sqrt{\frac{l}{g}}$, while the period of the revolution about O' is $\tau_2 = \frac{2\pi}{\omega_z} = \frac{2\pi}{\omega \sin\theta}$. For example, if $\theta = 45°$, $\tau_2 = 1.414$ days $= 33\,\text{h}\ 50\,\text{min}$.

This effect was discovered and studied by the French physicist *Jean Bernard Léon Foucault*. His most famous experiment was done in 1851, with a pendulum having $m = 28\,\text{kg}, l = 67$ m, under the cupola of the Paris Panthéon

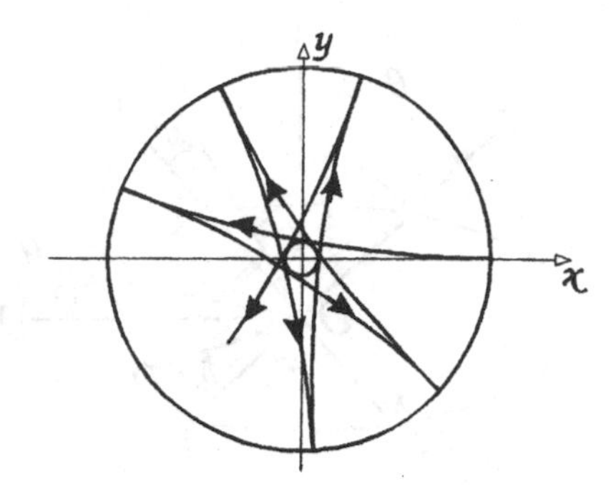

Fig. 4.7 A trajectory of the Foucault pendulum.

$(\theta = 48°50\,\text{min})$. He found $\tau_2 = 32$ h, in good agreement with the theoretical prediction.

Figure 4.7 shows the Foucault pendulum trajectory corresponding to our initial conditions. Different initial conditions give different trajectories.

Observation: An observer located in an inertial frame (e.g. the Sun) would see the pendulum oscillating in a plane, the Earth being in rotation relative to this plane. The Foucault pendulum shows, then, the motion of rotation of the Earth, *without* any other astronomical observations.

4.4 Euler's Angles

As we have seen in Sect. 4.1, the position of a free rigid body is fully determined if one knows the position – relative to S – of a certain point O' of the body and the angles between the two frames S and S', as well. Here it is assumed that S is fixed in space, while S' is fixed to the body. Since the definition of O' is an easy matter, in the following we shall consider the motion of the body about this point.

From the practical point of view, the choice of three direction cosines as independent parameters is not convenient. One has to look, then, for other solutions. The most efficient method was devised by *Leonhard Euler*. He defined a system of three angular parameters, attached to a group of three successive rotations about three conveniently chosen directions. In this way, the transition from $S(Oxyz)$ to $S'(O'x'y'z')$ is realized. There are three successive steps in the transition, as follows:

(a) A direct (i.e. counterclockwise) rotation of angle φ, in the xy-plane about the Oz-axis, until the new axis Op (Fig. 4.8a) is orthogonal to Oz'. (The orientation of S' relative to S is *given*!) Thus, we go from the frame S, of unit vectors $\mathbf{i}$, $\mathbf{j}$, $\mathbf{k}$, to the frame $Opqz$, of unit vectors $\mathbf{t}_1$, $\mathbf{t}_2$, $\mathbf{k}$ (Fig. 4.8b). The transformation relations are:

$$\begin{aligned} \mathbf{t}_1 &= \mathbf{i}\cos\varphi + \mathbf{j}\sin\varphi, \\ \mathbf{t}_2 &= -\mathbf{i}\sin\varphi + \mathbf{j}\cos\varphi, \\ \mathbf{k} &= \mathbf{k}. \end{aligned} \tag{4.4.1}$$

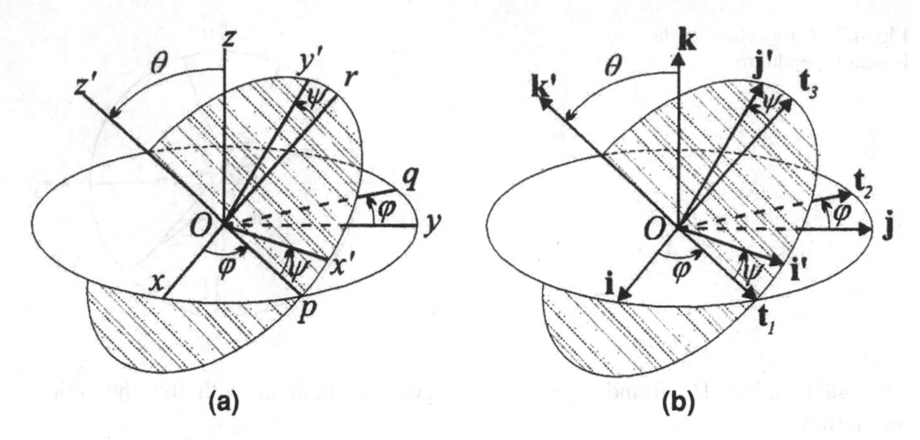

Fig. 4.8 Geometrical representation of the Euler angles: φ, θ and ψ.

The Op axis is known as the *line of nodes* and φ – as the *precession angle*, varying from 0 to 2π.

(b) A direct rotation of angle θ, in the Oqz-plane, about the line of nodes Op, until Oz coincides with Oz', i.e. the transition from $Opqz$ to $Oprz'$, of unit vectors $\mathbf{t}_1$, $\mathbf{t}_3$, $\mathbf{k}'$. The transformation formulas are:

$$\begin{aligned} \mathbf{t}_1 &= \mathbf{t}_1, \\ \mathbf{t}_3 &= \mathbf{t}_2 \cos\theta + \mathbf{k}\sin\theta, \\ \mathbf{k}' &= -\mathbf{t}_2 \sin\theta + \mathbf{k}\cos\theta. \end{aligned} \tag{4.4.2}$$

The angle θ is called *angle of nutation* and takes values from 0 to π.

(c) A direct rotation of angle ψ, in the Opr plane, about the Oz'-axis, until Op coincides with Ox'. If $\mathbf{i}'$, $\mathbf{j}'$, $\mathbf{k}'$ are the unit vectors of S', then the transition from $Oprz'$ to $Ox'y'z'$ is given by:

$$\begin{aligned} \mathbf{i}' &= \mathbf{t}_1 \cos\psi + \mathbf{t}_3 \sin\psi, \\ \mathbf{j}' &= -\mathbf{t}_1 \sin\psi + \mathbf{t}_3 \cos\psi, \\ \mathbf{k}' &= \mathbf{k}'. \end{aligned} \tag{4.4.3}$$

The angle ψ is the *angle of self-rotation* and takes values from 0 to 2π.

The angles φ, θ, ψ are called *Euler's angles*. Their names will be justified later in this chapter.

Let us write the direction cosines a_{ik} in terms of Euler's angles. To this end, we define the column matrices:

$$\mathrm{T}_1 = \begin{pmatrix} \mathbf{t}_1 \\ \mathbf{t}_2 \\ \mathbf{k} \end{pmatrix}, \quad \mathrm{T}_2 = \begin{pmatrix} \mathbf{t}_1 \\ \mathbf{t}_3 \\ \mathbf{k}' \end{pmatrix}, \quad \mathrm{T}_3 = \begin{pmatrix} \mathbf{i}' \\ \mathbf{j}' \\ \mathbf{k}' \end{pmatrix}, \quad \mathrm{T} = \begin{pmatrix} \mathbf{i} \\ \mathbf{j} \\ \mathbf{k} \end{pmatrix}. \tag{4.4.4}$$

The transformation formulas (4.4.1)–(4.4.3) can be then written in the matrix form

$$\mathbf{T}_1 = D\mathbf{T}, \quad \mathbf{T}_2 = C\mathbf{T}_1, \quad \mathbf{T}_3 = B\mathbf{T}_2,$$

where D, C, B are the three transformation matrices:

$$D = \begin{pmatrix} \cos\varphi & \sin\varphi & 0 \\ -\sin\varphi & \cos\varphi & 0 \\ 0 & 0 & 1 \end{pmatrix}, \quad C = \begin{pmatrix} 1 & 0 & 0 \\ 0 & \cos\theta & \sin\theta \\ 0 & -\sin\theta & \cos\theta \end{pmatrix},$$

$$B = \begin{pmatrix} \cos\psi & \sin\psi & 0 \\ -\sin\psi & \cos\psi & 0 \\ 0 & 0 & 1 \end{pmatrix}. \tag{4.4.5}$$

The transition from the coordinates x_i to x_i' $(i = 1, 2, 3)$ is given by the matrix relation $x' = Ax$, where $A = BCD$. Any element a_{ik} of the matrix A is calculated according to the rule

$$a_{ik} = (BCD)_{ik} = b_{il}\, c_{lm}\, d_{mk}. \tag{4.4.6}$$

Performing the necessary calculations, we obtain the transformation matrix A in terms of the Euler angles:

$$A = (a_{ik})$$
$$= \begin{pmatrix} \cos\psi\cos\varphi - \cos\theta\sin\varphi\sin\psi & \cos\psi\sin\varphi + \cos\theta\cos\varphi\sin\psi & \sin\theta\sin\psi \\ -\sin\psi\cos\varphi - \cos\theta\sin\varphi\cos\psi & -\sin\psi\sin\varphi + \cos\theta\cos\varphi\cos\psi & \sin\theta\cos\psi \\ \sin\theta\sin\varphi & -\sin\theta\cos\varphi & \cos\theta \end{pmatrix}. \tag{4.4.7}$$

4.5 Motion of a Rigid Body About a Fixed Point

In this section we shall use Latin indices $(i,\ j,\ k, \ldots)$ to indicate the number of particles and Greek indices $(\alpha,\ \beta,\ \gamma, \ldots)$ for the vector components and generalized coordinates. The summation convention over repeated (Greek) indices running from 1 to 3 will also be used.

As we already know (see Sect. 4.1), a free rigid body has six degrees of freedom: three of them associated with the three Cartesian coordinates of a certain point of the rigid body, describing its translation, and three independent angular parameters (e.g. Euler's angles) which define the rotation about this point. If the origin O' of the frame S' is chosen in the centre of mass G of the body, then according to (4.2.10) both angular momentum and kinetic energy will each be composed of two terms: one term containing *only* the Cartesian coordinates of the centre of mass and the other written *only* in terms of angular coordinates, describing the rotation. This decomposition also occurs when we deal with the potential energy. For instance, the potential energy of the electric dipole placed in a uniform field depends only on its orientation, while the gravitational potential energy of a body depends only on the coordinates of its centre of mass. Going further, we may say that the Lagrangian of such a mechanical system has the same property. As a result, the translation and rotation motions can be studied independently of each other.

In view of these considerations, in the following we shall analyze the motion of a rigid body about a fixed point, chosen so that $O \equiv O' \equiv G$. At the end of this section, we shall give the method used when both translation and rotation motions are taken into consideration.

In order to present the dynamical approach to the motion of the rigid body about a fixed point, we shall first introduce some elements which are necessary for the derivation of the equations of motion.

4.5.1 Kinematic Preliminaries

Since the fixed point coincides with the common origin of the two frames, $O \equiv O'$ (Fig. 4.9), the instantaneous velocity of a certain point P of the rigid body is (see (4.3.14)) $\mathbf{v} = \boldsymbol{\omega} \times \mathbf{r}$, which means that at any moment the body is engaged in a motion of rotation about an axis passing through O. This axis is oriented along the vector $\boldsymbol{\omega}$ and it is called *instantaneous axis of rotation*. All particles of the rigid body situated on this axis have zero velocity ($\mathbf{r} = \lambda\boldsymbol{\omega}$, so that $\mathbf{v} = \lambda\boldsymbol{\omega} \times \boldsymbol{\omega} = 0$).

We observe that the instantaneous vector $\boldsymbol{\omega}$ can be written as the resultant of three vectors, each of them corresponding to successive rotations about Oz, Op and Oz'. Therefore, using the definition of Euler's angles, we have:

$$\boldsymbol{\omega} = \dot{\varphi}\mathbf{k} + \dot{\theta}\mathbf{t}_1 + \dot{\psi}\mathbf{k}'. \tag{4.5.1}$$

In order to find the components $\omega_{x'}, \omega_{y'}, \omega_{z'}$ of $\boldsymbol{\omega}$ on the axes of the mobile frame S', we successively multiply (4.5.1) by $\mathbf{i}'$, $\mathbf{j}'$, $\mathbf{k}'$. In view of (4.4.1)–(4.4.3), we get:

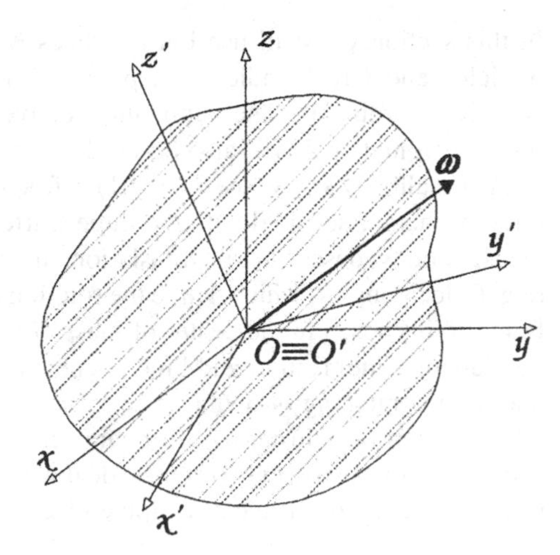

Fig. 4.9 Rigid body with a fixed point.

$$\begin{aligned} \omega_{x'} &= \omega'_x = \omega'_1 = \dot{\varphi} \sin\theta \sin\psi + \dot{\theta}\cos\psi, \\ \omega_{y'} &= \omega'_y = \omega'_2 = \dot{\varphi} \sin\theta \cos\psi - \dot{\theta}\sin\psi, \\ \omega_{z'} &= \omega'_z = \omega'_3 = \dot{\varphi}\cos\theta + \dot{\psi}. \end{aligned} \tag{4.5.2}$$

The components of the vector $\boldsymbol{\omega}$ on the axes of the frame S are obtained in the same manner, multiplying (4.5.1) by **i, j, k**:

$$\begin{aligned} \omega_x &= \omega_1 = \dot{\theta}\cos\varphi + \dot{\psi}\sin\theta\sin\varphi, \\ \omega_y &= \omega_2 = \dot{\theta}\sin\varphi - \dot{\psi}\sin\theta\cos\varphi, \\ \omega_z &= \omega_3 = \dot{\varphi} + \dot{\psi}\cos\theta. \end{aligned} \tag{4.5.3}$$

4.5.2 Angular Momentum

In view of (1.3.35), the angular momentum $\mathbf{L}$ of the rigid body relative to $O' \equiv O$ is:

$$\mathbf{L} = \sum_{i=1}^{N} m_i\,(\mathbf{r}_i \times \mathbf{v}_i) = \sum_{i=1}^{N} m_i\,[\mathbf{r}_i \times (\boldsymbol{\omega} \times \mathbf{r}_i)], \tag{4.5.4}$$

where $\mathbf{r}_i$ and $\mathbf{v}_i$ are the radius-vector and the velocity of the particle P_i of the mass m_i of the body relative to the fixed point.

Using the convention adopted in the beginning of this section, the projection of (4.5.4) on the axes of the frame S' yields:

$$\begin{aligned} L'_\alpha &= \sum_{i=1}^{N} m_i\Big[|\mathbf{r}_i|^2(\boldsymbol{\omega}\cdot\mathbf{u}'_\alpha) - (\mathbf{r}_i\cdot\boldsymbol{\omega})(\mathbf{r}_i\cdot\mathbf{u}'_\alpha)\Big] = \sum_{i=1}^{N} m_i\Big[|\mathbf{r}_i|^2\omega'_\alpha - (x'_{i\beta}\omega'_\beta)x'_{i\alpha}\Big] \\ &= \omega'_\beta \sum_{i=1}^{N} m_i(x'_{i\gamma}x'_{i\gamma}\delta_{\alpha\beta} - x'_{i\alpha}x'_{i\beta}) \qquad (\alpha,\beta,\gamma = 1,2,3), \end{aligned}$$

where $x'_{i\alpha}$ are the components of $\mathbf{r}'_i$ on the axes of S'. If we denote

$$I'_{\alpha\beta} = \sum_{i=1}^{N} m_i(x'_{i\gamma}\,x'_{i\gamma}\,\delta_{\alpha\beta} - x'_{i\alpha}\,x'_{i\beta}), \tag{4.5.5}$$

we can write:

$$L'_\alpha = I'_{\alpha\beta}\,\omega'_\beta \quad (\alpha,\beta = 1,2,3). \tag{4.5.6}$$

The components L'_α of the angular momentum on the axes of the frame S' are therefore obtained by means of a linear transformation. The quantities (4.5.5) are the components of a tensor, called the *inertia tensor*. From (4.5.5) it follows that the inertia tensor is symmetric. Its diagonal components $I'_{11}, I'_{22}, I'_{33}$, e.g.

$$I'_{11} = \sum_{i=1}^{N} m_i \left(|\mathbf{r}_i|^2 - x'^2_i\right) = \sum_{i=1}^{N} m_i \left(y'^2_i + z'^2_i\right) \text{ etc.,} \tag{4.5.7}$$

are the *axial moments of inertia*, while the non-diagonal components I'_{12}, I'_{23}, I'_{31}, e.g.

$$I'_{12} = I'_{21} = -\sum_{i=1}^{N} m_i x'_i y'_i \text{ etc.,} \tag{4.5.8}$$

are the *centrifugal momenta* or *inertia products*.

Observation: Although the inertia tensor $I'_{\alpha\beta}$ was defined relative to the centre of mass O, sometimes it can be found in a more convenient way by first calculating its components with respect to some other point. For instance, let us now choose a reference frame $\tilde{S}$, invariable with respect to S' and with the origin at $\tilde{O}$, defined by the radius-vector $\mathbf{R}(X', Y', Z')$ relative to O'. In order to determine the relationship between the components of the tensor $I_{\alpha\beta}$, given in the two frames $\tilde{S}$ and S', we denote $\tilde{\mathbf{r}}_i = \mathbf{r}_i - \mathbf{R}$. Then,

$$\begin{aligned}\tilde{I}_{\alpha\beta} &= \sum_{i=1}^{N} m_i (\tilde{x}_{i\gamma}\tilde{x}_{i\gamma}\delta_{\alpha\beta} - \tilde{x}_{i\alpha}\tilde{x}_{i\beta}) \\ &= \sum_{i=1}^{N} m_i \left[(x'_{i\gamma} - X'_\gamma)(x'_{i\gamma} - X'_\gamma)\delta_{\alpha\beta} - (x'_{i\alpha} - X'_\alpha)(x'_{i\beta} - X'_\beta)\right].\end{aligned}$$

But, in view of (1.3.54), since $O' \equiv G$, we have:

$$\sum_{i=1}^{N} m_i x'_{i\gamma} = 0;$$

consequently

$$\tilde{I}_{\alpha\beta} = I'_{\alpha\beta} + M(R^2\delta_{\alpha\beta} - X'_\alpha X'_\beta), \tag{4.5.9}$$

where M is the mass of the body. If $\tilde{I}_{\alpha\beta}$ is known, we can immediately determine $I'_{\alpha\beta}$.

4.5.3 Kinetic Energy

Using (4.2.14), we find the kinetic energy:

$$T = \frac{1}{2}\sum_{i=1}^{N} m_i |\mathbf{v}_i|^2 = \frac{1}{2}\sum_{i=1}^{N} m_i \mathbf{v}_i \cdot (\boldsymbol{\omega} \times \mathbf{r}_i) = \frac{1}{2}\boldsymbol{\omega} \cdot \sum_{i=1}^{N} m_i \mathbf{r}_i \times \mathbf{v}_i,$$

or, by virtue of (4.5.4) and (4.5.6),

$$T = \frac{1}{2}\boldsymbol{\omega} \cdot \mathbf{L} = \frac{1}{2} I'_{\alpha\beta}\omega'_\alpha\omega'_\beta. \tag{4.5.10}$$

The tensor $I'_{\alpha\beta}$ being given by (4.5.5), we can also write:

$$T = \frac{1}{2}\sum_{i=1}^{N} m_i \left[|\mathbf{r}_i|^2 \omega^2 - (\mathbf{r}'_i \cdot \boldsymbol{\omega})^2 \right],$$

or, if $\mathbf{s}$ is the unit vector of $\boldsymbol{\omega}$,

$$T = \frac{1}{2} I \omega^2, \tag{4.5.11}$$

where

$$I = \sum_{i=1}^{N} m_i \left[|\mathbf{r}_i|^2 - (\mathbf{r}_i \cdot \mathbf{s})^2 \right] \tag{4.5.12}$$

is a scalar, called the *moment of inertia* of the rigid body relative to the axis of rotation. As one can see (Fig. 4.10), $d_i^2 = r_i^2 - (\mathbf{r}_i \cdot \mathbf{s})^2$, i.e.

$$I = \sum_{i=1}^{N} m_i d_i^2. \tag{4.5.13}$$

Comparing (4.5.6) and (4.5.10), we obtain

$$L'_\alpha = \frac{\partial T}{\partial \omega'_\alpha} \quad (\alpha = 1, 2, 3). \tag{4.5.14}$$

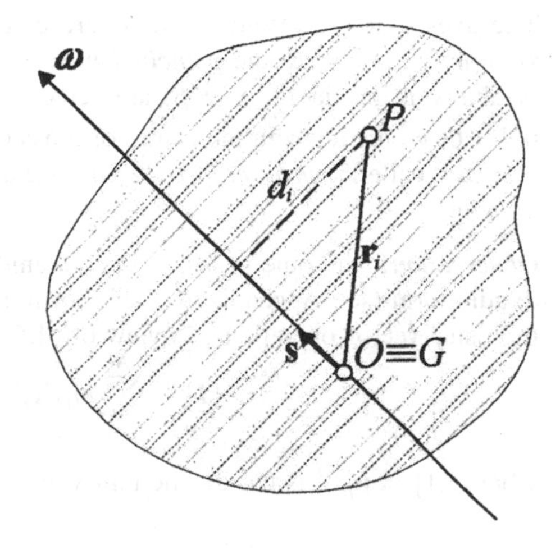

Fig. 4.10 Auxiliary construction used to determine the specific quantities associated with a rigid body: the kinetic energy and the ellipsoid of inertia.

4.5.4 Ellipsoid of Inertia

The moment of inertia (4.5.12) can be expressed in a form leading to a remarkable and useful geometrical interpretation. If s'_α $(\alpha = 1, 2, 3)$ are the direction cosines of $\mathbf{s}$ relative to S', we may write $\mathbf{s} = s'_\alpha \mathbf{u}'_\alpha$, such that

$$|\mathbf{s}|^2 = 1 = s'_\alpha s'_\beta \mathbf{u}'_\alpha \cdot \mathbf{u}'_\beta = s'_\alpha s'_\beta \delta_{\alpha\beta},$$

leading to

$$I = \sum_{i=1}^{N} m_i (|\mathbf{r}_i|^2 s'_\alpha s'_\beta \delta_{\alpha\beta} - s'_\alpha s'_\beta x'_{i\alpha} x'_{i\beta}) = I'_{\alpha\beta} s'_\alpha s'_\beta. \quad (4.5.15)$$

We take now a point M on the axis of rotation, given by

$$\vec{OM} = \frac{\mathbf{s}}{\sqrt{I}}.$$

If X'_α $(\alpha = 1, 2, 3)$ are the components of $\vec{OM}$ on the axes of S', then (4.5.15) leads to

$$I'_{\alpha\beta} X'_\alpha X'_\beta = 1. \quad (4.5.16)$$

This formula tells us that the geometric locus of the point M, when the direction of the axis of rotation varies in time, is a quadric surface with its centre in O'. Since $I' > 0$, the segment OM is always finite; in other words, the quadric surface does not have points at infinity. This means that our quadric is an ellipsoid, called *ellipsoid of inertia*. If the coordinate axes coincide with the symmetry axes of the ellipsoid of inertia, then its equation has the *canonical* form:

$$I'_{11} X'^2_1 + I'_{22} X'^2_2 + I'_{33} X'^2_3 = 1.$$

The axes of the ellipsoid of inertia are called *principal axes of inertia*, while its symmetry planes are the *principal planes of inertia*. Relative to these axes, the products of inertia $I'_{\alpha\beta}$ $(\alpha \neq \beta)$ are zero, i.e. the inertia tensor is diagonal. If the point O' coincides with the centre of mass of the body (our case), the ellipsoid of inertia is called *central ellipsoid of inertia* and its axes – *central principal axes of inertia*.

Observation: In some cases, it is convenient to write the tensor of inertia in a dyadic form (see Appendix A). This form is useful because it allows us to utilize the usual vector operations. In view of (4.5.5), we define the dyadic vector $\{I\}$ by

$$\{I\} = \sum_{i=1}^{N} m_i (|\mathbf{r}_i|^2 \{1\} - \mathbf{r}_i \mathbf{r}_i),$$

where $\{1\} = \mathbf{u}'_i \mathbf{u}'_i$ is the dyadic unit vector. Then,

$$\{I\} \cdot \boldsymbol{\omega} = \sum_{i=1}^{N} m_i \left[|\mathbf{r}_i|^2 \boldsymbol{\omega} - (\mathbf{r}_i \cdot \boldsymbol{\omega}) \mathbf{r}_i \right] = \mathbf{L},$$

$$T = \frac{1}{2} \boldsymbol{\omega} \cdot \mathbf{L} = \frac{1}{2} \boldsymbol{\omega} \cdot \{I\} \cdot \boldsymbol{\omega} = \frac{1}{2} \omega^2 \mathbf{s} \cdot \{I\} \cdot \mathbf{s}.$$

Since

$$\mathbf{s} \cdot \{I\} \cdot \mathbf{s} = \mathbf{s} \cdot \sum_{i=1}^{N} \left[|\mathbf{r}_i|^2 \mathbf{s} - (\mathbf{r}_i \cdot \mathbf{s}) \mathbf{r}_i \right] = \sum_{i=1}^{N} m_i [|\mathbf{r}_i|^2 - (\mathbf{r}_i \cdot \mathbf{s})^2] = I,$$

we finally obtain

$$T = \frac{1}{2} I \omega^2.$$

4.5.5 Euler's Equations of Motion

First, we notice that the coordinates $x'_{i\alpha}(i = \overline{1, N}; \alpha = \overline{1, 3})$ of the particles of the rigid body relative to S' do not change in time, meaning that the components $I'_{\alpha\beta}$ of the inertia tensor are also constant. Second, we recall that $\boldsymbol{\omega}$ is a function of time only: $\boldsymbol{\omega} = \boldsymbol{\omega}(t)$. In order to derive the differential equations of motion of the rigid body having one point fixed, we shall use the Lagrange equations (2.5.13):

$$\frac{d}{dt} \left(\frac{\partial T}{\partial \dot{q}_\gamma} \right) - \frac{\partial T}{\partial q_\gamma} = Q_\gamma \quad (\gamma = 1, 2, 3), \tag{4.5.17}$$

where the role of general coordinates is played by Euler's angles: $q_1 = \theta$, $q_2 = \varphi$, $q_3 = \psi$. Such a choice is dictated by the fact that the unit vectors $\mathbf{t}_1$, $\mathbf{k}$, $\mathbf{k}'$ (in this order!) form a clockwise orthogonal frame. By virtue of (4.5.10), we can write:

$$\frac{\partial T}{\partial \dot{q}_\gamma} = \frac{1}{2} I'_{\eta\beta} \omega'_\eta \frac{\partial \omega'_\beta}{\partial \dot{q}_\gamma} + \frac{1}{2} I'_{\eta\beta} \frac{\partial \omega'_\eta}{\partial \dot{q}_\gamma} \omega'_\beta = I'_{\eta\beta} \omega'_\beta \frac{\partial \omega'_\eta}{\partial \dot{q}_\gamma},$$

$$\frac{d}{dt} \left(\frac{\partial T}{\partial \dot{q}_\gamma} \right) = I'_{\eta\beta} \dot{\omega}'_\beta \frac{\partial \omega'_\eta}{\partial \dot{q}_\gamma} + I'_{\eta\beta} \omega'_\beta \frac{d}{dt} \left(\frac{\partial \omega'_\eta}{\partial \dot{q}_\gamma} \right),$$

$$\frac{\partial T}{\partial q_\gamma} = I'_{\eta\beta} \omega'_\beta \frac{\partial \omega'_\eta}{\partial q_\gamma}$$

and Eqs. (4.5.17) lead to:

$$I'_{\eta\beta} \dot{\omega}'_\beta \frac{\partial \omega'_\eta}{\partial \dot{q}_\gamma} + I'_{\eta\beta} \omega'_\beta \left[\frac{d}{dt} \left(\frac{\partial \omega'_\eta}{\partial \dot{q}_\gamma} \right) - \frac{\partial \omega'_\eta}{\partial q_\gamma} \right] = Q_\gamma, \tag{4.5.18}$$

where Q_γ are the components of the generalized force along the directions θ, φ, ψ. Using the definition (2.4.12), we obtain:

$$Q_\gamma = \sum_{i=1}^{N} \mathbf{F}_i \cdot \frac{\partial \mathbf{r}_i}{\partial q_\gamma} = \sum_{i=1}^{N} F'_{i\alpha} \frac{\partial x'_{i\alpha}}{\partial q_\gamma} = \sum_{i=1}^{N} F'_{i\alpha} \frac{\partial \dot{x}'_{i\alpha}}{\partial \dot{q}_\gamma}.$$

But

$$\dot{x}'_{i\alpha} = (\boldsymbol{\omega} \times \mathbf{r}_i)_\alpha = \epsilon_{\alpha\beta\gamma}\, \omega'_\beta\, x'_{i\gamma},$$

such that

$$Q_\gamma = \sum_{i=1}^{N} \epsilon_{\alpha\beta\xi} F'_{i\alpha} x'_{i\xi} \frac{\partial \omega'_\beta}{\partial \dot{q}_\gamma}. \tag{4.5.19}$$

On the other hand, the components $\mathcal{M}'_\beta$ $(\beta = 1, 2, 3)$ of the resultant moment of exterior forces relative to S' are, by definition,

$$\mathcal{M}'_\beta = \sum_{i=1}^{N} (\mathbf{r}_i \times \mathbf{F}_i)_\beta = \sum_{i=1}^{N} \epsilon_{\beta\xi\alpha}\, x'_{i\xi}\, F'_{i\alpha}. \tag{4.5.20}$$

Comparing the relations (4.5.19) and (4.5.20), we arrive at

$$Q_\gamma = \mathcal{M}'_\beta \frac{\partial \omega'_\beta}{\partial \dot{q}_\gamma}. \tag{4.5.21}$$

Before going further, we must remember that in (4.5.21) Q_γ are the components $Q_\theta,\ Q_\varphi,\ Q_\psi$, while $\mathcal{M}'_\beta$ stands for $\mathcal{M}_{x'},\ \mathcal{M}_{y'},\ \mathcal{M}_{z'}$.

It can be proved that the vector $\boldsymbol{\omega}$ defined by (4.5.1) satisfies the identity

$$\frac{d}{dt}\left(\frac{\partial \boldsymbol{\omega}}{\partial \dot{q}_\gamma}\right) = \mathbf{u}'_\beta \frac{\partial \omega'_\beta}{\partial q_\gamma}. \tag{4.5.22}$$

If we denote

$$\mathbf{t}_1 = \mathbf{s}_1, \quad \mathbf{k} = \mathbf{s}_2, \quad \mathbf{k}' = \mathbf{s}_3,$$

then we may write $\boldsymbol{\omega} = \dot{q}_\alpha \mathbf{s}_\alpha$ and (4.5.22) leads to

$$\frac{d\mathbf{s}_\gamma}{dt} = \mathbf{u}'_\beta \frac{\partial \omega'_\beta}{\partial q_\gamma}. \tag{4.5.23}$$

Since $\mathbf{u}'_\beta$ $(\beta = 1, 2, 3)$ do not depend on $\dot{q}_\gamma$, from (4.2.7) and (4.5.23) we obtain:

$$\mathbf{u}'_\beta \frac{d}{dt}\left(\frac{\partial \omega'_\beta}{\partial \dot{q}_\gamma}\right) + \boldsymbol{\omega} \times \mathbf{u}'_\beta \frac{\partial \omega'_\beta}{\partial \dot{q}_\gamma} = \mathbf{u}'_\beta \frac{\partial \omega'_\beta}{\partial q_\gamma},$$

or, in projection on the axes of the frame S',

$$\frac{d}{dt}\left(\frac{\partial \omega'_\eta}{\partial \dot{q}_\gamma}\right) - \frac{\partial \omega'_\eta}{\partial q_\gamma} = \epsilon_{\eta\xi\alpha}\, \omega'_\alpha \frac{\partial \omega'_\xi}{\partial \dot{q}_\gamma}. \tag{4.5.24}$$

Replacing the square bracket in (4.5.18) by the r.h.s. of (4.5.24) and using (4.5.21), we arrive at

$$\left(I'_{\eta\beta}\, \dot{\omega}'_\beta + \epsilon_{\eta\xi\alpha}\, \omega'_\xi\, \omega'_\beta\, I'_{\alpha\beta} - \mathcal{M}'_\eta\right) \frac{\partial \omega'_\eta}{\partial \dot{q}_\gamma} = 0, \tag{4.5.25}$$

where the summation indices have been conveniently changed. It is easy to show that

$$\det\left(\frac{\partial \omega'_\beta}{\partial \dot{q}_\gamma}\right) = \sin\theta \neq 0 \quad (0<\theta<\pi),$$

meaning that the system of homogeneous linear equations (4.5.25), in which the number of equations equals the number of unknowns (i.e. the brackets) admits only the trivial solution, which is expressed as

$$I'_{\eta\beta}\, \dot{\omega}'_\beta + \epsilon_{\eta\xi\alpha}\, \omega'_\xi\, \omega'_\beta\, I'_{\alpha\beta} = \mathcal{M}'_\eta \quad (\alpha, \beta, \eta, \xi = 1, 2, 3). \tag{4.5.26}$$

These are the equations of motion of the rigid body, called *Euler's equations.*

Equations (4.5.2), together with (4.5.26), form a system of six first-order differential equations, with six unknowns $\omega'_1, \omega'_2, \omega'_3, \theta, \varphi, \psi$, leading to the expressions for Euler's angles as functions of time. The general integral of the system (4.5.26) depends on six arbitrary constants, which are determined from the initial conditions: at the initial time $t = 0$, the values of the variables are fixed to $\omega'_1(0), \omega'_2(0), \omega'_3(0), \theta(0), \varphi(0), \psi(0)$.

Using (4.5.2), we can obtain the components Q_γ of the generalized force in terms of Euler's angles:

$$\begin{aligned}
Q_\theta &= \mathcal{M}'_x \cos\psi - \mathcal{M}'_y \sin\psi, \\
Q_\varphi &= \mathcal{M}'_x \sin\theta \sin\psi + \mathcal{M}'_y \sin\theta \cos\psi + \mathcal{M}'_z \cos\theta, \\
Q_\psi &= \mathcal{M}'_z.
\end{aligned} \tag{4.5.27}$$

We remind the reader that $\mathcal{M}'_x, \mathcal{M}'_y, \mathcal{M}'_z$ mean, in fact, $\mathcal{M}_{x'}, \mathcal{M}_{y'}, \mathcal{M}_{z'}$.

If, in particular, the axes of the frame S' coincide with the symmetry axes of the ellipsoid of inertia, then the angular momentum, the kinetic energy and Euler's equations are given respectively by:

$$\mathbf{L} = \sum_{\alpha=1}^{3} I'_\alpha \omega'_\alpha \mathbf{u}'_\alpha, \tag{4.5.28}$$

$$T = \frac{1}{2} \sum_{\alpha=1}^{3} I'_\alpha \omega'^2_\alpha, \tag{4.5.29}$$

$$\begin{aligned} I'_1\dot{\omega}'_1 - (I'_2 - I'_3)\omega'_2\omega'_3 &= \mathcal{M}'_1, \\ I'_2\dot{\omega}'_2 - (I'_3 - I'_1)\omega'_3\omega'_1 &= \mathcal{M}'_2, \\ I'_3\dot{\omega}'_3 - (I'_1 - I'_2)\omega'_1\omega'_2 &= \mathcal{M}'_3, \end{aligned} \tag{4.5.30}$$

where the following notations have been used:

$$I'_{11} = I'_1, \quad I'_{22} = I'_2, \quad I'_{33} = I'_3. \tag{4.5.31}$$

If no torque is applied to the rigid body, or if **F** is permanently directed towards the fixed point O', then Euler's equations (4.5.26) reduce to

$$I'_{\eta\beta}\,\dot{\omega}'_\beta + \epsilon_{\eta\xi\alpha}\,\omega'_\xi\,\omega'_\beta\,I'_{\alpha\beta} = 0.$$

Euler's equations do not change their form if all quantities are expressed in the inertial frame S, i.e.

$$I_{\alpha\beta}\,\dot{\omega}_\beta + \epsilon_{\alpha\beta\gamma}\,\omega_\beta\,\omega_\eta\,I_{\gamma\eta} = \mathcal{M}_\alpha, \tag{4.5.32}$$

where the components $I_{\alpha\beta}$ of the inertia tensor are functions of time. To prove this statement, let us apply the angular momentum theorem (1.3.38). Since the unit vectors $\mathbf{u}_\alpha$ of the frame S are constant, we have:

$$\frac{dL_\alpha}{dt} = \mathcal{M}_\alpha.$$

On the other hand, since $L_\alpha = I_{\alpha\beta}\omega_\beta$, where

$$I_{\alpha\beta} = \sum_{i=1}^{N} m_i(x_{i\gamma}\,x_{i\gamma}\,\delta_{\alpha\beta} - x_{i\alpha}\,x_{i\beta}),$$

we may write:

$$\dot{L}_\alpha = \dot{I}_{\alpha\beta}\omega_\beta + I_{\alpha\beta}\dot{\omega}_\beta.$$

But (see Appendix A)

$$I_{\alpha\beta} = a_{\theta\alpha}\,a_{\xi\beta}\,I'_{\theta\xi},$$

therefore

$$\begin{aligned} \dot{I}_{\alpha\beta} &= (\dot{a}_{\theta\alpha}\,a_{\xi\beta} + a_{\theta\alpha}\,\dot{a}_{\xi\beta})I'_{\theta\xi} = (\dot{a}_{\theta\alpha}\,a_{\xi\beta}\,a_{\theta\gamma}\,a_{\xi\eta} + a_{\theta\alpha}\,\dot{a}_{\xi\beta}\,a_{\theta\gamma}\,a_{\xi\eta})I_{\gamma\eta} \\ &= \dot{a}_{\theta\alpha}\,a_{\theta\gamma}\,I_{\gamma\beta} + \dot{a}_{\xi\beta}\,a_{\xi\eta}\,I_{\alpha\eta}. \end{aligned}$$

On the other hand, since

$$\mathbf{u}'_\alpha = a_{\alpha\beta}\,\mathbf{u}_\beta,$$

we have:

$$\dot{\mathbf{u}}'_\alpha = \dot{a}_{\alpha\beta}\,\mathbf{u}_\beta = \dot{a}_{\alpha\beta}\,a_{\gamma\beta}\,\mathbf{u}'_\gamma.$$

A comparison of the last relation with (4.2.4) yields:

$$\omega'_{\alpha\gamma} = \dot{a}_{\alpha\beta}\,a_{\gamma\beta}$$

and, in view of (4.2.6),

$$\dot{I}_{\alpha\beta} = \omega_{\gamma\alpha}\,I_{\gamma\beta} + \omega_{\eta\beta}\,I_{\alpha\eta} = \epsilon_{\gamma\alpha\xi}\,\omega_\xi\,I_{\gamma\beta} + \epsilon_{\eta\beta\xi}\,\omega_\xi\,I_{\alpha\eta},$$

from which we obtain:

$$\dot{L}_\alpha = \epsilon_{\gamma\alpha\xi}\,\omega_\beta\,\omega_\xi\,I_{\gamma\beta} + I_{\alpha\beta}\,\dot{\omega}_\beta,$$

which completes the proof.

Observations:

(a) A rigid body whose principal moments of inertia are all different ($I'_1 \neq I'_2 \neq I'_3$, $I'_1 \neq I'_3$) is called *asymmetrical top*. If any two principal moments of inertia coincide (e.g. $I'_1 = I'_2 \neq I'_3$), we have a *symmetrical top*. In the case $I'_1 = I'_2 = I'_3$ we deal with a *spherical top*.

The determination of the principal moments and principal axes of inertia is facilitated if the rigid body has planes or axes of symmetry. If there is a plane of symmetry, then the centre of mass and two principal axes of inertia lie in this plane, the third being orthogonal to the symmetry plane. For instance, if a discrete system of N particles is distributed in the plane $O'x'y'$, we have:

$$I'_1 = \sum_{i=1}^{N} m_i\,y_i'^2,\ I'_2 = \sum_{i=1}^{N} m_i\,x_i'^2,\ I'_3 = \sum_{i=1}^{N} m_i(x_i'^2 + y_i'^2) = I'_1 + I'_2. \qquad (4.5.33)$$

If the rigid body possesses an axis of symmetry, this coincides with one of the principal axes of inertia and the centre of mass lies on it. If, in particular, the system of particles is distributed along a straight line (say, the $O'z'$-axis), then

$$I'_1 = I'_2 = \sum_{i=1}^{N} m_i\,z_i'^2,\ I'_3 = 0. \qquad (4.5.34)$$

Such a system is called *rotator* and has only two degrees of freedom.

(b) If the rigid body is subject to potential forces only, then Euler's equations (4.5.26) can be obtained by means of the Lagrange equations in the form (2.5.17). In this case,

$$L = T - V = \frac{1}{2} I'_{\alpha\beta}\,\omega'_\alpha\,\omega'_\beta - V(\theta, \varphi, \psi). \qquad (4.5.35)$$

(c) If we consider the general motion of a rigid body relative to the frame S, i.e. both translation and rotation are considered, then the Lagrangian will be composed of two parts:

$$L = L_{trans} + L_{rot}.$$

The kinetic energy can be calculated by using (4.3.10), with $\mathbf{v}_0 = \mathbf{v}_G$:

$$T = \frac{1}{2}M|\mathbf{v}_G|^2 + \frac{1}{2}I'_{\alpha\beta}\,\omega'_\alpha\,\omega'_\beta \tag{4.5.36}$$

and we finally can write

$$L = \frac{1}{2}M|\mathbf{v}_G|^2 + \frac{1}{2}I'_{\alpha\beta}\,\omega'_\alpha\,\omega'_\beta - V(x_G,\, y_G,\, z_G,\, \theta,\, \varphi,\, \psi). \tag{4.5.37}$$

We mention, once again, that this analysis is valid *only* if the origin O' of the reference frame S', rigidly fixed to the body, is chosen in its centre of mass.

4.6 Applications

The purpose of this section is to study, by use of the analytical mechanics formalism and the theory developed in this chapter, both some classical applications and several extensions of this formalism to non-mechanical systems. This will be done by virtue of the analogies that can be identified between mechanical models and other physical representations.

4.6.1 Physical Pendulum

A rigid body able to oscillate about a fixed horizontal axis, which does not pass through the centre of mass, is a *physical pendulum*, or a *compound pendulum* (Fig. 4.11). Since the only applied force acting on the body is the force of gravity, the projection of the equation $M\mathbf{g} = -\mathrm{grad}\, V$ on the axes of the frame S gives by integration the potential energy of the pendulum:

$$V = Mgl(1 - \cos\theta).$$

Here, $l = OG$, M is the mass of the body, while the integration constant is determined from the condition: $V = 0$, for $x = l$.

The instantaneous vector of rotation $\boldsymbol{\omega}$ is directed along the fixed axis and has the magnitude $|\boldsymbol{\omega}| = \omega_z = \dot{\theta}$. In view of (4.5.11), the kinetic energy is $T = \frac{1}{2}I\dot{\theta}^2$; thus, the Lagrangian of the physical pendulum reads:

$$L = \frac{1}{2}I\dot{\theta}^2 - Mgl(1 - \cos\theta). \tag{4.6.1}$$

Let θ be the generalized coordinate associated to the only degree of freedom of the body. Therefore we have a single equation of motion,

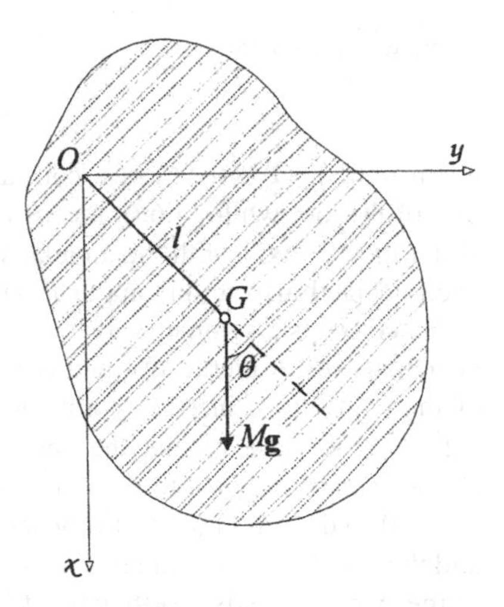

Fig. 4.11 Physical pendulum.

$$I\ddot{\theta} + Mgl\sin\theta = 0, \tag{4.6.2}$$

formally identical with the equation of a simple pendulum. For small amplitudes ($\sin\theta \simeq \theta$), the period of oscillation is

$$\tau = 2\pi\sqrt{\frac{I}{Mgl}}. \tag{4.6.3}$$

The physical pendulum is used in determining the acceleration of gravity, in the calculation of the moments of inertia of a rigid body, etc.

4.6.2 Symmetrical Top

Keeping the index 'prime' associated with the quantities in the frame S', let the symmetrical top be such that $I'_1 = I'_2 \neq I'_3$. In the following, we shall study the motion of the symmetrical top in two different cases: the force-free motion and the motion in the gravitational field of the Earth.

4.6.2.1 Force-Free Motion

The equations of motion of the rigid body are in this case Euler's equations (4.5.30), in which $\mathcal{M}'_\alpha = 0$. Since $I'_1 = I'_2$, the last equation gives $\omega_z = \text{const.}$, while the first two equations yield:

$$\dot{\omega}'_1 = -\Omega'\omega'_z, \quad \dot{\omega}'_2 = \Omega'\omega'_1, \tag{4.6.4}$$

where

$$\Omega' = \frac{I'_3 - I'_1}{I'_1}\omega'_3. \tag{4.6.5}$$

Then we can write

$$\dot{\boldsymbol{\omega}} = \boldsymbol{\Omega} \times \boldsymbol{\omega}. \tag{4.6.6}$$

From (4.6.5), $\boldsymbol{\Omega}(0, 0, \Omega')$ is a constant vector, directed along the symmetry axis Oz' of the top, while (4.6.6) means that the vector $\boldsymbol{\omega}$ performs a uniform rotation about the Oz'-axis, with the angular velocity Ω'. The convention $O \equiv O' \equiv G$ has been adopted also in this application.

Since Oz' is a symmetry axis, it is also a principal axis of inertia. The other two principal axes of inertia Ox' and Oy' can be arbitrarily chosen, because the section of the ellipsoid of inertia by a plane orthogonal to the axis Oz' is a circle. Since $\mathcal{M}'_\alpha = 0$, we observe that the angular momentum $\mathbf{L}$ is a constant vector. If we choose Ox' orthogonal to the plane determined by $\mathbf{L}$ and Oz', it results that $L'_1 \equiv L_{x'} = 0$, and from (4.5.28) it follows that $\omega'_1 = 0$. In other words, the vectors $\boldsymbol{\omega}, \mathbf{L}$ and the axis Oz' are coplanar. This shows that the velocity $\mathbf{v}_i$ of a certain particle P_i of the symmetry axis is orthogonal to this plane ($\mathbf{v}_i = \boldsymbol{\omega} \times \mathbf{r}_i$), meaning that the z'-axis performs a uniform rotation about $\mathbf{L}$. This motion is called *regular precession*.

Our result acquires a more suggestive interpretation if we make use of Euler's angles. Let the axis Oz of the fixed frame be directed along $\mathbf{L} = \text{const.}$, the Oz' axis along the top axis (as above), and the Ox'-axis along the line of nodes, i.e. $\psi = 0$ (Fig. 4.12). The components of the vector $\boldsymbol{\omega}$ in S' are then given by (4.5.2):

$$\omega'_1 = \dot{\theta}, \quad \omega'_2 = \dot{\varphi} \sin\theta, \quad \omega'_3 = \dot{\varphi}\cos\theta + \dot{\psi}. \tag{4.6.7}$$

The angular momentum is in the $Oy'z'$-plane, so that

$$L'_1 = 0, \quad L'_2 = L \sin\theta, \quad L'_3 = L\cos\theta, \tag{4.6.8}$$

where $L = L_z$. In view of (4.5.28), we also have:

$$\begin{aligned} L'_1 &= I'_1\omega'_1 = I'_1\dot{\theta}, \quad L'_2 = I'_2\omega'_2 = I'_1\dot{\varphi}\sin\theta, \\ L'_3 &= I'_3\omega'_3 = I'_3(\dot{\varphi}\cos\theta + \dot{\psi}). \end{aligned} \tag{4.6.9}$$

A comparison between (4.6.8) and (4.6.9) gives:

$$\dot{\theta} = 0, \quad \dot{\varphi} = \frac{L}{I'_1}, \quad \omega'_3 = \dot{\varphi}\cos\theta + \dot{\psi} = \frac{L\cos\theta}{I'_3}. \tag{4.6.10}$$

The first relation shows that the angle θ between the directions of the top axis and the angular momentum is constant; the second, that the precession angular velocity $\dot{\varphi}$ of the top axis about Oz is also constant; the third relation defines the angular velocity of the top about its axis of symmetry, which is constant as well.

The same result is obtained by integrating Euler's equations (4.5.30). In this respect, we see that the system (4.6.4) admits the solutions

$$\omega'_1 = A\sin\Omega' t, \quad \omega'_2 = A\cos\Omega' t,$$

meaning that the solutions of Euler's equations are:

$$\omega_1' = A \sin \Omega' t = \dot{\varphi} \sin \theta \sin \psi,$$
$$\omega_2' = A \cos \Omega' t = \dot{\varphi} \sin \theta \cos \psi,$$
$$\omega_3' = \dot{\varphi} \cos \theta + \dot{\psi} = \text{const.},$$

leading to the already known result:

$$A = \dot{\varphi} \sin \theta = \text{const.}, \quad \psi = \Omega' t. \tag{4.6.11}$$

Observation: The geometric locus of the instantaneous axes of rotation with respect to S' is a cone with its top at O, called *polhodic cone*; the geometric locus of the instantaneous axes of rotation relative to S is also a cone with the top at the same point, named *herpolhodic cone*. It can be shown that the two cones are tangent, and the polhodic cone is rolling without slipping over the herpolhodic cone (Fig. 4.12).

4.6.2.2 Motion in the Gravitational Field

Suppose that the fixed point of the top is on its axis of symmetry. We choose the origins O and O' in the fixed point, and the axis Oz' directed along the symmetry axis of the top. As one can see, in this case the centre of mass and the fixed point do not coincide anymore (Fig. 4.13). Then, if the principal moments of inertia $\tilde{I}_1, \tilde{I}_2, \tilde{I}_3$ relative to a frame fixed with respect to the rigid and having its origin in G are known, from (4.5.9) we obtain:

$$I_1' = I_2' = \tilde{I}_1 + Ml^2, \quad I_3' = \tilde{I}_3, \tag{4.6.12}$$

where M is the mass of the top and l is the distance between the fixed point O and the centre of mass G.

The kinetic energy of the moving top is found by means of (4.5.29):

$$T = \frac{1}{2} \sum_{\alpha=1}^{3} I_\alpha' \omega_\alpha'^2 = \frac{1}{2} I_1' (\omega_1'^2 + \omega_2'^2) + \frac{1}{2} I_3' \omega_3'^2,$$

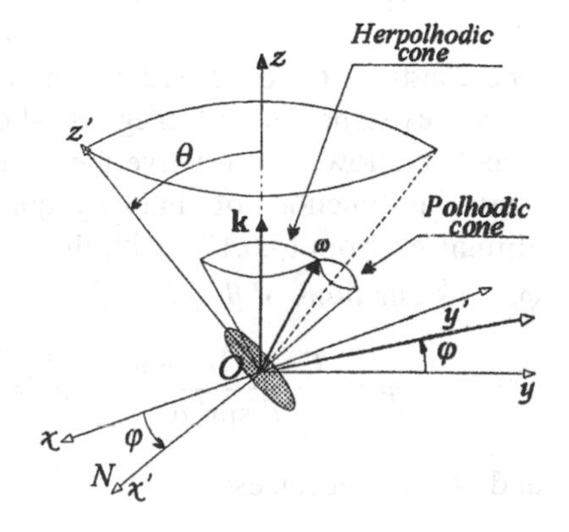

Fig. 4.12 Symmetrical top. *Herpolhodic* and *polhodic* cones.

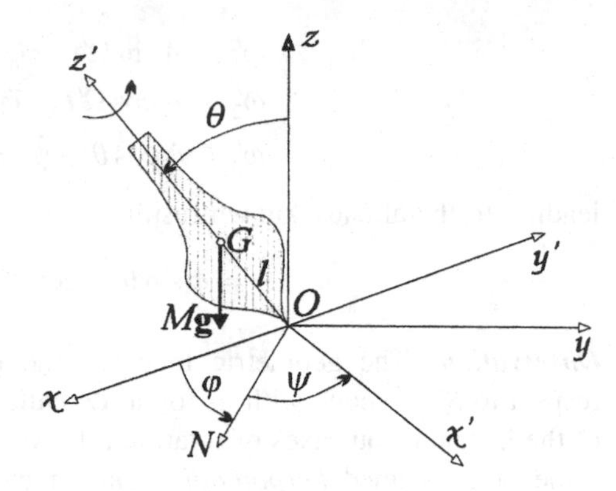

Fig. 4.13 Motion of the symmetrical top in the gravitational field.

or, in view of (4.5.2),

$$T = \frac{1}{2} I_1' (\dot{\varphi}^2 \sin^2 \theta + \dot{\theta}^2) + \frac{1}{2} I_3' (\dot{\varphi} \cos \theta + \dot{\psi})^2. \tag{4.6.13}$$

The potential energy of the body situated in the terrestrial gravitational field is

$$V = Mgl \cos \theta.$$

The Lagrangian is then:

$$L = \frac{1}{2} I_1' (\dot{\varphi}^2 \sin^2 \theta + \dot{\theta}^2) + \frac{1}{2} I_3' (\dot{\varphi} \cos \theta + \dot{\psi})^2 - Mgl \cos \theta. \tag{4.6.14}$$

The two cyclic coordinates φ, ψ lead to the following first integrals:

$$p_\varphi = \frac{\partial L}{\partial \dot{\varphi}} = (I_1' \sin^2 \theta + I_3' \cos^2 \theta) \dot{\varphi} + I_3' \dot{\psi} \cos \theta = C_1, \tag{4.6.15}$$

$$p_\psi = \frac{\partial L}{\partial \dot{\psi}} = I_3' (\dot{\varphi} \cos \theta + \dot{\psi}) = I_3' \omega_3' = C_2. \tag{4.6.16}$$

Since the physical system is conservative, there also exists the energy first integral:

$$\frac{1}{2} I_1' (\dot{\varphi}^2 \sin^2 \theta + \dot{\theta}^2) + \frac{1}{2} I_3' (\dot{\varphi} \cos \theta + \dot{\psi})^2 + Mgl \cos \theta = E. \tag{4.6.17}$$

The constants C_1, C_2, E are determined from the initial conditions.

The existence of first integrals whose number is equal to that of the degrees of freedom, allows us to derive the finite equations of motion (in our case, Euler's angles as functions of time) by quadratures. The cyclic variables φ, ψ can be eliminated from (4.6.17), with the help of (4.6.15) and (4.6.16). Then we obtain $\dot{\varphi}, \dot{\psi}$ as functions of θ:

$$\dot{\varphi} = \frac{C_1 - C_2 \cos \theta}{I_1' \sin^2 \theta}, \quad \dot{\psi} = \frac{C_2}{I_3'} - \frac{C_1 - C_2 \cos \theta}{I_1' \sin^2 \theta} \cos \theta \tag{4.6.18}$$

and (4.6.17) becomes:

$$E = \frac{1}{2} I_1' \dot{\theta}^2 + \frac{1}{2} \frac{C_2^2}{I_3'} + \frac{1}{2I_1' \sin^2\theta} (C_1 - C_2 \cos\theta)^2 + Mgl \cos\theta. \tag{4.6.19}$$

Introducing the notation

$$E - \frac{1}{2} \frac{C_2^2}{I_3'} = E_1 = \text{const.}, \tag{4.6.20}$$

$$\frac{1}{2I_1' \sin^2\theta} (C_1 - C_2 \cos\theta)^2 + Mgl \cos\theta = V_{eff}(\theta), \tag{4.6.21}$$

we have:

$$E_1 = \frac{1}{2} I_1' \dot{\theta}^2 + V_{eff}(\theta), \tag{4.6.22}$$

or, by integration,

$$\int_{t_0}^{t} dt = \int_{\theta_0}^{\theta} \frac{d\theta}{\sqrt{\frac{2}{I_1'} [E_1 - V_{eff}(\theta)]}}. \tag{4.6.23}$$

This equation gives $\theta = \theta(t)$. Introducing then $\theta(t)$ into (4.6.18), we find $\varphi = \varphi(t)$ and $\psi = \psi(t)$. This means that the problem of the motion of the rigid body – at least in principle – is determined. But the study of the potential energy $V_{eff}(\theta)$ and the resolution of the elliptic integral (4.6.23) encounter great difficulties, since $V_{eff}(\theta)$ depends on two parameters, C_1 and C_2, whose values are not *a priori* known. In the following, we shall give a brief account of the problem, considering only the most interesting cases.

First, we see that the integral on the r.h.s. of (4.6.23) acquires a more convenient form by the substitution $u = \cos\theta$. Second, let us denote

$$\alpha = \frac{2E_1}{I_1'}, \quad \beta = \frac{2Mgl}{I_1'} \tag{4.6.24}$$

and choose the constants a, b given by

$$C_1 = I_1' b, \quad C_2 = I_1' a, \tag{4.6.25}$$

instead of C_1, C_2. Then, (4.6.23) becomes:

$$t - t_0 = \int_{u(t_0)}^{u(t)} \frac{du}{\sqrt{(1-u^2)(\alpha - \beta u) - (b - au)^2}} = \int_{u(t_0)}^{u(t)} \frac{du}{\sqrt{f(u)}}, \tag{4.6.26}$$

where $f(u)$ is the polynomial under the square root.

The polynomial $f(u)$ is of third degree in u, meaning that the integral occurring in (4.6.26) is an elliptic integral. We shall use, however, a method to solve the problem without resorting to elliptic functions. Let us write

$$\dot{u}^2 = f(u) = (1-u^2)(\alpha - \beta u) - (b - au)^2. \quad (4.6.27)$$

In order that Eq. (4.6.27) has a real solution, it is necessary that $f(u) \geq 0$. This condition must also be fulfilled by $u_0 = u(t_0)$, i.e. we must have $f(u_0) \geq 0$. Let us analyze the two cases, $f(u_0) > 0$ and $f(u_0) = 0$.

(a) $f(u_0) > 0$. Let u in (4.6.27) vary in time. We find:

$$f(-\infty) < 0, \quad f(-1) < 0, \quad f(+1) < 0, \quad f(+\infty) > 0.$$

Since $u_0 = \cos\theta_0$ takes values in the interval $(-1, +1)$, the polynomial $f(u)$ changes its sign in the intervals

$$(-1, u_0), \quad (u_0, +1), \quad (+1, +\infty).$$

The three real roots of $f(u)$ will then be situated in these intervals:

$$u_1 \in (-1, u_0), \quad u_2 \in (u_0, +1), \quad u_3 \in (+1, +\infty).$$

Since the root u_3 does not make sense, we shall leave it out. The graphic representation of the case (a) is given in Fig. 4.14.

(b) $f(u_0) = 0$. In this case, we have $u_0 = u_1 = u_2$, the solutions $u = +1$ and $u = -1$ giving the vertical position of the top. It follows, then, that $u = \cos\theta$ is permanently within the interval (u_1, u_2), i.e. the inclination limit angles of the axis of the top with respect to the vertical line are $\theta_1 = \arccos u_1$ and $\theta_2 = \arccos u_2$. These two limit angles determine two circles on a sphere with the centre in the fixed point, so that the trajectory described by the intersection point of the top axis and the sphere lies between the two circles. One observes that

$$\dot{u}^2 = f(u) = \dot{\theta}^2 \sin\theta,$$

meaning that $\dot{\theta} = 0$ on the two circles, which correspond to the turning points. In order to determine the shape of the curve described by the point produced by the intersection of the top axis and the sphere with its centre in O, let us calculate the tangent of the angle α made by this curve and the meridian circle defined on the sphere by the plane $\varphi = \text{const}$. Thus, we have (see Appendix B)

$$ds_\theta = r\, d\theta, \quad ds_\varphi = r\, \sin\theta\, d\varphi;$$

consequently (see Fig. 4.15),

$$\tan\alpha = \frac{d\varphi}{d\theta} \sin\theta.$$

But

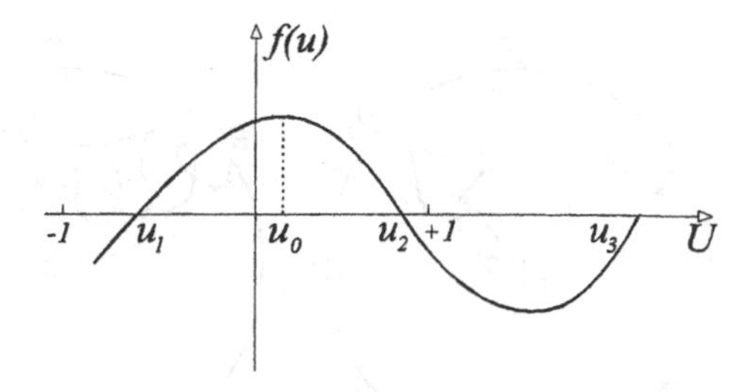

Fig. 4.14 Graphical representation of the case (a) $f(u_0) > 0$.

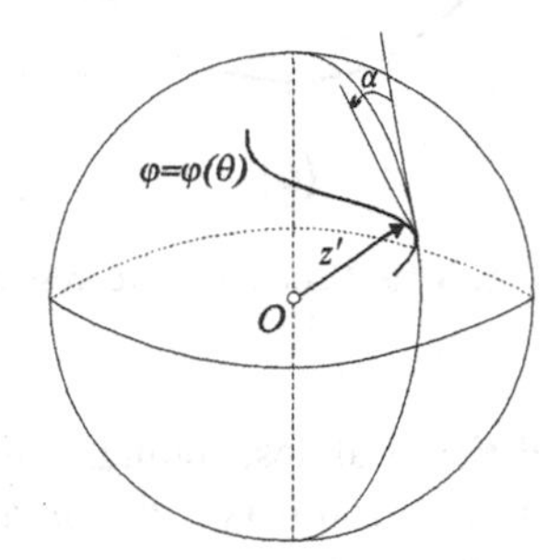

Fig. 4.15 Parameters describing the case (b) $f(u_0) = 0$.

$$\frac{d\varphi}{d\theta} = \frac{d\varphi}{du}\frac{du}{d\theta} = -\frac{d\varphi}{du}\sin\theta.$$

On the other hand, in view of (4.6.18), (4.6.25) and (4.6.27), we can write

$$\frac{d\varphi}{du} = \frac{\dot{\varphi}}{\dot{u}} = \frac{b - au}{\pm(1 - u^2)\,[f(u)]^{\frac{1}{2}}},$$

leading to

$$\tan\alpha = \frac{b - au}{\pm\,[f(u)]^{\frac{1}{2}}}. \qquad (4.6.28)$$

Let u' be the root of the equation $b - au = 0$. Then, depending on the initial conditions, one of the following situations can occur:

(1) u' is outside the interval $[u_1, u_2]$;
(2) u' is inside this interval (u_1, u_2);
(3) u' coincides with either u_1, or u_2.

In the first case, if we write $(4.6.18)_1$ in the form

$$\dot{\varphi} = \frac{b - au}{1 - u^2},$$

we realize that $\dot{\varphi}$ does not change its sign, meaning that the axis of the top performs a *precession motion* about the vertical line passing through O. If $\dot{\varphi} = 0$, the precession is *monotonic*. When u reaches the values u_1 or u_2, in view of (4.6.28), $\tan\,\alpha = \infty$, i.e. the curve $\varphi = \varphi(\theta)$ on the sphere is tangent to the parallel circles $\theta = \theta_1,\ \theta = \theta_2$. The axis of the top performs a periodic motion of

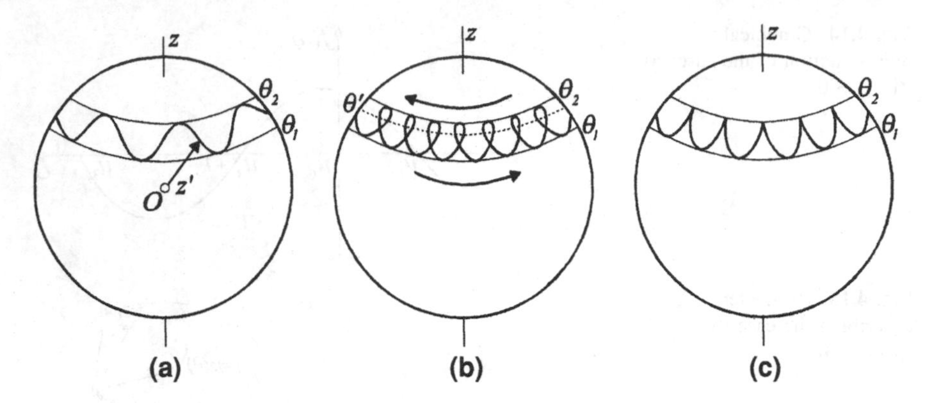

Fig. 4.16 Geometrical representation of *nutation* for various conditions.

lifting and descending. This motion is done on the background of the precession motion and it is called *nutation* (Fig. 4.16a).

In the second case, the curve $\varphi = \varphi(\theta)$ is also tangent to the parallel circles $\theta = \theta_1$, $\theta = \theta_2$, but $\dot{\varphi}$ changes its sign for $u = u'$, meaning that the directions of the precession on the circles $\theta = \theta_1$, $\theta = \theta_2$ are opposite (Fig. 4.16b).

Let us, finally, analyze the third case. Putting $b = au'$ in (4.6.27), we have:

$$f(u) = (1 - u^2)(\alpha - \beta u) - a^2(u' - u)^2 \tag{4.6.29}$$

and, since either $u' = u_1$, or $u' = u_2$, it follows that

$$f(u') = (1 - u'^2)(\alpha - \beta u') = 0.$$

This leads to $\alpha = \beta u'$, and (4.6.29) takes the form:

$$f(u) = (u' - u)[\beta(1 - u^2) - a^2(u' - u)]. \tag{4.6.30}$$

Writing (4.6.30) for the two roots u_1 and u_2, we have:

$$f(u_1) = (u' - u_1)[\beta(1 - u_1^2) - a^2(u' - u_1)] = 0,$$
$$f(u_2) = (u' - u_2)[\beta(1 - u_2^2) - a^2(u' - u_2)] = 0.$$

As we can see, if in one of these equations the parenthesis vanishes, in the other the bracket becomes zero. Since $1 - u^2 > 0$, it results that either $u' > u_1$, or $u' > u_2$. But $u' > u_2$ does not have any sense, so that the only remaining possibility is $u' > u_1$. This inequality is satisfied only by $u' = u_2$, i.e. u' may coincide *only* with the biggest of the two roots u_1, u_2.

At the same time, (4.6.28) leads to $\tan \alpha = 0$, for $u = u' = u_2$, which means that the angle α, made by the curve $\varphi = \varphi(\theta)$ at the intersection points with the parallel circle $\theta = \theta_2$, is zero. Since $\dot{\varphi}$ does not change its sign in the interval (u_1, u_2), we conclude that these points are *turning points* (Fig. 4.16c).

In order to explain the motion of nutation, let us take as initial conditions:

$$\theta(0) = \theta_0 \neq 0, \pi, \quad \dot{\theta}(0) = 0, \quad \dot{\varphi}(0) = 0. \tag{4.6.31}$$

The relations (4.6.16), (4.6.19) and (4.6.20) then lead to

$$(E_1)_{t=0} = \left(E - \frac{1}{2} I'_3 \omega'^2_3 \right)_{t=0} = Mgl \cos\theta_0 = \text{const.} \tag{4.6.32}$$

In the light of this result, let us now give a physical interpretation of the equation of conservation (4.6.17). It shows that, if at the initial moment $t = 0$, the quantity E_1 satisfies Eq. (4.6.32), then the potential energy must diminish with the growth of $\dot{\varphi}, \dot{\theta}$. The angle θ_0 is then precisely the minimum value θ_2 of θ, i.e. the top always begins its motion with a falling tendency. The axis of self-rotation inclines until θ reaches its maximum value θ_1, then the top gradually recovers, the angle θ diminishes up to its minimum value θ_2, and the motion repeats itself in this manner, periodically. As we have already mentioned, the nutation is interwoven with the motion of precession.

4.6.3 *Fast Top. The Gyroscope*

A symmetrical top with a fixed point on its axis of symmetry, having a rapid motion of rotation about this axis, is called *gyroscope*. The initial conditions (4.6.18) and (4.6.31) yield

$$b - au_0 = 0. \tag{4.6.33}$$

Similarly, the relation (4.6.19) with the notations (4.6.20), (4.6.24) and (4.6.25) leads to

$$\alpha - \beta u_0 = 0. \tag{4.6.34}$$

By substituting b and α given by (4.6.33) and (4.6.34), respectively, into (4.6.27), we find that

$$f(u) = (u_0 - u)[\beta(1 - u^2) - a^2(u_0 - u)]. \tag{4.6.35}$$

We recall that u' is the root of the equation $b - au = 0$, so that (4.6.33) gives $u' = u_0$. For $u = u_1$ and $u = u_2$, we have $f(u) = 0$, i.e. u_0 equals either u_1, or u_2. Since $u_0 = u'$, the only possibility is $u_0 = u_2$ (the case (3) above). Then u may vary between the limits u_1 and $u_2 = u_0$.

If $u = u_1$, (4.6.35) yields:

$$u_0 - u_1 = \frac{\beta}{a^2}(1 - u^2) > 0.$$

Since $1 - u^2 > 0$, we can write

$$0 < u_0 - u_1 < \frac{\beta}{a^2}. \tag{4.6.36}$$

Suppose now that the angular velocity of the top about its axis is very high. In this case, $\frac{\beta}{a^2} \to 0$ and (4.6.36) shows that the domain of variation of u is very small. Then $u \simeq u_0$, the two parallel circles $\theta = \theta_1$ and $\theta = \theta_2$ coincide, and the top seems to move *without nutation*. From $(4.6.18)_1$ and (4.6.36), we obtain:

$$\dot{\varphi} = \frac{a(u_0 - u)}{1 - u^2} < \frac{\beta}{|a|} \frac{u_0 - u_1}{1 - u^2}. \tag{4.6.37}$$

Since $|a|$ is large and the difference $u_0 - u_1$ is very small, it follows that the motion of precession is *very slow*. The formula (4.6.37) also shows that $\dot{\varphi}$ and a have the same sign, i.e. the motions of precession and spinning about the top axis have the same sense.

Observation: This analysis justifies the names given to Euler's angles: θ – angle of nutation, φ – angle of precession and ψ – angle of self-rotation.

4.6.4 Motion of a Rigid Body Relative to a Non-inertial Frame. The Gyrocompass

Consider a rigid body moving with respect to a non-inertial frame $S'(O'x'y'z')$, which in its turn moves relative to an inertial frame $S(Oxyz)$. Let $S''(O''x''y''z'')$ be the orthogonal frame invariably related to the body, $\boldsymbol{\omega}$ – the instantaneous vector of rotation of the body about its own axis and $\boldsymbol{\Omega}$ – the instantaneous vector of rotation of S' relative to S. If $O'' \equiv G$ and we take into account the notations used in Fig. 4.17, we can write:

$$\mathbf{r}'_i = \mathbf{r}_G + \mathbf{r}''_i. \tag{4.6.38}$$

In order to study the motion of the body relative to S', we must first find the form the Lagrangian. To this end, we shall take advantage of the formula (4.3.4), furnishing the Lagrangian L_i of a certain particle P_i of the body with respect to the non-inertial frame S':

$$L_i = \frac{1}{2} m_i |\mathbf{v}_r|^2 + \frac{1}{2} m_i |\boldsymbol{\omega} \times \mathbf{r}'_i|^2 + m_i \mathbf{v}_r \cdot (\boldsymbol{\omega} \times \mathbf{r}'_i) - V_i. \tag{4.6.39}$$

Here, the term $-m_i \mathbf{r}'_i \cdot \mathbf{a}_0$ has been included in the potential energy V_i, because $\mathbf{a}_0$ can always be considered as due to a uniform gravitational field acting on the body, while $\mathbf{v}_r = \mathbf{v}_G + \boldsymbol{\omega} \times \mathbf{r}'_i$ is the velocity of P_i relative to O'. In view of (1.3.58) and (4.6.39), the Lagrangian separates into two groups of terms, one corresponding to the motion of the centre of mass and the other giving the rotation about it:

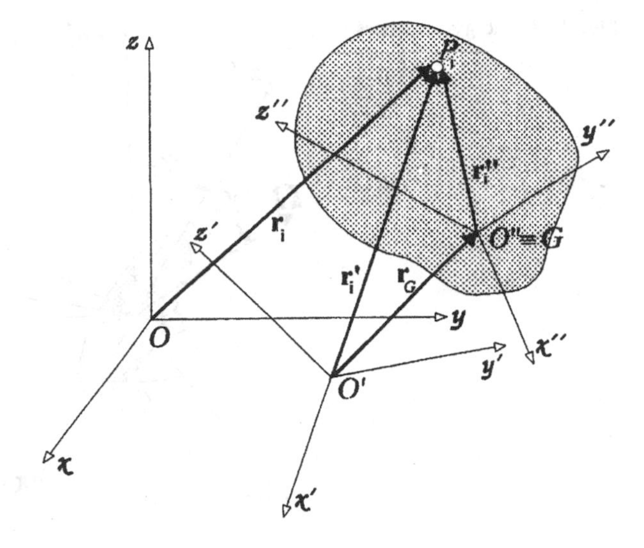

Fig. 4.17 Choice of the coordinate systems in order to study the motion of a rigid body relative to a non-inertial frame.

$$
\begin{aligned}
L = \sum_{i=1}^{N} L_i = T_G + \frac{1}{2} M(|\mathbf{r}_G|^2 \delta_{\alpha\beta} - x_{G\alpha}\, x_{G\beta})\Omega_\alpha\, \Omega_\beta \\
+ \boldsymbol{\Omega} \cdot \mathbf{L}_G + T_{rot} + \frac{1}{2} I_{\alpha\beta}\Omega_\alpha\Omega_\beta + \boldsymbol{\Omega} \cdot \mathbf{L}_{rot} - V,
\end{aligned}
\tag{4.6.40}
$$

where

$$
\begin{aligned}
I_{\alpha\beta} &= \sum_{i=1}^{N} m_i (x''_{i\gamma}\, x''_{i\gamma}\, \delta_{\alpha\beta} - x''_{i\alpha}\, x''_{i\beta}), \\
\mathbf{L}_{rot} &= \sum_{i=1}^{N} m_i \mathbf{r}''_i \times (\boldsymbol{\omega} \times \mathbf{r}''_i), \\
T_{rot} &= \frac{1}{2} I_{\alpha\beta}\, \omega_\alpha\, \omega_\beta,
\end{aligned}
$$

while V depends both on the centre of mass coordinates and on the parameters describing the motion about this point (Euler's angles). If a point belonging to the body, e.g. its centre of mass is fixed, then the Lagrangian associated with the motion about G is:

$$
L = T_{rot} + \frac{1}{2} I_{\alpha\beta}\, \Omega_\alpha\, \Omega_\beta + \boldsymbol{\omega} \cdot \mathbf{L}_{rot} - V, \tag{4.6.41}
$$

where the quantities T_{rot}, $I_{\alpha\beta}$, and $\mathbf{L}_{rot}$ are defined relative to the fixed point. The second term in (4.6.41) is called *centrifugal term* and the third – *Coriolis term.* Since the Lagrangian (4.6.41) is invariant under rotations, it will keep its form relative to the frames S and S'.

The aforementioned analysis represents the theoretical basis of the construction of the *gyrocompass*. This is a very rapidly rotating gyroscope, its symmetry axis

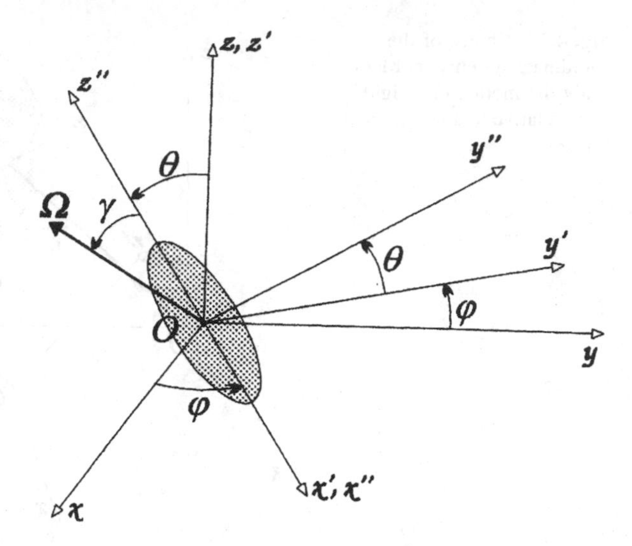

Fig. 4.18 The gyrocompass.

lying permanently in the horizontal plane. Without reducing the generality of the problem, we presume that the origins of the frames S, S', S'' coincide with the fixed point. Next, we take the frame S invariable with respect to the Earth, with its x-axis pointing North and its y-axis pointing West. The x'-axis is taken along the gyroscope axis, while Euler's angles φ, θ are chosen as shown in Fig. 4.18. Since the x'-axis remains permanently in the horizontal plane, we have $\psi = 0$.

It is more convenient to express all quantities relative to the frame S' (e.g. the ship, or the airplane where the gyrocompass is installed). In this frame, the vector $\boldsymbol{\omega} = \dot{\varphi}\mathbf{k} + \dot{\theta}\mathbf{i}'$ has the components:

$$\omega_1' = \dot{\theta}, \quad \omega_2' = 0, \quad \omega_3' = \dot{\varphi}. \tag{4.6.42}$$

Since Ox' is a principal axis of inertia, we have $I_2' = I_3' \neq I_1'$. The angular momentum and the kinetic energy of the gyroscope are given by (4.5.28) and (4.5.29), respectively:

$$L_1' = I_1'\dot{\theta}, \quad L_2' = 0, \quad L_3' = I_3'\dot{\varphi}, \tag{4.6.43}$$

$$T = \frac{1}{2}I_1'\dot{\theta}^2 + \frac{1}{2}I_2'\dot{\varphi}^2. \tag{4.6.44}$$

Let γ be the angle between the z-axis and the instantaneous vector of rotation $\boldsymbol{\Omega}$ of the Earth (Fig. 4.18). Then,

$$\Omega_1' = \Omega' \sin\gamma \cos\varphi, \quad \Omega_2' = -\Omega' \sin\gamma \sin\varphi, \quad \Omega_3' = \Omega' \cos\gamma,$$

such that

$$I'_{\alpha\beta}\,\Omega'_\alpha\,\Omega'_\beta = I'_2\,\Omega'^2\cos^2\gamma + \Omega'^2\sin^2\gamma(I'_1\cos^2\varphi + I'_2\sin^2\varphi),$$
$$\boldsymbol{\Omega}\cdot\mathbf{L}_{rot} = I'_1\,\dot\theta\,\Omega'\sin\gamma\cos\varphi + I'_2\,\dot\varphi\,\Omega'\cos\gamma.$$

Therefore, the corresponding Lagrangian is

$$L = \frac{1}{2}I'_1\dot\theta^2 + \frac{1}{2}I'_2\dot\varphi^2 + \frac{1}{2}\Omega'^2\sin^2\gamma(I'_1\cos^2\varphi + I'_2\sin^2\varphi)$$
$$+ I'_2\,\Omega'^2\cos^2\gamma + I'_1\,\dot\theta\,\Omega'\sin\gamma\cos\varphi + I'_2\,\dot\varphi\,\Omega'\cos\gamma, \qquad (4.6.45)$$

where the constant potential energy has been omitted.

We observe that the generalized coordinate θ is cyclic, so that we have the first integral

$$p_\theta = \frac{\partial L}{\partial\dot\theta} = I'_1(\dot\theta + \Omega'\sin\gamma\cos\varphi) = C_1. \qquad (4.6.46)$$

Another first integral follows from the fact that the energy is conserved, since the Lagrangian L does not depend explicitly on time. In view of (2.8.30), we may write:

$$p_\alpha\dot q_\alpha - L = \frac{\partial L}{\partial\dot\theta}\dot\theta + \frac{\partial L}{\partial\dot\varphi}\dot\varphi - L$$
$$= \frac{1}{2}I'_1\dot\theta^2 + \frac{1}{2}I'_2\dot\varphi^2 - \frac{1}{2}\Omega'^2\sin^2\gamma(I'_1\cos^2\varphi + I'_2\sin^2\varphi) - I'_2\Omega'^2\cos^2\gamma = C_2.$$
$$(4.6.47)$$

Using these two first integrals, let us write the differential equation of motion corresponding to the variable φ. To this end, we eliminate $\dot\theta$ from (4.6.46) and (4.6.47) and then take the time derivative of the result:

$$I'_2\ddot\varphi + C_1\Omega'\sin\gamma\sin\varphi - I'_2\Omega'^2\sin^2\gamma\sin\varphi\cos\varphi = 0,$$

where $\dot\varphi \neq 0$ has been simplified. If we neglect the term in Ω'^2 (recall that $\Omega' \simeq 7.27\times10^{-5}\,\mathrm{s}^{-1}$), we finally obtain:

$$\ddot\varphi + \frac{C_1\Omega'\sin\gamma}{I'_2}\sin\varphi = 0. \qquad (4.6.48)$$

This non-linear second-order differential equation is similar to the equation of a simple pendulum. Consequently, if $C_1 > 0$ and the gyroscope is initially oriented to show North, this is a position of stable equilibrium. Once directed towards North ($\varphi = 0$), the gyroscope axis tends to remain in this direction. Such a system is called *gyroscopic compass* or *gyrocompass*.

4.6.5 Motion of Rigid Bodies in Contact

Let us consider a moving rigid body, constrained to remain permanently in contact with another rigid body. In this case, there occurs a constraint force in each of the contact points. This force can be decomposed into two vector components: the *normal reaction* and the *force of friction*. The first is normal to the contact surface, while the second lies in the plane tangent to this surface.

The relative displacement of the two rigid bodies can be performed either by sliding, or by rolling. If the contact surfaces are perfectly smooth, we are dealing with a *pure sliding*, while in the case of perfectly rough surfaces, the displacement is a *pure rolling*. In practice, there is a mixture of these two cases.

The constraints imposed by the contact between bodies are, in general, non-holonomic, and can be expressed by relations of the form (see (2.1.58)):

$$\sum_{i=1}^{N} \mathbf{g}_i^a(\mathbf{r}_1, \ldots, \mathbf{r}_N, t) \cdot \dot{\mathbf{r}}_i + g_0^a(\mathbf{r}_1, \ldots, \mathbf{r}_N, t) = 0 \quad (a = \overline{1, s}).$$

Since $\mathbf{r}_i = \mathbf{r}_i\ (q, t)$ and $\dot{\mathbf{r}}_i = \sum_{\alpha=1}^{n} \frac{\partial \mathbf{r}_i}{\partial q_\alpha} \dot{q}_\alpha + \frac{\partial \mathbf{r}_i}{\partial t}$, we can write:

$$\sum_{\alpha=1}^{n} b_\alpha^a \dot{q}_\alpha + b_0^a = 0, \tag{4.6.49}$$

where

$$b_\alpha^a = \sum_{i=1}^{N} \mathbf{g}_i^a \cdot \frac{\partial \mathbf{r}_i}{\partial q_\alpha}, \quad b_0^a = g_0^a + \sum_{i=1}^{N} \mathbf{g}_i^a \cdot \frac{\partial \mathbf{r}_i}{\partial t}.$$

The elementary displacements δq_α must be compatible with the constraints (4.6.49), i.e. they must satisfy the relations

$$\sum_{\alpha=1}^{n} b_\alpha^a \delta q_\alpha = 0. \tag{4.6.50}$$

On the other hand, by D'Alembert's principle (see Sect. 2.5), we have:

$$\sum_{\alpha=1}^{n} \left[\frac{d}{dt}\left(\frac{\partial T}{\partial \dot{q}_\alpha}\right) - \frac{\partial T}{\partial q_\alpha} - Q_\alpha \right] \delta q_\alpha = 0. \tag{4.6.51}$$

In order to derive the equations of motion, we multiply (4.6.50) by the arbitrary parameters λ_a (Lagrange multipliers) and subtract the result from (4.6.51). We obtain:

$$\frac{d}{dt}\left(\frac{\partial T}{\partial \dot{q}_\alpha}\right) - \frac{\partial T}{\partial q_\alpha} = Q_\alpha + \sum_{a=1}^{s} \lambda_a b_\alpha^a \quad (\alpha = \overline{1, n}), \tag{4.6.52}$$

the infinitesimal displacements δq_α being now arbitrary. Equations (4.6.52) are called *Lagrange equations with multipliers*. These equations, together with the equations of non-holonomic constraints (4.6.49), form a system of $n + s$ equations with $n + s$ unknowns $q_1, \ldots, q_n, \lambda_1, \ldots, \lambda_s$.

Let us analyze, as an example, the case of a homogeneous sphere of radius R, rolling (with pivoting) on a horizontal plane. If $\mathbf{v}_G$ is the velocity of the centre of the sphere and $\boldsymbol{\omega}$ is the instantaneous vector of rotation, then the constraint is expressed by the condition that the velocity of the contact point P is zero, i.e. (see (4.2.10)):

$$\mathbf{v}_G + \boldsymbol{\omega} \times \mathbf{R}_I = \mathbf{v}_G + \boldsymbol{\omega} \times \vec{GP} = 0. \tag{4.6.53}$$

Let the xy-plane of the fixed frame $S(Oxyz)$ be the horizontal plane, the z-axis being oriented along the ascending vertical, while the frame $S'(Gx'y'z')$ – as usual – is invariably related to the sphere. In this case, the point G will always lie in the plane $z = R$. Projecting the equation (4.6.53) on the axes of the frame S and using (4.5.3), we arrive at:

$$\begin{aligned} \dot{x}_G - R(\dot{\theta}\sin\varphi - \dot{\psi}\sin\theta\cos\varphi) &= 0, \\ \dot{y}_G + R(\dot{\theta}\cos\varphi + \dot{\psi}\sin\theta\sin\varphi) &= 0. \end{aligned} \tag{4.6.54}$$

Thus, the generalized coordinates $x_G, y_G, \theta, \varphi, \psi$ satisfy two non-integrable constraints. This means that we have a non-holonomic, scleronomous system, with three degrees of freedom.

To solve the problem, we shall make use of the Lagrange equations with multipliers (4.6.52). Since

$$I_1' = I_2' = I_3' = I' = \frac{2}{5}MR^2$$

and recalling (4.5.36), the kinetic energy of the body is:

$$T = \frac{1}{2}M(\dot{x}_G^2 + \dot{y}_G^2) + \frac{1}{2}I'(\dot{\theta}^2 + \dot{\varphi}^2 + \dot{\psi}^2 + 2\dot{\varphi}\dot{\psi}\cos\theta).$$

Using (4.6.52), we obtain the following system of equations:

$$\begin{aligned} M\ddot{x}_G &= \lambda_1, \\ M\ddot{y}_G &= \lambda_2, \\ I'(\ddot{\theta} + \dot{\varphi}\dot{\psi}\sin\theta) &= R(\lambda_2\cos\varphi - \lambda_1\sin\varphi), \\ \dot{\varphi} + \dot{\psi}\cos\theta &= \text{const.}, \\ I'\frac{d}{dt}(\dot{\psi} + \dot{\varphi}\cos\theta) &= R(\lambda_1\cos\varphi + \lambda_2\sin\varphi)\sin\theta. \end{aligned} \tag{4.6.55}$$

In addition to these equations, we have the energy first integral:

$$\frac{1}{2}MR^2(\dot{\theta}^2 + \dot{\psi}^2\sin^2\theta) + \frac{1}{2}I'(\dot{\theta}^2 + \dot{\varphi}^2 + \dot{\psi}^2 + 2\dot{\varphi}\dot{\psi}\cos\theta) = \text{const.},$$

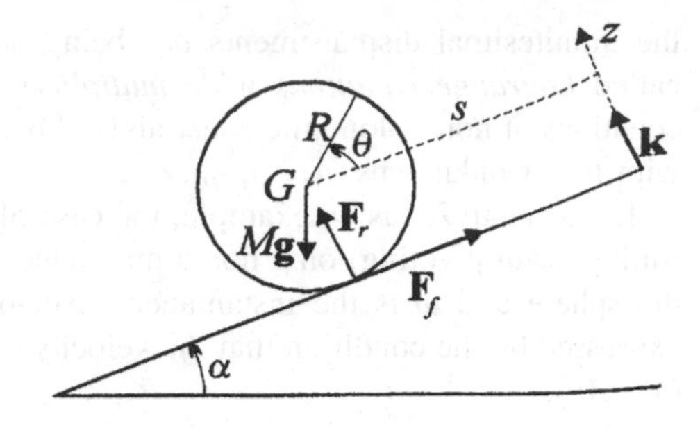

Fig. 4.19 A homogeneous cylinder rolling without sliding on an inclined plane.

where we have used (4.6.54) and took $V = 0$, because of the position of the centre of mass.

Another first integral is obtained from the remaining equations by eliminating the unknown parameters λ_1, λ_2. To this end, one introduces (4.6.54) into $(4.6.55)_{1,2}$, the result is substituted into $(4.6.55)_3$ and then one eliminates φ between the obtained equation and $(4.6.55)_4$. Consequently,

$$\dot{\psi} \sin^2 \theta = C_1 + C_2 \cos \theta,$$

where C_1 and C_2 are two constants. Since we have obtained three first integrals, our problem has been reduced to quadratures.

If the axis of rotation of the rolling sphere does not change its direction, then the constraint (4.6.53) becomes holonomic. This is the case, for instance, of a homogeneous cylinder, rolling without sliding on an inclined plane (Fig. 4.19). At the point of contact between the plane and the circle of transversal section with its centre in G, we have $\mathbf{R} = -R\mathbf{k}$, such that:

$$\mathbf{v}_G = R\boldsymbol{\omega} \times \mathbf{k}, \quad |\mathbf{v}_G| = \dot{s} = R|\boldsymbol{\omega} \times \mathbf{k}| = R\omega = R\dot{\theta}.$$

If we choose $s(0) = 0$, $\theta(0) = 0$, we find $s = R\theta$. A survey of the forces acting on the cylinder (see Fig. 4.19) leads to the Lagrangian:

$$L = \frac{1}{2}(I_G + MR^2)\dot{\theta}^2 + RMg\theta \sin \alpha,$$

where I_G is the moment of inertia of the cylinder relative to its axis. Then the Lagrange equations (2.5.17) yield:

$$\ddot{\theta} - \frac{2g}{3R} \sin \alpha = 0.$$

This equation, together with the initial conditions, determine uniquely the motion of the cylinder.

4.6.6 *Mechanical–Electromagnetic Analogies*

4.6.6.1 Larmor Precession

Let us consider a system of charged particles (electrons), performing a finite motion in a central electric field, created by a fixed point charge. We also assume that the whole system is in an external, constant, homogeneous magnetic field **B**. An atom placed in a magnetic field can be considered as such a system: the electrons are point charges, moving in the central field of the nucleus. In the following, we suppose that all particles of the system have the same charge e and the same mass m.

In order to study the action of the magnetic field on the particles, we first write the Lagrangian of the system relative to a frame having the origin in the centre of the electric field (nucleus) and performing a rotation about an axis passing through the fixed centre. The rotation is considered relative to a fixed frame with the same origin. Then (4.3.4) yields:

$$L = \frac{m}{2}\sum_{i=1}^{N}|\mathbf{v}_i|^2 + \frac{m}{2}\sum_{i=1}^{N}|\omega \times \mathbf{r}_i|^2 + m\sum_{i=1}^{N}\mathbf{v}_i \cdot (\omega \times \mathbf{r}_i) - V, \qquad (4.6.56)$$

where $\mathbf{v}_i$ is the velocity of the particle P_i relative to the mobile frame and V is the potential energy of the system of particles.

On the other hand, the Lagrangian of a system of N charged particles, moving in the electric field produced by the nucleus and the magnetic field **B** (see (2.5.30)), is:

$$L = \frac{m}{2}\sum_{i=1}^{N}|\mathbf{v}_i|^2 + e\sum_{i=1}^{N}\mathbf{v}_i \cdot \mathbf{A} - e\phi,$$

where **A** is the vector potential of the field **B**, while the potential ϕ can include a term proportional to $\sum_{ij} e^2/r_{ij}$. If **B** is constant and homogeneous, it is easy to prove that the equation $\mathbf{B} = \operatorname{curl}\mathbf{A}$ has the solution $\mathbf{A} = \frac{1}{2}\mathbf{B} \times \mathbf{r}$. Then, we may write:

$$L = \frac{m}{2}\sum_{i=1}^{N}|\mathbf{v}_i|^2 + \frac{e}{2}\sum_{i=1}^{N}\mathbf{v}_i \cdot (\mathbf{B} \times \mathbf{r}_i) - e\phi. \qquad (4.6.57)$$

We observe that the two Lagrangians (4.6.56) and (4.6.57) are equivalent if we set

$$\omega = \frac{e}{2m}\mathbf{B} \qquad (4.6.58)$$

and if the second term in the r.h.s. of (4.6.56) can be neglected. The second condition is fulfilled if **B** is weak ($B^2 \simeq 0$), while the first shows that the motion of the system of particles relative to the frame rotating with the angular velocity $\omega = -\frac{e\mathbf{B}}{2m}$ does not differ from the motion of the system relative to the fixed frame, when **B** is absent.

We then conclude that a weak homogeneous and constant magnetic field $\mathbf{B}$ gives rise to a motion of rotation of the system of charged particles about the direction of the field (precession), with the angular velocity $|\omega_L| = \frac{eB}{2m}$. This is known as *Larmor's theorem*, while the corresponding angular velocity ω_L is usually referred to as the *Larmor frequency*.

Larmor's theorem is still valid if there are rigid connections among the particles of the system. Let $\boldsymbol{\mu}$ be the resultant of magnetic moments of the closed currents produced by the charges and $\mathbf{L}$ be the angular momentum of the rigid system relative to one of its points, considered to be fixed (e.g. the centre of mass). As it is well known from the general courses on electricity and magnetism, the resultant moment of the magnetic forces acting on the particles of the rigid body is

$$\mathcal{M} = \boldsymbol{\mu} \times \mathbf{B},$$

while $\boldsymbol{\mu}$ and $\mathbf{L}$ are related by

$$\boldsymbol{\mu} = \frac{e}{2m}\mathbf{L}.$$

Then the angular momentum theorem (1.3.35) yields:

$$\frac{d\mathbf{L}}{dt} = \boldsymbol{\mu} \times \mathbf{B} = -\frac{e\mathbf{B}}{2m} \times \mathbf{L} = \omega_L \times \mathbf{L}. \quad (4.6.59)$$

This result shows that the vector of constant magnitude $\mathbf{L}$ performs a uniform motion of precession about $\mathbf{B}$ (Larmor's precession), just like in the case of the symmetrical top, with angular velocity ω_L. The same motion is performed by the vector $\boldsymbol{\mu} = \lambda\mathbf{L}$, where $\lambda = \frac{e}{2m}$. The sense of precession depends on the sign of the electrical charge of the particles.

4.6.6.2 Gyroscopic Forces

If we compare the Lagrangian (4.3.4) of a particle of mass m moving relative to the non-inertial frame S':

$$L = \frac{1}{2}m|\mathbf{v}_r|^2 + \frac{1}{2}m|\omega \times \mathbf{r}'|^2 + m\mathbf{v}_r \cdot (\omega \times \mathbf{r}') - m\mathbf{a}_0 \cdot \mathbf{r}' - V,$$

with the Lagrangian of a charged particle subject to an electromagnetic force (2.5.30):

$$L = \frac{1}{2}m|\mathbf{v}|^2 - e\phi + e\mathbf{v} \cdot \mathbf{A},$$

as well as the equations obtained by the use of these two Lagrangian functions, we can see some remarkable similarities.

First, both the Coriolis force $\mathbf{F}_c = 2m\mathbf{v}_r \times \boldsymbol{\omega}$ and the Lorentz force $\mathbf{F} = e\mathbf{v} \times \mathbf{B}$ are gyroscopic forces. Since we are dealing with a motion in a non-inertial frame, we shall keep the notation $\mathbf{v}$ instead of $\mathbf{v}_r$ and $\mathbf{r}$ instead of $\mathbf{r}'$.

Second, each of the two Lagrangians contains two velocity-dependent terms. The remaining terms are only position-dependent.

Third, if we assume that the field $\mathbf{B}$ is constant and homogeneous, then $\mathbf{A} = \frac{1}{2}\mathbf{B} \times \mathbf{r}$. This means that the terms $m\mathbf{v} \cdot (\boldsymbol{\omega} \times \mathbf{r})$, belonging to the first Lagrangian, and $e\mathbf{v}\cdot\mathbf{A}$, which occurs in the second, are equivalent if we make the correspondence $m \leftrightarrow e$ and choose

$$\mathbf{A}' = \frac{1}{2}\boldsymbol{\omega} \times \mathbf{r}, \tag{4.6.60}$$

meaning that

$$\boldsymbol{\omega} = \operatorname{curl} \mathbf{A}'. \tag{4.6.61}$$

This analogy leads to the following Lagrangian of the particle, relative to the non-inertial frame:

$$L = \frac{1}{2}m|\mathbf{v}|^2 - m\phi' + 2m\mathbf{v} \cdot \mathbf{A}' - V(x_i), \tag{4.6.62}$$

where $\mathbf{A}'$ is defined by (4.6.60) and ϕ' by

$$\phi' = \mathbf{r} \cdot \mathbf{a}_0 - \frac{1}{2}|\boldsymbol{\omega} \times \mathbf{r}|^2. \tag{4.6.63}$$

Note that the potentials $\mathbf{A}'$ and ϕ' are functions of the coordinates x_i and the time t, while $V(x_i)$ yields the potential force (the force of gravity in our case). Also note that the quantities $\dot{\mathbf{r}} = \mathbf{v}$ and $\ddot{\mathbf{r}} = \mathbf{a}$ are relative to the non-inertial frame S'.

The equation of motion is obtained by applying the Lagrange equations (2.5.17). Performing calculations similar to those leading to (4.3.6), we arrive at:

$$m\ddot{\mathbf{r}} + 2m\dot{\mathbf{A}}' + m \operatorname{grad} \phi' - 2m \operatorname{grad}(\mathbf{v} \cdot \mathbf{A}') + \operatorname{grad} V = 0,$$

or, if we make some rearrangements of the terms and use the results of Appendix B,

$$m\ddot{\mathbf{r}} = m(\mathbf{E}' + \mathbf{v} \times \mathbf{B}') + \mathbf{F}, \tag{4.6.64}$$

where $\mathbf{E}'$ and $\mathbf{B}'$ are given by:

$$\mathbf{E}' = -\operatorname{grad} \phi' - \frac{\partial}{\partial t}(2\mathbf{A}'), \tag{4.6.65}$$

$$\mathbf{B}' = \operatorname{curl}(2\mathbf{A}'). \tag{4.6.66}$$

Equation (4.6.64) shows that the terms $\mathbf{E}'$ and $\mathbf{v} \times \mathbf{B}'$ have the units of acceleration. It also tells us that the particle moves in an applied ($\mathbf{F}$) and an inertial fields of force, the latter being defined by means of the potentials $\mathbf{A}', \phi'$, as can be seen from the definitions (4.6.65) and (4.6.66).

If the frame S' becomes inertial and $\mathbf{F}$ is absent, then $\mathbf{a}_0 = 0$, $\omega = 0$, and equation (4.6.64) reduces to $\ddot{\mathbf{r}} = 0$, as expected. We reach the same result if the charged particle is neither accelerated by the electric field $\mathbf{E}$, nor engaged in a motion of rotation about the magnetic field $\mathbf{B}$.

This analogy can be further developed. For instance, remark that $\mathbf{E}'$ and $\mathbf{B}'$ obey the well-known source-free Maxwell equations:

$$\operatorname{curl} \mathbf{E}' = -\frac{\partial \mathbf{B}'}{\partial t}, \quad \operatorname{div} \mathbf{B}' = 0. \tag{4.6.67}$$

In conclusion, the study of a heavy non-charged particle in a non-inertial frame can be accomplished by using the same Lagrangian formalism as for a charged particle, moving in a velocity-dependent field of force, the generalized potential being given by

$$V = m(\phi' - 2\mathbf{v} \cdot \mathbf{A}'). \tag{4.6.68}$$

4.7 Problems

1. Determine the principal moments of inertia of the following homogeneous rigid objects:

   (a) A sphere of radius R;
   (b) A circular cylinder of radius R and height h;
   (c) A circular cone of base radius R and height h;
   (d) A rectangular parallelepiped of sides a, b, c;
   (e) An ellipsoid of semi-axes a, b, c.

2. Making an abstraction for the oscillation and vibration motions, determine the principal moments of inertia of the following molecules, considered as rigid systems:

   (a) A diatomic molecule AB;
   (b) A triatomic molecule A_2B, with the atoms disposed as an isosceles triangle;
   (c) A tetratomic molecule A_3B, with the atoms disposed as a tetrahedron, the basis being an equilateral triangle.

3. Assuming that the Earth is a homogeneous sphere of radius R, uniformly rotating about the Poles-axis, determine the acceleration of gravity g at some point on the Earth in terms of the angular velocity ω and the radius R. For which latitude λ does the deviation from the geocentric vertical attain its maximum?

4. A heavy particle of mass m falls without initial velocity from the height $h > 0$. Using the successive approximation method, determine the influence of the Coriolis force on the falling particle.
5. Determine the deviation of the plane of motion of a body thrown from the surface of the Earth with the initial velocity $\mathbf{v}_0$.
6. Show that the trajectory of a particle constrained to move in a horizontal plane (but otherwise free) is always deviated to the right in the Northern hemisphere and to the left in the Southern hemisphere (Baer's rule).
7. Check the validity of the relations (4.5.2) and (4.5.3), expressing the components of the vector $\boldsymbol{\omega}$ on the axes of the frames S and S'.
8. A homogeneous disk of mass M and radius R can rotate about its centre. Study the motion of the disk, assuming that at time $t = 0$ the angular velocity of the disk is ω_0, while the angle between the instantaneous axis of rotation and the normal to the disk is α.
9. A homogeneous sphere of mass M and radius R rolls without sliding on an inclined plane of angle α. Using the Lagrange equations with multipliers, determine the law of motion. The initial velocity is assumed to be zero.
10. Find the kinetic energy of a cylinder of radius R rolling on a plane. One assumes that the principal axis of inertia is parallel to the axis of the cylinder. The moment of inertia relative to this axis is I, while the moment between the two axes is a.
11. Find the kinetic energy of a homogeneous cylinder of radius a, rolling within a cylinder of radius R.
12. Determine the kinetic energy of a homogeneous cone rolling on a plane.
13. Using Euler's equations, show that the torque of constraint for a body rotating about one of its principal axes is zero.
14. A thin rod AB of mass M and length l is pinned at one end to a rotating shaft. Find the equilibrium value of the angle α, for a given constant angular velocity ω of rotation of the shaft.

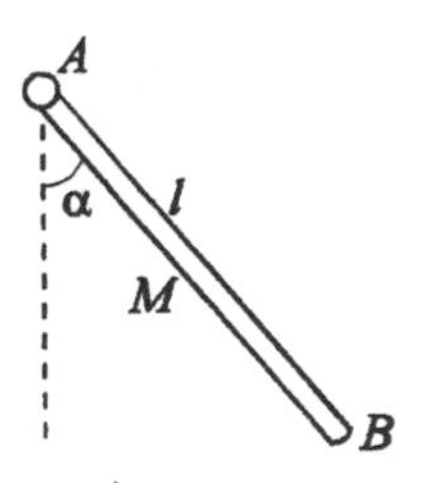

15. Find Euler's angles as functions of time, for the free rotation of a symmetrical top.
16. Under which conditions does a rigid body with axial symmetry behave like a symmetric top? Apply the result to a homogeneous cylinder.
17. The mass density of a rigid ellipsoid of semi-axes a, b, c $(a \neq b \neq c)$ varies according to

$$\rho(x_1, x_2, x_3) = \rho_0\left(1 + \alpha\frac{x_3^2}{c^2}\right),$$

where ρ_0 and a are two constants. Prove that α can be chosen in such a way that the rigid body behaves like a symmetric top.

18. Determine the condition under which the rotation of the symmetric top about a vertical axis is stable.
19. Study the motion of a symmetric top for which the kinetic energy of self-rotation is bigger than the gravitation energy (rapid top).
20. A right homogeneous cylinder of mass M, radius R and height h rotates about a vertical axis z passing through its centre of mass G, with a constant angular velocity ω. If the angle α between the vertical axis and the cylinder's axis of symmetry is also constant, find the lateral pressure at the two bearings A and B.

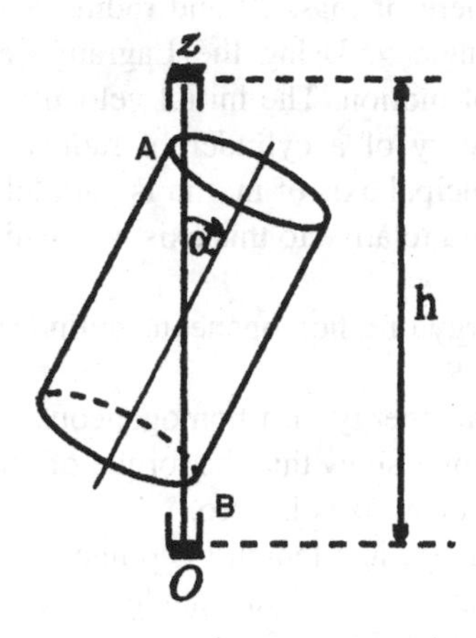

Chapter 5
Hamiltonian Formalism

5.1 Hamilton's Canonical Equations

The previous three chapters have been primarily concerned with the use of the Lagrangian formalism in the study of various mechanical problems. Assuming again that our field of investigation concerns only *natural* systems (i.e. systems possessing either a simple, or a generalized potential), we remind the reader that the Lagrangian approach to mechanical systems with a finite number of degrees of freedom consists of the definition of the Lagrangian L, as a scalar function of the generalized coordinates q_j $(j=\overline{1,n})$, generalized velocities $\dot{q}_j$ $(j=\overline{1,n})$ and the time t:

$$L(q,\dot{q},t)=T(q,\dot{q},t)-V(q,\dot{q},t), \tag{5.1.1}$$

and the integration of the Lagrange equations:

$$\frac{d}{dt}\left(\frac{\partial L}{\partial \dot{q}_j}\right)-\frac{\partial L}{\partial q_j}=0 \quad (j=\overline{1,n}), \tag{5.1.2}$$

which means the determination of q_j as functions of the time:

$$q_j=q_j(t) \quad (j=\overline{1,n}).$$

These are the finite equations of motion of the system in the configuration space and, at the same time, the parametric equations of the generalized trajectory described by the representative point. The two sets of variables q_j and $\dot{q}_j$ completely determine, at any time, the position and the velocity of the system in the configuration space.

As it has been shown, the definition (2.8.6) of the generalized momenta,

$$p_j=\frac{\partial L}{\partial \dot{q}_j} \quad (j=\overline{1,n}), \tag{5.1.3}$$

M. Chaichian et al., *Mechanics*, DOI: 10.1007/978-3-642-17234-2_5,

leads to a slightly different form of the Lagrange equations:

$$\dot{p}_j = \frac{\partial L}{\partial q_j} \quad (j = \overline{1,n}). \tag{5.1.4}$$

On the other hand, the theory of differential equations shows that a system of n second-order differential equations with n variables of the type (5.1.2) can be put in a *normal* form, i.e. it can be expressed relative to the derivatives of the highest order $\ddot{q}_j$:

$$\ddot{q}_j = g_j(q, \dot{q}, t) \quad (j = \overline{1,n}), \tag{5.1.5}$$

if the Hessian determinant is non-zero, that is, in our case,

$$\left| \frac{\partial^2 L}{\partial \dot{q}_j \partial \dot{q}_k} \right| = \left| \frac{\partial p_k}{\partial \dot{q}_j} \right| \neq 0. \tag{5.1.6}$$

Recalling the meaning of the functions T and V, we conclude that, in our case, this condition is fulfilled. Equation (5.1.3) can be solved with respect to $\dot{q}_j$:

$$\dot{q}_j = h_j(q, p, t) \quad (j = \overline{1,n}), \tag{5.1.7}$$

and then (5.1.4) takes the form:

$$\dot{p}_j = \frac{\partial}{\partial q_j} L(q_1, \dots, q_n, h_1, \dots, h_n, t)\Big|_{h_1,\dots,h_n=\text{const.}} \quad (j = \overline{1,n}). \tag{5.1.8}$$

The system of n second-order differential equations (5.1.5), written in normal form, is equivalent to the system of $2n$ first-order differential equations (5.1.7) and (5.1.8), which are not yet in normal form. Let us show that these $2n$ equations can be written in a symmetric form, customarily called *canonical*. To do this, we pass from the set of independent variables $(q_j, \dot{q}_j)$ to the set (q_j, p_j), $j = \overline{1,n}$. In this new representation, our system is completely determined by n generalized coordinates q_j and n generalized momenta p_j. This implies the substitution of the Lagrangian function $L(q, \dot{q}, t)$ by another characteristic function, in which q_j and p_j appear as independent variables.

The mathematical procedure that gives the transition from the old set $(q_j, \dot{q}_j)$ to the new set (q_j, p_j) is called *Legendre transformation*. To illustrate the method, let us write the total differential of a two-variable function $f(x, y)$:

$$df = \frac{\partial f}{\partial x} dx + \frac{\partial f}{\partial y} dy = \varphi dx + \psi dy. \tag{5.1.9}$$

Here,

$$\varphi = \left(\frac{\partial f}{\partial x} \right)_y, \quad \psi = \left(\frac{\partial f}{\partial y} \right)_x \tag{5.1.10}$$

and the subscripts show which variable is kept constant at the derivation. We realize that, if (x, y) are chosen as independent parameters, then (φ, ψ) are obtained by taking the partial derivatives of the function f with respect to the independent variables.

If, instead of y, we take ψ as an independent variable,

$$(x, y) \rightarrow (x, \psi),$$

then, obviously, the characteristic function will be different. To find it, we subtract $d(y\psi)$ from both sides of (5.1.9). The result is:

$$d\eta = \varphi\, dx - y\, d\psi, \tag{5.1.11}$$

where the new characteristic function $\eta(x, \psi)$ is defined by

$$\eta = f - y\psi. \tag{5.1.12}$$

Then, (5.1.11) yields:

$$\varphi = \left(\frac{\partial\eta}{\partial x}\right)_\psi, \quad y = -\left(\frac{\partial\eta}{\partial\psi}\right)_x. \tag{5.1.13}$$

The Legendre transformation is widely used in thermodynamics. To show how it works, let us write the fundamental thermodynamic equation for reversible processes:

$$dU = T\, dS - p\, dV, \tag{5.1.14}$$

where $U(S, V)$ is the internal energy, T – the absolute temperature, S – the entropy, p – the pressure and V – the volume. If we subtract $d(TS)$ from both sides of (5.1.14), the result is

$$dF = -S\, dT - p\, dV, \tag{5.1.15}$$

where the state function F,

$$F = U - TS, \tag{5.1.16}$$

was defined by *Hermann von Helmholtz* and has the meaning of *free energy*. Now the new independent variables are (T, V), while S and p are given by:

$$S = -\left(\frac{\partial F}{\partial T}\right)_V, \quad p = -\left(\frac{\partial F}{\partial V}\right)_T. \tag{5.1.17}$$

Going back to our formalism, we first notice that the change of the independent variables $(q, \dot{q}) \rightarrow (q, p)$ does not affect the variable t, which remains the same in both representations. Differentiating (5.1.1), we may write:

$$dL = \sum_{j=1}^{n}\left(\frac{\partial L}{\partial q_j} dq_j + \frac{\partial L}{\partial \dot{q}_j} d\dot{q}_j\right) + \frac{\partial L}{\partial t} dt = \sum_{j=1}^{n}(\dot{p}_j dq_j + p_j d\dot{q}_j) + \frac{\partial L}{\partial t} dt, \tag{5.1.18}$$

where (5.1.3) and (5.1.4) have been used. Now, if we subtract the quantity $d\sum_{j=1}^{n} p_j\dot{q}_j$ from both sides, we get:

$$-dH = \sum_{j=1}^{n}(\dot{p}_j\,dq_j - \dot{q}_j\,dp_j) + \frac{\partial L}{\partial t}dt, \tag{5.1.19}$$

where

$$H(q,p,t) = \sum_{j=1}^{n} p_j\dot{q}_j - L(q,\dot{q},t)|_{\dot{q}_j=h_j(q,p,t)} \tag{5.1.20}$$

and the velocities $\dot{q}_j$ have to be replaced according to (5.1.7). The function $H(p, q, t)$ is the already known *Hamiltonian function* or, as it is usually called, the *Hamiltonian*. Equation (5.1.19) shows that the independent variables are now (q_j, p_j). Thus,

$$\dot{q}_j = \left(\frac{\partial H}{\partial p_j}\right)_q, \quad \dot{p}_j = -\left(\frac{\partial H}{\partial q_j}\right)_p \quad (j = \overline{1,n}), \tag{5.1.21}$$

$$\frac{\partial L}{\partial t} = -\frac{\partial H}{\partial t}. \tag{5.1.22}$$

The system of $2n$ first-order differential equations (5.1.21) is equivalent to the system of n second-order differential equations (5.1.2). By integration, one obtains the set of independent variables (q_j, p_j) as functions of time and $2n$ arbitrary constants. In order to uniquely determine the motion of the system in the space defined by the variables (q_j, p_j), one must know $2n$ independent initial conditions, e.g. $q_j(0),\ p_j(0)\ (j = \overline{1,n})$. Remark that Eqs. (5.1.21) are written in a normal and symmetric form. They were established by *William Rowan Hamilton* and are known as *Hamilton's canonical equations*. The function H, as well as the canonical equations, play a fundamental role in analytical mechanics, with many applications in physics, chemistry, mathematics, etc.

The independent variables (q_j, p_j) are called *canonical variables*, or *conjugate variables*. Each generalized momentum p_k (k fixed) is canonically conjugated to a generalized coordinate q_k. The variables $q_1, \dots, q_n, p_1, \dots, p_n$ can be considered as the coordinates of a *generalized* or *representative* point in a $2n$ dimensional space, introduced by the American physicist *Josiah Willard Gibbs* and called the *phase space*. In this space, any solution $q_j(t), p_j(t)$ of the canonical equations (5.1.21) is represented by a generalized curve, which is the *generalized trajectory* of the representative point.

Before going further, we wish to deduce Hamilton's equations in a different way. To this end, we remember that the Lagrange equations (5.1.2) can be derived by means of Hamilton's principle,

$$\delta\int_{t_1}^{t_2} L(q,\dot{q},t)\,dt = 0. \tag{5.1.23}$$

The equivalence between Lagrange's and Hamilton's systems of equations gives us the idea that the canonical equations can also be obtained using Hamilton's principle. Indeed, by substituting (5.1.20) into (5.1.23) and performing the variation indicated by the operator δ, we have:

$$\int_{t_1}^{t_2} \sum_{j=1}^{n} \left(p_j \delta \dot{q}_j + \dot{q}_j \delta p_j - \frac{\partial H}{\partial q_j} \delta q_j - \frac{\partial H}{\partial p_j} \delta p_j \right) dt = 0, \tag{5.1.24}$$

where we used the fact that the virtual displacements $\delta q_j, \delta p_j$ in the phase space are taken at $t = \text{const.}$ ($\delta t = 0$). Integrating the first term on the l.h.s. of (5.1.24) by parts, we obtain:

$$\int_{t_1}^{t_2} \sum_{j=1}^{n} p_j \, \delta \dot{q}_j = \sum_{j=1}^{n} p_j \delta q_j \Bigg|_{t_1}^{t_2} - \int_{t_1}^{t_2} \sum_{j=1}^{n} \dot{p}_j \delta q_j \, dt,$$

or, since $(\delta q_j)_{t_1} = (\delta q_j)_{t_2} = 0$,

$$\int_{t_1}^{t_2} \sum_{j=1}^{n} \left[\left(\dot{q}_j - \frac{\partial H}{\partial p_j} \right) \delta p_j - \left(\dot{p}_j + \frac{\partial H}{\partial q_j} \right) \delta q_j \right] dt = 0. \tag{5.1.25}$$

For arbitrary and independent δq_j, δp_j, this equality holds true if and only if each of the $2n$ parentheses is zero. As a result, we obtain again Hamilton's equations (5.1.21).

Observations:

(a) If one of the n generalized coordinates $q_1, ..., q_n$, say q_k (k fixed), does not explicitly appear in the Lagrangian L, i.e. if q_k is cyclic, then it remains cyclic in the new representation. Indeed, $(5.1.21)_2$ yields

$$p_k = \text{const.}, \tag{5.1.26}$$

showing that the *general momentum theorem holds true*. If all the generalized coordinates are cyclic:

$$p_1 = C_1, \ldots, p_n = C_n,$$

the Hamiltonian H becomes a function of time only,

$$H = H(C_1, \ldots, C_n, t),$$

and so the cyclic coordinates are obtained by quadratures:

$$q_j = \int \frac{\partial H}{\partial p_j} dt + q_j^0 \quad (j = \overline{1, n}), \tag{5.1.27}$$

where the integration constants q_j^0 are determined from the initial conditions.

(b) The relation

$$f(q_1, \ldots, q_n, p_1, \ldots, p_n, t) = \text{const.} \tag{5.1.28}$$

is a *first integral* of Hamilton's equations (5.1.21) if f is a constant for any solution of these equations. The first integrals are also called *constants of motion*. For example, the relation (5.1.26) is a first integral. If the Hamiltonian does not explicitly depend on the time, then

$$H(q_1, \ldots, q_n, p_1, \ldots, p_n) = \text{const.} \tag{5.1.29}$$

is also a first integral, since

$$\frac{dH}{dt} = \sum_{j=1}^{n} \left(\frac{\partial H}{\partial q_j} \dot{q}_j + \frac{\partial H}{\partial p_j} \dot{p}_j \right) = \sum_{j=1}^{n} \left(\frac{\partial H}{\partial q_j} \frac{\partial H}{\partial p_j} - \frac{\partial H}{\partial p_j} \frac{\partial H}{\partial q_j} \right) \equiv 0.$$

In view of (5.1.22), we also have:

$$\frac{dH}{dt} = \frac{\partial H}{\partial t} = -\frac{\partial L}{\partial t}. \tag{5.1.30}$$

To know h cyclic coordinates means to know h distinct first integrals of the canonical equations (5.1.21). If we are able to find $2n$ distinct first integrals of Hamilton's equations, then the system is integrated.

(c) We recall (see Chap. 2, Sect. 2.8) that the Hamiltonian of a natural scleronomous system represents its total energy. If the time t does not explicitly occur in the potential function V, then (5.1.29) reads:

$$H = T + V = \text{const.}, \tag{5.1.31}$$

i.e. the energy first integral. This result also holds true for the case of a generalized potential.

The Hamiltonian for rheonomous systems is given by (2.8.33):

$$H = T_2 - T_0 + V = T + V - (T_1 + 2T_0), \tag{5.1.32}$$

meaning that in this case the function H is not identical to the total energy. If the time t does not explicitly appear in (5.1.32), H is a first integral, without being the total energy. It is also possible for H to be the total energy, without being a constant of the motion. Finally, there are cases with H being neither a first integral, nor the total energy.

In the examples to follow, we shall show how to find the Hamiltonian and how to determine the differential equations of motion by means of the Hamiltonian formalism.

5.1.1 Motion of a Particle in a Plane

Consider a particle of mass m, moving without friction in the plane $z = 0$, subject to the conservative force $\mathbf{F} = -\text{grad } V$. Choosing $q_1 = x,\ q_2 = y$, we can write:

$$H = T + V = \frac{1}{2}(\dot{x}^2 + \dot{y}^2) + V(x, y) = \text{const.} \tag{5.1.33}$$

In order to make the use of the Hamiltonian formalism possible, it is necessary to express H as a function of the generalized coordinates x, y and the generalized momenta p_x, p_y. To this end, we use (5.1.3):

$$p_x = \frac{\partial L}{\partial \dot{x}} = m\dot{x}, \quad p_y = \frac{\partial L}{\partial \dot{y}} = m\dot{y}, \tag{5.1.34}$$

such that

$$H = \frac{1}{2m}(p_x^2 + p_y^2) + V(x, y). \tag{5.1.35}$$

According to the canonical equations (5.1.21), we have:

$$\dot{x} = \frac{p_x}{m}, \quad \dot{y} = \frac{p_y}{m}, \quad \dot{p}_x = -\frac{\partial V}{\partial x}, \quad \dot{p}_y = -\frac{\partial V}{\partial y}, \tag{5.1.36}$$

or, finally,

$$m\ddot{x} = -\frac{\partial V}{\partial x} = F_x, \quad m\ddot{y} = -\frac{\partial V}{\partial y} = F_y, \tag{5.1.37}$$

which are projections on the x- and y-axis of the vector equation

$$m\ddot{\mathbf{r}} = -\text{grad}\, V = \mathbf{F}. \tag{5.1.38}$$

Sometimes, as we already know, it is more convenient to choose the polar coordinates $q_1 = r$, $q_2 = \varphi$. Then,

$$H = \frac{1}{2}m(\dot{r}^2 + r^2\dot{\varphi}^2) + V(r, \varphi). \tag{5.1.39}$$

Since

$$p_r = \frac{\partial L}{\partial \dot{r}} = \frac{\partial T}{\partial \dot{r}} = m\dot{r}, \quad p_\varphi = \frac{\partial L}{\partial \dot{\varphi}} = \frac{\partial T}{\partial \dot{\varphi}} = mr^2\dot{\varphi}, \tag{5.1.40}$$

the Hamiltonian reads:

$$H = \frac{1}{2m}\left(p_r^2 + \frac{1}{r^2}p_\varphi^2\right) + V(r, \varphi). \tag{5.1.41}$$

The canonical equations (5.1.21) yield then:

$$\dot{r} = \frac{\partial H}{\partial p_r} = \frac{p_r}{m}, \quad \dot{\varphi} = \frac{\partial H}{\partial p_\varphi} = \frac{p_\varphi}{mr^2}, \tag{5.1.42}$$

$$\dot{p}_r = -\frac{\partial H}{\partial r} = \frac{p_\varphi}{mr^3} - \frac{\partial V}{\partial r}, \quad \dot{p}_\varphi = -\frac{\partial H}{\partial \varphi} = -\frac{\partial V}{\partial \varphi}, \tag{5.1.43}$$

leading to the expected equations of motion:

$$m(\ddot{r} - r\dot{\varphi}^2) = -\frac{\partial V}{\partial r} = F_r, \quad m(2\dot{r}\dot{\varphi} + r\ddot{\varphi}) = -\frac{1}{r}\frac{\partial V}{\partial \varphi} = F_\varphi. \tag{5.1.44}$$

If V is independent of φ (cyclic variable), the associated generalized momentum is the area first integral:

$$p_\varphi = mr^2\dot{\varphi} = \text{const.} \tag{5.1.45}$$

5.1.2 *Motion of a Particle Relative to a Non-inertial Frame*

The Lagrangian corresponding to a particle moving with respect to a non-inertial frame is (see (4.3.4)):

$$L = \frac{1}{2}m|\mathbf{v}_r|^2 + \frac{1}{2}m|\omega \times \mathbf{r}'|^2 + m\mathbf{v}_r \cdot (\omega \times \mathbf{r}') - m\mathbf{r}' \cdot \mathbf{a}_0 - V, \tag{5.1.46}$$

where $\mathbf{r}' = x'_k\mathbf{u}'_k$ is the radius-vector of the particle relative to the non-inertial frame and $\mathbf{v}_r = \dot{x}'_k\mathbf{u}'_k$ is its relative velocity. To shorten the calculation, we take $\mathbf{r}'$ as a vector generalized coordinate. The vector generalized momentum

$$\mathbf{p}' = \frac{\partial L}{\partial \mathbf{v}_r} = m\mathbf{v}_r + m\omega \times \mathbf{r}' \tag{5.1.47}$$

is then introduced into (5.1.20), yielding:

$$H = \mathbf{p}' \cdot \mathbf{v}_r - L = \frac{1}{2}m|\mathbf{v}_r|^2 - \frac{1}{2}m|\omega \times \mathbf{r}'|^2 + m\mathbf{r}' \cdot \mathbf{a}_0 + V. \tag{5.1.48}$$

If the origin of the non-inertial frame has the acceleration $\mathbf{a}_0 = \mathbf{g}$, meaning that V is not connected with the gravitational force, we have:

$$H = \frac{1}{2m}|\mathbf{p}'|^2 - \mathbf{p}' \cdot (\omega \times \mathbf{r}') + m\mathbf{r}' \cdot \mathbf{a}_0 + V. \tag{5.1.49}$$

From (5.1.21), we obtain:

$$\mathbf{v}_r = \frac{\partial H}{\partial \mathbf{p}'} = \frac{1}{m}\mathbf{p}' - \omega \times \mathbf{r}', \tag{5.1.50}$$

$$\dot{\mathbf{p}}' = -\frac{\partial H}{\partial \mathbf{r}'} = -\omega \times \mathbf{p}' - m\mathbf{a}_0 - \frac{\partial V}{\partial \mathbf{r}'}. \tag{5.1.51}$$

Equations (5.1.50) and (5.1.51) yield the expected equation of motion of the particle relative to an accelerated frame:

$$m\mathbf{a}_r = \mathbf{F} - m\mathbf{a}_0 - m\dot{\omega} \times \mathbf{r} - m\omega \times (\omega \times \mathbf{r}') - 2m\omega \times \mathbf{v}_r = \mathbf{F} + \mathbf{F}_{tr} + \mathbf{F}_c, \tag{5.1.52}$$

where

$$\mathbf{a}_r = \ddot{x}'_k \mathbf{u}'_k, \quad \mathbf{F} = -\frac{\partial V}{\partial \mathbf{r}'}. \tag{5.1.53}$$

5.1.3 Motion of a Charged Particle in an Electromagnetic Field

Let us choose $q_1 = x,\ q_2 = y,\ q_3 = z$ as generalized coordinates. The Lagrangian associated to a particle of mass m and charge e, moving in the electromagnetic field $\mathbf{E}, \mathbf{B}$ is (see (2.5.30)):

$$L = \frac{1}{2} m v_j v_j - e\phi + e v_j A_j. \tag{5.1.54}$$

The generalized momenta

$$p_j = \frac{\partial L}{\partial v_j} = m v_j + e A_j \quad (j = 1, 2, 3) \tag{5.1.55}$$

are then introduced in the Hamiltonian, which reads:

$$H = p_j v_j - L = \frac{1}{2} m v_j v_j + e\phi, \tag{5.1.56}$$

or, in terms of x_j, p_j,

$$H = \frac{1}{2m}(p_j - eA_j)(p_j - eA_j) + e\phi. \tag{5.1.57}$$

Hamilton's canonical equations (5.1.21) yield:

$$\dot{x}_k = v_k = \frac{\partial H}{\partial p_k} = \frac{1}{m}(p_j - eA_j)\delta_{jk} = \frac{1}{m}(p_k - eA_k), \tag{5.1.58}$$

$$\dot{p}_k = -\frac{\partial H}{\partial x_k} = \frac{e}{m}(p_j - eA_j)\frac{\partial A_j}{\partial x_k} - e\frac{\partial \phi}{\partial x_k}. \tag{5.1.59}$$

The total time derivative of $A_k(x, y, z, t)$ is:

$$\dot{A}_k = \frac{\partial A_k}{\partial t} + v_j \frac{\partial A_k}{\partial x_j}; \tag{5.1.60}$$

consequently,

$$m\dot{v}_k = m\ddot{x}_k = -e\frac{\partial \phi}{\partial x_k} - e\frac{\partial A_k}{\partial t} + e v_j \left(\frac{\partial A_j}{\partial x_k} - \frac{\partial A_k}{\partial x_j} \right). \tag{5.1.61}$$

Since

$$\frac{\partial A_j}{\partial x_k} - \frac{\partial A_k}{\partial x_j} = \epsilon_{kjm}(\text{curl}\,\mathbf{A})_m,$$

we notice that

$$v_j\left(\frac{\partial A_j}{\partial x_k} - \frac{\partial A_k}{\partial x_j}\right) = \epsilon_{kjm} v_j (\text{curl}\,\mathbf{A})_m = (\mathbf{v} \times \text{curl}\,\mathbf{A})_k$$

and finally

$$m\ddot{x}_k = e(\mathbf{E} + \mathbf{v} \times \mathbf{B})_k, \tag{5.1.62}$$

which is the x_k-component of the equation of motion.

An interesting and useful special case concerns a stationary electromagnetic field with axial symmetry, characterized by

$$\mathbf{E} = -\text{grad}\,\phi, \quad \mathbf{B} = \text{curl}\,\mathbf{A},$$

where the potentials $\mathbf{A}, \phi$ do not depend explicitly on time. Let us choose a cylindrical system of coordinates ρ, φ, z, direct the z-axis along the symmetry axis and assume that $A_\rho = 0,\ A_\varphi \neq 0,\ A_z = 0$. Since by definition $B_\varphi = 0$, the non-zero components of $\mathbf{B}$ are:

$$B_\rho = -\frac{\partial A_\varphi}{\partial z}, \quad B_z = \frac{1}{\rho}\frac{\partial}{\partial \rho}(\rho A_\varphi).$$

Noting that $\mathbf{A}$ and ϕ do not depend on φ, we may write the Hamiltonian as

$$H = \frac{1}{2m}\left\{p_\rho^2 + \frac{1}{\rho^2}[p_\varphi - e\rho A_\varphi(\rho, z)]^2 + p_z^2\right\} + e\phi(\rho, z). \tag{5.1.63}$$

Hamilton's equations for the conjugate variables φ and p_φ are:

$$\dot{\varphi} = \frac{\partial H}{\partial p_\varphi} = \frac{1}{m\rho^2}(p_\varphi - e\rho A_\varphi), \quad \dot{p}_\varphi = -\frac{\partial H}{\partial \varphi} = 0,$$

and lead to the first integral

$$p_\varphi = m\rho^2\dot{\varphi} + e\rho A_\varphi = \text{const.} \tag{5.1.64}$$

Since H does not explicitly depend on time, we also have the energy first integral:

$$E = \frac{1}{2}m(\dot{\rho}^2 + \rho^2\dot{\varphi}^2 + \dot{z}^2) + e\phi = \text{const.} \tag{5.1.65}$$

The differential equations of motion for ρ and z are obtained either by using the canonical equations, or directly from (5.1.62):

$$m(\ddot{\rho} - \rho\dot{\varphi}^2) = eE_\rho + e\dot{\rho}B_z, \tag{5.1.66}$$

$$m\ddot{z} = eE_z - e\rho\dot{\varphi}B_\rho. \tag{5.1.67}$$

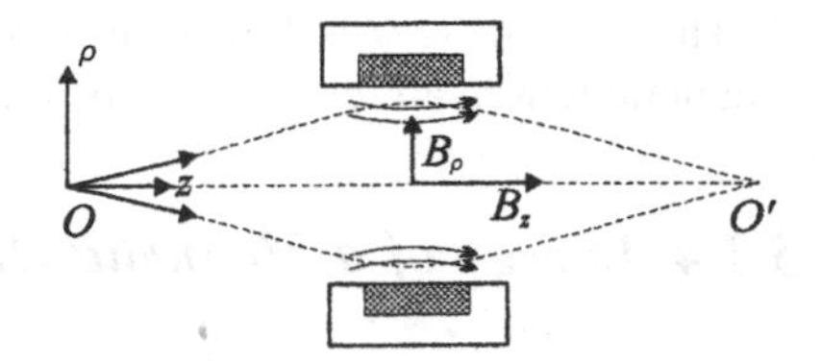

Fig. 5.1 A magnetic lens.

Assume now that $\mathbf{E} = 0$, while $\mathbf{B}$ is constant and homogeneous. Then (see Chap. 4, Sect. 4.6.6):

$$A_\varphi = \frac{1}{2}(\mathbf{B} \times \mathbf{r})_\varphi = \frac{1}{2}\rho B_z,$$

and the first integral (5.1.64) reads:

$$m\rho^2\dot{\varphi} + \frac{1}{2}e\rho^2 B_z = C(\text{const.}), \tag{5.1.68}$$

known as *Busch's relation*. If at the time $t = 0$ the particle is at the origin O of the coordinate system, where $B_z = 0,\ v_\varphi = 0$, then $C = 0$, and the last relation yields:

$$\dot{\varphi} = -\frac{eB_z}{2m}, \tag{5.1.69}$$

meaning that the particle performs a motion of precession about the z-axis (Fig. 5.1).

This effect is applied in the construction of magnetic focusing devices, called *magnetic lenses*. The substitution of (5.1.69) into (5.1.66) leads to the second order differential equation

$$\ddot{\rho} + \left(\frac{eB_z}{2m}\right)^2 \rho = 0. \tag{5.1.70}$$

Assume that the particles move close to the z-axis and that the magnetic field acts only on a small portion of the path of the beam (paraxial beam). Then the components of the velocity along ρ and φ are negligible in comparison with the component along z. If $\dot{z} = v = \text{const.}$, we have:

$$\ddot{\rho} = \dot{z}^2\frac{d^2\rho}{dz^2} = v^2\frac{d^2\rho}{dz^2}$$

and (5.1.70) yields:

$$\frac{d^2\rho}{dz^2} + \left(\frac{eB_z}{2mv}\right)^2 \rho = 0. \tag{5.1.71}$$

Since $\frac{d^2\rho}{dz^2} < 0$, the magnetic lens is *converging*, independent of the sign of the charged particles.

The *electric lens* is based on a similar focusing principle. Both electric and magnetic lenses are used in electronic microscopy, television devices, etc.

5.1.4 Energy of a Magnetic Dipole in an External Field

Let us consider a charged particle describing a circular trajectory in the constant and homogeneous magnetic field $\mathbf{B}$. This is a closed circuit, equivalent to a magnetic dipole. Recall that for such a field the equation $\mathbf{B} = \operatorname{curl}\mathbf{A}$ admits the solution

$$\mathbf{A} = \frac{1}{2}\mathbf{B} \times \mathbf{r}. \tag{5.1.72}$$

Substitute this relation into the Hamiltonian (5.1.57) and assume that $\mathbf{B}$ is weak enough so that we can neglect its second power. Thus,

$$H = \frac{1}{2m}|\mathbf{p}|^2 + e\phi - \frac{e}{2}\mathbf{v}\cdot(\mathbf{B}\times\mathbf{r}) = H_0 + H', \tag{5.1.73}$$

where H_0 is the energy of the dipole when $\mathbf{B}$ is absent and H' is the supplementary energy of the dipole due to the existence of $\mathbf{B}$:

$$H_0 = \frac{1}{2m}|\mathbf{p}|^2 + e\phi, \quad H' = -\frac{e}{2}\mathbf{v}\cdot(\mathbf{B}\times\mathbf{r}). \tag{5.1.74}$$

The area $\Delta\mathbf{S}$ swept by the radius-vector of the particle during the time Δt is:

$$\Delta\mathbf{S} = \frac{1}{2}\mathbf{r}\times\Delta\mathbf{r},$$

and the areal velocity reads:

$$\frac{\Delta\mathbf{S}}{\Delta t} = \frac{1}{2}\mathbf{r}\times\mathbf{v},$$

such that

$$H' = -e\mathbf{B}\cdot\frac{\Delta\mathbf{S}}{\Delta t} = -\mathbf{B}\cdot\left(\frac{e}{\Delta t}\Delta\mathbf{S}\right) = -\mathbf{B}\cdot(I\Delta\mathbf{S}).$$

But $\boldsymbol{\mu} = I\Delta\mathbf{S}$ is the magnetic moment of the dipole and we arrive at the expected result:

$$H' = -\boldsymbol{\mu}\cdot\mathbf{B}. \tag{5.1.75}$$

5.2 Routh's Equations

The Lagrangian and Hamiltonian methods are distinct from each other by the choice of the independent variables: the generalized coordinates and velocities $(q_j, \dot{q}_j)$ $(j = \overline{1,n})$ in the Lagrangian approach and the canonical variables (q_j, p_j) in the Hamiltonian formulation.

In some circumstances, as in that of cyclic coordinates, it is convenient to choose as the set of independent variables a mixture of Lagrangian and Hamiltonian parameters. Such a method was given by the English mechanicist *Edward John Routh*.

Assume a dynamic system with n degrees of freedom, defined by two sets of independent variables: the Lagrangian variables $(q_j, \dot{q}_j), j = \overline{1,s}$, and the Hamiltonian variables $(q_\alpha, p_\alpha), \alpha = \overline{s+1,n}$. The variables

$$q_j, \dot{q}_j, q_\alpha, p_\alpha, t \tag{5.2.1}$$

are called *Routhian variables*. The transition from the Lagrangian set $(q_j,\ \dot{q}_j,\ t)$, $j = \overline{1,n}$, to the Routhian variables (5.2.1) is performed by a Legendre transformation. For infinitesimal and arbitrary variations of the parameters, we have:

$$dL(q_1, \ldots, q_s, q_{s+1}, \ldots, q_n, \dot{q}_1, \ldots, \dot{q}_s, \dot{q}_{s+1}, \ldots, \dot{q}_n, t)$$
$$= \sum_{j=1}^{s} \frac{\partial L}{\partial q_j} dq_j + \sum_{\alpha=s+1}^{n} \frac{\partial L}{\partial q_\alpha} dq_\alpha + \sum_{j=1}^{s} \frac{\partial L}{\partial \dot{q}_j} d\dot{q}_j + \sum_{\alpha=s+1}^{n} \frac{\partial L}{\partial \dot{q}_\alpha} d\dot{q}_\alpha + \frac{\partial L}{\partial t} dt.$$

Subtracting the quantity $d\left(\sum_{\alpha=s+1}^{n} \frac{\partial L}{\partial \dot{q}_\alpha} \dot{q}_\alpha\right)$ from both sides, we obtain:

$$dR = -\sum_{j=1}^{s} \frac{\partial L}{\partial q_j} dq_j - \sum_{j=1}^{s} \frac{\partial L}{\partial \dot{q}_j} d\dot{q}_j + \sum_{\alpha=s+1}^{n} \dot{q}_\alpha dp_\alpha - \sum_{\alpha=s+1}^{n} \dot{p}_\alpha dq_\alpha - \frac{\partial L}{\partial t} dt, \tag{5.2.2}$$

where

$$R(q_1, \ldots, q_s, \dot{q}_1, \ldots, \dot{q}_s, q_{s+1}, \ldots, q_n, p_{s+1}, \ldots, p_n, t) = \sum_{\alpha=s+1}^{n} p_\alpha \dot{q}_\alpha - L \tag{5.2.3}$$

is *Routh's function*, or, simply the *Routhian*. Equation (5.2.2) yields, on the one hand,

$$\frac{\partial L}{\partial q_j} = -\frac{\partial R}{\partial q_j}, \quad \frac{\partial L}{\partial \dot{q}_j} = -\frac{\partial R}{\partial \dot{q}_j},$$

or, by means of the Lagrange equations (5.1.2),

$$\frac{d}{dt}\left(\frac{\partial R}{\partial \dot{q}_j}\right) - \frac{\partial R}{\partial q_j} = 0 \quad (j = \overline{1,s}). \tag{5.2.4}$$

On the other hand, we have:

$$\dot{q}_\alpha = \frac{\partial R}{\partial p_\alpha}, \quad \dot{p}_\alpha = -\frac{\partial R}{\partial q_\alpha} \quad (\alpha = \overline{s+1,n}), \tag{5.2.5}$$

as well as

$$\frac{\partial R}{\partial t} = -\frac{\partial L}{\partial t}. \tag{5.2.6}$$

We conclude that the characteristic function R generates simultaneously two sets of equations: s second-order Lagrange-type differential equations (5.2.4) and $2(n - s)$ first-order Hamilton-type differential equations (5.2.5). In the first set R plays the role of the Lagrangian, while in the second it stands for the Hamiltonian. The equations (5.2.4) and (5.2.5) are called *Routh's equations.*

Combining (5.2.6) with (5.1.22), we obtain:

$$\frac{\partial L}{\partial t} = -\frac{\partial H}{\partial t} = -\frac{\partial R}{\partial t}, \tag{5.2.7}$$

meaning that if one of the three functions L, H, R does not explicitly depend on time, neither do the other two.

The definition (5.2.3) of R shows that if a coordinate q_k (k fixed) is absent from L, it will likewise not occur in the Routhian. Suppose that all the $(n - s)$ variables q_α $(\alpha = \overline{s+1, n})$ are cyclic. Then (5.1.4) gives $(n - s)$ first integrals:

$$p_\alpha = h_\alpha \quad (\alpha = \overline{s+1, n}), \tag{5.2.8}$$

where h_α are some arbitrary constants, such that

$$R = R(q_1, \ldots, q_s, \dot{q}_1, \ldots, \dot{q}_s, h_{s+1}, \ldots, h_n, t). \tag{5.2.9}$$

This means that the Lagrange-type equations (5.2.4) will be expressed in the non-cyclic coordinates $q_1, \ldots, q_s$. Integrating these equations, we obtain the non-cyclic coordinates $q_j = q_j(t)$ and the associated velocities $\dot{q}_j = \dot{q}_j(t)$. These quantities are then introduced into R and, by means of $(5.2.5)_1$, the cyclic coordinates are found by quadratures:

$$q_\alpha = \int \frac{\partial R}{\partial h_\alpha}\, dt + q_\alpha^0 \quad (\alpha = \overline{s+1, n}), \tag{5.2.10}$$

where $\frac{\partial R}{\partial h_\alpha}$ are functions of time only. The constants q_α^0 are determined from the initial conditions. In this way, the existence of the $(n - s)$ cyclic variables reduces the number of Lagrange equations from n to $s < n$.

Application. Let us study, using the Routhian formalism, the motion of a particle subject to a central force. The already known Lagrangian

$$L = \frac{1}{2} m(\dot{r}^2 + r^2\dot{\varphi}^2) - V(r) \tag{5.2.11}$$

displays the cyclic coordinate φ, such that

$$p_\varphi = \frac{\partial L}{\partial \dot{\varphi}} = mr^2\dot{\varphi} = h_\varphi(\text{const.}). \tag{5.2.12}$$

The Routh function is then

$$R = p_\varphi\dot{\varphi} - L = \frac{1}{2}mr^2\dot{\varphi}^2 - \frac{1}{2}m\dot{r}^2 + V(r)$$

or, in view of (5.2.12),

$$R(r, \dot{r}, h_\varphi) = \frac{h_\varphi^2}{2mr^2} - \frac{1}{2} m\dot{r}^2 + V(r). \tag{5.2.13}$$

The non-cyclic coordinate r is obtained by integrating

$$\frac{d}{dt}\left(\frac{\partial R}{\partial \dot{r}}\right) - \frac{\partial R}{\partial r} = 0,$$

which, by means of (5.2.13), becomes:

$$m\ddot{r} - \frac{h_\varphi^2}{mr^3} = -\frac{\partial V}{\partial r} = F. \tag{5.2.14}$$

The solution $r = r(t)$ of this equation is then introduced into (5.2.10) to determine the cyclic coordinate φ:

$$\varphi = \int \frac{\partial R}{\partial h_\varphi} dt + \varphi_0 = \frac{h_\varphi}{m} \int \frac{dt}{[r(t)]^2} + \varphi_0. \tag{5.2.15}$$

The last step is the substitution in (5.2.15) of $r(t)$ by its value earlier determined. As one can see, the variable $\varphi = \varphi(t)$ has been found by a quadrature.

5.3 Poisson Brackets

A mathematical device, which proved to be very useful in the analytical study of mechanical and non-mechanical systems, is the so-called *Poisson bracket formalism.* Let $f(q, p, t)$ and $g(q, p, t)$ be any two functions of the canonical variables $q_1, \ldots, q_n, p_1, \ldots, p_n$ and of the time t. The functions are supposed to be continuous and differentiable with respect to the variables. The expression

$$\{f, g\} = \sum_{j=1}^{n} \left(\frac{\partial f}{\partial q_j} \frac{\partial g}{\partial p_j} - \frac{\partial f}{\partial p_j} \frac{\partial g}{\partial q_j} \right) \tag{5.3.1}$$

is called the *Poisson bracket of f and g.*

By means of this abbreviation and of the canonical equations (5.1.21), we can write the total time derivative of any function $f(q, p, t)$ as

$$\frac{df}{dt} = \frac{\partial f}{\partial t} + \sum_{j=1}^{n} \left(\frac{\partial f}{\partial q_j} \dot{q}_j + \frac{\partial f}{\partial p_j} \dot{p}_j \right) = \frac{\partial f}{\partial t} + \sum_{j=1}^{n} \left(\frac{\partial f}{\partial q_j} \frac{\partial H}{\partial p_j} - \frac{\partial f}{\partial p_j} \frac{\partial H}{\partial q_j} \right),$$

or

$$\frac{df}{dt} = \frac{\partial f}{\partial t} + \{f, H\}. \tag{5.3.2}$$

Then, the relation

$$\frac{\partial f}{\partial t} + \{f, H\} = 0 \tag{5.3.3}$$

expresses the necessary and sufficient condition that $f(q, p, t) =$ const. be a first integral of Hamilton's equations (5.1.21). Indeed, since $f =$ const. is a first integral of (5.1.21), it satisfies (5.3.3). Reciprocally, the function f deduced from (5.3.3) is a first integral of the canonical equations, because the characteristic system associated with (5.3.3) is the canonical system. If f does not explicitly depend on time, (5.3.3) reduces to:

$$\{f, H\} = 0. \tag{5.3.4}$$

The following properties result immediately from the definition (5.3.1) of the Poisson bracket:

1°. $\{C, f\} = 0, \quad C = \text{const.};$
2°. $\{f, Cg\} = C\{f, g\};$
3°. $\{f, g\} = -\{g, f\};$
4°. $\{f, -g\} = -\{f, g\};$
5°. $\{f_1 + f_2, g\} = \{f_1, g\} + \{f_2, g\};$
6°. $\{f_1 f_2, g\} = f_1\{f_2, g\} + \{f_1, g\} f_2;$
7°. $\mathcal{D}\{f, g\} = \{\mathcal{D}f, g\} + \{f, \mathcal{D}g\}$, where $\mathcal{D}$ is any scalar or vector differential operator, like $\frac{\partial}{\partial t}, \frac{\partial}{\partial x}, \nabla$, etc.;
8°. $\{q_i, q_j\} = \{p_i, p_j\} = 0;$
9°. $\{q_i, p_k\} = \delta_{ik};$
10°. $\{f, \{g, h\}\} + \{h, \{f, g\}\} + \{g, \{h, f\}\} = 0$ (Jacobi's identity);
11°. $\{f, g(y_1, \ldots, y_n)\} = \sum_{i=1}^{n} \frac{\partial g}{\partial y_i}\{f, y_i\}.$

Property 3° shows the antisymmetry of the Poisson bracket and 5° its linearity relative to both functions. Property 7° is called the Leibniz rule. The brackets given by 8° and 9° are known as the *fundamental Poisson brackets*. Any pair of conjugate variables q_k, p_k (k fixed) obeys the relation

$$\{q_k, p_k\} = 1.$$

The proof of each property is very easy, except for Jacobi's identity 10°. To verify its validity, let us calculate:

$$\{f, \{g, h\}\} - \{g, \{f, h\}\} = \left\{ f, \sum_{j=1}^{n} \left(\frac{\partial g}{\partial q_j} \frac{\partial h}{\partial p_j} - \frac{\partial g}{\partial p_j} \frac{\partial h}{\partial q_j} \right) \right\} - \left\{ g, \sum_{j=1}^{n} \left(\frac{\partial f}{\partial q_j} \frac{\partial h}{\partial p_j} - \frac{\partial f}{\partial p_j} \frac{\partial h}{\partial q_j} \right) \right\}$$

or, in view of property 6°,

$$\{f,\{g,h\}\} - \{g,\{f,h\}\}$$
$$= \sum_{j=1}^{n} \frac{\partial h}{\partial p_j}\left[\left\{f, \frac{\partial g}{\partial q_j}\right\} + \left\{\frac{\partial f}{\partial q_j}, g\right\}\right] - \sum_{j=1}^{n} \frac{\partial h}{\partial q_j}\left[\left\{f, \frac{\partial g}{\partial p_j}\right\} + \left\{\frac{\partial f}{\partial p_j}, g\right\}\right]$$
$$+ \sum_{j=1}^{n}\left[\frac{\partial g}{\partial q_j}\left\{f, \frac{\partial h}{\partial p_j}\right\} - \frac{\partial g}{\partial p_j}\left\{f, \frac{\partial h}{\partial q_j}\right\} - \frac{\partial f}{\partial q_j}\left\{g, \frac{\partial h}{\partial p_j}\right\} + \frac{\partial f}{\partial p_j}\left\{g, \frac{\partial h}{\partial q_j}\right\}\right].$$

The third sum vanishes and, using the property 7°, we finally arrive at:

$$\{f,\{g,h\}\} - \{g,\{f,h\}\} = -\sum_{j=1}^{n}\left[\frac{\partial h}{\partial q_j}\frac{\partial}{\partial p_j}\{f,g\} - \frac{\partial h}{\partial p_j}\frac{\partial}{\partial q_j}\{f,g\}\right]$$
$$= -\{h,\{f,g\}\},$$

which completes the proof.

Poisson's Theorem. *If $f(q, p, t) =$ const. and $g(q, p, t) =$ const. are two first integrals of Hamilton's equations, then $\{f, g\} =$ const. is also a first integral of the same system.*

To prove the theorem, we first note that, by hypothesis,

$$\frac{\partial f}{\partial t} + \{f,H\} = 0, \quad \frac{\partial g}{\partial t} + \{g,H\} = 0. \tag{5.3.5}$$

We now write Jacobi's identity for f, g, H:

$$\{f,\{g,H\}\} + \{H,\{f,g\}\} + \{g,\{H,f\}\} = 0$$

which, by (5.3.5), becomes:

$$\left\{\frac{\partial f}{\partial t}, g\right\} + \left\{f, \frac{\partial g}{\partial t}\right\} + \{\{f,g\},H\} = 0,$$

or, in view of property 7°:

$$\frac{\partial}{\partial t}\{f,g\} + \{\{f,g\},H\} = 0,$$

and thus Poisson's theorem is proved.

The procedure indicated by Poisson's theorem does not always lead to a new (i.e. independent) first integral. Indeed, the number of the independent first integrals is limited by the number of canonical equations. Besides, the Poisson bracket of, say, f and g, can be either a linear combination of these functions, or even zero; in both situations Poisson's theorem will not lead to any independent first integral.

***Observations*:**

(a) If the Hamiltonian H does not explicitly depend on time, $H = H(q, p)$, then by (5.3.2), we have:

$$\frac{dH}{dt} = \{H, H\} = 0.$$

By taking the partial derivative with respect to time of (5.3.3), we also obtain:

$$\frac{\partial}{\partial t}\left(\frac{\partial f}{\partial t}\right) + \left\{\frac{\partial f}{\partial t}, H\right\} = 0,$$

meaning that, if $f(q, p, t)$ is a first integral of Hamilton's equations, then so is $\frac{\partial f}{\partial t}$.

(b) Using definition (5.3.1), we can give another form of the canonical equations (5.1.21). To this end, we calculate the Poisson brackets $\{q_j, H\}$ and $\{p_j, H\}$:

$$\{q_j, H\} = \sum_{k=1}^{n}\left(\frac{\partial q_j}{\partial q_k}\frac{\partial H}{\partial p_k} - \frac{\partial q_j}{\partial p_k}\frac{\partial H}{\partial q_k}\right) = \frac{\partial H}{\partial p_k}\delta_{jk} = \frac{\partial H}{\partial p_j},$$

$$\{p_j, H\} = \sum_{k=1}^{n}\left(\frac{\partial p_j}{\partial q_k}\frac{\partial H}{\partial p_k} - \frac{\partial p_j}{\partial p_k}\frac{\partial H}{\partial q_k}\right) = -\frac{\partial H}{\partial q_k}\delta_{jk} = -\frac{\partial H}{\partial q_j},$$

leading to Hamilton's equations (5.1.21) in the new form:

$$\dot{q}_j = \{q_j, H\}, \quad \dot{p}_j = \{p_j, H\}. \tag{5.3.6}$$

5.3.1 Poisson Brackets for Angular Momentum

1. Assume the choice:

$$x_1 = x, \quad x_2 = y, \quad x_3 = z, \quad p_1 = p_x, \quad p_2 = p_y, \quad p_3 = p_z. \tag{5.3.7}$$

Let us show that, if l_i and l_j are any two components of the angular momentum of a particle,

$$\mathbf{l} = \mathbf{r} \times \mathbf{p}, \quad l_i = \epsilon_{ijk} x_j p_k, \tag{5.3.8}$$

then

$$\{l_i, l_j\} = \epsilon_{ijk} l_k, \tag{5.3.9}$$

where the summation convention has been used.

To prove (5.3.9), we use (5.3.8) and the definition (5.3.1):

$$\begin{aligned}\{l_i, l_j\} &= \frac{\partial l_i}{\partial x_k}\frac{\partial l_j}{\partial p_k} - \frac{\partial l_i}{\partial p_k}\frac{\partial l_j}{\partial x_k} \\ &= \epsilon_{ism}\epsilon_{juv}(\delta_{sk}\delta_{vk}x_u p_m - \delta_{mk}\delta_{uk}x_s p_v),\end{aligned}$$

which, after some index manipulation, leads to

$$\{l_i, l_j\} = (\delta_{mj}\delta_{iu} - \delta_{mu}\delta_{ij})(x_u p_m - x_m p_u).$$

Performing the summation over u and v, we obtain:

$$\{l_i, l_j\} = x_i p_j - x_j p_i. \tag{5.3.10}$$

The antisymmetric tensor $L_{ij} = x_i p_j - x_j p_i$ can be uniquely associated with the axial vector l_k (see Appendix A), such that

$$l_k = \frac{1}{2}\,\epsilon_{kij} L_{ij}, \quad L_{ij} = \epsilon_{ijk} l_k, \tag{5.3.11}$$

which yields (5.3.9). The cyclic permutation of indices leads to three relations for the three components of the angular momentum $\mathbf{l}$. The significance of (5.3.9) is that, in view of the fundamental brackets 8° and 9°, no two components of the angular momentum can simultaneously play the role of conjugate momenta in a given system of reference. This relation also shows that, if $l_1 = \text{const.}$, $l_2 = \text{const.}$ are two first integrals of the canonical equations, then there exists a third first integral $l_3 = \text{const.}$, i.e. only two out of the three components are independent.

Let us now apply the theory to the isotropic oscillator, considered as a conservative system. The Hamiltonian

$$H = \frac{1}{2m}\,p_i p_i + \frac{1}{2}\,k x_i x_i \tag{5.3.12}$$

and the canonical equations (5.1.21) yield:

$$\dot{x}_j = \frac{1}{m}\,p_j, \quad \dot{p}_j = -k x_j \quad (j = 1, 2, 3). \tag{5.3.13}$$

Three first integrals (two of them independent) are:

$$l_i = \text{const.} \quad (i = 1, 2, 3). \tag{5.3.14}$$

Another first integral can be deduced from (5.3.12), by any choice of the index j, say $j = 1$:

$$\frac{1}{m}\,p_1^2 + k x_1^2 = E_1. \tag{5.3.15}$$

The Poisson bracket of l_3 and E_1 gives:

$$\{l_3, E_1\} = 2k x_1 x_2 + \frac{2}{m}\,p_1 p_2 = \alpha_2. \tag{5.3.16}$$

The independent first integrals are then: H, l_2, l_3, E_1, α_2. Poisson's theorem cannot furnish any new independent constant of motion. Otherwise, all the six coordinates and momenta would be equal to some constants and the integration of the canonical system would reduce to an algebraic exercise.

Nevertheless, a time-dependent sixth integral can be obtained from the equations of motion (5.3.13):

$$\omega_0 x_1 \cos \omega_0 t - \frac{1}{m}\,p_1 \sin \omega_0 t = \alpha_1 \qquad \left(\omega_0^2 = \frac{k}{m}\right).$$

Since we have already found six independent first integrals, it can be proved easily that neither Poisson's theorem, nor any other method can furnish any more independent constants of motion.

2. Let us now show that, if l^2 is the squared angular momentum of a particle, then

$$\{l_i, l^2\} = 0. \tag{5.3.17}$$

In view of property 6° and Eq. (5.3.9), we have:

$$\{l_i, l^2\} = \{l_i, l_j l_j\} = 2l_j\{l_i, l_j\} = 2\epsilon_{ijk} l_j l_k = 0,$$

since ϵ_{ijk} is antisymmetric and the product $l_j l_k$ is symmetric in the summation indices j and k. We conclude that l^2 and any component of $\mathbf{l}$ can be simultaneously chosen as conjugate momenta.

3. If x_i is an arbitrary Cartesian coordinate and l_i is an arbitrary component of the angular momentum, then

$$\{x_i, l_j\} = \epsilon_{ijk} x_k. \tag{5.3.18}$$

This relation is verified if one makes use of the properties 6°, 8° and 9°:

$$\{x_i, l_j\} = \epsilon_{jms}\{x_i, x_m p_s\} = \epsilon_{jms} x_m \delta_{is} = \epsilon_{ijm} x_m.$$

In a similar way it can be proven that

$$\{p_i, l_j\} = \epsilon_{ijk} p_k. \tag{5.3.19}$$

4. A useful Poisson bracket is that of the Hamiltonian of a particle of mass m, subject to a conservative force $\mathbf{F} = -\text{grad}\, V(x, y, z)$, with any component l_i of its angular momentum. Let us prove that

$$\{l_i, H\} = (\mathbf{r} \times \mathbf{F})_i = \mathcal{M}_i, \tag{5.3.20}$$

where

$$H = \frac{1}{2m} p_j p_j + V(x, y, z). \tag{5.3.21}$$

Indeed,

$$\{l_i, H\} = \frac{1}{2m}\{l_i, p_j p_j\} + \{l_i, V\} = \frac{1}{m} p_j\{l_i, p_j\} + \frac{\partial l_i}{\partial x_k}\frac{\partial V}{\partial p_k} - \frac{\partial l_i}{\partial p_k}\frac{\partial V}{\partial x_k},$$

or, using (5.3.8) and (5.3.19),

$$\{l_i, H\} = -\epsilon_{isk} x_s \frac{\partial V}{\partial x_k} = (\mathbf{r} \times \mathbf{F})_i,$$

which proves (5.3.20). If the force $\mathbf{F}$ is central, then

$$\{l_i, H\} = 0, \tag{5.3.22}$$

i.e. in this case l_i and H can be simultaneously regarded as conjugate momenta. The relation (5.3.22) also shows that, if l_i is time-independent, then l_i is a first integral of the canonical equations (areas theorem).

The result is the same if the particle carries the electric charge e and moves in the static electromagnetic field $\mathbf{E}$, $\mathbf{B}$. In such a case,

$$H = \frac{1}{2m}(p_j - eA_j)(p_j - eA_j) + e\phi, \tag{5.3.23}$$

$$l_i = \epsilon_{ijk} x_j (p_k - eA_k), \tag{5.3.24}$$

and the Poisson bracket reads:

$$\begin{aligned}
\{l_i, H\} &= \left\{\epsilon_{ijk} x_j (p_k - eA_k), \frac{1}{2m}(p_j - eA_j)(p_j - eA_j) + e\phi\right\} \\
&= \frac{1}{m}\left[\epsilon_{isk}(p_k - eA_k) - e\epsilon_{ijk} x_j \frac{\partial A_k}{\partial x_s}\right](p_s - eA_s) \\
&\quad - \epsilon_{ijs} x_j \left[-\frac{e}{m}(p_k - eA_k)\frac{\partial A_k}{\partial x_s} + e\frac{\partial \phi}{\partial x_s}\right] \\
&= \epsilon_{ijs} x_j \left[-e\frac{\partial \phi}{\partial x_s} + e v_k \left(\frac{\partial A_k}{\partial x_s} - \frac{\partial A_s}{\partial x_k}\right)\right].
\end{aligned}$$

Thus, by (5.1.61) and (5.1.62), the relation (5.3.20) follows immediately.

5.3.2 Poisson Brackets and Commutators

In all the aforementioned examples the physical quantities occurring in the Poisson brackets had the property of *commutativity*, meaning that for any two quantities A and B, the relation

$$AB = BA \tag{5.3.25}$$

is valid. In other words, in classical (non-quantum) mechanics we do not care about the order of the factors in a product. But, if $\hat{A}$ and $\hat{B}$ are two *operators* associated with the physical quantities A and B then, in general, the two operators are *non-commutative*, i.e.

$$\hat{A}\hat{B} \neq \hat{B}\hat{A}. \tag{5.3.26}$$

This property lies at the basis of quantum mechanics, where to each physical quantity we associate a linear Hermitian operator. If we apply (5.3.26) to some function f, then

$$\hat{A}(\hat{B}\, f) \neq \hat{B}(\hat{A}\, f), \tag{5.3.27}$$

which means that the *order* of action of the two operators is not arbitrary. The difference

$$\hat{A}\hat{B} - \hat{B}\hat{A} = [\hat{A}, \hat{B}] \tag{5.3.28}$$

is called *commutator*. Therefore, if $[\hat{A}, \hat{B}] = 0$, we say that the two operators *commute* and if $[\hat{A}, \hat{B}] \neq 0$, the operators *do not commute*. In the language of quantum mechanics, the commutativity of any two operators expresses the fact that the associated physical quantities can simultaneously be measured with arbitrary accuracy; if the operators do not commute, the simultaneous measurement of the associated physical quantities cannot be made more precise than a certain minimal value, given by Heisenberg's uncertainty principle.

In the Hamiltonian formulation of quantum mechanics, the Poisson brackets are replaced by commutators of operators, and in the following we shall establish the correlation between the Poisson bracket of any two physical quantities and the commutator of their associated operators. To this end, we assume that properties 2°–7° are valid, except for the arbitrary order of the operators appearing in property 6°. Let us denote by $\{\hat{A}, \hat{B}\}$ the quantum analog of the Poisson bracket of the two operators. In view of 6°, for any three operators $\hat{A}, \hat{B}, \hat{C}$ we can write

$$\{\hat{A}, \hat{B}\hat{C}\} = \hat{B}\{\hat{A}, \hat{C}\} + \{\hat{A}, \hat{B}\}\hat{C}, \tag{5.3.29}$$

as well as

$$\{\hat{A}\hat{B}, \hat{C}\} = \hat{A}\{\hat{B}, \hat{C}\} + \{\hat{A}, \hat{C}\}\hat{B}. \tag{5.3.30}$$

In the same way, for any four operators $\hat{A}, \hat{B}, \hat{C}, \hat{D}$ we have, on the one hand:

$$\begin{aligned}\{\hat{A}\hat{B}, \hat{C}\hat{D}\} &= \hat{A}\{\hat{B}, \hat{C}\hat{D}\} + \{\hat{A}, \hat{C}\hat{D}\}\hat{B}\\ &= \hat{A}\hat{C}\{\hat{B}, \hat{D}\} + \hat{A}\{\hat{B}, \hat{C}\}\hat{D} + \hat{C}\{\hat{A}, \hat{D}\}\hat{B} + \{\hat{A}, \hat{C}\}\hat{D}\hat{B},\end{aligned}$$

and on the other hand:

$$\begin{aligned}\{\hat{A}\hat{B}, \hat{C}\hat{D}\} &= \hat{C}\{\hat{A}\hat{B}, \hat{D}\} + \{\hat{A}\hat{B}, \hat{C}\}\hat{D}\\ &= \hat{C}\hat{A}\{\hat{B}, \hat{D}\} + \hat{C}\{\hat{A}, \hat{D}\}\hat{B} + \hat{A}\{\hat{B}, \hat{C}\}\hat{D} + \{\hat{A}, \hat{C}\}\hat{B}\hat{D}.\end{aligned}$$

The last two relations yield, by subtracting one from the other:

$$\frac{\{\hat{A}, \hat{C}\}}{[\hat{A}, \hat{C}]} = \frac{\{\hat{D}, \hat{B}\}}{[\hat{D}, \hat{B}]},$$

meaning that for any pair of operators $\hat{f}$, $\hat{g}$ we may write:

$$\{\hat{f}, \hat{g}\} = C[\hat{f}, \hat{g}], \tag{5.3.31}$$

where C is a constant. In quantum mechanics, it is shown that the operator $\hat{\varphi}$ corresponding to an observable physical quantity has to be *Hermitian*: $\hat{\varphi}^\dagger = \hat{\varphi}$. This property is satisfied if we take $C^* = -C$. The units of C follow from the definition of the Poisson bracket:

$$[C] = (Energy \times Time)^{-1}.$$

All these conditions are fulfilled by the choice

$$C = -\frac{i}{\hbar} \quad \left(\hbar = \frac{h}{2\pi}\right), \tag{5.3.32}$$

where h is the Planck constant. Thus, finally we can write:

$$\{\hat{f}, \hat{g}\} = \frac{1}{i\hbar}[\hat{f}, \hat{g}]. \tag{5.3.33}$$

Using property 9°, let us write (5.3.33) for the operators $\hat{x}_i$ and $\hat{p}_j$, associated with the conjugate variables x_i and p_j, respectively:

$$\{\hat{x}_i, \hat{p}_j\} = -\frac{i}{\hbar}\ [\hat{x}_i, \hat{p}_j] = \delta_{ij},$$

or

$$[\hat{x}_i, \hat{p}_j] = i\hbar\delta_{ij} \quad (i, j = 1, 2, 3), \tag{5.3.34}$$

which is the well-known *Heisenberg commutation relation*. It is clear that (5.3.34) is satisfied by the choice

$$\hat{x}_i = x_i, \quad \hat{p}_j = \frac{\hbar}{i}\frac{\partial}{\partial x_j} \quad (i, j = 1, 2, 3). \tag{5.3.35}$$

Here, $\hat{x}$ is the *coordinate operator* and $\hat{p}$ is the *momentum operator*. They play an important role in quantum mechanics.

Let us now transpose in the quantum language some results obtained in this section. By (5.3.2) and (5.3.33), the operator $\hat{A}$ associated with a certain physical quantity A obeys the relation:

$$\frac{d\hat{A}}{dt} = \frac{\partial\hat{A}}{\partial t} + \frac{1}{i\hbar}[\hat{A}, \hat{H}], \tag{5.3.36}$$

where the *Hamiltonian operator* is defined by

$$\hat{H} = \frac{1}{2m}\ \hat{p}_j\hat{p}_j + \hat{V}(x) \tag{5.3.37}$$

or, using the representation (5.3.35),

$$\hat{H} = -\frac{\hbar^2}{2m}\ \Delta + \hat{V}(x), \tag{5.3.38}$$

where $\Delta = \frac{\partial^2}{\partial x_j \partial x_j}$ is the Laplacian operator. Applying the operator $\hat{H}$ to the state function ψ associated with the microparticle, we find the Schrödinger equation for non-stationary states:

$$\left(-\frac{\hbar^2}{2m}\Delta + V\right)\psi = \hat{E}\psi, \quad \hat{E} = -\frac{\hbar}{i}\frac{\partial}{\partial t}. \tag{5.3.39}$$

The commutation relations for the operators associated with the components of the angular momentum are obtained in the same way, from the corresponding Poisson brackets:

$$[\hat{l}_i, \hat{l}_j] = i\hbar\epsilon_{ijk}\hat{l}_k, \tag{5.3.40}$$

$$[\hat{l}_i, \hat{l}^2] = 0. \tag{5.3.41}$$

5.3.3 Lagrange Brackets

Let $\varphi(q, p, t)$ and $\psi(q, p, t)$ be two arbitrary functions, continuous and derivable with respect to all arguments. Their Poisson bracket is:

$$\{\varphi, \psi\} = \sum_{j=1}^{n} \left(\frac{\partial \varphi}{\partial q_j} \frac{\partial \psi}{\partial p_j} - \frac{\partial \varphi}{\partial p_j} \frac{\partial \psi}{\partial q_j} \right). \tag{5.3.42}$$

As we can see, the r.h.s. is a sum of n functional determinants (Wronskians),

$$\frac{\partial(\varphi, \psi)}{\partial(q_k, p_k)} = \begin{vmatrix} \frac{\partial \varphi}{\partial q_k} & \frac{\partial \varphi}{\partial p_k} \\ \frac{\partial \psi}{\partial q_k} & \frac{\partial \psi}{\partial p_k} \end{vmatrix} \quad (k \text{ fixed}), \tag{5.3.43}$$

and we may write:

$$\{\varphi, \psi\} = \sum_{j=1}^{n} \frac{\partial(\varphi, \psi)}{\partial(q_j, p_j)} = \frac{\partial(\varphi, \psi)}{\partial(q_1, p_1)} + \cdots + \frac{\partial(\varphi, \psi)}{\partial(q_n, p_n)}. \tag{5.3.44}$$

If the Wronskians do not vanish,

$$\frac{\partial(\varphi, \psi)}{\partial(q_j, p_j)} \neq 0, \tag{5.3.45}$$

then the transformation $(\varphi, \psi) \to (q, p)$ is locally reversible and we have:

$$q_j = q_j(\varphi, \psi), \quad p_j = p_j(\varphi, \psi). \tag{5.3.46}$$

Let us now introduce the notation

$$(\varphi, \psi) = \sum_{j=1}^{n} \left(\frac{\partial q_j}{\partial \varphi} \frac{\partial p_j}{\partial \psi} - \frac{\partial q_j}{\partial \psi} \frac{\partial p_j}{\partial \varphi} \right), \tag{5.3.47}$$

called the *Lagrange bracket* of the two functions, φ and ψ.

The Lagrange brackets obey properties similar to 1°–11°. For example, the choice $\varphi \to q_k$, $\psi \to p_s$ $(k, s = \overline{1, n})$ yields:

$$(q_k, p_s) = -(p_s, q_k) = \delta_{ks}. \tag{5.3.48}$$

Similarly,

$$(q_k, q_s) = (p_k, p_s) = 0. \tag{5.3.49}$$

The brackets (5.3.48) and (5.3.49) are the *fundamental Lagrange brackets*. Since

$$(\varphi, \psi) = \sum_{k=1}^{n} \frac{\partial(q_k, p_k)}{\partial(\varphi, \psi)}, \tag{5.3.50}$$

we have:

$$\begin{aligned}(\varphi, \psi)\{\varphi, \psi\} &= \sum_{j=1}^{n} \sum_{k=1}^{n} \frac{\partial(q_k, p_k)}{\partial(\varphi, \psi)} \cdot \frac{\partial(\varphi, \psi)}{\partial(q_j, p_j)} \\ &= \sum_{j=1}^{n} \sum_{k=1}^{n} \frac{\partial(q_k, p_k)}{\partial(q_j, p_j)} = \sum_{j=1}^{n} \sum_{k=1}^{n} \delta_{jk} = n. \end{aligned} \tag{5.3.51}$$

Let us now prove that, if

$$\varphi_a = \varphi_a(q, p, t) \quad (a = \overline{1, 2n}) \tag{5.3.52}$$

are $2n$ independent and invertible functions, then

$$\sum_{a=1}^{2n} (\varphi_a, \varphi_j)\{\varphi_a, \varphi_k\} = \delta_{jk} \quad (j, k = \overline{1, n}). \tag{5.3.53}$$

From the definitions (5.3.1) and (5.3.47), we have:

$$\sum_{a=1}^{2n} (\varphi_a, \varphi_j)\{\varphi_a, \varphi_k\} = \sum_{a=1}^{2n} \left[\sum_{s=1}^{n} \left(\frac{\partial q_s}{\partial \varphi_a} \frac{\partial p_s}{\partial \varphi_j} - \frac{\partial q_s}{\partial \varphi_j} \frac{\partial p_s}{\partial \varphi_a} \right) \cdot \sum_{m=1}^{n} \left(\frac{\partial \varphi_a}{\partial q_m} \frac{\partial \varphi_k}{\partial p_m} - \frac{\partial \varphi_a}{\partial p_m} \frac{\partial \varphi_k}{\partial q_m} \right) \right].$$

But

$$\begin{aligned} &\sum_{a=1}^{2n} \frac{\partial q_s}{\partial \varphi_a} \frac{\partial \varphi_a}{\partial q_m} = \sum_{a=1}^{2n} \frac{\partial p_s}{\partial \varphi_a} \frac{\partial \varphi_a}{\partial p_m} = \delta_{sm}, \\ &\sum_{a=1}^{2n} \frac{\partial q_s}{\partial \varphi_a} \frac{\partial \varphi_a}{\partial p_m} = \sum_{a=1}^{2n} \frac{\partial p_s}{\partial \varphi_a} \frac{\partial \varphi_a}{\partial q_m} = 0, \end{aligned}$$

and thus, finally,

$$\sum_{a=1}^{2n} (\varphi_a, \varphi_j)\{\varphi_a, \varphi_k\} = \sum_{s=1}^{n} \sum_{m=1}^{n} \left(\frac{\partial \varphi_k}{\partial p_m} \frac{\partial p_s}{\partial \varphi_j} \delta_{sm} + \frac{\partial \varphi_k}{\partial q_m} \frac{\partial q_s}{\partial \varphi_j} \delta_{sm} \right) = \frac{\partial \varphi_k}{\partial \varphi_j} = \delta_{jk}.$$

As an immediate application of (5.3.53), we shall obtain the fundamental Poisson brackets, starting from the Lagrange fundamental brackets. To this end, we split the sum into two groups of terms:

$$\sum_{a=1}^{2n}(\varphi_a,\varphi_j)\{\varphi_a,\varphi_k\} = \sum_{i=1}^{n}(\varphi_i,\varphi_j)\{\varphi_i,\varphi_k\} + \sum_{i=n+1}^{2n}(\varphi_i,\varphi_j)\{\varphi_i,\varphi_k\}.$$

Now we choose $\varphi_i = q_i$ in the first sum on the r.h.s. and $\varphi_{n+i} = p_i$ in the second. Then, for $\varphi_j = q_j, \varphi_k = p_k$, we find:

$$\sum_{i=1}^{n}(q_i,q_j)\{q_i,p_k\} + \sum_{i=1}^{n}(p_i,q_j)\{p_i,p_k\} = 0,$$

or, by (5.3.48) and (5.3.49),

$$\{p_i,p_k\} = 0.$$

Using the same procedure, an appropriate choice of φ_j, φ_k yields the rest of the fundamental Poisson brackets:

$$\{q_j,q_k\} = 0, \quad \{q_j,p_k\} = \delta_{jk}.$$

5.4 Canonical Transformations

As we have seen, the choice of generalized coordinates within the analytical formalism is not unique, but it should be made appropriately, so that to obtain the maximum information about the physical system in the simplest possible way. For example, the coordinates r, φ are more useful than x, y in the study of central force problems.

In the three-dimensional space, the transition from the Cartesian coordinates x_i $(i = 1, 2, 3)$ to, for example, the curvilinear orthogonal coordinates q_j $(j = 1, 2, 3)$ is given by:

$$x_i = x_i(q_1,q_2,q_3) \quad (i = 1,2,3). \tag{5.4.1}$$

This transformation can be generalized in the configuration space. Here, the position of the representative point can be determined by any appropriate choice of Lagrangian variables. Let q_j and Q_j $(j = \overline{1,n})$ be two such sets of variables. The relations

$$Q_j = Q_j(q_1,\ldots,q_n,t) \quad (j = \overline{1,n}), \tag{5.4.2}$$

where it is assumed that the variable t appears explicitly, represent a *point transformation* in the configuration space. If the Jacobian of the transformation (5.4.2) is non-zero, we also have the inverse transformation,

$$q_k = q_k(Q_1,\ldots,Q_n,t) \quad (k = \overline{1,n}). \tag{5.4.3}$$

Passing now to the Hamiltonian formalism, we first recall that the generalized coordinates and the generalized momenta are independent variables. This means that in the phase space the most general transformation is

$$Q_j = Q_j(q, p, t), \quad P_j = P_j(q, p, t) \quad (j = \overline{1, n}), \tag{5.4.4}$$

where, as before, we have left out the indices inside the parentheses. This is, obviously, a point transformation in the $2n$-dimensional phase space. We also assume that the transformation (5.4.4) is locally invertible, i.e.

$$\frac{\partial(Q, P)}{\partial(q, p)} = \frac{\partial(Q_1, \ldots, Q_n, P_1, \ldots, P_n)}{\partial(q_1, \ldots, q_n, p_1, \ldots, p_n)} \neq 0. \tag{5.4.5}$$

If the system of canonical equations (5.1.21) keep their form under the transformation (5.4.4), that is if the new variables Q_j, P_j obey

$$\dot{Q}_j = \frac{\partial \mathcal{H}}{\partial P_j}, \quad \dot{P}_j = -\frac{\partial \mathcal{H}}{\partial Q_j} \quad (j = \overline{1, n}), \tag{5.4.6}$$

where

$$\mathcal{H} = \mathcal{H}(Q_1, \ldots, Q_n, P_1, \ldots, P_n, t) \tag{5.4.7}$$

is the Hamiltonian of the system in the new representation, the transformation (5.4.4) preserving the form of Hamiltonian equation of motion is called *canonical*. In mathematics, such a transformation is called *contact* transformation, whereas the concept of *canonical* or *symplectic* transformation refers to the case when none of the physical quantities explicitly depends on the time t, i.e., there is no explicit t-dependence in (5.4.2)–(5.4.4) and similarly $H = H(q, p)$ and $\mathcal{H} = \mathcal{H}(Q, P)$ are time-independent. However, the contact geometry is related but different from the symplectic one.

Since not *all* the transformations of the form (5.4.4) have this property, let us find the condition of canonicity. To this end, we use Hamilton's principle (5.1.23) in the form:

$$\delta \int_{t_1}^{t_2} \left[\sum_{j=1}^{n} p_j \, dq_j - H(q, p, t) \, dt \right] = 0. \tag{5.4.8}$$

The reader is already acquainted (see Sect. 5.1) with the fact that the canonical equations can be derived by means of Hamilton's principle. In order that the new variables Q_j, P_j also satisfy the canonical equations, they must verify Hamilton's principle:

$$\delta \int_{t_1}^{t_2} \left[\sum_{j=1}^{n} P_j \, dQ_j - \mathcal{H}(Q, P, t) \, dt \right] = 0. \tag{5.4.9}$$

The equations deriving from (5.4.8) and (5.4.9) must describe the same motion, so that we must have

$$\int_{t_1}^{t_2}\left(\sum_{j=1}^{n} p_j\, dq_j - H\, dt\right) = c\int_{t_1}^{t_2}\left(\sum_{j=1}^{n} P_j\, dQ_j - \mathcal{H}\, dt\right), \tag{5.4.10}$$

where the constant c is called the *valence* of the canonical transformation. If $c = 1$, the transformation is *univalent*. Here and hereafter we shall deal with univalent canonical transformations only.

The integrands in (5.4.10) can differ only by a total differential of some scalar function F, because

$$\delta\int_{t_1}^{t_2} dF = \delta F(t_2) - \delta F(t_1) = 0.$$

Consequently, we obtain:

$$\sum_{j=1}^{n} p_j dq_j - Hdt = \sum_{j=1}^{n} P_j dQ_j - \mathcal{H}dt + dF, \tag{5.4.11}$$

known as the *canonicity condition.*

The function F is called the *generating function* of the canonical transformation (5.4.4). As one can see, it is a function of $2n + 1$ independent variables: $q_1, ..., q_n, Q_1, ..., Q_n, t$. Let us name it $F_1(q, Q, t)$. Then,

$$dF_1 = \sum_{j=1}^{n}\left(\frac{\partial F_1}{\partial q_j}\, dq_j + \frac{\partial F_1}{\partial Q_j} dQ_j\right) + \frac{\partial F_1}{\partial t} dt, \tag{5.4.12}$$

and, in view of (5.4.11),

$$p_j = \frac{\partial F_1}{\partial q_j}, \quad P_j = -\frac{\partial F_1}{\partial Q_j}, \quad \mathcal{H} = H + \frac{\partial F_1}{\partial t} \quad (j = \overline{1,n}). \tag{5.4.13}$$

Equations $(5.4.13)_1$ give Q_j in terms of $q_1, \ldots, q_n, p_1, \ldots, p_n, t$, i.e. the first set $(5.4.4)_1$ of the canonical transformations. These are then introduced into $(5.4.13)_2$, which gives the second set $(5.4.4)_2$ of the canonical transformations. Finally, $(5.4.13)_3$ expresses the connection between the two Hamiltonians, H and $\mathcal{H}$.

The generating function F makes possible the transition from the old variables q_j, p_j to the new variables Q_j, P_j. It must depend on n old variables (either q_j or p_j), n new variables (either Q_j or P_j) and, of course, on the time t. Thus, besides our previous case, there are still three other possibilities:

$$F_2(q, P), \quad F_3(p, Q), \quad F_4(p, P).$$

The transition from q_j, Q_j to q_j, P_j, as independent variables, is performed by means of a Legendre transformation. Adding the quantity $d\left(\sum_{j=1}^{n} P_j Q_j\right)$ to both sides of (5.4.11), we find

$$\sum_{j=1}^{n} p_j dq_j - Hdt = -\sum_{j=1}^{n} Q_j dP_j - \mathcal{H}dt + dF_2, \tag{5.4.14}$$

where

$$F_2(q, P, t) = F_1(q, Q, t) + \sum_{j=1}^{n} P_j Q_j \tag{5.4.15}$$

is the new generating function. Following the procedure given in the previous case, we obtain:

$$p_j = \frac{\partial F_2}{\partial q_j}, \quad Q_j = \frac{\partial F_2}{\partial P_j}, \quad \mathcal{H} = H + \frac{\partial F_2}{\partial t} \quad (j = \overline{1,n}). \tag{5.4.16}$$

Another possible Legendre transformation is realized by subtracting $d\left(\sum_{j=1}^{n} p_j q_j\right)$ from both sides of (5.4.11). The result is:

$$-\sum_{j=1}^{n} q_j dp_j - Hdt = \sum_{j=1}^{n} P_j dQ_j - \mathcal{H}dt + dF_3, \tag{5.4.17}$$

with

$$F_3(p, Q, t) = F_1(q, Q, t) - \sum_{j=1}^{n} p_j q_j. \tag{5.4.18}$$

The transformation equations are:

$$q_j = -\frac{\partial F_3}{\partial p_j}, \quad P_j = -\frac{\partial F_3}{\partial Q_j}, \quad \mathcal{H} = H + \frac{\partial F_3}{\partial t}. \tag{5.4.19}$$

The last possible choice is the transition from q_j, Q_j to p_j, P_j as independent variables. This is done by a double Legendre transformation. Adding $d\sum_{j=1}^{n}(P_j Q_j - p_j q_j)$ to both sides of (5.4.11), we arrive at:

$$-\sum_{j=1}^{n} q_j dp_j - H\,dt = -\sum_{j=1}^{n} Q_j dP_j - \mathcal{H}dt + dF_4, \tag{5.4.20}$$

where we have denoted

$$F_4(p, P, t) = F_1(q, Q, t) + \sum_{j=1}^{n}(P_j Q_j - p_j q_j). \tag{5.4.21}$$

The corresponding equations of transformation are:

$$q_j = -\frac{\partial F_4}{\partial p_j}, \quad Q_j = \frac{\partial F_4}{\partial P_j}, \quad \mathcal{H} = H + \frac{\partial F_4}{\partial t}. \tag{5.4.22}$$

Observations:

(a) The differential equation (5.4.11) expresses the *necessary* and *sufficient* condition of canonicity. The first part of this assertion has already been proved. To prove the sufficiency, we must show that, if (5.4.13) and the canonical equations (5.1.21) written for q_j, p_j are satisfied, then the canonical equations in terms of Q_j, P_j are also valid. The total time derivative of $(5.4.13)_1$ yields:

$$\dot{p}_j = \frac{\partial^2 F_1}{\partial t \partial q_j} + \sum_{k=1}^{n} \frac{\partial^2 F_1}{\partial q_k \partial q_j} \dot{q}_k + \sum_{k=1}^{n} \frac{\partial^2 F_1}{\partial Q_k \partial q_j} \dot{Q}_k$$
$$= \frac{\partial}{\partial q_j}(\mathcal{H} - H) + \sum_{k=1}^{n} \frac{\partial p_k}{\partial q_j} \dot{q}_k - \sum_{k=1}^{n} \frac{\partial P_k}{\partial q_j} \dot{Q}_k. \tag{5.4.23}$$

But, in view of (5.4.13),

$$\mathcal{H}(Q, P, t) = \mathcal{H}[Q, P(q, Q, t), t], \quad H(q, p, t) = H[q, p(q, Q, t), t],$$

and thus

$$\frac{\partial}{\partial q_j}(\mathcal{H} - H) = \sum_{k=1}^{n} \left(\frac{\partial \mathcal{H}}{\partial P_k} \frac{\partial P_k}{\partial q_j} - \frac{\partial H}{\partial p_k} \frac{\partial p_k}{\partial q_j} \right) - \frac{\partial H}{\partial q_j}. \tag{5.4.24}$$

Substituting (5.4.24) into (5.4.23) and taking into account (5.1.21), we arrive at

$$\sum_{k=1}^{n} \left(\dot{Q}_k - \frac{\partial \mathcal{H}}{\partial P_k} \right) \frac{\partial P_k}{\partial q_j} = 0 \quad (j = \overline{1, n}). \tag{5.4.25}$$

If

$$\frac{\partial P_k}{\partial q_j} \neq 0, \tag{5.4.26}$$

then we obtain the first set of canonical equations (5.4.6). The second set is deduced in a similar way, by taking the total time derivative of $(5.4.13)_2$ and following the same procedure. This last step is left to the reader as an exercise.

(b) Irrespective of the type of the generating function F, the new Hamiltonian $\mathcal{H}(Q, P, t)$ is obtained from the old one $H(q, p, t)$ by adding the partial derivative with respect to time of the function F. If the generating function does not explicitly depend on time, then

$$\mathcal{H} = H. \tag{5.4.27}$$

In this case we have a *completely canonical* transformation.

5.4.1 Extensions and Applications

The choice of the set of independent variables is indicated by the characteristics of the problem. The following examples will give the reader a better idea of the utility of canonical transformations.

1. Consider the generating function

$$F = F_2 = \sum_{j=1}^{n} q_j P_j. \tag{5.4.28}$$

Then the transformation equations (5.4.16) yield:

$$p_j = P_j, \quad Q_j = q_j, \quad \mathcal{H} = H \quad (j = \overline{1,n}), \tag{5.4.29}$$

meaning that (5.4.28) generates an *identity transformation*. Reciprocally, it can be shown that any identity transformation is canonical.

Now, if instead we choose

$$F = F_2 = -\sum_{j=1}^{n} q_j P_j, \tag{5.4.30}$$

then

$$p_j = -P_j, \quad Q_j = -q_j, \quad \mathcal{H} = H, \tag{5.4.31}$$

hence any inversion in the phase space is a canonical transformation.

2. Consider the completely canonical transformation

$$Q_j = Q_j(q, p), \quad P_j = P_j(q, p) \quad (j = \overline{1,n}). \tag{5.4.32}$$

Since in this case $\mathcal{H} = H$, Eq. (5.4.11) reads:

$$\sum_{j=1}^{n} p_j dq_j = \sum_{j=1}^{n} P_j dQ_j + dF(q, Q), \tag{5.4.33}$$

or, in view of (5.4.32),

$$\sum_{j=1}^{n} p_j dq_j = \sum_{j=1}^{n} \sum_{k=1}^{n} P_j \left(\frac{\partial Q_j}{\partial q_k} dq_k + \frac{\partial Q_j}{\partial p_k} dp_k \right) + \sum_{j=1}^{n} \left(\frac{\partial F}{\partial q_j} dq_j + \frac{\partial F}{\partial p_j} dp_j \right),$$

which yields:

$$p_j = \frac{\partial F}{\partial q_j} + \sum_{k=1}^{n} P_k \frac{\partial Q_k}{\partial q_j}, \quad 0 = \frac{\partial F}{\partial p_j} + \sum_{k=1}^{n} P_k \frac{\partial Q_k}{\partial p_j} \quad (j = \overline{1,n}). \tag{5.4.34}$$

Next, take partial derivative of $(5.4.34)_1$ with respect to p_s, then change the index j to s in $(5.4.34)_2$ and take its partial derivative with respect to q_j and finally subtract one equation from the other, obtaining

$$\sum_{k=1}^{n}\left(\frac{\partial Q_k}{\partial q_j}\frac{\partial P_k}{\partial p_s}-\frac{\partial Q_k}{\partial p_s}\frac{\partial P_k}{\partial q_j}\right)=(q_j,p_s)=\delta_{js}, \tag{5.4.35}$$

i.e. one of the fundamental Lagrange brackets. Suitable derivatives of (5.4.33) give the remaining fundamental Lagrange brackets:

$$(q_j,q_s)=0, \quad (p_j,p_s)=0. \tag{5.4.36}$$

This analysis shows that the transformation (5.4.32) is canonical, if the Lagrange fundamental brackets hold true. If this transformation is invertible, a similar procedure leads to the Lagrange brackets in the new variables:

$$(Q_m,P_k)=\delta_{km}, \quad (Q_m,Q_k)=0, \quad (P_m,P_k)=0. \tag{5.4.37}$$

3. Let us now show that the Poisson bracket of any two functions of the canonical variables is invariant under the canonical transformation (5.4.4), i.e.

$$\{\varphi,\psi\}_{Q,P}=\{\varphi,\psi\}_{q,p}. \tag{5.4.38}$$

We begin by proving the invariance of the fundamental Poisson brackets:

$$\begin{gathered}\{q_j,p_k\}_{Q,P}=\{q_j,p_k\}_{q,p}, \quad \{q_j,q_k\}_{Q,P}=\{q_j,q_k\}_{q,p},\\ \{p_j,p_k\}_{Q,P}=\{p_j,p_k\}_{q,p}.\end{gathered} \tag{5.4.39}$$

Taking the partial derivative of $(5.4.13)_1$ and $(5.4.13)_2$ relative to Q_s and q_j, respectively, and choosing suitably the indices, we have:

$$\frac{\partial p_j}{\partial Q_s}=\frac{\partial^2 F}{\partial q_j\partial Q_s}, \quad \frac{\partial P_s}{\partial q_j}=-\frac{\partial^2 F}{\partial Q_s\partial q_j},$$

and thus

$$\frac{\partial p_j}{\partial Q_s}=-\frac{\partial P_s}{\partial q_j} \quad (j,s=\overline{1,n}). \tag{5.4.40}$$

Similarly, by means of (5.4.16), (5.4.19) and (5.4.22), we obtain:

$$\frac{\partial p_j}{\partial P_s}=\frac{\partial Q_s}{\partial q_j}, \quad \frac{\partial q_j}{\partial Q_s}=\frac{\partial P_s}{\partial p_j}, \quad \frac{\partial q_j}{\partial P_s}=-\frac{\partial Q_s}{\partial p_j} \quad (j,s=\overline{1,n}). \tag{5.4.41}$$

On the other hand, the Poisson bracket of q_j and p_k in the new variables Q_j, P_j, is:

$$\{q_j,p_k\}_{Q,P}=\sum_{s=1}^{n}\left(\frac{\partial q_j}{\partial Q_s}\frac{\partial p_k}{\partial P_s}-\frac{\partial q_j}{\partial P_s}\frac{\partial p_k}{\partial Q_s}\right) \quad (j,k=\overline{1,n}), \tag{5.4.42}$$

or, using (5.4.40) and (5.4.41),

$$\{q_j, p_k\}_{Q,P} = \sum_{s=1}^{n} \left(\frac{\partial q_j}{\partial Q_s} \frac{\partial Q_s}{\partial q_k} + \frac{\partial q_j}{\partial P_s} \frac{\partial P_s}{\partial q_k} \right) = \frac{\partial q_j}{\partial q_k} = \delta_{jk} = (q_j, p_k)_{q,p}.$$

Also,

$$\{q_j, q_k\}_{Q,P} = \sum_{s=1}^{n} \left(\frac{\partial q_j}{\partial Q_s} \frac{\partial q_k}{\partial P_s} - \frac{\partial q_j}{\partial P_s} \frac{\partial q_k}{\partial Q_s} \right)$$

$$= -\sum_{s=1}^{n} \left(\frac{\partial q_j}{\partial Q_s} \frac{\partial Q_s}{\partial p_k} + \frac{\partial q_j}{\partial P_s} \frac{\partial P_s}{\partial p_k} \right) = -\frac{\partial q_j}{\partial p_k} = 0 = \{q_j, q_k\}_{q,p},$$

as well as

$$\{p_j, p_k\}_{Q,P} = 0 = \{p_j, p_k\}_{q,p},$$

which completes the first part of our proof.

Now, going back to the Poisson bracket of φ and ψ, we have:

$$\{\varphi, \psi\}_{Q,P} = \sum_{k=1}^{n} \left(\frac{\partial \varphi}{\partial Q_k} \frac{\partial \psi}{\partial P_k} - \frac{\partial \varphi}{\partial P_k} \frac{\partial \psi}{\partial Q_k} \right)$$

$$= \sum_{k=1}^{n} \left[\frac{\partial \varphi}{\partial Q_k} \left(\frac{\partial \psi}{\partial q_j} \frac{\partial q_j}{\partial P_k} + \frac{\partial \psi}{\partial p_j} \frac{\partial p_j}{\partial P_k} \right) - \frac{\partial \varphi}{\partial P_k} \left(\frac{\partial \psi}{\partial q_j} \frac{\partial q_j}{\partial Q_k} + \frac{\partial \psi}{\partial p_j} \frac{\partial p_j}{\partial Q_k} \right) \right],$$

which becomes, after some manipulation,

$$\{\varphi, \psi\}_{Q,P} = \sum_{j=1}^{n} \frac{\partial \psi}{\partial q_j} \{\varphi, q_j\}_{Q,P} + \sum_{j=1}^{n} \frac{\partial \psi}{\partial p_j} \{\varphi, p_j\}_{Q,P}. \tag{5.4.43}$$

Apply now this relation to the brackets $\{q_j, \varphi\}_{Q,P}$ and $\{p_j, \varphi\}_{Q,P}$. Since

$$\{q_j, \varphi\}_{Q,P} = \sum_{k=1}^{n} \frac{\partial \varphi}{\partial q_k} \{q_j, q_k\}_{Q,P} + \sum_{k=1}^{n} \frac{\partial \varphi}{\partial p_k} \{q_j, p_k\}_{Q,P} = \sum_{k=1}^{n} \frac{\partial \varphi}{\partial p_k} \delta_{jk} = \frac{\partial \varphi}{\partial p_j},$$

$$\{p_j, \varphi\}_{Q,P} = \sum_{k=1}^{n} \frac{\partial \varphi}{\partial q_k} \{p_j, q_k\}_{Q,P} + \sum_{k=1}^{n} \frac{\partial \varphi}{\partial p_k} \{p_j, p_k\}_{Q,P} = -\sum_{k=1}^{n} \frac{\partial \varphi}{\partial q_k} \delta_{jk} = -\frac{\partial \varphi}{\partial q_j},$$

equation (5.4.43) simplifies to

$$\{\varphi, \psi\}_{Q,P} = \sum_{j=1}^{n} \left(\frac{\partial \varphi}{\partial q_j} \frac{\partial \psi}{\partial p_j} - \frac{\partial \varphi}{\partial p_j} \frac{\partial \psi}{\partial q_j} \right) = \{\varphi, \psi\}_{q,p},$$

which completes the proof of (5.4.38).

This property enables us to show that the successive application of two canonical transformations is also a canonical transformation. Let us assume that the first transformation is given by (5.4.4), while the second is

$$Q'_j = Q'_j(Q, P, t), \quad P'_j = P'_j(Q, P, t) \qquad (j = \overline{1, n}). \tag{5.4.44}$$

The transformation (5.4.4) is canonical if, according to (5.4.38),

$$\{Q'_j, P'_k\}_{q,p} = \{Q'_j, P'_k\}_{Q,P}, \tag{5.4.45}$$

while (5.4.44) is canonical if it satisfies the fundamental Poisson bracket,

$$\{Q'_j, P'_k\}_{Q,P} = \delta_{jk}. \tag{5.4.46}$$

The last two relations then yield

$$\{Q'_j, P'_k\}_{q,p} = \delta_{jk}. \tag{5.4.47}$$

In a similar way we find:

$$\{Q'_j, Q'_k\}_{q,p} = 0, \quad \{P'_j, P'_k\}_{q,p} = 0. \tag{5.4.48}$$

The relations (5.4.47) and (5.4.48) give the proof of the aforementioned property. Combining this result with the existence of the identity and inverse transformations as canonical transformations, eqs. (5.4.28)–(5.4.31), we conclude that the class of canonical transformations has a *group structure*.

Using a similar procedure, it is not difficult to prove that the Lagrange bracket (φ, ψ) is also invariant with respect to the canonical transformation (5.4.4):

$$(\varphi, \psi)_{Q,P} = (\varphi, \psi)_{q,p}. \tag{5.4.49}$$

This proof is left to the reader.

4. Let us show that the absolute value of the Jacobian of any canonical transformation is equal to one. We first observe that the Jacobian

$$J = \frac{\partial(Q_j, P_k)}{\partial(q_s, p_m)} \quad (j, k, s, m = \overline{1, n}) \tag{5.4.50}$$

is a $2n \times 2n$ determinant:

$$J = \begin{vmatrix} \frac{\partial Q_j}{\partial q_s} \cdots & \cdots \frac{\partial Q_j}{\partial p_m} \\ \vdots & \vdots \\ \frac{\partial P_k}{\partial q_s} \cdots & \cdots \frac{\partial P_k}{\partial p_m} \end{vmatrix}. \tag{5.4.51}$$

Using (5.4.40) and (5.4.41) in (5.4.51), we have:

$$J = \begin{vmatrix} \frac{\partial p_s}{\partial P_j} \cdots & \cdots -\frac{\partial q_m}{\partial P_j} \\ \vdots & \vdots \\ -\frac{\partial p_s}{\partial Q_k} \cdots & \cdots \frac{\partial q_m}{\partial Q_k} \end{vmatrix} = \frac{\partial(q_m, p_s)}{\partial(Q_k, P_j)} = J^{-1}, \tag{5.4.52}$$

and thus $J^2 = 1$ and $|J| = 1$.

5. In some applications, the generating function F is given and we are supposed to find the associated transformation and show whether it is canonical. In other cases, the transformation is given and we have to show whether it is canonical and find the generating function.

Let us consider an example in which the generating function is

$$F \equiv F_2(q, P) = qP + \frac{1}{6}\frac{P^3}{m^2 g}, \tag{5.4.53}$$

where m is the mass and q is the vertical coordinate of a freely falling particle in the constant gravitational field specified by the acceleration of gravity g. Since in our case the independent variables are q and P, using (5.4.16) we obtain:

$$p = \frac{\partial F_2}{\partial q} = P, \quad Q = \frac{\partial F_2}{\partial P} = q + \frac{1}{2}\frac{P^2}{m^2 g}. \tag{5.4.54}$$

It is very easy to show that this transformation is canonical. Indeed, calculating the Jacobian, which in our case coincides with the Poisson bracket $\{Q, P\}$, we have:

$$\{Q, P\} = \frac{\partial Q}{\partial q}\frac{\partial P}{\partial p} - \frac{\partial P}{\partial q}\frac{\partial Q}{\partial p} = 1.$$

Since the generating function does not explicitly depend on time, the Hamiltonian of our application is found by using $(5.4.16)_3$ and (5.4.54):

$$\mathcal{H} = H = \frac{p^2}{2m} + mgq = mgQ. \tag{5.4.55}$$

Hamilton's canonical equations (5.4.6) then yield:

$$\dot{Q} = \frac{\partial \mathcal{H}}{\partial P} = 0, \quad \dot{P} = -\frac{\partial \mathcal{H}}{\partial Q} = -mg,$$

which, by integration, give:

$$Q = C_1, \quad P = -mgt + C_2,$$

where C_1, C_2 are constants. Therefore the expected solution of the problem, written in the variables q, p, is:

$$q = Q - \frac{P^2}{2m^2 g} = -\frac{1}{2}gt^2 + C_3 t + C_4, \quad p = P = -mgt + C_2, \tag{5.4.56}$$

with $C_3 = C_2/m, C_4 = C_1 - C_2^2/2m^2 g$. The integration constants C_1 and C_2 are determined from the initial conditions.

The reader is advised to solve the same problem using the remaining possible forms of the generating function: $F_1(q, Q), F_3(p, Q), F_4(p, P)$. Are all these functions consistent with the theory?

6. Consider now the following generating function:

$$F \equiv F_1 = \frac{1}{2} m\omega_0 q^2 \cot Q, \tag{5.4.57}$$

where m and ω_0 are two constants. Then, using (5.4.13), we find the variables p, P, and $\mathcal{H}$ as functions of the independent variables q, Q:

$$p = \frac{\partial F_1}{\partial q} = m\omega_0 q \cot Q, \tag{5.4.58}$$

$$P = -\frac{\partial F_1}{\partial Q} = \frac{m\omega_0}{2\sin^2 Q} q^2, \tag{5.4.59}$$

$$\mathcal{H} = H. \tag{5.4.60}$$

The desired transformation is then:

$$q = \left(\frac{2P}{m\omega_0}\right)^{\frac{1}{2}} \sin Q, \quad p = (2m\omega_0 P)^{\frac{1}{2}} \cos Q, \tag{5.4.61}$$

or, conversely,

$$Q = \arctan\frac{m\omega_0 q}{p}, \quad P = \frac{1}{2m\omega_0} p^2 + \frac{1}{2} m\omega_0 q^2. \tag{5.4.62}$$

Since

$$\{Q, P\} = \frac{\partial Q}{\partial q}\frac{\partial P}{\partial p} - \frac{\partial Q}{\partial p}\frac{\partial P}{\partial q} = J = 1, \tag{5.4.63}$$

the transformation (5.4.61) is canonical.

Assume now that m and ω_0 are the mass and the angular frequency of a harmonic oscillator. In this case,

$$\mathcal{H} = H = \frac{p^2}{2m} + \frac{1}{2} m\omega_0 q^2$$

and, according to (5.4.61),

$$\mathcal{H} = \omega_0 P. \tag{5.4.64}$$

The coordinate Q is cyclic, which means that we have the first integral

$$P = \text{const.}, \tag{5.4.65}$$

while the canonical equations (5.4.6) give:

$$Q = \omega_0 t + \beta, \tag{5.4.66}$$

where β is a constant of integration. Using (5.4.61) and (5.4.66), we arrive at

$$q = \left(\frac{2P}{m\omega_0}\right)^{\frac{1}{2}} \sin(\omega_0 t + \beta) = A\ \sin(\omega_0 t + \beta).$$

The constants A and β are determined from the initial conditions.

7. Suppose, this time, that the transformation

$$Q = \sqrt{2q}\, e^t \cos p, \quad P = \sqrt{2q}\, e^{-t} \sin p \tag{5.4.67}$$

is given and we have to show that it is canonical and to find the generating function. To show the first part of the problem, we can either prove that the absolute value of the Jacobian is one, as above, or use the canonicity condition (5.4.11), which yields:

$$p\,dq - P\,dQ = (p - \sin p \cos p)dq + 2q\ \sin^2 p\ dp, \tag{5.4.68}$$

or, after a convenient grouping of terms,

$$p\,dq - P\,dQ = d(qp - q \sin p \cos p),$$

and consequently

$$F = q(p - \sin p \cos p), \tag{5.4.69}$$

The last step is to express F in terms of q (or p) and Q (or P). Using (5.4.67), we obtain the generating function in the form:

$$F(q, Q, t) = q \cos^{-1} \frac{Q\ e^{-t}}{\sqrt{2q}} - \frac{1}{2}\ Q\ e^{-t}\sqrt{2q - Q^2\ e^{-2t}}. \tag{5.4.70}$$

Note that we arrived at the result (5.4.68) by a straightforward calculation, but in general one uses mathematical formalism to prove that a differential expression is an exact differential. In our case, in view of (5.4.68), we must have

$$\frac{\partial}{\partial q}\,(2q \sin^2 p) = \frac{\partial}{\partial p}\,(p - \sin\ p \cos\ p),$$

which is true. Since

$$\frac{\partial F}{\partial q} = p - \sin\ p \cos\ p, \quad \frac{\partial F}{\partial p} = 2q\ \sin^2 p,$$

by integration we obviously obtain the same result (5.4.69).

5.4.2 Mechanical–Thermodynamical Analogy. Thermodynamic Potentials

In some applications, like in the study of the motion of a fluid, the parameters characterizing the thermodynamical behaviour of the system must also be taken into account. Since the principles of mechanics do not provide the equations describing thermodynamical processes, we 'borrow' from thermodynamics the equation of state, usually written in the form:

$$f(p, V, T) = 0. \tag{5.4.71}$$

Here, p is the pressure, V – the volume and T – the absolute temperature of the system. The principles of thermodynamics connect these three parameters and two other thermodynamic functions, the entropy S and the internal energy U, by the relation

$$T\,dS \geq dU + p\,dV, \tag{5.4.72}$$

where the equality corresponds to the equilibrium processes. Assuming that we are concerned only with equilibrium transformations, let us compare

$$T\,dS = dU + p\,dV \tag{5.4.73}$$

with $(5.4.1)_1$, written for a complete canonical transformation:

$$\sum_{j=1}^{n} p_j dq_j = \sum_{j=1}^{n} P_j dQ_j + dF(q, Q). \tag{5.4.74}$$

It is obvious that (5.4.73) expresses the canonical transformation of the variables S, T to the parameters V, p, the generating function being $U(S, V)$. Since S, V are independent variables, the functions p, T are given by (5.4.13):

$$T = \left(\frac{\partial U}{\partial S}\right)_V, \quad p = -\left(\frac{\partial U}{\partial V}\right)_S. \tag{5.4.75}$$

Here, the volume V plays the role of a generalized coordinate conjugated to the pressure p, while the entropy S is a generalized coordinate conjugate to the temperature T. Then, p and T are generalized momenta, while the phase space reduces to two dimensions: T, S in the old variables and p, V in the new ones. This gives the possibility of graphic representations of the thermodynamic transformations.

The function $U(S, V)$ is called *thermodynamic potential*. The name shows that T and p are expressed in terms of U by (5.4.75), very much like a force $\mathbf{F}$ is obtained from a given potential energy.

The choice of $U(S, V)$ as a thermodynamic potential is not useful, because the independent variable S cannot be directly determined. Then, for practical reasons, instead of U are used some other thermodynamic potentials: the enthalpy H, the Helmholtz free energy F and Gibbs' free energy (sometimes called free enthaply) G. The transition from U to the new characteristic functions is realized by some appropriate Legendre transformations. A comparative table for the canonical variables and the generating functions is given below:

$$\begin{aligned}
&T = T(S, V), \quad p = p(S, V), \quad F_1(S, V) = U(S, V), \\
&T = T(S, p), \quad V = V(S, p), \quad F_2(S, p) = H(S, p) = U + pV, \\
&S = S(T, V), \quad p = p(T, V), \quad F_3(T, V) = F(T, V) = U - TS, \\
&p = p(T, S), \quad V = V(T, S), \quad F_4(T, S) = G(T, S) = U + pV - TS.
\end{aligned} \tag{5.4.76}$$

The choice of one or another of the potentials U, H, F, G depends on the physical problem at hand. In their turn, these functions can be taken as independent variables, leading to new sets of generating functions. Nonetheless, not all the state functions obtained this way have a practical utility.

The use of the Legendre transformation method yields:

$$dH = T\,dS + V\,dp, \quad dF = -S\,dT - p\,dV, \quad dG = -S\,dT + V\,dp, \tag{5.4.77}$$

as well as:

$$\begin{aligned}
T &= \left(\frac{\partial H}{\partial S}\right)_p, \quad V = \left(\frac{\partial H}{\partial p}\right)_S, \\
S &= -\left(\frac{\partial F}{\partial T}\right)_V, \quad p = \left(\frac{\partial F}{\partial V}\right)_T, \\
S &= -\left(\frac{\partial G}{\partial T}\right)_p, \quad V = \left(\frac{\partial G}{\partial p}\right)_T.
\end{aligned} \tag{5.4.78}$$

By means of (5.4.76) and (5.4.78), we find:

$$\begin{aligned}
U &= F - T\left(\frac{\partial F}{\partial T}\right)_V, \\
H &= G - T\left(\frac{\partial G}{\partial T}\right)_p,
\end{aligned} \tag{5.4.79}$$

which are the well-known *Gibbs–Helmholtz relations.*

5.5 Infinitesimal Canonical Transformations

We call *infinitesimal canonical transformation* any canonical transformation that differs infinitesimally from the identity transformation (5.4.29).

In the previous section, we assumed that the transition from the set q_j, p_j of canonical variables to the set Q_j, P_j is performed by a finite canonical transformation. Since the canonical transformations have group structure then, supposing that the generating function of the canonical transformation (5.4.4) is continuous and derivable, any finite canonical transformation can be regarded as a succession of infinitesimal canonical transformations. The fundamental property of an infinitesimal canonical transformation is that it can be expressed in an explicit form. This gives rise, in turn, to an intimate connection between some infinitesimal canonical transformations and the symmetry properties of mechanical systems.

Recall that

$$F_2(q, P) = \sum_{j=1}^{n} q_j P_j$$

is the generating function of the identity transformation

$$Q_j = q_j, \quad P_j = p_j \quad (j = \overline{1,n}).$$

If the canonical variables Q_j, P_j differ from q_j, p_j by some arbitrary infinitesimal variations δq_j, δp_j, then

$$Q_j = q_j + \delta q_j, \quad P_j = p_j + \delta p_j \quad (j = \overline{1,n}) \tag{5.5.1}$$

is an infinitesimal canonical transformation. Its generating function is

$$F_2' = \sum_{j=1}^{n} q_j P_j + \varepsilon W(q, P), \tag{5.5.2}$$

where W is an arbitrary function of q_j, P_j and ε is an infinitesimal parameter, independent of the canonical variables. Using (5.4.16) we obtain

$$p_j = \frac{\partial F_2'}{\partial q_j} = P_j + \varepsilon \frac{\partial W(q, P)}{\partial q_j}, \quad Q_j = \frac{\partial F_2'}{\partial P_j} = q_j + \varepsilon \frac{\partial W(q, P)}{\partial P_j},$$

$$\mathcal{H} = H + \varepsilon \frac{\partial W(q, P)}{\partial t}, \tag{5.5.3}$$

and the infinitesimal variations in (5.5.1) become:

$$Q_j - q_j = \delta q_j = \varepsilon \frac{\partial W(q, P)}{\partial P_j}, \quad P_j - p_j = \delta p_j = -\varepsilon \frac{\partial W(q, P)}{\partial q_j} \quad (j = \overline{1,n}). \tag{5.5.4}$$

Since P_j is only slightly different from p_j, we can take $W(q, p)$ instead of $W(q, P)$ and $\frac{\partial W}{\partial p_j}$ instead of $\frac{\partial W}{\partial P_j}$. Thus,

$$\delta q_j = \varepsilon \frac{\partial W}{\partial p_j}, \quad \delta p_j = -\varepsilon \frac{\partial W}{\partial q_j} \quad (j = \overline{1,n}). \tag{5.5.5}$$

Here, $W(q, p)$ plays the role of the generating function of the infinitesimal canonical transformation, while δq_j, δp_j stand for the canonical variables.

Let $f(q, p, t)$ be any function of the canonical variables. Its variation δf due to the infinitesimal transformation (5.5.1) can be written using (5.5.5) in terms of Poisson brackets as

$$\delta f = \varepsilon \{f, W\}. \tag{5.5.6}$$

For δq_j, δp_j and δH, this yields:

$$\delta q_j = \varepsilon \{q_j, W\}, \quad \delta p_j = \varepsilon \{p_j, W\}, \quad \delta H = \varepsilon \{H, W\}. \tag{5.5.7}$$

As we already know, if $W(q, p)$ is a constant of motion, the Poisson bracket $\{H, W\}$ vanishes. As a result, the first integrals of motion are the infinitesimal generators of the canonical transformations that leave the Hamiltonian invariant. This conclusion implies a connection between the symmetry properties and the constants of motion.

Let us exemplify this property in the case of an isolated and constraint-free system of N particles.

5.5.1 Total Momentum as Generator of Translations

The infinitesimal generator of the spatial translations, along a direction of unit vector **s**, is the projection of the total momentum on that direction:

$$W = \mathbf{s} \cdot \sum_{i=1}^{N} \mathbf{p}_i. \tag{5.5.8}$$

Then, by (5.5.7):

$$\delta x_{i\alpha} = \varepsilon\{x_{i\alpha}, W\} = \varepsilon \sum_{j=1}^{N} \sum_{\beta=1}^{3} s_\beta \{x_{i\alpha}, p_{j\beta}\} = \varepsilon (s_\beta \delta_{\alpha\beta})_i = \varepsilon (s_\alpha)_i,$$

$$\delta p_{i\alpha} = \varepsilon\{p_{i\alpha}, W\} = \varepsilon \sum_{j=1}^{N} \sum_{\beta=1}^{3} s_\beta \{p_{i\alpha}, p_{j\beta}\} = 0,$$

or, in vector form,

$$\delta \mathbf{r}_i = (\delta a)\mathbf{s}_i; \quad \delta \mathbf{p}_i = 0, \tag{5.5.9}$$

where $\delta a = \varepsilon$.

The Hamiltonian of the system,

$$H = \frac{1}{2} \sum_{i=1}^{N} \frac{1}{m_i} \mathbf{p}_i^2 + V, \tag{5.5.10}$$

is invariant under the transformation (5.5.9). Thus W given by (5.5.8) is a constant for any **s**, which shows the conservation of the total momentum $\mathbf{P} = \sum_{i=1}^{N} \mathbf{p}_i$ (see Chap. 1 and Chap. 2).

5.5.2 Total Angular Momentum as Generator of Rotations

The infinitesimal generator of the spatial rotations of angle $\delta\theta$ about an axis of unit vector **s** is the projection of the total angular momentum on that direction:

$$W = \mathbf{s} \cdot \mathbf{L} = \mathbf{s} \cdot \sum_{i=1}^{N} \mathbf{r}_i \times \mathbf{p}_i. \tag{5.5.11}$$

Following the same procedure as in the previous case, we have:

$$\delta x_{i\alpha} = \varepsilon \left\{ x_{i\alpha}, \sum_{j=1}^{N} \sum_{\beta,\gamma,\eta=1}^{3} \epsilon_{\beta\gamma\eta} s_\beta x_{j\gamma} p_{j\eta} \right\} = \varepsilon \sum_{\beta,\gamma=1}^{3} \epsilon_{\alpha\beta\gamma} s_\beta x_{i\gamma},$$

$$\delta p_{i\alpha} = \varepsilon \left\{ p_{i\alpha}, \sum_{j=1}^{N} \sum_{\beta,\gamma,\eta=1}^{3} \epsilon_{\beta\gamma\eta} s_\beta x_{j\gamma} p_{j\eta} \right\} = \varepsilon \sum_{\beta,\gamma=1}^{3} \epsilon_{\alpha\beta\gamma} s_\beta p_{i\gamma},$$

or, in vector form,

$$\delta \mathbf{r}_i = \delta \boldsymbol{\theta} \times \mathbf{r}_i, \quad \delta \mathbf{p}_i = \delta \boldsymbol{\theta} \times \mathbf{p}_i, \tag{5.5.12}$$

where $\delta\boldsymbol{\theta} = \mathbf{s}\delta\theta$ and $\delta\theta = \varepsilon$. Since the Hamiltonian (5.5.10) is invariant under the transformation (5.5.12), the quantity (5.5.11) is a first integral of motion and shows the conservation of the total angular momentum $\mathbf{L} = \sum_{i=1}^{N} \mathbf{r}_i \times \mathbf{p}_i$ (see Chap. 1 and Chap. 2).

5.5.3 Hamiltonian as Generator of Time-Evolution

Let us now show that the time-evolution of the canonical variables is a succession of infinitesimal canonical transformations, with the Hamiltonian as the generator of the transformation. Suppose that at the time t_0 we have

$$q_j^0 = q_j(t_0), \quad p_j^0 = p_j(t_0) \quad (j = \overline{1,n}),$$

while at the time $t = t_0 + \delta t$,

$$q_j(t) = q_j(q^0, p^0, t), \quad p_j(t) = p_j(q^0, p^0, t) \quad (j = \overline{1,n}).$$

Using the canonical equations (5.1.21), we obtain:

$$q_j = q_j^0 + \dot{q}_j \delta t = q_j^0 + \frac{\partial H(q^0, p^0, t_0)}{\partial p_j^0} \delta t, \tag{5.5.13}$$

$$p_j = p_j^0 + \dot{p}_j \delta t = p_j^0 - \frac{\partial H(q^0, p^0, t_0)}{\partial q_j^0} \delta t, \tag{5.5.14}$$

as well as

$$\begin{aligned}\mathcal{H}(q, p, t) &= H[q(t_0 + \delta t), p(t_0 + \delta t), t_0 + \delta t] \\ &= H(q^0, p^0, t_0) + \left[\frac{\partial H}{\partial t} + \{H, H\}\right]_0 \delta t,\end{aligned}$$

or

$$\mathcal{H}(q, p, t) = H(q^0, p^0; t_0) + \frac{\partial H(q^0, p^0, t_0)}{\partial t_0} \delta t. \tag{5.5.15}$$

In general, any mechanical quantity f obeys the rule:

$$f[q(t_0 + \delta t), p(t_0 + \delta t), t_0 + \delta t] = f(q^0, p^0, t_0) + \left[\frac{\partial f}{\partial t} + \{f, H\}\right]_0 \delta t. \tag{5.5.16}$$

If f does not depend explicitly on time, then

$$f[q(t_0 + \delta t), p(t_0 + \delta t)] = f(q^0, p^0) + \{f, H\}_0 \delta t. \tag{5.5.17}$$

Compare now (5.5.13)–(5.5.15) with (5.5.3). This shows that H is the infinitesimal generator of the transformation of parameter $\delta t = \varepsilon$. If we take

$$\delta t = \frac{t - t_0}{n} \quad (n \to \infty)$$

and use the group property of the canonical transformations, then

$$q_j(t) = q_j(q^0, p^0, t), \quad p_j(t) = (q^0, p^0, t) \quad (j = \overline{1, n}) \tag{5.5.18}$$

is a canonical transformation. The proof is therefore complete.

5.6 Integral Invariants

We have emphasized, in various approaches, the importance of the invariance properties of mechanical systems. The invariance is expressed either as a principle (e.g. Hamilton's principle), as a theorem (e.g. Noether's theorem), or in other ways still to be encountered. In each case, the invariance leads to important results, some of them being fundamental in physics.

The development of analytical mechanics shows that the study of canonical equations necessarily leads to some integral expressions which are invariant with respect to certain transformations. *Henri Poincaré* called these expressions *integral invariants*. This notion finds many applications both in physics and in many other branches of science.

Let us consider the system of $2n$ first-order differential equations

$$\dot{x}_s = X_s(x_1, ..., x_{2n}, t) = X_s(x, t) \quad (s = \overline{1, 2n}), \tag{5.6.1}$$

where $X_1, ..., X_{2n}$ are $2n$ functions, derivable in a certain domain D. Equations (5.6.1) are the equations of motion of a point in the $2n$-dimensional space R_{2n}. If we take as initial conditions $x_s(t_0) = x_s^0$ $(s = \overline{1, 2n})$, then the general solution of (5.6.1) is:

$$x_s = x_s(x^0, t) \quad (s = \overline{1, 2n}). \tag{5.6.2}$$

Suppose that all the points which obey (5.6.1) at time t_0 are in a p-dimensional $(p < 2n)$ manifold $V_p^0 \subset R_{2n}$. If the Jacobian of the transformation (5.6.2) satisfies the condition

$$J = \frac{\partial(x)}{\partial(x^0)} \neq 0, \tag{5.6.3}$$

then to any initial point $P_0 \in V_p^0$, it will correspond at the time $t > t_0$ a single point P of a manifold $V_p \subset R_{2n}$ $(p = \overline{1, 2n})$. In other words, this correspondence *conserves the dimension* of any manifold V_p $(p < 2n)$, i.e. the set of points which at the time t_0 form a p-dimensional manifold, will form, at time $t > t_0$, a manifold of the same dimension.

In general, if V_p $(p < 2n)$ is a p-dimensional manifold in R_{2n} and $d\tau_p^h$ $(h = 1, ..., \nu)$ is one of the $\nu = C_{2n}^p$ components of the 'volume element' that can be formed in R_{2n} (for example, $d\tau_p^1 = dx_1 dx_2 ... dx_p$), then

$$I_p = \int_{V_p} \sum_{k=1}^{\nu} f_k(x,t)\, d\tau_p^k \tag{5.6.4}$$

is an integral invariant of p-order of the system (5.6.1), if the integral is independent of time, i.e.

$$\frac{dI_p}{dt} = 0 \tag{5.6.5}$$

or, equivalently,

$$\int_{V_p^0} \sum_{k=1}^{\nu} f_k(x^0, t_0)\, d(\tau_p^k)_0 = \int_{V_p} \sum_{k=1}^{\nu} f_k(x,t)\, d\tau_p^k. \tag{5.6.6}$$

Here, $d(\tau_p^k)_0$ is the volume element of the manifold V_p^0.

The integral invariants are of the first order, second order, etc., as p takes the values 1,2... If V_p is an open manifold, the integral invariant is called *absolute*; if V_p is a closed manifold, we are dealing with a *relative* integral invariant. It can be shown that a relative integral invariant of order p is equivalent to an absolute integral invariant of order $p+1$. For example, by using the generalized Stokes' theorem (see Appendix B), we can see that

$$\oint_{\Gamma} \sum_{k=1}^{2n} f_k\, dx_k = \int_S \sum_{\substack{k,m=1\\ k<m}}^{2n} \left(\frac{\partial f_k}{\partial x_m} - \frac{\partial f_m}{\partial x_k} \right) dx_m dx_k, \tag{5.6.7}$$

where the surface S of arbitrary shape is bounded by the closed curve Γ.

Let us consider the absolute invariant integral of order $2n$:

$$\mathcal{I}_{2n} = \int_{V_{2n}} f(x,t)\, d\tau, \tag{5.6.8}$$

where $d\tau = dx_1 ... dx_{2n}$, and find the condition which must be obeyed by the function f, so that (5.6.8) is an integral invariant, i.e.

$$\frac{d\mathcal{I}_{2n}}{dt} = \frac{d}{dt} \int_{V_{2n}} f(x,t)\, d\tau = 0. \tag{5.6.9}$$

To perform the derivative in (5.6.9), recall that our domain is moving, i.e. $d\tau$ changes in time. Making use of (5.6.2), we have:

$$\frac{d\mathcal{I}_{2n}}{dt} = \int_{V_{2n}^0} \left\{ \frac{dJ}{dt} f[x(x^0,t)] + J\frac{df}{dt} \right\} d\tau^0, \tag{5.6.10}$$

where J is the Jacobian of the transformation (5.6.2) and

$$d\tau^0 = dx_1^0 \dots dx_{2n}^0$$

is the volume element of the manifold V_{2n}^0. The total time derivative of the Jacobian is performed by the usual rule:

$$\frac{dJ}{dt} = \sum_{i=1}^{2n} \frac{\partial(x_1, \dots, x_{i-1}, X_i, x_{i+1}, \dots, x_{2n})}{\partial(x_1^0, x_2^0, \dots, x_{2n}^0)}, \tag{5.6.11}$$

where (5.6.1) has also been used. But

$$\frac{\partial X_i}{\partial x_k^0} = \sum_{j=1}^{2n} \frac{\partial X_i}{\partial x_j} \frac{\partial x_j}{\partial x_k^0} \quad (i, k = \overline{1, 2n}),$$

and thus the only non-zero determinants in (5.6.11) are those obtained for $i = j$, i.e.

$$\frac{\partial(x_i)}{\partial(x_k^0)} \frac{\partial X_i}{\partial x_i} = J \frac{\partial X_i}{\partial x_i} \quad \text{(no summation)},$$

therefore

$$\frac{dJ}{dt} = J \sum_{i=1}^{2n} \frac{\partial X_i}{\partial x_i}, \tag{5.6.12}$$

known as *Euler's theorem*. Introducing this expression into (5.6.10), we obtain:

$$\frac{d\mathcal{I}_{2n}}{dt} = \int_{V_{2n}^0} \left(J \frac{df}{dt} + fJ \sum_{i=1}^{2n} \frac{\partial X_i}{\partial x_i} \right) d\tau^0 = \int_{V_{2n}} \left(\frac{df}{dt} + f \sum_{i=1}^{2n} \frac{\partial X_i}{\partial x_i} \right) d\tau.$$

Since the domain of integration V_{2n} is arbitrary, the function f must satisfy the following *condition of continuity*:

$$\frac{df}{dt} + f \sum_{i=1}^{2n} \frac{\partial X_i}{\partial x_i} = 0. \tag{5.6.13}$$

5.6.1 Integral Invariants of the Canonical Equations

Let us assume that R_{2n} is the phase space and that the $2n$ first-order differential equations (5.6.1) are precisely the canonical equations (5.1.21). Then, any integral invariant of the type (5.6.4) is associated with the system of canonical equations or, equivalently, with the canonical transformation (5.5.18). Consequently, if one finds some integral which proves to be invariant with respect to a canonical transformation, then this integral is an integral invariant of the canonical equations.

We can construct the following absolute integral invariant:

$$\mathcal{I}_{2n} = \int_{V_{2n}} \sum_{k=1}^{\nu} d\tau_n^k,$$

where $v = C_n^p$ $(p \leq n)$, V_{2n} is a $2n$-dimensional manifold in the phase space and

$$d\tau_n^k = \prod_{i=1}^{k} dq_{\alpha_i}\, dp_{\alpha_i}.$$

Here, $\alpha_1, \ldots, \alpha_k$ are k numbers out of 1, ..., n, taken in an arbitrary order. The relation between $\mathcal{I}_{2l}$ and the relative integral invariant I_{2l-1} is:

$$I_{2l-1} = \mathcal{I}_{2l}. \tag{5.6.13$'$}$$

The integral invariants in which the Hamiltonian H does not occur are called *universal*.

In the following, we shall discuss two cases: $l = 1$ and $l = n$.

5.6.2 The Relative Universal Invariant of Mechanics

In the case $l = 1$, formula (5.6.13′) gives:

$$I_1 = \oint_{\Gamma} \sum_{j=1}^{n} p_j\, dq_j = \int_{S} \sum_{j=1}^{n} dp_j\, dq_j = \mathcal{I}_2.$$

Here, Γ is a one-dimensional closed manifold, while S is a two-dimensional open manifold. We wish to prove that I_1 is indeed an invariant, associated with the transformations (5.5.18). In view of (5.6.6), the condition of invariance can be written as

$$\int_{S} \sum_{j=1}^{n} dp_j dq_j = \int_{S_0} \sum_{j=1}^{n} dp_j^0 dq_j^0. \tag{5.6.14}$$

To prove this identity, let us express q_j^0, p_j^0 as functions of two parameters v_1, v_2:

$$q_j^0 = q_j^0(v_1, v_2), \quad p_j^0 = p_j^0(v_1, v_2). \tag{5.6.15}$$

Using the definition of the Lagrange bracket, we can write:

$$\int_{S} \sum_{j=1}^{n} dp_j dq_j = \int_{\Sigma} \sum_{j=1}^{n} \frac{\partial(q_j, p_j)}{\partial(v_1, v_2)} dv_1 dv_2 = \int_{\Sigma} (v_1, v_2) dv_1 dv_2,$$

$$\int_{S_0} \sum_{j=1}^{n} dp_j^0 dq_j^0 = \int_{\Sigma} \sum_{j=1}^{n} \frac{\partial(q_j^0, p_j^0)}{\partial(v_1, v_2)} dv_1 dv_2 = \int_{\Sigma} (v_1, v_2)_0 dv_1 dv_2,$$

where Σ is the integration domain in the plane of the variables v_1, v_2 and the subscript 'zero' shows that the Lagrange bracket is calculated in terms of the canonical variables q_j^0, p_j^0. On the other hand,

$$\frac{\partial(q_j, p_j)}{\partial(v_1, v_2)} = \sum_{k,l=1}^{n} \left[\frac{\partial(q_j, p_j)}{\partial(q_k^0, q_l^0)} \frac{\partial(q_k^0, q_l^0)}{\partial(v_1, v_2)} + \frac{\partial(q_j, p_j)}{\partial(q_k^0, p_l^0)} \frac{\partial(q_k^0, p_l^0)}{\partial(v_1, v_2)} + \frac{\partial(q_j, p_j)}{\partial(p_k^0, p_l^0)} \frac{\partial(p_k^0, p_l^0)}{\partial(v_1, v_2)} \right],$$

or, if the summation over j is performed,

$$(v_1, v_2) = \sum_{k,l=1}^{n} \left[\frac{\partial(q_k^0, q_l^0)}{\partial(v_1, v_2)} (q_k^0, q_l^0) + \frac{\partial(q_k^0, p_l^0)}{\partial(v_1, v_2)} (q_k^0, p_l^0) + \frac{\partial(p_k^0, p_l^0)}{\partial(v_1, v_2)} (p_k^0, p_l^0) \right].$$

If we remember that the Lagrange bracket is invariant with respect to any canonical transformation, we are left with

$$(v_1, v_2) = \sum_{k,l=1}^{n} \frac{\partial(q_k^0, p_l^0)}{\partial(v_1, v_2)} \delta_{kl} = (v_1, v_2)_0, \tag{5.6.16}$$

and (5.6.14) follows immediately. We conclude that $I_1 (= \mathcal{I}_2)$ is an integral invariant associated with the transformation (5.5.18).

The contour Γ occurring in the definition of the integral invariant I_1 is composed by a set of representative points in the phase space, determined at the same time t. The integral invariant I_1 was introduced by *Henri Poincaré* and it is called the *relative universal invariant of mechanics*. Later, *Eli Cartan* gave a generalization of this invariant for closed paths considered at different times. In particular, the quantity:

$$I = \oint_{\Gamma} \left(\sum_{j=1}^{n} p_j dq_j - H dt \right) \tag{5.6.17}$$

is the *Poincaré–Cartan integral invariant.*

A suggestive explanation of the conditions related to the path of integration is given by the following geometric interpretation. Let us define the *state space*, or the *extended phase space* R_{2n+1}. Each point of this space corresponds to a *state* of the system. When referring to the invariant I_1, the paths Γ_0 and Γ lie in the hyperplanes $t_0 = \text{const.}$ and $t = \text{const.}$, respectively. In contrast, the closed contour Γ for the Poincaré–Cartan invariant I has no connection with any hyperplane $t = \text{const.}$ If the integration path in (5.6.17) corresponds to simultaneous states, then along the integration path we have $dt = 0$ and we get back the invariant I_1.

The importance of the Poincaré–Cartan integral invariant (5.6.17) is that through it one can reformulate the fundamental postulate of mechanics: *The motion of a mechanical system is described by the system of 2n first-order differential equations*

$$\dot{q}_j = A_j(q, p, t), \quad \dot{p}_j = B_j(q, p, t) \quad (j = \overline{1, n}), \tag{5.6.18}$$

with the initial conditions $q_j(t_0) = q_j^0, p_j(t_0) = p_j^0 (j = \overline{1, n})$, *if the system* (5.6.18) *admits the integral* (5.6.17) *as a relative integral invariant.*

Indeed, in order that (5.6.17) be an integral invariant, the following conditions must be fulfilled:

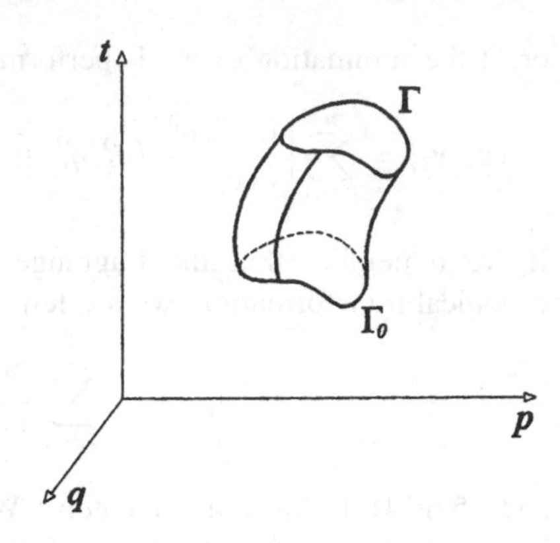

Fig. 5.2 The *extended phase space (state space).*

$$A_j = \frac{\partial H}{\partial p_j}, \quad B_j = -\frac{\partial H}{\partial q_j} \quad (j = \overline{1, n}), \tag{5.6.19}$$

which leads to the canonical equations (5.1.21).

To justify this statement, let Γ_0 and Γ be two neighbouring closed paths, Γ being the image of Γ_0 in the state space. The one-to-one correspondence between Γ and Γ_0 shows that the generalized trajectories connecting the points on Γ with the points on Γ_0 do not intersect with each other, giving rise to a tube-like geometric volume (Fig. 5.2). The variation of (5.6.17), when passing from Γ_0 to Γ, is:

$$\delta I = \oint_\Gamma \left[\sum_{j=1}^n \delta p_j dq_j + \sum_{j=1}^n p_j \delta(dq_j) - \delta H\, dt - H\delta(dt) \right].$$

The operators δ and d are independent; consequently,

$$\delta(dq_j) = d(\delta q_j), \quad \delta(dt) = d(\delta t),$$

and thus, integrating by parts and expanding δH, we obtain:

$$\delta I = \oint_\Gamma \left[\sum_{j=1}^n \left(dq_j - \frac{\partial H}{\partial p_j} dt \right) \delta p_j - \sum_{j=1}^n \left(dp_j + \frac{\partial H}{\partial q_j} dt \right) \delta q_j + \left(dH - \frac{\partial H}{\partial t} dt \right) \delta t \right].$$

The integral I is an invariant if $\delta I = 0$. Then, for arbitrary variations $\delta q_j, \delta p_j, \delta t$ we find the canonical equations (5.1.21), as well as the property (5.1.30) of the Hamiltonian.

5.6.3 Liouville's Theorem

In the case $l = n$, the integral invariant is:

$$\mathcal{I}_{2n} = \int_{V_{2n}} dq_1 \ldots dq_n dp_1 \ldots dp_n \tag{5.6.20}$$

and expresses the fact that the "volume" of any portion of the phase space remains invariant under the transformation (5.5.18). The proof of the invariance of (5.6.20) is simple. Let us write

$$dq_1 \dots dq_n dp_1 \dots dp_n = \frac{\partial(q, p)}{\partial(q_0, p_0)} dq_1^0 \dots dq_n^0 dp_1^0 \dots dp_n^0.$$

But the transformation (5.5.18) is canonical, meaning that $J = 1$, which proves the invariance of $\mathcal{I}_{2n}$.

If by V_{2n}^0 we mean a vicinity of the representative point P_0, at the time t_0, and by V_{2n} the vicinity of the point P, reached by the system at the time $t > t_0$, we have arrived at an important theorem: *The "volume" of any vicinity of the representative point does not change in time.* This is *Liouville's theorem.*

Liouville's theorem is fundamental in statistical physics, where it was introduced in connection with the notion of *statistical ensemble*. Let us imagine a large number of replicas of a system, which are macroscopically identical, but microscopically distinct. Assume, for instance, that our system consists of N identical molecules, enclosed in a container. These molecules are characterized by $q_j(t_0), p_j(t_0)$, at the time t_0, and by $q_j(t), p_j(t)$ at the time t. To each of these sets is associated a point in the phase space. Let dN be the number of particles which, at the time t, have their generalized coordinates in the infinitesimal interval $(q_j, q_j + dq_j)$, and their generalized momenta in the interval $(p_j, p_j + dp_j)$. Then we can write

$$dN = f(q, p, t) dq_1 \dots dq_n dp_1 \dots dp_n. \tag{5.6.21}$$

Here, $f(q, p, t)$ is the *distribution function*, which expresses the probability of finding the set of N particles in a *cell* of the phase space, whose 'volume' is

$$d\tau = dq_1 \dots dq_n dp_1 \dots dp_n.$$

Assuming that there are no collisions between particles, the number of particles which, at time $t + dt$, are in the same cell of the phase space, is:

$$dN' = f(q + dq, p + dp, t + dt) dq_1 \dots dq_n dp_1 \dots dp_n.$$

The variation δN of the number of particles in the elementary volume $d\tau$ is

$$\delta N = dN' - dN = [f(q + dq, p + dp, t + dt) - f(q, p, t)] dq_1 \dots dq_n dp_1 \dots dp_n.$$

If in the Taylor series expansion for δN we keep only those terms which are linear in $\delta q_j, \delta p_j, \delta t$, we get:

$$\delta N \approx \left[\sum_{j=1}^{n} \frac{\partial f}{\partial q_j} \dot{q}_j + \sum_{j=1}^{n} \frac{\partial f}{\partial p_j} \dot{p}_j + \frac{\partial f}{\partial t} \right] d\tau dt. \tag{5.6.22}$$

Since, by hypothesis, the collisions are absent, $\delta N = 0$, and (5.6.22) yields

$$\frac{df}{dt} = 0, \tag{5.6.23}$$

as a consequence of Liouville's theorem. This shows that the distribution function $f(q, p, t)$ is a first integral of the canonical equations.

The phase space may be imagined as a $2n$-dimensional fluid. Indeed, the position at time t of a certain molecule of the fluid is given by (5.6.2):

$$x_i = x_i(x_1^0, x_2^0, x_3^0, t) \quad (i = 1, 2, 3),$$

where x_i^0 and x_i are the Cartesian coordinates of the molecule at times t_0 and t, respectively. This hydrodynamic model can be compared with the phase space, the only difference being the number of coordinates that determine the position of the representative point. This $2n$-dimensional fluid is called the *phase fluid*.

The Hamiltonian of a conservative system is a constant of motion:

$$H(q_1, \ldots, q_n, p_1, \ldots, p_n) = E = \text{const.} \tag{5.6.24}$$

This relation leads to an interesting geometric interpretation. In the phase space R_{2n}, (5.6.24) represents a *hypersurface*, which is any s-dimensional manifold with $2 < s < 2n$. Thus, if at the time t_0 the state of the system lies on some energetic hypersurface, at any time $t > t_0$ it will lie on the same hypersurface. This analogy also shows that the phase fluid associated with the canonical equations behaves like an *incompressible fluid*. This property is easily proved by integrating the divergence-free condition

$$\sum_{i=1}^{2n} \frac{\partial X_i}{\partial x_i} = 0 \tag{5.6.25}$$

over some $2n$-dimensional manifold of R_{2n} and then applying the Green–Gauss generalized formula:

$$\int_{V_{2n}} \sum_{i=1}^{2n} \frac{\partial X_i}{\partial x_i} d\tau = \int_{V_{2n-1}} \sum_{i=1}^{2n} X_i dS_i = 0, \tag{5.6.26}$$

where

$$d\tau = dx_1 \ldots dx_{2n} = dS_i \, dx_i \quad (\text{no summation}), \tag{5.6.27}$$

with

$$dS_i = dx_1 \ldots dx_{i-1} dx_{i+1} \ldots dx_{2n}, \tag{5.6.28}$$

while V_{2n-1} is the boundary of V_{2n}. If we take $x_1, \ldots, x_n$ as the generalized coordinates $q_1, \ldots, q_n$, and $x_{n+1}, \ldots, x_{2n}$ as the generalized momenta $p_1, \ldots, p_n$, then, in view of (5.1.21) and (5.6.1),

$$\sum_{j=1}^{2n} \frac{\partial X_i}{\partial x_i} \to \sum_{j=1}^{n} \left(\frac{\partial \dot{q}_j}{\partial q_j} + \frac{\partial \dot{p}_j}{\partial p_j} \right) \equiv 0. \tag{5.6.29}$$

Thus, Eq. (5.6.26), in hydrodynamic notation (see Chap. 6), reads:

$$\int_{V_{2n}} \operatorname{div} \mathbf{v}\, d\tau = \int_{V_{2n-1}} \mathbf{v} \cdot d\mathbf{S} = 0, \tag{5.6.30}$$

meaning that *the flux of the phase fluid passing through a closed hypersurface of the phase space is zero*. This is a different way of expressing the incompressibility of the phase fluid. It also represents another formulation of Liouville's theorem.

5.6.4 Pfaff Forms

Let us consider the differential expression

$$\omega(d) = \sum_{i=1}^{n} X_i dx_i, \tag{5.6.31}$$

where $X_i = X_i(x_1, \ldots, x_n)$ are n functions defined in some domain of the n-dimensional space R_n. The expression (5.6.31) is called a *Pfaff form.* We notice that the integrand of a first-order integral invariant is a Pfaff form.

Consider now the Pfaff form $\omega(\delta)$, associated with the variation performed by the operator δ, independent of the operator d:

$$\omega(\delta) = \sum_{i=1}^{n} X_i \delta x_i. \tag{5.6.32}$$

The difference

$$\Delta\omega = \delta\omega(d) - d\omega(\delta) \tag{5.6.33}$$

is known as the *bilinear covariant* associated with the form ω. Since the operators d and δ commute, we can write:

$$\Delta\omega = \sum_{i,k=1}^{n} T_{ki} \delta x_k dx_i. \tag{5.6.34}$$

Here,

$$T_{ki} = \frac{\partial X_i}{\partial x_k} - \frac{\partial X_k}{\partial x_i} \tag{5.6.35}$$

is a the second-order antisymmetric tensor (if X_i are the components of a n-dimensional vector). If ω is a total differential, then Stokes' theorem implies

$$\Delta\omega = 0. \tag{5.6.36}$$

By using this property, it is not difficult to show that two Pfaff forms which differ from one another by a total differential of some twice-differentiable function $F(x_1, \ldots, x_n)$ have the same bilinear covariant. Indeed, since $\delta(dF) = d(\delta F)$, the two forms

$$\omega(d) = \sum_{i=1}^{n} X_i dx_i,$$

$$\omega'(d) = \omega(d) + dF(x_1, \ldots, x_n)$$

give

$$\Delta\omega = \Delta\omega'.$$

If we denote

$$\omega_k = \sum_{i=1}^{n} \left(\frac{\partial X_i}{\partial x_k} - \frac{\partial X_k}{\partial x_i} \right) dx_i \quad (k = \overline{1,n}), \tag{5.6.37}$$

the expression (5.6.34) reads:

$$\Delta\omega = \sum_{k=1}^{n} \omega_k \delta x_k. \tag{5.6.38}$$

For arbitrary and independent variations δx_k, we have $\Delta\omega = 0$ if

$$\omega_k = 0 \quad (k = \overline{1,n}). \tag{5.6.39}$$

The n equations (5.6.39) form a system associated with the Pfaff form (or contact 1-form) ω.

Let us show that the canonical equations (5.1.21) can be considered as a system associated with the Pfaff form

$$\omega(d) = \sum_{i=1}^{n} p_i dq_i - H dt, \tag{5.6.40}$$

which is nothing else but the integrand of the Cartan–Poincaré integral invariant (5.6.17). Using the definition (5.6.33), we can write:

$$\Delta\omega = \sum_{i=1}^{n} [\delta p_i dq_i + p_i \delta(dq_i) - \delta H \; dt - H\delta(dt)$$
$$- dp_i \delta q_i - p_i \; d(\delta q_i) + dH \delta t + H \; d(\delta t)].$$

But the operators δ and d commute, such that

$$\Delta\omega = \sum_{i=1}^{n} \left[\left(dq_i - \frac{\partial H}{\partial p_i} dt \right) \delta p_i - \left(dp_i + \frac{\partial H}{\partial q_i} dt \right) \delta q_i \right] + \left(dH - \frac{\partial H}{\partial t} dt \right) \delta t.$$

Therefore the condition $\Delta\omega = 0$, for arbitrary and independent $\delta q_j, \delta p_j, \delta t$, leads indeed to the canonical equations (5.1.21). Reciprocally, if q_j and p_j are canonical variables, the bilinear covariant of the corresponding Pfaff form is zero.

As a final example, let us consider the transformation (5.4.4):

$$q_j = q_j(Q, P, t), \quad p_j = p_j(Q, P, t) \quad (j = \overline{1,n}). \tag{5.6.41}$$

Since, as we have already proven, the canonical equations written for q_j, p_j form a system associated with the Pfaff form (5.6.40), the transformation (5.6.41) is canonical if Hamilton's equations written for Q_j, P_j are associated with the Pfaff form

$$\Omega = \sum_{i=1}^{n} P_i dQ_i - \mathcal{H} dt. \tag{5.6.42}$$

But the two forms ω and Ω have the same bilinear covariant, namely zero, and thus they can differ from one another only by a total differential of some function $F(q, Q, t)$, i.e.

$$\sum_{j=1}^{n} p_j dq_j - H\, dt = \sum_{j=1}^{n} P_j dQ_j - \mathcal{H} dt + dF(q, Q, t),$$

which is precisely the canonicity condition (5.4.11).

5.6.5 Quantum Mechanical Harmonic Oscillator

The quantum theory of the atom was initially based on the Bohr–Sommerfeld quantization rule:

$$J_i = \oint p_i dq_i = n_i h \quad \text{(no summation)}, \tag{5.6.43}$$

where the index i takes as many values, as there are degrees of freedom in the system, n_i are positive integers and h is the Planck constant. Since the units of the product pq are

$$[p \times q] = Energy \times Time,$$

the integral (5.6.43) is called the *action modulus*. One also observes that J_i is a relative, first-order integral invariant.

Let us use the rule (5.6.43) to derive some essential properties of the one-dimensional quantum harmonic oscillator. The system possesses a single degree of freedom, and so the phase space reduces to a phase plane (p, q). Since the system is conservative, the total energy is constant:

$$E = T + V = \frac{p^2}{2m} + \frac{1}{2} kq^2 = \text{const.}$$

Alternatively, dividing by E we get:

$$\frac{q^2}{\frac{2E}{k}} + \frac{p^2}{2mE} = 1. \tag{5.6.44}$$

This shows that the trajectory in the phase plane is an ellipse, with its centre at the origin of the coordinate system, the semi-axes a and b being given by

$$a = \sqrt{\frac{2E}{k}}, \quad b = \sqrt{2mE}. \tag{5.6.45}$$

The condition (5.6.43), corresponding to a single degree of freedom, reads:

$$\oint p dq = 2\pi n\hbar. \tag{5.6.46}$$

This integral is nothing else but the ellipse area in the phase plane:

$$\pi ab = \pi\sqrt{\frac{2E}{k}}\sqrt{2mE},$$

leading by the quantization condition (5.6.46) to

$$E_n = n\, h\nu_0 = n\hbar\omega_0. \tag{5.6.47}$$

Here, ω_0 is the angular frequency associated with the periodic motion of q.

The relation (5.6.47) shows that the quantum harmonic oscillator can exist only in certain energy states. Further developments in quantum mechanics showed that a better description is provided by the relation

$$E_n = \left(n + \frac{1}{2}\right) h\nu_0. \tag{5.6.48}$$

Going back to Eq. (5.6.44), one can see that to each quantized value of the energy E corresponds an ellipse. Then the energy hypersurfaces (5.6.24) are here reduced to some closed curves, being also current lines of the two-dimensional phasic fluid. The relation $p = m\dot{q}$ shows that, if $p > 0$, then q increases with time, and thus we find the sense of the representative point on the trajectory (Fig. 5.3).

The state space has three dimensions, associated with q, p, t. The trajectory of the representative point is a cylindrical helix, any cross section obtained by a plane t = const. being an ellipse (Fig. 5.4), corresponding to the energy E = const. of the harmonic oscillator.

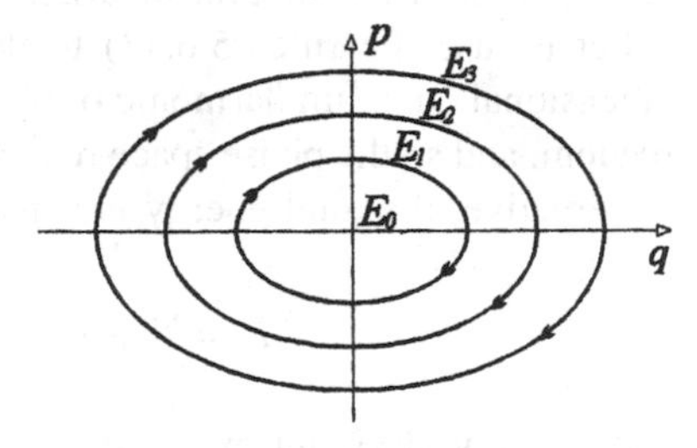

Fig. 5.3 Energetic representation of a quantum harmonic oscillator in the phase space.

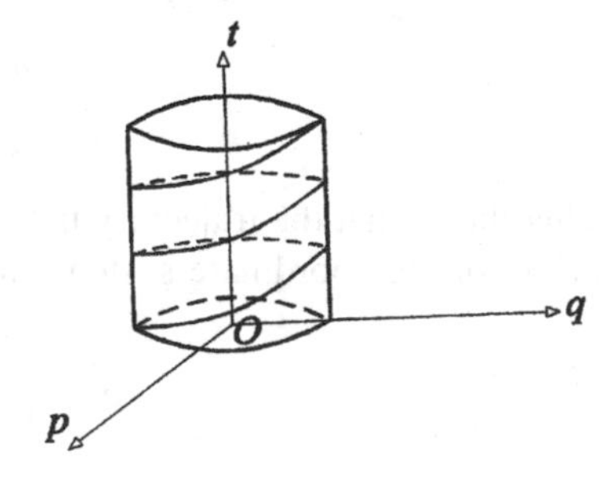

Fig. 5.4 Trajectory in the state space of the representative point associated to the harmonic oscillator.

5.7 Hamilton–Jacobi Formalism

5.7.1 Hamilton–Jacobi Equation

The canonical transformations studied in Sect. 5.4 are particularly useful in the derivation of the canonical equations (5.4.6) in terms of the new variables $Q_j, P_j\,(j=\overline{1,n})$ in the simplest possible form, i.e.:

$$\dot{Q}_j = 0, \quad \dot{P}_j = 0 \quad (j=\overline{1,n}), \tag{5.7.1}$$

which can be integrated immediately, to yield

$$Q_j = b_j, \quad P_j = a_j \quad (j=\overline{1,n}), \tag{5.7.2}$$

where a_j, b_j are $2n$ constants of integration. Equations (5.4.6) yield (5.7.1) if the Hamiltonian $\mathcal{H}$ does not depend on Q_j and P_j, or if it is a constant. Since only the derivatives of $\mathcal{H}$ occur in the canonical equations, we may take $\mathcal{H}=0$. Let us then find the canonical transformation leading to $\mathcal{H}=0$.

If we choose q_j, P_j as independent variables, the transformation formulas are given by (5.4.16):

$$p_j = \frac{\partial F_2}{\partial q_j}, \quad Q_j = b_j = \frac{\partial F_2}{\partial P_j}, \quad \frac{\partial F_2}{\partial t} + H\left(q_1,\ldots,q_n,\frac{\partial F_2}{\partial q_1},\ldots,\frac{\partial F_2}{\partial q_n},t\right) = 0 \quad (j=\overline{1,n}). \tag{5.7.3}$$

Denoting

$$F_2(q_1,\ldots,q_n,P_1,\ldots,P_n,t) = F_2(q_1,\ldots,q_n,a_1,\ldots,a_n,t) = S(q,t), \tag{5.7.4}$$

equations (5.7.3) become:

$$p_j = \frac{\partial S}{\partial q_j}, \quad b_j = \frac{\partial S}{\partial a_j} \quad (j=\overline{1,n}), \tag{5.7.5}$$

$$\frac{\partial S}{\partial t} + H\left(q_1,\ldots,q_n,\frac{\partial S}{\partial q_1},\ldots,\frac{\partial S}{\partial q_n},t\right) = 0. \tag{5.7.6}$$

The first-order partial differential equation (5.7.6) is called the *Hamilton–Jacobi equation*, while $S(q, t)$ is *Hamilton's principal function*. If we are able to integrate the Hamilton–Jacobi equation, then, by substituting the solution $S(q, t)$ into $(5.7.5)_2$, we find:

$$q_j = q_j(a_1,\ldots,a_n,b_1,\ldots,b_n,t) \quad (j=\overline{1,n}),$$

i.e. the generalized coordinates as functions of time and $2n$ constants of integration, which are determined from the initial conditions. This means that the integration of the canonical equations, on the one hand, and of the Hamilton–Jacobi equation, on the other, are equivalent. Let us more thoroughly investigate the solution of the Hamilton–Jacobi equation.

Jacobi's Theorem. *If $S(q, a, t)$ is any complete solution of the Hamilton–Jacobi equation* (5.7.6), *then the general solution of the canonical equations is provided by* (5.7.5).

Before proving the theorem, we remind the reader that a *complete solution* or *complete integral* of a first-order partial differential equation is a solution of this equation which contains as many independent arbitrary constants as there are independent variables. In particular, if we choose t as an independent parameter, then a complete solution of the Hamilton–Jacobi equation should contain $n+1$ independent constants. But, since only the partial derivatives of S appear in Eq. (5.7.6), one of the constants $a_1, \dots, a_{n+1}$, say a_{n+1}, is purely additive and disappears upon differentiation. Therefore, the most general form of a complete integral of the Hamilton–Jacobi equation is

$$S(q_1, \dots, q_n, a_1, \dots, a_n, t) + \text{const.} \tag{5.7.7}$$

The independent constants $a_1, \dots, a_n$ are called *essential*.

To prove the theorem, we start by taking the total time derivative of $(5.7.5)_2$:

$$\frac{d}{dt}\left(\frac{\partial S}{\partial a_j}\right) = \left(\frac{\partial}{\partial t} + \sum_{k=1}^{n} \dot{q}_k \frac{\partial}{\partial q_k}\right)\frac{\partial S}{\partial a_j} = \frac{\partial^2 S}{\partial t \partial a_j} + \sum_{k=1}^{n} \dot{q}_k \frac{\partial^2 S}{\partial q_k \partial a_j} = 0$$

and also the partial derivative with respect to a_j of the Hamilton–Jacobi equation:

$$\frac{\partial^2 S}{\partial a_j \partial t} + \sum_{k=1}^{n} \frac{\partial H}{\partial(\frac{\partial S}{\partial q_k})} \frac{\partial(\frac{\partial S}{\partial q_k})}{\partial a_j} = \frac{\partial^2 S}{\partial a_j \partial t} + \sum_{k=1}^{n} \frac{\partial H}{\partial p_k} \frac{\partial^2 S}{\partial q_k \partial a_j} = 0.$$

Subtracting these two equations one from the other, we find that

$$\sum_{k=1}^{n}\left(\dot{q}_k - \frac{\partial H}{\partial p_k}\right)\frac{\partial^2 S}{\partial q_k \partial a_j} = 0. \tag{5.7.8}$$

Assuming that

$$\left|\frac{\partial^2 S}{\partial a_j \partial q_k}\right| \neq 0, \tag{5.7.9}$$

we obtain:

$$\dot{q}_k = \frac{\partial H}{\partial p_k} \quad (k = \overline{1, n}), \tag{5.7.10}$$

which is the first group of Hamilton's equations.

Next, we take the total time derivative of $(5.7.5)_1$ and the partial derivative with respect to q_k of the Hamilton–Jacobi equation (5.7.6):

$$\frac{d}{dt}\left(\frac{\partial S}{\partial q_j}\right) = \frac{\partial^2 S}{\partial q_j \partial t} + \sum_{k=1}^{n} \dot{q}_k \frac{\partial^2 S}{\partial q_j \partial q_k} = \dot{p}_j,$$

$$\frac{\partial^2 S}{\partial q_j \partial t} + \frac{\partial H}{\partial q_j} + \sum_{k=1}^{n} \frac{\partial H}{\partial(\frac{\partial S}{\partial q_k})} \frac{\partial(\frac{\partial S}{\partial q_k})}{\partial q_j} = \frac{\partial^2 S}{\partial q_j \partial t} + \frac{\partial H}{\partial q_j} + \sum_{k=1}^{n} \frac{\partial H}{\partial p_k} \frac{\partial^2 S}{\partial q_j \partial q_k} = 0.$$

Using (5.7.10) and assuming that

$$\left|\frac{\partial^2 S}{\partial q_j \partial q_k}\right| \neq 0 \tag{5.7.11}$$

we obtain:

$$\dot{p}_j = -\frac{\partial H}{\partial q_j} \quad (j = \overline{1,n}), \tag{5.7.12}$$

i.e. the second set of Hamilton's equations.

Observation: Let us consider the action integral (2.7.17), taken on the generalized trajectory, and assume that only one of the two end-points is fixed:

$$S(q, \dot{q}, t) = \int_{t_0}^{t} L(q, \dot{q}, t) dt. \tag{5.7.13}$$

Take now the variation of the action for all possible neighbouring trajectories:

$$\delta S = \sum_{j=1}^{n} \left[\frac{\partial L}{\partial \dot{q}_j} \delta q_j\right]_{t_0}^{t} + \int_{t_0}^{t} \sum_{j=1}^{n} \left[\frac{\partial L}{\partial q_j} - \frac{d}{dt}\left(\frac{\partial L}{\partial \dot{q}_j}\right)\right] \delta q_j dt.$$

Since on any of these trajectories the Lagrange equations are satisfied, the second term above vanishes and we arrive at

$$\delta S = \sum_{j=1}^{n} \frac{\partial L}{\partial \dot{q}_j} \delta q_j = \sum_{j=1}^{n} p_j \delta q_j, \tag{5.7.14}$$

which yields:

$$p_j = \frac{\partial S}{\partial q_j} \quad (j = \overline{1, n}). \tag{5.7.15}$$

On the other hand, the total time derivative of S is

$$\frac{dS}{dt} = \frac{\partial S}{\partial t} + \sum_{j=1}^{n} \frac{\partial S}{\partial q_j} \dot{q}_j,$$

from which, using (5.7.15), follows

$$-\frac{\partial S}{\partial t} = \sum_{j=1}^{n} p_j \dot{q}_j - L = H(q, p, t). \tag{5.7.16}$$

We can also write

$$dS = \sum_{j=1}^{n} \frac{\partial S}{\partial q_j} dq_j + \frac{\partial S}{\partial t} dt = \sum_{j=1}^{n} p_j dq_j - H\, dt,$$

which shows that the action S satisfies the canonicity condition required for the generating function. For this reason, we used the same letter, S, to denote the generating function of the canonical transformation (5.7.1). We also have

$$S = \int L dt + \text{const.}, \tag{5.7.17}$$

showing that Hamilton's principal function differs from the indefinite time integral of L by at most a constant.

5.7.2 Methods for Solving the Hamilton–Jacobi Equation

We should mention from the very beginning that there are no general methods of finding a complete integral of the Hamilton–Jacobi equation. However, in some particular cases, a complete integral can be determined by separation of variables.

5.7.2.1 Separation of Time

If the Hamiltonian H does not explicitly depend on time, we look for a solution of the Hamilton–Jacobi equation of the form:

$$S(q, a, t) = W(q, a) + S_1(a, t). \tag{5.7.18}$$

Then, (5.7.6) becomes:

$$\frac{\partial S_1(a, t)}{\partial t} + H\left(q, \frac{\partial W(q, a)}{\partial q}\right) = 0. \tag{5.7.19}$$

Since the first term depends only on the time variable and the second only on the independent variables q_j, Eq. (5.7.19) is satisfied only if the first term equals a constant E and the second one equals $-E$. Thus,

$$\frac{\partial S_1}{\partial t} = -E, \tag{5.7.20}$$

and, by integration,

$$S_1(a, t) = -Et, \tag{5.7.21}$$

where the additive constant has been omitted. We have also

$$H\left(q_1,\ldots,q_n,\frac{\partial W}{\partial q_1},\ldots,\frac{\partial W}{\partial q_n}\right)=E, \tag{5.7.22}$$

called the *abbreviated* (or *restricted*) *Hamilton–Jacobi equation.* Since the Hamiltonian H does not explicitly depend on time, the constant E is usually the total energy of the system.

As one can see, the function W depends essentially on the constant E, which cannot be independent of the other constants $a_1,\ldots,a_n$. But the number of the independent constants must be n and it is customary to choose $a_n = E$. Then the complete integral (5.7.18) reads:

$$S(q_1,\ldots,q_n,a_1,\ldots,a_n,t)=-Et+W(q_1,\ldots,q_n,a_1,\ldots,a_{n-1},E), \tag{5.7.23}$$

while Eqs. (5.7.5) yield:

$$\frac{\partial W}{\partial a_k}=b_k \quad (k=\overline{1,n-1}), \tag{5.7.24}$$

$$\frac{\partial W}{\partial E}-t=\pm t_1(\text{const.}), \tag{5.7.25}$$

$$\frac{\partial W}{\partial q_j}=p_j \quad (j=\overline{1,n}). \tag{5.7.26}$$

The $n-1$ equations (5.7.24) determine the generalized trajectory in the configuration space, Eq. (5.7.25) gives the law of motion, i.e. the generalized coordinates as functions of time, while (5.7.26) define the generalized momenta. The constant t_1 has the units of time and only one sign is chosen.

5.7.2.2 Separation of Variables

The Hamilton–Jacobi formalism is useful when the variables are separable, meaning that S can be written as a sum of n functions, each function depending only on a single variable. In this case, the Hamilton–Jacobi equation leads to n first-order differential equations in each variable. There are also situations in which only part of the variables are separable.

A particular case of separable variables are the cyclic ones. Assume that the variable q_n is cyclic and look for the solution S of the form

$$S(q_1,\ldots,q_n,a_1,\ldots,a_n,t)=S'(q_1,\ldots q_{n-1},a_1,\ldots,a_n,t)+a_0q_n. \tag{5.7.27}$$

The generalized momenta are:

$$p_k=\frac{\partial S'}{\partial q_k} \quad (k=\overline{1,n-1}),$$
$$p_n=a_0, \tag{5.7.28}$$

and the Hamilton–Jacobi equation reads:

$$\frac{\partial S'}{\partial t} + H\left(q_1, \ldots, q_{n-1}, \frac{\partial S'}{\partial q_1}, \ldots, \frac{\partial S'}{\partial q_{n-1}}, a_0, t\right) = 0. \tag{5.7.29}$$

The constant a_0 must depend on the essential constants $a_1, \ldots, a_n$, so that the most convenient choice is $a_0 = a_n$. Consequently, $(5.7.5)_2$ leads to:

$$\begin{aligned} &\frac{\partial S'}{\partial a_k} = b_k \quad (k = \overline{1, n-1}), \\ &\frac{\partial S'}{\partial a_n} + q_n = b_n. \end{aligned} \tag{5.7.30}$$

If there are several cyclic variables, say q_α $(\alpha = \overline{r, n})$, the generalized complete integral is

$$S(q_1, \ldots, q_n, a_1, \ldots, a_n, t) = S'(q_1, \ldots, q_{r-1}, a_1, \ldots, a_n, t) + \sum_{\alpha=r}^{n} a_\alpha q_\alpha. \tag{5.7.31}$$

Example. Let us illustrate the separation of variables in the case of a particle moving in a conservative force field $\mathbf{F} = -\text{grad } V$, admitting that $V = V(r, \theta)$ (cylindrical symmetry). The corresponding Hamiltonian is:

$$H = \frac{1}{2m}\left(p_r^2 + \frac{1}{r^2}\,p_\theta^2 + \frac{1}{r^2 \sin^2\theta}\,p_\varphi^2\right) + V(r, \theta). \tag{5.7.32}$$

The variable φ is automatically separated due to its cyclicity, and the other two variables, r and θ, can be separated if the potential is of the form:

$$V(r, \theta) = u(r) + \frac{1}{r^2}\,v(\theta). \tag{5.7.33}$$

The Hamilton–Jacobi equation reads:

$$\frac{\partial S}{\partial t} + \frac{1}{2m}\left(p_r^2 + \frac{1}{r^2}\,p_\theta^2 + \frac{1}{r^2 \sin^2\theta}\,p_\varphi^2\right) + u(r) + \frac{1}{r^2}\,v(\theta) = 0. \tag{5.7.34}$$

Since the Hamiltonian does not depend explicitly on time, we can take

$$S = -Et + W(r, \theta, \varphi),$$

which leads to

$$\frac{1}{2m}\left(\frac{\partial W}{\partial r}\right)^2 + u(r) + \frac{1}{2mr^2}\left(\frac{\partial W}{\partial \theta}\right)^2 + \frac{1}{r^2}\,v(\theta) + \frac{1}{2mr^2 \sin^2\theta}\left(\frac{\partial W}{\partial \varphi}\right)^2 = E. \tag{5.7.35}$$

Since the variable φ is cyclic, we choose

$$W(r, \theta, \varphi) = \varphi p_\varphi + R(r) + \Theta(\theta). \tag{5.7.36}$$

This leads to the separation of variables in (5.7.35). Following the usual procedure, we find:

$$\frac{1}{2m}\left(\frac{dR}{dr}\right)^2 + u(r) + \frac{a}{2mr^2} = E,$$

$$\left(\frac{d\Theta}{d\theta}\right)^2 + 2mv(\theta) + \frac{p_\varphi^2}{\sin^2\theta} = a,$$

where a is a constant. The solutions $R(r)$ and $\Theta(\theta)$ of these equations, when introduced into W, deliver the complete integral:

$$S = -Et + W(r, \theta, \varphi, a, E, p_\varphi), \tag{5.7.37}$$

where

$$W(r, \theta, \varphi, a, E, p_\varphi) = \varphi p_\varphi + \int \sqrt{2mE - 2mu(r) - \frac{a}{r^2}}\, dr + \int \sqrt{a - 2mv(\theta) - \frac{p_\varphi^2}{\sin^2\theta}}\, d\theta. \tag{5.7.38}$$

With this solution, Eqs. (5.7.24) and (5.7.25) give, respectively, the trajectory and the equation of motion. The explicit form of $u(r)$ and $v(\theta)$ depend on the concrete conditions of the problem.

5.7.2.3 Separation of Pair-Variables

Suppose that the variables q_n and $p_n = \frac{\partial S}{\partial q_n}$ appear in the Hamilton–Jacobi equation only as the combination

$$f_n\left(q_n, \frac{\partial S}{\partial q_n}\right). \tag{5.7.39}$$

In this case, the Hamilton–Jacobi equation reads:

$$\frac{\partial S}{\partial t} + H\left[q_1, \ldots, q_{n-1}, \frac{\partial S}{\partial q_1}, \ldots, \frac{\partial S}{\partial q_{n-1}}, f_n\left(q_n, \frac{\partial S}{\partial q_n}\right), t\right] = 0. \tag{5.7.40}$$

We look for a complete integral of the form

$$S = S'(q_1, \ldots, q_{n-1}, t) + S_n(q_n), \tag{5.7.41}$$

which leads us to

$$\frac{\partial S'}{\partial t} + H\left[q_1, \ldots, q_{n-1}, \frac{\partial S'}{\partial q_1}, \ldots, \frac{\partial S'}{\partial q_{n-1}}, f_n\left(q_n, \frac{\partial S_n}{\partial q_n}\right), t\right] = 0. \tag{5.7.42}$$

This equation becomes an identity with respect to q_n if S_n satisfies the differential equation

$$f_n\left(q_n, \frac{\partial S_n}{\partial q_n}\right) = a_n(\text{const.}), \tag{5.7.43}$$

so that

$$\frac{\partial S'}{\partial t} + H\left(q_1, \ldots, q_{n-1}, \frac{\partial S'}{\partial q_1}, \ldots, \frac{\partial S'}{\partial q_{n-1}}, a_n, t\right) = 0, \tag{5.7.44}$$

which is a equation with n partial derivatives of S'. The function S_n can be found by reversing (5.7.43):

$$\frac{\partial S_n}{\partial q_n} = F_n(q_n, a_n)$$

and integrating the result:

$$S_n = \int F_n(q_n, a_n)dq_n. \tag{5.7.45}$$

If this procedure can be continued further, then we may, for example, have:

$$S(q_1, \ldots, q_n, a_1, \ldots, a_n, t) = S'(q_1, \ldots, q_{r-1}, a_1, \ldots, a_n, t) + \sum_{k=r}^{n} \int F_k(q_k, a_k)dq_k. \tag{5.7.46}$$

If all the variables $q_1, \ldots, q_n, t$ can be separated, the form of a complete integral of the Hamilton–Jacobi equation is:

$$S(q_1, \ldots, q_n, a_1, \ldots, a_n, t) = -E(a_1, \ldots, a_n)t + \sum_{k=1}^{n} S_k(q_k, a_k). \tag{5.7.47}$$

Each one of the functions F_k $(k = \overline{1, n})$ are obtained by inverting equations of the form

$$f_k\left(q_k, \frac{\partial S_k}{\partial q_k}\right) = a_k. \tag{5.7.48}$$

The dependence of the energy E upon $a_1, \ldots, a_n$ is obtained by introducing $W = \sum_k S_k$ into the abbreviated Hamilton–Jacobi equation (5.7.22). We notice that the separation of the cyclic variables can be discussed as a particular case of this procedure. Indeed, for a cyclic variable, say q_α, we have $\frac{\partial S_\alpha}{\partial q_\alpha} = f_\alpha$, then by taking $f_\alpha = a_\alpha$ and integrating the result, we arrive at $S_\alpha = a_\alpha q_\alpha$ (no summation).

5.7.3 Applications

5.7.3.1 Free Particle

Let us study, in the Hamilton–Jacobi formalism, the motion of a free particle of mass m. Since no force acts on the particle and $p_i = \frac{\partial S}{\partial q_i}$ by (5.7.5), the Hamilton–Jacobi equation is

$$\frac{\partial S}{\partial t} + \frac{1}{2m}(\text{grad } S)^2 = 0. \qquad (5.7.49)$$

The Hamiltonian does not explicitly depend on time and all the variables x, y, z are cyclic, therefore a completely separated solution is

$$S = -Et + \mathbf{a} \cdot \mathbf{r}. \qquad (5.7.50)$$

In order for (5.7.50) to be a solution of (5.7.49), we must have

$$E = \frac{1}{2m}\mathbf{a}^2;$$

consequently

$$S = -\frac{\mathbf{a}^2}{2m}t + \mathbf{a} \cdot \mathbf{r}. \qquad (5.7.51)$$

Next, using (5.7.5), we get:

$$\mathbf{b} = \mathbf{r} - \frac{1}{m}\mathbf{a}t,$$

which is the law of motion

$$\mathbf{r} = \frac{1}{m}\mathbf{a}t + \mathbf{b}, \qquad (5.7.52)$$

as well as the first integral:

$$\mathbf{p} = \text{grad } S = \mathbf{a}(\text{const.}). \qquad (5.7.53)$$

If we do not wish to separate the variables in the Hamilton–Jacobi equation (5.7.49), we can take as a complete integral

$$S = \frac{m}{2t}(\mathbf{r} - \mathbf{a})^2. \qquad (5.7.54)$$

Again using (5.7.5), we arrive at a similar result:

$$\mathbf{r} = -\frac{1}{m}\mathbf{b}t + \mathbf{a}, \qquad (5.7.55)$$

$$\mathbf{p} = \text{grad } S = -\mathbf{b}. \qquad (5.7.56)$$

5.7.3.2 Linear Harmonic Oscillator

Since the system possesses one degree of freedom and the Hamiltonian is time-independent, we search for a complete integral of the form

$$S = W(x, E) - Et. \tag{5.7.57}$$

The abbreviated Hamilton–Jacobi equation,

$$\frac{1}{2m}\left(\frac{dW}{dx}\right)^2 + \frac{1}{2}m\omega_0^2 x^2 = E, \tag{5.7.58}$$

has the solution

$$W(x, E) = \int \sqrt{2mE - m^2\omega_0^2 x^2}\,dx.$$

The integration is performed via the substitution $x = \sqrt{\frac{2E}{m\omega_0^2}}\sin u$, so that

$$W(x, E) = \frac{E}{\omega_0}\left(u + \frac{1}{2}\sin 2u\right),$$

and (5.7.57) finally reads:

$$S = \frac{x}{2}\sqrt{2mE - m^2\omega_0^2 x^2} + \frac{E}{\omega_0}\arcsin\left(\sqrt{\frac{m}{2E}}\omega_0 x\right) - Et.$$

Using (5.7.5), we obtain:

$$t_1 = \frac{\partial S}{\partial E} = -t + \frac{1}{\omega_0}\arcsin\left(\sqrt{\frac{m}{2E}}\omega_0 x\right),$$

$$p = \frac{\partial S}{\partial x} = \sqrt{(2mE - m^2\omega_0^2 x^2)}$$

and, finally, with the substitution $t_1\omega_0 = \varphi(\text{const.})$:

$$x = \frac{1}{\omega_0}\sqrt{\frac{2E}{m}}\sin(\omega_0 t + \varphi),$$

$$p = \sqrt{2mE}\cos(\omega_0 t + \varphi),$$

which is the expected result.

5.7.3.3 Newtonian Central Force

As we know (see Chap. 3), the trajectory of a particle subject to such a force lies in a plane. Let us choose $q_1 = r, q_2 = \varphi$ as the Lagrangian parameters. Since

$$F(r) = -\frac{k}{r^2} = -(\text{grad } V)_r = -\frac{dV}{dr},$$

by integrating from r to ∞, with $V(\infty) = 0$, we have:

$$V = -\frac{k}{r}.$$

The Hamiltonian is then:

$$H = \frac{1}{2m}\left(p_r^2 + \frac{1}{r^2}p_\varphi^2\right) - \frac{k}{r}. \tag{5.7.59}$$

The cyclic variable φ is associated with the first integral

$$p_\varphi = mr^2\dot{\varphi} = \text{const.} \tag{5.7.60}$$

The Hamilton–Jacobi equation,

$$\frac{\partial S}{\partial t} + \frac{1}{2m}\left[\left(\frac{\partial S}{\partial r}\right)^2 + \frac{1}{r^2}\left(\frac{\partial S}{\partial \varphi}\right)^2\right] - \frac{k}{r} = 0, \tag{5.7.61}$$

is a particular case of (5.7.34), with $\theta = \frac{\pi}{2}, p_\theta = 0, v(\theta) = 0, u(r) = -\frac{k}{r}$. Then, we can take

$$S = -Et + \varphi p_\varphi + R(r)$$

and so we get

$$\frac{1}{2m}\left[\left(\frac{dR}{dr}\right)^2 + \frac{1}{r^2}p_\varphi^2\right] - \frac{k}{r} = E. \tag{5.7.62}$$

Recalling that $p_\varphi = \text{const.}$, we can recast (5.7.62) in the form:

$$r^2\left(\frac{dR}{dr}\right)^2 - 2mEr^2 - 2mkr = -p_\varphi^2 = -2\gamma(\text{const.})$$

and then integrate, to find a complete integral of (5.7.62) with the expression

$$W(r, \varphi, E, \gamma) = \sqrt{2\gamma}\varphi + \int \frac{1}{r}\sqrt{2mEr^2 + 2mkr - 2\gamma}dr. \tag{5.7.63}$$

The trajectory is given by (5.7.24):

$$\frac{\partial W}{\partial \gamma} = \gamma_1(\text{const.}),$$

which, upon differentiation, yields:

$$\varphi - \varphi_0 = \int \frac{\sqrt{2\gamma}}{r\sqrt{2mEr^2 + 2mkr - 2\gamma}}dr, \quad \gamma_1\sqrt{2\gamma} = \varphi_0(\text{const.}). \tag{5.7.64}$$

The integral is easily worked out by using the following substitution:

$$x = \frac{\sqrt{2\gamma}}{r}, \quad p = \frac{2mk}{\sqrt{2\gamma}}, \quad q = 2mE, \tag{5.7.65}$$

which gives:

$$\varphi - \varphi_0 = \arccos \frac{x - \frac{p}{2}}{\sqrt{q + \frac{p^2}{4}}} = \arccos \frac{\frac{1}{r} - \frac{mk}{\sqrt{2\gamma}}}{\sqrt{\frac{mE}{\gamma} + \frac{m^2k^2}{4\gamma^2}}}$$

Using the notation

$$\frac{mk}{2\gamma} = \frac{1}{p}, \quad \sqrt{\frac{mE}{\gamma} + \frac{m^2k^2}{4\gamma^2}} = \frac{e}{p},$$

we finally arrive at the previously established result,

$$\frac{1}{r} = \frac{1}{p}[1 + e\cos(\varphi - \varphi_0)], \tag{5.7.66}$$

which says that, when subject to a Newtonian central force field, the particle describes a conic.

5.7.3.4 Symmetrical Top

As a final example, using the Hamilton–Jacobi formalism, let us find the finite equation of motion (4.6.26) of the symmetrical top. To this end, we use the Lagrangian (4.6.14):

$$L = \frac{1}{2}I_1'(\dot{\varphi}^2 \sin^2\theta + \dot{\theta}^2) + \frac{1}{2}I_3'(\dot{\varphi}\cos\theta + \dot{\psi})^2 - Mgl\cos\theta.$$

Since the variables φ, ψ are cyclic, the associated conjugate momenta $p_\varphi,\ p_\psi$ are first integrals and if we choose $q_1 = \varphi,\ q_2 = \theta,\ q_3 = \psi$, we find:

$$p_\varphi = (I_1' \sin^2\theta + I_3' \cos^2\theta)\dot{\varphi} + I_3'\dot{\psi}\cos\theta = C_1,$$

$$p_\theta = I_1'\dot{\theta},$$

$$p_\psi = I_3'(\dot{\varphi}\cos\theta + \dot{\psi}) = C_2.$$

The Hamilton–Jacobi equation then reads:

$$\frac{\partial S}{\partial t} + \frac{1}{2}\left[\frac{p_\theta^2}{I_1'} + \frac{p_\psi^2}{I_3'} + \frac{1}{I_1'}\frac{(p_\varphi - p_\psi\cos\theta)^2}{\sin^2\theta}\right] + Mgl\cos\theta = 0. \tag{5.7.67}$$

Since H does not depend explicitly on t, φ, or ψ, we can take

$$S = -Et + W(\varphi, \theta, \psi) = -Et + C_1\varphi + C_2\psi + \Theta(\theta). \tag{5.7.68}$$

The substitution of (5.7.68) into (5.7.67) leads to an equation with θ as single variable:

$$\frac{1}{2}\left[\frac{1}{I_1'}\left(\frac{d\Theta}{d\theta}\right)^2+\frac{C_2^2}{I_3'}+\frac{1}{I_1'}\frac{(C_1-C_2\cos\theta)^2}{\sin^2\theta}\right]+Mgl\cos\theta=E,$$

which can be integrated to give

$$\Theta(\theta)=\int\frac{\sqrt{2I_1'\sin^2\theta(E-A-Mgl\cos\theta)-(C_1-C_2\cos\theta)^2}}{\sin\theta}d\theta,$$

where $A=\frac{C_2^2}{2I_3'}$. Therefore, a complete integral of the equation (5.7.67) is:

$$W(\varphi,\theta,\psi,C_1,C_2,E)=C_1\varphi+C_2\psi + \int\frac{\sqrt{2I_1'\sin^2\theta(E-A-Mgl\cos\theta)-(C_1-C_2\cos\theta)^2}}{\sin\theta}d\theta.$$

Following the general procedure, the finite equation of motion for the coordinate θ is found by applying (5.7.25):

$$t-t_1=-\frac{\int I_1'\sin\theta\,d\theta}{\sqrt{2I_1'\sin^2\theta(E-A-Mgl\cos\theta)-(C_1-C_2\cos\theta)^2}}.$$

With the change of variable $\cos\theta=u$ and the notations (see (4.6.24))

$$\alpha=\frac{2(E-A)}{I_1'}=\frac{2E_1}{I_1'},\quad \beta=\frac{2Mgl}{I_1'},\quad C_1=I_1'b,\quad C_2=I_1'a,$$

we arrive at:

$$t-t_1=\int\frac{du}{\sqrt{(1-u^2)(\alpha-\beta u)-(b-au)^2}},$$

which is precisely Eq. (4.6.26).

Conclusion. The Hamilton–Jacobi formalism provides an ingenious and efficient method of integrating the canonical system of equations. This formalism consists in the determination of a complete integral of a single first-order partial derivative equation. This procedure is most useful when the possibility of the separation of variables exists.

5.7.4 Action–Angle Variables

In many physical problems we deal with systems performing a periodic motion. A useful procedure of studying these problems derives from the Hamilton–Jacobi formalism and consists in using some new canonical variables: J_α – *action variables* and w_α – *angle variables*, instead of the constant canonical parameters (5.7.2).

5.7.4.1 Systems with One Degree of Freedom

To make the procedure as clear as possible, let us first consider a conservative system with one degree of freedom. Then we can write

$$H(q, p) = E, \tag{5.7.69}$$

which is the implicit equation of a curve in the phase plane and represents the generalized trajectory corresponding to the evolution of the given system possessing the constant energy E.

The properties of the periodical motion are given by the type of the generalized trajectory. We can speak about two types of periodical motion:

1. If the generalized trajectory is a *closed* curve, the motion is called *vibration*. (In most books on mechanics the term *libration* is used instead; but, since the linear harmonic oscillator falls into this category, the word 'vibration' would be more appropriate for a physicist.) In this case, q oscillates between two constant values, both q and p being periodical functions of time with *the same period* (Fig. 5.5a).
2. If by solving Eq. (5.7.69) we obtain

$$p = p(q, E) \tag{5.7.70}$$

as a periodical function of q with period q_0, i.e. $p(q + kq_0, E) = p(q, E)$, with k integer, the periodical motion is termed as *rotation* or, sometimes, *revolution*. In this case the coordinate q can take any value (Fig. 5.5b). A very simple example for such a system is a rigid body rotating about a fixed axis, the generalized coordinate q being the angle of rotation.

It is important to note that a certain system can perform, in certain conditions, either vibration or rotation motions. For example, let us consider again the case of the simple pendulum discussed in Chap. 3, Sect. 3.4. Using the Lagrangian (3.4.1), we can construct the Hamiltonian and (5.7.69) yields:

$$\frac{1}{2mR^2}p^2 - mgR\cos\theta = E,$$

where p is the momentum conjugated to the coordinate θ. The equation of the generalized trajectory in the phase plane (θ, p) is then

$$p = \pm\sqrt{2mR^2(E + mgR\cos\theta)}. \tag{5.7.71}$$

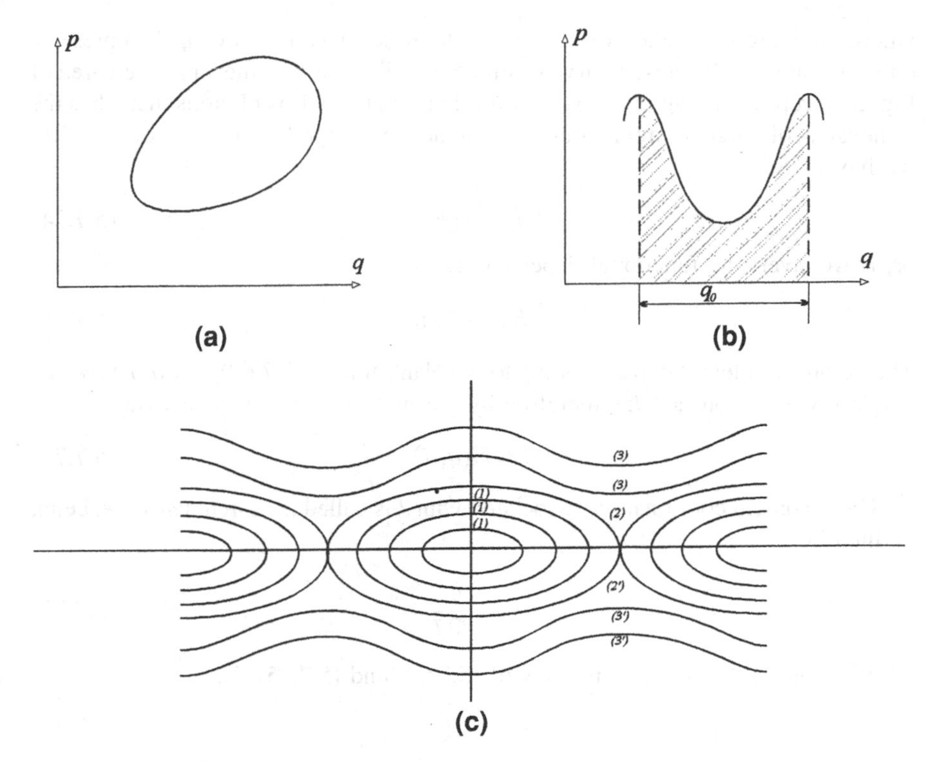

Fig. 5.5 Geometrical representations of the *action variable* in the phase space under various conditions.

If $E < mgR$, the pendulum performs periodical motions as the angle θ varies between $-\theta_0$ and θ_0, where θ_0 is defined by

$$\cos\theta_0 = -\frac{E}{mgR}. \tag{5.7.72}$$

In this case we have a *vibration*-type motion, the generalized trajectories in the phase plane (θ, p) being the closed curves (1) as shown in Fig. 5.5c. But if $E > mgR$, there are no limitations for the angle θ, and the pendulum *rotates* around the suspension point. The corresponding generalized trajectories are the curves (3) in Fig. 5.5c. A special situation appears in the limit case $E = mgR$, marked on Fig. 5.5c by the curves (2) or (2′). These curves correspond to the situation when the pendulum reaches the positions with $\theta = \pm\pi$ (in general $\theta = (2k+1)\pi$), where $p = 0$, therefore these are positions of unstable equilibrium. A small perturbation can remove the pendulum from such a position, and the representative point may trace out either curve (2), or curve (2′).

We define the *action variable* by

$$J = \oint p\, dq, \tag{5.7.73}$$

where the integral is taken over a complete cycle of variation of q. It represents either the area of the closed curve of Fig. 5.5a (vibration), or the shadowed area of Fig. 5.5b corresponding to a period of motion (rotation). It is obvious that J has the dimension of angular momentum (or of action integral). In view of (5.7.70), we have:

$$J = J(E), \tag{5.7.74}$$

or, if we invert the functional dependence,

$$E = E(J). \tag{5.7.75}$$

The complete integral corresponding to the Hamiltonian (5.7.69) is $S(q, E)$ (with a single *essential* constant E), therefore by means of (5.7.75) we can write:

$$S = \mathcal{S}(q, J). \tag{5.7.76}$$

The canonical coordinate w associated with J is called the *angle variable*, being defined by

$$w = \frac{\partial \mathcal{S}}{\partial J}, \tag{5.7.77}$$

while the new Hamiltonian, in view of (5.7.69) and (5.7.75), is:

$$H = \mathcal{H}(J). \tag{5.7.78}$$

Since the coordinate w is conjugated to an angular momentum, its dimension is that of an angle.

With this choice, the canonical equations yield:

$$\dot{J} = 0, \quad \dot{w} = \frac{\partial \mathcal{H}}{\partial J} = \nu(J), \tag{5.7.79}$$

meaning that J is a constant (which is already known from (5.7.74)) and, consequently, ν is also a constant (depending on J). Integrating $(5.7.79)_2$, we obtain:

$$w = \nu t + \varphi. \tag{5.7.80}$$

To find the significance of the constant ν, let us determine the variation of the angle variable w when q performs a complete cycle of variation (either vibration or rotation). We have

$$\Delta w = \oint \frac{\partial w}{\partial q} dq,$$

or, using (5.7.77) and the definition $\frac{\partial \mathcal{S}}{\partial q} = p$,

$$\Delta w = \oint \frac{\partial^2 \mathcal{S}}{\partial q \partial J}\, dq = \oint \frac{\partial p}{\partial J}\, dq = \frac{d}{dJ} \oint p\, dq = 1. \tag{5.7.81}$$

If we denote by τ the period corresponding to a complete cycle, from (5.7.80) and (5.7.81) we obtain $\Delta w = \nu\tau = 1$, hence:

$$\nu = \frac{1}{\tau}. \tag{5.7.82}$$

This result means that ν is the *frequency* of the periodical variation of q. It then follows that we can determine the period of the motion if we know the dependence of the Hamiltonian on the variable J, without solving the equations of motion. Inverting relation (5.7.76) and using (5.7.80), we can obtain the time-dependence of the coordinate q. If the coordinate q is cyclic, the corresponding momentum p is conserved, p = const. The representative point then traces a straight line in the (q, p)-plane. This can be regarded as a limiting case of rotation (see Fig. 5.5c, curve 3) with arbitrary period. Since w is an angle variable, it is natural to choose its period as 2π. Consequently, for cyclic coordinates, the corresponding action variable is always $J = 2\pi p$.

Example. Let us now apply the aforementioned formalism to the motion of vibration of a simple pendulum. Using (5.7.71), the action variable J reads:

$$J = -2\int_{\theta_0}^{-\theta_0} [2mR^2(E + mgR\cos\theta)]^{\frac{1}{2}}\, d\theta = 4\int_0^{\theta_0} [2mR^2(E + mgR\cos\theta)]^{\frac{1}{2}}\, d\theta.$$

By virtue of (5.7.69), we can write $(5.7.79)_2$ in the form

$$\nu(J) = \frac{\partial \mathcal{H}}{\partial J} = \left(\frac{\partial J}{\partial E}\right)^{-1}.$$

Then, according to (5.7.82), the period can be written as

$$\tau = \frac{dJ}{dE},$$

or, after the derivative is performed,

$$\tau = 4\int_0^{\theta_0} \frac{2g}{R}(\cos\theta - \cos\theta_0)^{-\frac{1}{2}}\, d\theta,$$

which is an already known formula (see Chap. 3, Sect. 3.4). In the case of small oscillations, using the Hamiltonian

$$H = \frac{1}{2mR^2}p^2 + \frac{1}{2}mgR^2\theta = E,$$

we easily obtain

$$J = 2\pi E\sqrt{\frac{R}{g}}, \tag{5.7.83}$$

yielding the well-known relation

$$\tau = 2\pi\sqrt{\frac{R}{g}}. \tag{5.7.84}$$

5.7.4.2 Systems with Many Degrees of Freedom

Next, we shall generalize this formalism for conservative systems with many degrees of freedom. Assuming that the variables are completely separable, a complete integral W of the abbreviated Hamilton–Jacobi equation can be written as

$$W(q,a) = \sum_{j=1}^{n} S_j(q_j,a) \quad (a \equiv a_1,\ldots,a_n) \tag{5.7.85}$$

and Eqs. $(5.7.5)_1$ give:

$$p_j = \frac{\partial S_j(q_j,a)}{\partial q_j} \quad (j=\overline{1,n};\ \text{no summation}),$$

or

$$p_j = p_j(q_j,a) \quad (j=\overline{1,n}). \tag{5.7.86}$$

For a given index j, this is the equation of a curve in the phase plane $(q_j,\ p_j)$. Therefore we can introduce the action–angle variables only if, for any j, these curves are either closed (vibration), or correspond to some periodical functions of q_j (rotation).

Thus, the *action variables* J_j are defined by

$$J_j = \oint p_j\, dq_j = \oint \frac{\partial S_j(q_j,a)}{\partial q_j}\, dq_j \quad (\text{no summation}). \tag{5.7.87}$$

These integrals are calculated for a complete cycle of variation of q_j and can be considered as the increase of the generating functions S_j during this cycle. The last relation yields

$$J_j = J_j(a) \quad (j=\overline{1,n}), \tag{5.7.88}$$

or, by inverting the functional dependence,

$$a_j = a_j(J) \quad (j=\overline{1,n}), \tag{5.7.89}$$

which is possible because the Jacobian $\partial(J)/\partial(a)$ is always non-zero. Here, by J we mean the set of variables $J_1,\ldots,J_n$.

Now, we can write:

$$S_j\big(q_j,a(J)\big) = \mathcal{S}(q_j,J), \tag{5.7.90}$$

as well as $\mathcal{S}(a,J) = \sum_{j=1}^{n} S_j$. Using again (5.7.5), we obtain the canonical transformation from the set of variables $(q,\ p)$ to the set $(J,\ w)$:

$$w_j = \frac{\partial \mathcal{S}}{\partial J_j}, \quad p_j = \frac{\partial \mathcal{S}}{\partial q_j} = \frac{\partial \mathcal{S}_j}{\partial q_j} \quad (j=\overline{1,n}). \tag{5.7.91}$$

The first n relations (5.7.91) define the angle variables w_j. In order to define the new Hamiltonian, we recall that the system is conservative and, therefore, we can write $H(q, p) = E(a)$. In view of (5.7.89), this yields:

$$H = \mathcal{H}(J),$$

which shows that $\mathcal{H}$ does not depend on the new canonical variables w.

The canonical equations, written in terms of the new variables J and w, are:

$$\dot{J}_j = \frac{\partial \mathcal{H}}{\partial w_j} = 0, \quad \dot{w}_j = \frac{\partial \mathcal{H}}{\partial J_j} = \nu_j(J) \quad (j = \overline{1,n}), \tag{5.7.92}$$

and, by integration,

$$J_j = \text{const.}, \quad w_j = \nu_j t + \varphi_j \quad (j = \overline{1,n}). \tag{5.7.93}$$

Let us now show that the quantities ν_j have the significance of frequencies of the motion. One first observes that, according to the definition (5.7.91), the variables w_j depend on coordinates q. To realize the meaning of this dependence, we shall calculate the variation of w_j while one coordinate – say q_k – performs a cycle of variation, the other ones being fixed. Obviously, during the variation all J_j remain unchanged. Denoting this variation by $\Delta_k w_j$, we have:

$$\Delta_k w_j = \oint \frac{\partial w_j}{\partial q_k}\, dq_k = \oint \frac{\partial^2 \mathcal{S}}{\partial q_k \partial J_j}\, dq_k = \frac{\partial}{\partial J_j} \oint \frac{\partial \mathcal{S}}{\partial q_k}\, dq_k = \frac{\partial J_k}{\partial J_j} = \delta_{jk}. \tag{5.7.94}$$

This relation shows that w_k is a monotonic periodical function of the coordinates q_k $(k \neq j)$ (it increases by 1 after each complete cycle). But this fact does not allow us to conclude on the behaviour of ν_j, as we did in problems with one degree of freedom, because all the coordinates change during the motion.

Using Eqs. (5.7.91), we can write the canonical transformations $(q, p) \to (J, w)$ and their inverse as well:

$$q_j = q_j(J, w), \quad p_j = p_j(J, w) \quad (j = \overline{1,n}). \tag{5.7.95}$$

According to (5.7.94), in the case of *vibration*, we have:

$$q_j(J, w + m) = q_j(J, w), \tag{5.7.96}$$

where $w + m$ means the set $(w_1 + m_1, \ldots, w_n + m_n)$, with m_k integers $(k = \overline{1,n})$. In the case of *rotation*, we have:

$$q_j(J, w + m) = q_j(J, w) + q_j^0 n_j,$$

where q_j^0 are the periods of the functions (5.7.95). One observes that the quantities $q_j' = q_j - q_j^0 w_j$ satisfy the relations (5.7.96), which means that we can use in the case of rotation, for the quantities q_j', the formalism developed for vibration.

Assume that the frequencies ν_j are *commensurable*, that is there exist integers m_j and m_k so that

$$\frac{\nu_j}{m_j} = \frac{\nu_k}{m_k}, \tag{5.7.97}$$

for any $j, k = \overline{1, n}$. Therefore, after a time interval

$$\tau = \frac{m_j}{\nu_j} = \frac{m_k}{\nu_k} \tag{5.7.98}$$

all coordinates q_j recover their values:

$$q_j(t + \tau) = q_j(t), \tag{5.7.99}$$

meaning that the real motion is periodical with *period* τ. Indeed, using (5.7.93) and (5.7.98), we have:

$$w_j(t + \tau) = w_j(t) + m_j \quad (j = \overline{1, n}) \tag{5.7.100}$$

and (5.7.96) leads to the above statement. It can be shown that, starting from (5.7.99), we arrive at the condition of commensurable frequencies (5.7.97).

If condition (5.7.97) is satisfied for any j, k, and hence (5.7.99) is fulfilled for any j, we call the system *completely degenerated* and the real motion is a truly periodical motion. Even in this situation, the frequencies ν_j cannot be identified with the individual frequencies of the motion. Indeed, if (5.7.99) are satisfied for any j, the only thing we can say is that there exist the quantities τ_j $(j = \overline{1, n})$ so that the conditions (5.7.98) can be written as

$$\tau = m_j \tau_j = m_k \tau_k.$$

In this case τ_j is the shortest time interval for which the relation $q_j(t + \tau) = q_j(t)$ holds true, for any j and t, i.e. τ_j is the period of $q_j(t)$ only.

If relation (5.7.100) holds true only for certain values of j, the system is called *simply degenerated*. If the frequencies are not commensurable for any pair ν_j, ν_k, the system is said to be *periodically conditioned*. In this case the system does not come back to its initial state, and the generalized trajectory in the configuration space fills up a certain domain in this space. The action–angle formalism proves to be a powerful tool in such cases, since it provides all the frequencies of the individual motions, without the complete solutions being known.

Example. Let us discuss the Kepler problem within the frame of the action–angle formalism. The problem is described by the Hamiltonian (5.7.32) with the potential (5.7.33), in which $u(r) = -|k|/r$, $v(\theta) = 0$. We denote $p_\varphi = l_3, a = l^2$, where l is the magnitude of the angular momentum. The complete integral corresponding to our problem is then (see (5.7.38)):

$$S = -Et + \varphi l_3 + \int \sqrt{2mE + 2m\frac{|k|}{r} - \frac{l^2}{r^2}}\, dr + \int \sqrt{l^2 - \frac{l_3^2}{\sin^2\theta}}\, d\theta,$$

leading to the action variables:

$$J_\varphi = \oint \frac{\partial S_\varphi}{\partial \varphi}\, d\varphi = \oint l_3\, d\varphi, \tag{5.7.101}$$

$$J_\theta = \oint \frac{\partial S_\theta}{\partial \theta}\, d\theta = \oint \left[l^2 - \frac{l_3^2}{\sin^2\theta} \right]^{\frac{1}{2}} d\theta, \tag{5.7.102}$$

$$J_r = \oint \frac{\partial S_r}{\partial r}\, dr = \oint \left[2mE + \frac{2m|k|}{r} - \frac{l^2}{r^2} \right]^{\frac{1}{2}} dr. \tag{5.7.103}$$

The first integral yields:

$$J_\varphi = 2\pi l_3, \tag{5.7.104}$$

as we expected, since φ is a cyclic variable.

In a cycle, the variable θ increases from $-\theta_0$ to θ_0, then decreases from θ_0 to $-\theta_0$, where

$$\theta_0 = \arcsin \frac{l_3}{l},$$

therefore

$$J_\theta = 4 \int_0^{\theta_0} \left[l^2 - \frac{l_3^2}{\sin^2\theta} \right]^{\frac{1}{2}} d\theta.$$

The integral is easily worked out with the substitution

$$\sin\theta = \left[\frac{1 + \frac{l_3^2}{l^2} x^2}{1 + x^2} \right]^{\frac{1}{2}},$$

which gives:

$$J_\theta = 4l \left(1 - \frac{l_3^2}{l^2} \right) \int_0^\infty \left[(1 + x^2) \left(1 + \frac{l_3^2}{l^2} x^2 \right) \right]^{-1} dx,$$

and finally:

$$J_\theta = 2\pi (l - l_3). \tag{5.7.105}$$

The motion is periodical only for $E < 0$. In this case, r varies cyclically from r_m to r_M and back to r_m, where r_m and r_M are the roots of the equation $2mEr^2 + 2m|k|r - l^2 = 0$. Then, the integral (5.7.103) can be written as

$$J_r = 2\int_{r_m}^{r_M} [2mEr^2 + 2m|k|r - l^2]^{\frac{1}{2}}\, dr.$$

Using the substitution

$$[2mE(r - r_m)(r - r_M)]^{\frac{1}{2}} = y(r - r_M),$$

we finally obtain:

$$J_r = \pi|k|\sqrt{\frac{2m}{|E|}} - 2\pi l. \qquad (5.7.106)$$

Relations (5.7.104) and (5.7.105) yield $J_\theta + J_\varphi = 2\pi l$ and therefore (5.7.106) leads to

$$\mathcal{H} \equiv E = -2\pi^2 mk^2 (J_r + J_\theta + J_\varphi)^{-2}. \qquad (5.7.107)$$

In view of (5.7.92), the frequencies are:

$$\nu_r = \nu_\theta = \nu_\varphi = 4\pi^2 mk^2 (J_r + J_\theta + J_\varphi)^{-3}, \qquad (5.7.108)$$

which means that the system is *completely degenerated*. The motion is periodical, with the period $\tau = \frac{1}{\nu_r} = \frac{1}{\nu_\theta} = \frac{1}{\nu_\varphi}$, that can be expressed, using (5.7.107), as

$$\tau = \pi|k|\sqrt{\frac{m}{2|E|^3}},$$

which is nothing else but Kepler's third law (see Chap. 3, Sect. 3.2.2).

Observation: Since the integrals (5.7.101) and (5.7.102) do not depend on $V(r)$, the relation $2\pi l = J_\theta + J_\varphi$ holds true for any central field. Consequently, by integrating (5.7.103) for any given $V(r)$ we obtain E as a function of J_r and l, which means that E depends on J_θ and J_φ only through the sum $J_\theta + J_\varphi$. Consequently, the frequencies ν_θ and ν_φ are always equal. This simple degeneracy of the motion in a central field is a consequence of the fact that the motion takes place in a plane, orthogonal to the angular momentum $\mathbf{l}$, and thus the variations of the angles θ and φ are related (a variation of 2π in φ corresponds to a variation of $4|\theta_0|$ in θ).

5.7.5 Adiabatic Invariants

Consider a mechanical system whose Hamiltonian depends on some real parameter λ: $H = H(q, p, \lambda)$. This parameter can be *internal* (i.e. characterizes the system itself) or *external* (e.g. defines the external field in which the system is placed). We shall also assume that the problem of motion of the system can be solved by using the action–angle variables for all fixed values of λ and that there is no relation of type (5.7.97) (the system is not degenerated).

If λ depends on time, the system is not conservative and, therefore, the action variables J_j are not constant. Nevertheless, it can be shown that, if λ varies slowly in time, that is if

$$\tau\dot{\lambda} \ll \lambda,$$

where τ is the period of the motion, then the action variables J_j still remain unchanged. Such quantities are called *adiabatic invariants*.

The adiabatic invariants have been originally defined by *Paul Ehrenfest* in connection with the earliest research on quantum theory of the atom. For example, the Bohr–Sommerfeld quantization rules were postulated precisely for the adiabatic invariants J_j. During the recent years, the study of adiabatic invariants has been resumed, in connection with their use in plasma physics, thermonuclear processes, particle accelerators, etc.

To simplify the discussion, we consider a system with only one degree of freedom, for which we rewrite (5.7.69) and (5.7.70) as follows:

$$H(q, p, \lambda) = E, \tag{5.7.109}$$

and

$$p = p(q, E, \lambda), \tag{5.7.110}$$

with the parameter λ varying slowly in time. Introducing (5.7.110) into (5.7.73), it results that J depends on time through λ. To determine the time variation of J over a period of motion, we first calculate its derivative with respect to time:

$$\frac{dJ}{dt} = \oint \left(\frac{\partial p}{\partial E} \frac{\mathrm{d}E}{\mathrm{d}t} + \frac{\partial p}{\partial \lambda} \dot{\lambda} \right) dq, \tag{5.7.111}$$

and then take the average for one period of motion. By virtue of (5.7.109), we have:

$$\frac{dE}{dt} = \frac{\partial H}{\partial t} = \frac{\partial H}{\partial \lambda} \dot{\lambda}. \tag{5.7.112}$$

By averaging this relation and assuming that during the time interval $\tau, \dot{\lambda}$ is practically constant, we obtain:

$$\overline{\frac{dE}{dt}} = \dot{\lambda}\overline{\frac{\partial H}{\partial \lambda}} = \frac{1}{\tau}\dot{\lambda}\int_0^\tau \frac{\partial H}{\partial \lambda}\,dt. \tag{5.7.113}$$

The dependence of the rate of change of the energy on the time variation of λ is taken into account in (5.7.113) through the factor $\dot{\lambda}$, therefore $\frac{\partial H}{\partial \lambda}$ under the integral can be considered only as a function of the varying q and p, for fixed λ. Using the canonical equation $\dot{q} = \partial H/\partial p$, we can replace the integration with respect to time by one with respect to the coordinate q. Since

$$dt = \left(\frac{\partial H}{\partial p}\right)^{-1} dq \quad \text{and} \quad \tau = \int_0^\tau dt,$$

we have:

$$\overline{\frac{dE}{dt}} = \dot{\lambda}\frac{\oint \frac{\partial H}{\partial \lambda}\left(\frac{\partial H}{\partial p}\right)^{-1} dq}{\oint \left(\frac{\partial H}{\partial p}\right)^{-1} dq}, \tag{5.7.114}$$

with the integrals taken for fixed λ. Consequently, we regard λ and E as constant independent parameters in (5.7.109) and (5.7.110). Thus, H depends on λ explicitly, and implicitly through p, therefore the derivative with respect to λ of (5.7.109) is:

$$\frac{\partial H}{\partial \lambda} + \frac{\partial H}{\partial p}\frac{\partial p}{\partial \lambda} = 0.$$

This relation serves to obtain

$$\frac{\partial H}{\partial \lambda}\left(\frac{\partial H}{\partial p}\right)^{-1} = -\frac{\partial p}{\partial \lambda}.$$

Since, according to (5.7.109) and (5.7.110), we also have:

$$\left(\frac{\partial H}{\partial p}\right)^{-1} = \frac{\partial p}{\partial E},$$

introducing the last two relations into (5.7.114), we arrive at

$$\overline{\frac{dE}{dt}} = -\dot{\lambda}\frac{\oint \frac{\partial p}{\partial \lambda}\,dq}{\oint \frac{\partial p}{\partial E}\,dq}.$$

As neither $\dot{\lambda}$, nor $\overline{\frac{dE}{dt}}$ depend on q, the last relation can be written as

$$\oint \left(\frac{\partial p}{\partial E}\overline{\dot{E}} + \frac{\partial p}{\partial \lambda}\dot{\lambda}\right) dq = 0.$$

Comparing now this equation with (5.7.111), we finally obtain:

$$\overline{\frac{dJ}{dt}} = 0,$$

where $J = \oint pdq$, meaning that in this approximation the action variable J remains unchanged during the variation of the parameter λ. Therefore, J is an adiabatic invariant.

For oscillatory motion, the adiabatic invariant $J = \oint pdq$ represents the area of the surface bounded by the closed generalized trajectory in phase space. In the case of a one-dimensional oscillator, for example, as we know (see (5.6.45)), the path in the phase plane q, p is an ellipse with semi-axes $a = \sqrt{2E/m\omega^2}, b = \sqrt{2mE}$ which has the area

$$\pi ab = 2\pi \frac{E}{\omega},$$

and thus $J = 2\pi E/\omega$ is an adiabatic invariant. In other words, if the parameters of the linear oscillator vary slowly, the energy and the frequency are proportional.

We also mention, as another example, the case of the simple pendulum. Here we can take as the slowly varying parameter either its length l, or the gravitational acceleration g. In view of (5.7.83) and (5.7.84) it then results that the ratio of the energy E to the frequency $\nu = \sqrt{\frac{g}{l}}$ is practically constant.

As a final remark, it should be mentioned that one must also know the physical conditions which have to be fulfilled in order to have an adiabatic invariance. Indeed, we may vary very slowly but periodically the length of the simple pendulum, such that the period of the variation of the parameter is $n/2$ times (n integer) bigger than the period of oscillation of the pendulum. Then we can reach the so-called phenomenon of *parametric resonance*, in which the amplitude (and, consequently, the energy) increases rapidly with time, although the frequency performs a slow variation.

5.8 Problems

1. Determine the Hamiltonian of the double coplanar pendulum and write the equations of motion.
2. Determine the motion of a charged particle of mass m and charge e, moving in the static electromagnetic field $(\mathbf{E}, \mathbf{B})$. The angle between $\mathbf{E}$ and $\mathbf{B}$ is arbitrary and the initial velocity $\mathbf{v}_0$ of the particle is orthogonal to the plane determined by the $(\mathbf{E}, \mathbf{B})$ field.
3. A heavy bead of mass m slides without friction in a straight pipe, which rotates at a constant angular velocity ω in a vertical plane about one of its points, considered fixed. Find the finite equation of motion.
4. A homogeneous straight rod moves in a vertical plane, with its ends sliding without friction on two perpendicular walls, one on the horizontal plane, and the other on the vertical. Using the Hamiltonian formalism, determine the

trajectory of some point P of the rod. Also, find the motion of P if its mass is m and the mass of the rod can be neglected.

5. Write the differential equations of motion of a spherical pendulum of mass m, sliding without friction on a fixed sphere of radius R.
6. The point of support P of a simple pendulum of mass m and length l moves horizontally according to the law $Y(t) = A \sin \omega t$. Find the law of motion in the following two frames:

   (a) The reference frame is fixed, with the origin at O;
   (b) The origin of the frame is at the point P.

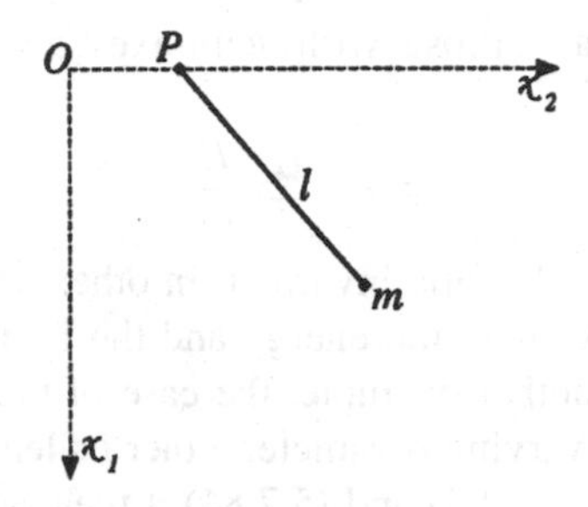

7. A particle of mass m is subjected to a central force. Compare the Hamiltonian H of the particle relative to an inertial (say, fixed) frame S, with the Hamiltonian H' relative to a frame S', rotating about the centre of force at a constant angular velocity ω. Write the canonical equations in both representations.
8. Construct the Hamiltonian of a dipole whose opposite charges have masses m_1 and m_2, and which is located in the homogeneous electric field $\mathbf{E}$.
9. Using the Routhian formalism, obtain the integrals of motion of a spherical pendulum of mass m and length l.
10. Find the Poisson bracket $\{\varphi, l_i\}$, where φ is any function, spherically symmetric about the origin, and depending on the coordinates and momentum of a particle, while l_i ($i = 1, 2, 3$) is any one of the three components of its angular momentum.
11. Show that $\{\mathbf{f}, L_z\} = \mathbf{k} \times \mathbf{f}$, where $\mathbf{f}$ is a vector function of $\mathbf{r}$ and $\mathbf{p}$, while $\mathbf{k}$ is the unit vector of the z-axis.
12. Use the Poisson bracket, the Lagrange bracket and the bilinear covariant methods to show that the transformation

$$Q = \sqrt{e^{-2q} - p^2}, \quad P = \cos^{-1}(pe^q)$$

is canonical.

13. Prove that the transformation

$$Q = \log \frac{\sin p}{q}, \quad P = q \cot p$$

is canonical and find all possible versions of the generating function.

14. Show that under the transformation

$$Q_j = q_j e^{-\beta t}, \quad P_j = p_j e^{-\gamma t} \quad (j = \overline{1,n})$$

the system

$$\dot{q}_j = \frac{\partial H}{\partial p_j} + \beta q_j, \quad \dot{p}_j = -\frac{\partial H}{\partial q_j} + \gamma p_i$$

becomes canonical and find the corresponding Hamiltonian $\mathcal{H}(Q, P, t)$.

15. Show that the transformation

$$Q_1 = q_1 - v_0 t, \quad P_1 = p_1 - v_0,$$

$$Q_2 = \sqrt{2p_2} e^{-t} \sin q_2, \quad P_2 = \sqrt{2p_2} e^{t} \cos q_2$$

is canonical. If the Hamiltonian in the original variables is $H = \frac{1}{2}(p_1^2 + p_2^2 + q_1^2)$, find the generating function $F_2(q, P, t)$, the new Hamiltonian $\mathcal{H}(Q, P, t)$ and the canonical equations in terms of Q_1, Q_2, P_1, P_2.

16. The Lagrangian associated with a damped linear harmonic oscillator subject to the forces $F_1 = -kx$, $F_2 = -r\dot{x}$ is

$$L = e^{\frac{rt}{m}} \left(\frac{1}{2} m \dot{x}^2 - \frac{1}{2} k x^2 \right).$$

Using the Hamilton–Jacobi formalism, find the integral of motion.

17. Determine the values of a and b, so that the transformation

$$Q = (2q)^a \cos^b p, \quad P = (2q)^a \sin^b q$$

is canonical and find the generating function associated to this transformation.

18. Show that the line integral

$$I = \int_\gamma \sum_{j=1}^{n} q_j dp_j$$

is a relative integral invariant associated with the canonical system.

19. Find the conditions which must be satisfied by the functions $A_j(q, p)$, $B_j(q, p)$ $(j = \overline{1,n})$, in order that

$$I = \oint \sum_{j=1}^{n} (A_j dq_j + B_j dp_j)$$

be an integral invariant of Hamilton's canonical equations.

20. A particle moves without friction on a fixed sphere of radius R. Using the Hamilton–Jacobi formalism, determine the trajectory of the particle and find the finite equation of motion. The initial conditions are:

$$\theta(0) = \varphi(0) = 0, \quad \dot{\varphi} = 0, \quad \dot{\theta}(0) = \frac{v_0}{R}.$$

Chapter 6
Mechanics of Continuous Deformable Media

6.1 General Considerations

By a *continuous medium* we mean a material body which fills a certain spatial domain in such a way that there is a mass point of the medium in each geometric point of the domain. But, since the substance possesses a granular structure, this definition has to be more accurate.

Consider a portion of our medium, of mass Δm and volume $\Delta\tau$, and assume that the number of molecules contained in this volume is ΔN. In order that the medium satisfies the property of continuity, there must exist the limit

$$\rho = \lim_{\Delta\tau \to 0} \frac{\Delta m}{\Delta\tau}, \tag{6.1.1}$$

meaning that no matter how small the volume $\Delta\tau$ is, it must contain a sufficiently big number of particles. Since this division is limited by the atomic structure of the medium, to give a *physical* sense to the definition (6.1.1) we introduce the notion of *physically small infinity*. So, by a *physically infinitely small volume* we mean an elementary volume, small by comparison with macroscopic volume inhomogeneities, but big enough so as to contain a large number of molecules. The number ΔN of molecules contained in a physically infinitely small volume produces a *physically infinitely small particle*. From now on, in the study of continuous deformable media, by *particle* we shall mean a physically infinitely small particle. Thus, formula (6.1.1) defines the *mass density* at some point P of the medium, with the aforementioned restrictions regarding the limit.

In general, the mass density ρ is a non-negative continuous function of position and time:

$$\rho = \rho(\mathbf{r}, t), \tag{6.1.2}$$

where $\mathbf{r}$ is the radius-vector of the particle of mass Δm. In other words, the mass density is a *scalar field*.

M. Chaichian et al., *Mechanics*, DOI: 10.1007/978-3-642-17234-2_6,

A continuous medium can be either at rest or in motion relative to an arbitrary reference frame. If the distance between any two points of the medium does not change during the motion, the medium is a *rigid body* (see Chap. 4), while in the opposite case we have a *deformable medium*. This chapter is concerned with the mechanical study of *continuous deformable media* (CDM).

A deformable medium can be in either a *solid* or a *fluid* state. A solid medium can have either *elastic* or *inelastic* properties, while a fluid can be either *compressible* $\left(\frac{\partial \rho}{\partial t} \neq 0\right)$ or *incompressible* $\left(\frac{\partial \rho}{\partial t} = 0\right)$. A special type of continuous deformable medium is the *plasma*, which is a mixture of neutral and excited atoms, ions, electrons and photons.

A macroscopic volume of a continuous medium contains a big (practically, infinite) number of molecules. To describe the motion of each molecule, one must characterize it by a finite number of Lagrangian parameters. It then results that a continuous deformable medium has a huge, practically *infinite number of degrees of freedom*. As we shall see, this gives rise to a special analytical treatment of the motion of such media.

6.2 Kinematics of Continuous Deformable Media

6.2.1 Lagrange's Method

Consider a portion of a continuous deformable medium (CDM) which, at time t_0, occupies the domain $D_0(t_0)$, of volume V_0 in the physical, Euclidean space E_3. Due to the motion, the form and the volume of the domain occupied by the medium vary. Let V be the volume of the domain $D(t)$ occupied by the medium at time $t > t_0$ (Fig. 6.1).

According to *Lagrange*, to know the motion of the medium means to follow and determine the motion of each particle. Consider a particle which, at the time t_0, is at the point $P_0(x_1^0, x_2^0, x_3^0)$ and at the time $t > t_0$ is at the point $P(x_1, x_2, x_3)$. Since the position of the particle depends both on the time t and its initial position, we can write:

$$x_i = x_i(x_1^0, x_2^0, x_3^0, t) \quad (i = 1, 2, 3), \tag{6.2.1}$$

or, in vector form,

$$\mathbf{r} = \mathbf{r}(\mathbf{r}_0, t), \tag{6.2.1$'$}$$

where $\mathbf{r}$ and $\mathbf{r}_0$ are the position vectors of the points P and P_0, respectively. The quantities x_1^0, x_2^0, x_3^0, t are called *Lagrange variables*. Equations (6.2.1) define the law of motion of the particle P which, at the time t_0, was at the point P_0. In order to find the motion of the medium, we must find the equation of motion of each particle.

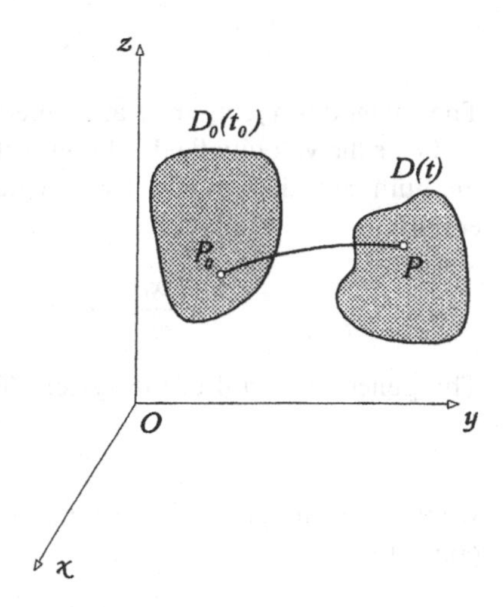

Fig. 6.1 Evolution in time of a domain occupied by a continuous deformable medium.

Assume that the vector function $\mathbf{r}$ is continuous with respect to the Lagrange variables and at least twice differentiable, and that there is a one-to-one correspondence between the domains D_0 and D, i.e.

$$J = \frac{\partial(x_1, x_2, x_3)}{\partial(x_1^0, x_2^0, x_3^0)} \neq 0, \tag{6.2.2}$$

where J is the Jacobian of the transformation (6.2.1).

The trajectory of the particle is given by Eqs. (6.2.1), where the initial position $\mathbf{r}_0$ of the particle is fixed, while the time t varies. Using (6.2.1′), we can write the velocity of the particle as

$$\mathbf{v} = \frac{\partial \mathbf{r}}{\partial t}(x_1^0, x_2^0, x_3^0, t), \tag{6.2.3}$$

while its acceleration is

$$\mathbf{a} = \frac{\partial^2 \mathbf{r}}{\partial t^2}(x_1^0, x_2^0, x_3^0, t). \tag{6.2.4}$$

6.2.2 Euler's Method

Suppose that, instead of following the particles in motion, we choose a fixed point $P(x_1, x_2, x_3)$ and look for the characteristic quantities (velocity, acceleration) at this point. For example, an observer fixed with respect to the banks of a river determines the velocity of the water at his observation point.

If we use (6.2.1) to express x_1^0, x_2^0, x_3^0 in terms of x_1, x_2, x_3, and then introduce them in (6.2.3), we obtain the velocity field at the time t:

$$\mathbf{v} = \mathbf{v}(x_1, x_2, x_3, t). \tag{6.2.5}$$

The quantities x_1, x_2, x_3, t are called *Euler variables*.

Once the velocity field is known, the parametric equations of the particles of the medium are determined by integrating the system of first-order differential equations

$$\frac{dx_i}{dt} = v_i(x_1, x_2, x_3, t) \quad (i = 1, 2, 3). \tag{6.2.6}$$

The general integral of the system (6.2.6) is

$$x_i = x_i(C_1, C_2, C_3, t), \tag{6.2.7}$$

where the integration constants C_1, C_2, C_3 are determined from the initial conditions,

$$x_i^0 = x_i(C_1, C_2, C_3, t_0). \tag{6.2.8}$$

Therefore, the expressions (6.2.7) become:

$$x_i = x_i(x_1^0,\, x_2^0,\, x_3^0, t),$$

which are precisely Eqs. (6.2.1).

The acceleration field in Euler variables is determined by calculating the derivative

$$\mathbf{a} = \frac{d}{dt}\mathbf{v}(x_1, x_2, x_3, t), \tag{6.2.9}$$

which yields

$$\mathbf{a} = \mathbf{a}(x_1, x_2, x_3, t). \tag{6.2.10}$$

Trajectories and Streamlines

By eliminating the time t between the three parametric equations of the trajectory (6.2.1), we obtain two equations in x_i, $x_i^0\ (i = 1, 2, 3)$, i.e. a family of curves depending on three parameters, x_1^0, x_2^0, x_3^0. Each trajectory is tangent to the velocity in *each* point and at *any* moment. The differential equations of trajectories can also be written as

$$\frac{dx_1}{v_1(\mathbf{r}, t)} = \frac{dx_2}{v_2(\mathbf{r}, t)} = \frac{dx_3}{v_3(\mathbf{r}, t)}. \tag{6.2.11}$$

The same system of equations is obtained by projecting on axes the obvious vector relation

$$\mathbf{v} \times d\mathbf{r} = 0.$$

If in Eqs. (6.2.11) we take a fixed value for the time t, we obtain a family of curves tangent to the velocity at that moment. These curves are called *streamlines* (or *current lines*).

Substantial and Local Derivatives

Let $f(x_1, x_2, x_3, t)$ be a function defined on the domain $D \subset E_3$ for $t_0 \leq t \leq t_1$, and let us calculate its derivative with respect to time in two cases:

(a) If x_i $(i = 1, 2, 3)$ are the coordinates of a moving particle of the medium, then

$$\frac{df}{dt} = \frac{\partial f}{\partial t} + \dot{x}_j \frac{\partial f}{\partial x_j} = \frac{\partial f}{\partial t} + (\mathbf{v} \cdot \nabla) f \tag{6.2.12}$$

is called *substantial* or *total* derivative of the function f.

(b) If x_i $(i = 1, 2, 3)$ are the coordinates of a fixed point, we have

$$\frac{df}{dt} = \frac{\partial f}{\partial t}, \tag{6.2.13}$$

which is the *space* or *local* derivative.

As an example, the substantial acceleration $\frac{d\mathbf{v}}{dt}$ and the local acceleration $\frac{\partial \mathbf{v}}{\partial t}$ are related by:

$$\frac{d\mathbf{v}}{dt} = \frac{\partial \mathbf{v}}{\partial t} + (\mathbf{v} \cdot \nabla)\mathbf{v}. \tag{6.2.14}$$

Observation: In the case of small displacements and deformations, the difference between the Euler and the Lagrange variables disappears. The first are mostly used in fluid mechanics and the latter – in the theory of elasticity.

6.3 Dynamics of Continuous Media

6.3.1 Equation of Continuity

Consider a continuous deformable medium in motion and delimit an arbitrary portion of the medium which, at the time t_0, occupies the domain $D_0(t_0)$ and at the time $t > t_0$ – the domain $D(t)$. The mass of the medium contained in the domain is

$$m = \int_V \rho(\mathbf{r}, t) d\tau. \tag{6.3.1}$$

If there are no mass transformations during the motion, this mass conserves. The conservation of mass is expressed by the *equation of continuity.*

This equation is easily obtained by means of the theory of integral invariants developed in Sect. 5.6. The mass expressed by (6.3.1) is an integral invariant of the transformation

$$x_i = x_i(\mathbf{r}_0, t) \quad (i = 1, 2, 3) \tag{6.3.2}$$

if the mass density ρ satisfies the equation

$$\frac{d\rho}{dt} + \rho \operatorname{div} \mathbf{v} = 0. \tag{6.3.3}$$

Using (6.2.12), this can also be written as

$$\frac{\partial\rho}{\partial t} + \rho \operatorname{div} \mathbf{v} + \mathbf{v} \cdot \operatorname{grad} \rho = 0,$$

and finally

$$\frac{\partial\rho}{\partial t} + \operatorname{div}(\rho\mathbf{v}) = 0. \tag{6.3.4}$$

This is the *equation of continuity* in the form given by *Euler*. It is a first-order partial differential equation which connects the velocity field $\mathbf{v}(\mathbf{r}, t)$ and the density field $\rho(\mathbf{r}, t)$. Using the convention of summation, we can also write it as

$$\frac{\partial\rho}{\partial t} + \frac{\partial}{\partial x_i}(\rho v_i) = 0. \tag{6.3.5}$$

To give a more suggestive physical interpretation of the equation of continuity (6.3.4), let us integrate it over an arbitrary, fixed volume V, inside the medium:

$$\int_V \frac{\partial\rho}{\partial t} d\tau = -\int_V \operatorname{div}(\rho\mathbf{v}) d\tau,$$

or, in view of the Green–Gauss theorem (see Appendix B),

$$\frac{\partial}{\partial t}\int_V \rho\, d\tau = -\oint_S \rho\mathbf{v} \cdot d\mathbf{S}, \tag{6.3.6}$$

where $d\mathbf{S}$ is an oriented element of the surface S which bounds the volume V. The particles of the medium can enter and leave the volume V through the boundary surface S. The l.h.s. clearly represents the rate of variation of the mass within the volume V. To emphasize the physical significance of the r.h.s. of Eq. (6.3.6), let $\mathbf{v}$ be the velocity of a particle passing through S during the time interval dt. The mass dm of the particle is proportional to the volume of an infinitesimal cylinder, of base $d\mathbf{S}$ and generatrix $\mathbf{v}dt$ (Fig. 6.2):

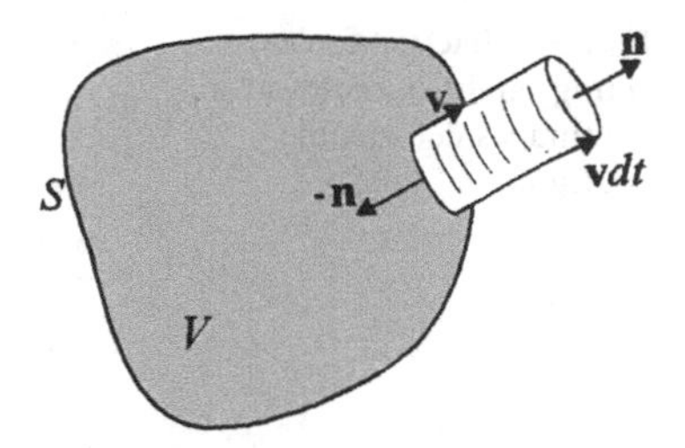

Fig. 6.2 Intuitive representation of the conservation of mass in the mechanics of continuous deformable media.

$$dm = \rho(\mathbf{v} \cdot d\mathbf{S}dt).$$

The quantity $\rho\mathbf{v} \cdot d\mathbf{S}$ is then the mass which flows in unit time through the surface element $d\mathbf{S}$. Choosing the outward normal, this quantity is positive if the mass leaves the volume V ($\mathbf{v} \cdot \mathbf{n} > 0$), and negative if the mass enters the volume ($\mathbf{v} \cdot \mathbf{n} < 0$). Consequently, the quantity $\oint_V \rho\mathbf{v} \cdot d\mathbf{S}$ represents the total mass which leaves or enters the volume V per unit time. The vector quantity $\rho\mathbf{v}$ is termed *mass current density*. Its direction coincides with that of $\mathbf{v}$, while its magnitude represents the mass passing in unit time a unit area orthogonal to $\mathbf{v}$.

An alternative form of the equation of continuity was given by *Jean-Baptiste le Rond D'Alembert*. Using again the theory of integral invariants (see (5.6.6)), the invariance of the mass contained in the volume V can be written as

$$\int_{V_0} \rho(x_1^0, x_2^0, x_3^0, t_0)d\tau_0 = \int_V \rho(x_1, x_2, x_3, t)d\tau. \tag{6.3.7}$$

But, in view of (6.2.1),

$$d\tau = dx_1dx_2dx_3 = \frac{\partial(x_1, x_2, x_3)}{\partial(x_1^0, x_2^0, x_3^0)}dx_1^0dx_2^0dx_3^0 = Jd\tau_0,$$

leading to

$$\int_{V_0} (\rho_0 - J\rho)d\tau_0 = 0, \tag{6.3.8}$$

where we denoted

$$\rho_0 = \rho(\mathbf{r}_0, t_0), \quad \rho = \rho(\mathbf{r}, t). \tag{6.3.9}$$

Since the integration volume V_0 is arbitrary, relation (6.3.8) yields D'Alembert's version of the equation of continuity:

$$J\rho = \rho_0. \tag{6.3.10}$$

It is easy to prove that Eqs. (6.3.6) and (6.3.10) are equivalent. Indeed, taking the total time-derivative of (6.3.10) and using formula (5.6.12), we obtain:

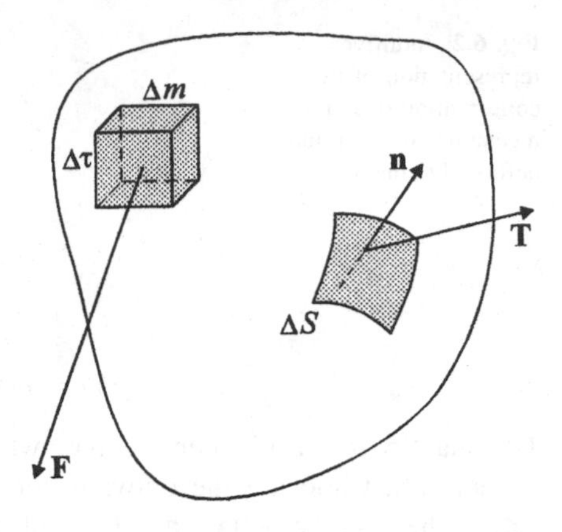

Fig. 6.3 Interior forces acting on the particles of a continuous deformable medium.

$$J\frac{d\rho}{dt} + \rho J \operatorname{div} \mathbf{v} = 0,$$

leading straightforwardly to (6.3.6) after simplification by $J \neq 0$.

The equation of continuity expresses fundamental conservation laws of physics, for example conservation of mass in mechanics, conservation of charge in electrodynamics, or conservation of probability in quantum mechanics, etc.

Observation: If the medium is incompressible and homogeneous, then $\rho(\mathbf{r}, t) = \text{const.}$, and the equation of continuity (6.3.4) reduces to

$$\operatorname{div} \mathbf{v} = 0, \tag{6.3.11}$$

which means that, in this case, the velocity field is *solenoidal* (or *source-free*).

6.3.2 Forces Acting upon a Continuous Deformable Medium

Unlike the case of rigid systems, the action of a force upon a continuous deformable medium depends on whether the particle on which it acts lies on the boundary of the body, or in its volume. Consider a portion of the medium in motion, which at the time t occupies the volume $V(t)$, bounded by the surface $S(t)$. The forces acting on it fall into two categories:

(a) *Body forces*, which act on three-dimensional particles of the domain containing the medium. These forces are proportional to the mass Δm of the particle contained in the volume $\Delta\tau$. Denoting by $\mathbf{F}$ the *specific body force* (force acting on unit mass), the body forces can be written as $\mathbf{F}\Delta m$ (Fig. 6.3). In general, the vector quantity $\mathbf{F}$ is a function of the position of the particle, its velocity and the time: $\mathbf{F} = \mathbf{F}(\mathbf{r}, \mathbf{v}, t)$ (see Chap. 1).
(b) *Superficial forces*, acting on the particles which form the boundary surface of the medium. These forces are proportional to the area ΔS of the surface element,

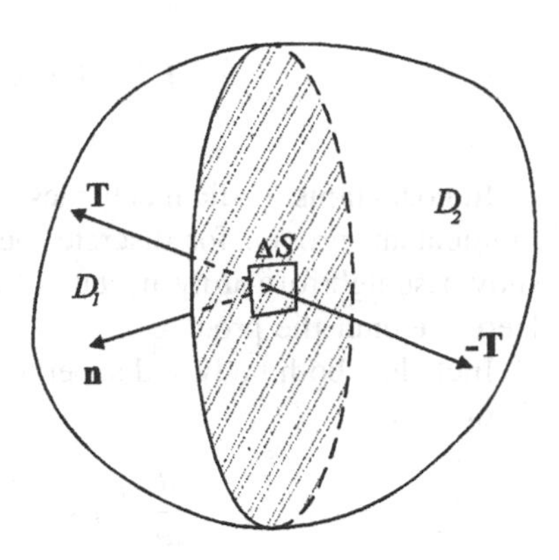

Fig. 6.4 The internal forces obey the action and reaction principle.

and we shall denote them by $\mathbf{T}\,\Delta\, S$, where the force $\mathbf{T}$ acting on unit area is known as the *stress* or *tension*. In general, $\mathbf{T}$ is a function of the position of the surface element, its orientation and time: $\mathbf{T} = \mathbf{T}(\mathbf{r}, \mathbf{n}, t)$, where $\mathbf{n}$ is the unit vector of the outward normal to ΔS. Let $T = |\mathbf{T}|$. If $\mathbf{T} \cdot \mathbf{n} = -T$, the stress is a *pressure*; if $\mathbf{T} \cdot \mathbf{n} = T$, the stress is a *traction*; finally, if $\mathbf{T} \cdot \mathbf{n} = 0$, we have a *shear stress*. In practice, we usually find mixtures of these particular cases. In the study of fluid media, superficial forces occur as pressures, while in the case of solid deformable media we meet all possible kinds of tensions.

Both body and superficial forces can be grouped, in their turn, into two classes:

(a) *Body (superficial) exterior forces*, coming from bodies outside our medium,
(b) *Body (superficial) interior forces*, due to the mutual interactions of the particles of the medium. According to the action and reaction principle, any pair of these forces are equal and directly opposed to each other. Consider, for example, two portions of the medium, which at the time t occupy the domains D_1 and D_2, being in contact through the surface S (Fig. 6.4). Then, one must have:

$$\mathbf{T}(\mathbf{r}, \mathbf{n}, t) = -\mathbf{T}(\mathbf{r}, -\mathbf{n}, t). \tag{6.3.12}$$

6.3.3 General Theorems

By definition, the *linear momentum* of a CDM, which occupies the domain $D(t)$ of volume V, bounded by the surface S, is

$$\mathbf{P} = \int_V \mathbf{v}\, dm = \int_V \rho \mathbf{v}\, d\tau, \tag{6.3.13}$$

while the *angular momentum* of this system, relative to the origin O of a Cartesian system of coordinates, is

$$\mathbf{L} = \int_V \mathbf{r} \times \mathbf{v}\, dm = \int_V \rho \mathbf{r} \times \mathbf{v}\, d\tau. \tag{6.3.14}$$

In continuous media mechanics, the theorems of variation of linear and angular momentum, proven for discrete mechanical systems, remain valid. This can be shown straightforwardly by taking the continuum limit of (1.3.34) and (1.3.38). Here we omit the proof.

Including both body and superficial forces, we shall write the linear momentum theorem as

$$\frac{d}{dt}\int_V \rho \mathbf{v}\, d\tau = \int_V \rho \mathbf{F}\, d\tau + \oint_S \mathbf{T}\, dS, \tag{6.3.15}$$

while the angular momentum theorem becomes:

$$\frac{d}{dt}\int_V \rho \mathbf{r} \times \mathbf{v}\, d\tau = \int_V \rho \mathbf{r} \times \mathbf{F}\, d\tau + \oint_S \mathbf{r} \times \mathbf{T}\, dS. \tag{6.3.16}$$

Here all momenta are taken with respect to the same point, the origin O.

In a similar way, one can define the kinetic energy:

$$T = \frac{1}{2}\int_V \rho \mathbf{v}^2 d\tau \tag{6.3.17}$$

and the radius-vector of the centre of mass:

$$\mathbf{r}_G = \frac{1}{M}\int_V \rho \mathbf{r}\, d\tau, \quad M = \int_V \rho\, d\tau. \tag{6.3.18}$$

Using these definitions, the reader can easily prove that in continuous mechanics the centre of mass theorem (1.3.55) and König's theorems (1.3.60), (1.3.61) remain valid.

6.3.4 Equations of Motion of a CDM. Cauchy Stress Tensor

Let us choose an arbitrary point P of our CDM and construct the Cartesian orthogonal frame $Px_1x_2x_3$. A plane defined by the normal vector $\mathbf{n}$ intersects the axes at the points P_1, P_2 and P_3, forming the elementary tetrahedron $PP_1P_2P_3$ shown in Fig. 6.5, in which

$$PP_i = \epsilon l_i \quad (i = 1, 2, 3), \tag{6.3.19}$$

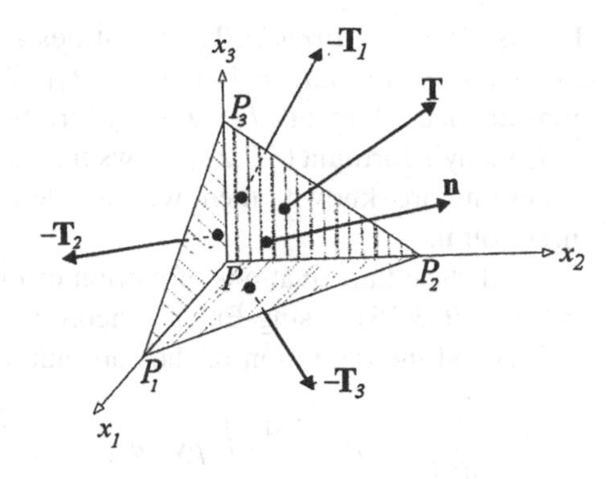

Fig. 6.5 Cauchy's tetrahedron. Sign convention: the tension is positive if it acts on a surface whose normal points in the positive direction of the axes of coordinates and negative otherwise.

where $\epsilon > 0$ is a small parameter ($\epsilon^2 \approx 0$). This is known as *Cauchy's tetrahedron*.

Let dS be the area of the basis $P_1P_2P_3$ and denote by n_1, n_2, n_3 the direction cosines of the normal to dS. Then the surfaces of the orthogonal faces of the tetrahedron, which are the projections of the oblique face on the corresponding planes of coordinates, can be expressed as

$$dS_i = (\mathbf{n}\, dS)_i = n_i dS \quad (i = 1, 2, 3). \tag{6.3.20}$$

Also, let $\mathbf{T}$ be the average stress on dS and $\mathbf{T}_i$ the average stress on dS_i ($i = 1, 2, 3$). Then T_j, T_{ij} are the components of $\mathbf{T}$ and $\mathbf{T}_i$, respectively, on the axes of the Cartesian frame.

With our sign convention (see Fig. 6.5), D'Alembert's principle (2.5.3) for the forces acting on Cauchy's tetrahedron reads:

$$\mathbf{a}\, dm - \mathbf{F}\, dm + \mathbf{T}\, dS - \mathbf{T}_i\, dS_i = 0, \tag{6.3.21}$$

where the summation convention has been used and $\mathbf{F}$ represents the body forces acting on the tetrahedron. Since the tetrahedron is infinitesimally small, we can assume that the tensions on the four faces are actually applied all at the point P. If dm is the mass of the elementary tetrahedron and $d\tau$ its volume, we have:

$$(\mathbf{a} - \mathbf{F})\, dm = (\mathbf{a} - \mathbf{F})\rho\, d\tau = \frac{1}{6}(\mathbf{a} - \mathbf{F})\rho\epsilon^3 l_1 l_2 l_3.$$

Since this expression is proportional to ϵ^3, it can be neglected with respect to the rest of the terms in (6.3.21), and we are left with

$$\mathbf{T}\, dS = \mathbf{T}_i\, dS_i,$$

or, in view of (6.3.20),

$$\mathbf{T} = n_i \mathbf{T}_i \quad \text{or} \quad T_k = n_i T_{ik} \quad (i, k = 1, 2, 3). \tag{6.3.22}$$

This is *Cauchy's formula*. The quantities T_{ik} define a second-rank tensor, called the *stress tensor*. Its diagonal elements T_{11}, T_{22}, T_{33} are the *normal stresses* and the non-diagonal elements T_{12}, T_{23}, T_{31} are the *tangent* or *shear stresses*.

Cauchy's formula (6.3.22) shows that, if the stresses at P along three orthogonal directions are known, then we can determine the stress at P relative to *any* direction $\mathbf{n}$.

To deduce the equations of motion of our CDM, we shall apply the momentum theorem (6.3.15). Using Euler's theorem (5.6.12), and the equation of continuity (6.3.4) and the definition of the substantial derivative (6.2.12), we have:

$$\frac{d}{dt}\int_V \rho\mathbf{v}\,d\tau = \frac{d}{dt}\int_{V_0} \rho\mathbf{v}J\,d\tau_0 = \int_{V_0}\left[J\frac{d}{dt}(\rho\mathbf{v}) + \rho\mathbf{v}J\,\operatorname{div}\mathbf{v}\right]d\tau_0$$

$$= \int_V\left[\frac{d}{dt}(\rho\mathbf{v}) + \rho\mathbf{v}\,\operatorname{div}\mathbf{v}\right]d\tau = \int_V \rho\frac{d\mathbf{v}}{dt}\,d\tau = \int_V \rho\mathbf{a}\,d\tau.$$

Therefore, the linear momentum theorem taken for a finite domain D of volume V and boundary S, reads:

$$\int_V \rho(\mathbf{a}-\mathbf{F})\,d\tau = \oint_S \mathbf{T}\,dS.$$

Recalling that $\mathbf{n}$ is the unit vector of the outward normal to $d\mathbf{S}$ and using Cauchy's formula (6.3.22) and the Green–Gauss theorem, the r.h.s. of the above equation becomes:

$$\oint_S \mathbf{T}\,dS = \oint_S n_k\mathbf{T}_k\,dS = \oint_S \mathbf{T}_k\,dS_k = \int_V \frac{\partial\mathbf{T}_k}{\partial x_k}\,d\tau,$$

hence

$$\int_V\left[\rho(\mathbf{a}-\mathbf{F}) - \frac{\partial\mathbf{T}_k}{\partial x_k}\right]d\tau = 0.$$

Since the integration volume is arbitrary, we obtain

$$\rho\mathbf{a} = \rho\mathbf{F} + \frac{\partial\mathbf{T}_k}{\partial x_k} \quad \text{or} \quad \rho a_i = \rho F_i + \frac{\partial T_{ki}}{\partial x_k} \quad (i = 1,2,3). \tag{6.3.23}$$

These are the *equations of motion* the CDM. They have been first deduced by *Augustin-Louis Cauchy*. Because the model of CDM is not specified, it is said that this is the *non-definite* form of the equations of motion. Using the derivation rule (6.2.14), we can recast Cauchy's equations in the form:

$$\rho\left(\frac{\partial v_i}{\partial t} + v_k \frac{\partial v_i}{\partial v_k}\right) = \rho F_i + \frac{\partial T_{ki}}{\partial x_k}. \tag{6.3.24}$$

Let us now show that the stress tensor T_{ki} is symmetric. To this end, we shall apply the angular momentum theorem (6.3.16) to the domain $D(t)$, bounded by the surface $S(t)$:

$$\int_V \rho \mathbf{r} \times (\mathbf{a} - \mathbf{F})\, d\tau = \oint_S \mathbf{r} \times \mathbf{T}\, dS,$$

or, in components,

$$\int_V \rho \epsilon_{ijk} x_j (a_k - F_k)\, d\tau = \oint_S \epsilon_{ijk} x_j T_k\, dS.$$

Using Cauchy's formula and the Green–Gauss theorem, we obtain:

$$\begin{aligned}\oint_S \epsilon_{ijk} x_j T_k\, dS &= \int_S \epsilon_{ijk} x_j T_{lk}\, dS_l \\ &= \int_V \epsilon_{ijk} \frac{\partial}{\partial x_l}(x_j T_{lk})\, d\tau = \int_V \epsilon_{ijk}\left(\delta_{jl} T_{lk} + x_j \frac{\partial T_{lk}}{\partial x_l}\right) d\tau\end{aligned}$$

and with this result, the expression for the angular momentum theorem can be put in the form:

$$\int_V \epsilon_{ijk} x_j \left[\rho(a_k - F_k) - \frac{\partial T_{lk}}{\partial x_l}\right] d\tau = \int_V \epsilon_{ijk} T_{jk}\, d\tau.$$

But, according to (6.3.23), the l.h.s. is identically zero. Then, since V is arbitrary, we have:

$$\epsilon_{ijk} T_{jk} = 0.$$

Since the pseudo-tensor ϵ_{ijk} is antisymmetric in the summation indices j and k and the product is zero, it follows that the stress tensor T_{jk} has to be symmetric:

$$T_{jk} = T_{kj}. \tag{6.3.25}$$

The equations of motion (6.3.23), together with the equation of continuity (6.3.4), represent a system of four partial differential equations with ten unknowns: the density ρ, the components v_i $(i = 1, 2, 3)$ of the velocity and the six independent components of the stress tensor T_{ik} $(i, k = 1, 2, 3)$. Since the principles of mechanics do not furnish any other equation connecting these variables, in order to determine the motion of the medium one must know some supplementary data, like: the radius-vectors and the velocity fields at the initial time, the constraint

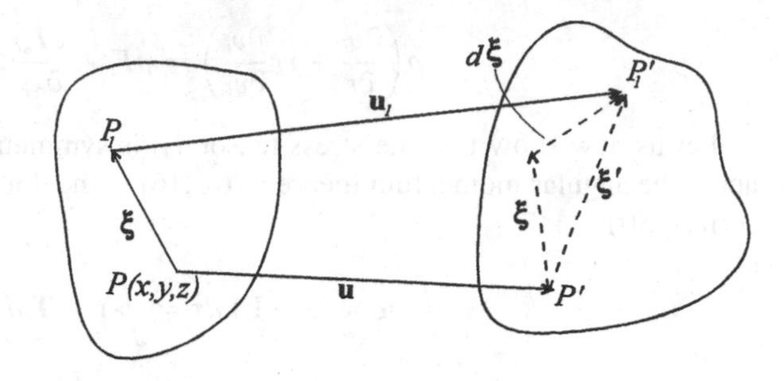

Fig. 6.6 Deformation of a continuous deformable medium about a point.

forces, the relations between the applied forces and the resulting deformations, etc. A part of these data is determined experimentally. In this way, one defines certain *models* of CDM, like: the elastic medium, the perfect fluid, the viscous fluid, etc. We shall present some of these models, but we must first introduce some new concepts, necessary in the study of *any* CDM.

6.4 Deformation of a Continuous Deformable Medium About a Point. Linear Approximation

6.4.1 Rotation Tensor and Small-Strain Tensor

Consider a particle of a CDM which, at the time t, occupies the position P, and let $D(t)$ be a vicinity of the point P. Assume that during the time interval dt the particle has moved from P to P', while the vicinity $D(t)$ of P has become the vicinity $D(t+dt)$ of P'. Our purpose is to study the deformation of the domain D during the motion, i.e. how the relative distances between its points change.

Take another particle at the point P_1 of $D(t)$ and let P_1' be its image at time $t+dt$ (Fig. 6.6). If we denote by $\mathbf{u} = d\mathbf{r}$ the infinitesimal displacement vector $\vec{PP'}$ and by $\mathbf{v}$ the velocity of the point P, we can write:

$$d\mathbf{r} = \mathbf{u} = \mathbf{v}\,dt. \tag{6.4.1}$$

Our continuous medium is deformable, which means that the elementary displacement is a function of the coordinates of P: $\mathbf{u} = \mathbf{u}(x_1, x_2, x_3)$. The distance between the points P and P_1 at the time t is given by the vector ξ. At $t+dt$, the distance between P' and P_1' will be ξ'. From Fig. 6.6, the variations of $\mathbf{u}$ and ξ during the motion are:

$$\delta\mathbf{u} = \vec{P_1P_1'} - \vec{PP'} = \mathbf{u}_1 - \mathbf{u}, \quad d\xi = \vec{P'P_1'} - \vec{PP_1} = \xi' - \xi. \tag{6.4.2}$$

But $\xi + \mathbf{u}_1 - \xi' - \mathbf{u} = 0$, therefore

$$d\xi = \delta\mathbf{u}. \tag{6.4.3}$$

Assuming that the displacement-gradient matrix $\frac{\partial u_i}{\partial x_k}$ has infinitesimal elements, we can write:

$$\delta \mathbf{u} = \mathbf{u}_1(\mathbf{r}) - \mathbf{u}(\mathbf{r}) = \mathbf{u}(\mathbf{r} + \boldsymbol{\xi}) - \mathbf{u}(\mathbf{r}) = \mathbf{u}(\mathbf{r}) + \xi_k \frac{\partial \mathbf{u}}{\partial x_k} + \cdots - \mathbf{u}(\mathbf{r}) \simeq \xi_k \frac{\partial \mathbf{u}}{\partial x_k},$$

and with this linearization, formula (6.4.3) becomes:

$$d\boldsymbol{\xi} = \frac{\partial \mathbf{u}}{\partial x_k} \xi_k \quad \text{or} \quad d\xi_i = \frac{\partial u_i}{\partial x_k} \xi_k \quad (i = 1, 2, 3). \tag{6.4.4}$$

In the case of orthogonal coordinates, the quantities $\frac{\partial u_i}{\partial x_k}$ are the components of a second-rank Cartesian tensor. Denoting $A_{ki} = \frac{\partial u_i}{\partial x_k}$, we have:

$$d\xi_i = A_{ki} \xi_k. \tag{6.4.4$'$}$$

We know (see Appendix A) that any second-rank Cartesian tensor A_{ki} can be written as the sum of an antisymmetric and a symmetric tensor:

$$\begin{aligned} A_{ki} &= \frac{1}{2}(A_{ki} - A_{ik}) + \frac{1}{2}(A_{ki} + A_{ik}) \\ &= \omega_{ki} + e_{ki}. \end{aligned} \tag{6.4.5}$$

The antisymmetric tensor

$$\omega_{ki} = \frac{1}{2}(A_{ki} - A_{ik}) = \frac{1}{2}\left(\frac{\partial u_i}{\partial x_k} - \frac{\partial u_k}{\partial x_i}\right) \tag{6.4.6}$$

is called the *rotation tensor*, while the symmetric tensor

$$e_{ki} = \frac{1}{2}(A_{ki} + A_{ik}) = \frac{1}{2}\left(\frac{\partial u_i}{\partial x_k} + \frac{\partial u_k}{\partial x_i}\right) \tag{6.4.7}$$

is called the *small-strain tensor*. Let us find their physical significance.

Beginning with ω_{ki}, let $\boldsymbol{\omega}$ be the axial vector associated with it:

$$\omega_{ki} = \epsilon_{kij} \omega_j. \tag{6.4.8}$$

Therefore, if we ignore the second term in (6.4.5), we find:

$$d\xi_i = \epsilon_{ijk} \omega_j \xi_k = (\boldsymbol{\omega} \times \boldsymbol{\xi})_i, \tag{6.4.9}$$

which corresponds to a *rigid rotation* about a fixed axis. Examining the defining relation (6.4.6), we notice that the *rotation vector* $\boldsymbol{\omega}$ is given by

$$\boldsymbol{\omega} = \frac{1}{2} \operatorname{curl} \mathbf{u}. \tag{6.4.10}$$

From (6.4.2) and (6.4.9), we obtain:

$$\boldsymbol{\xi}' = \boldsymbol{\xi} + d\boldsymbol{\xi} = \boldsymbol{\xi} + \boldsymbol{\omega} \times \boldsymbol{\xi}, \tag{6.4.11}$$

meaning that, when its components are very small, the antisymmetric tensor ω_{ki} produces an infinitesimal *rigid rotation*, without deformation (see Sect. 4.3).

Passing now to the small-strain tensor e_{ik}, it is seen that if we omit the first term in (6.4.5), we can write:

$$d\xi_i = e_{ki}\xi_k = \frac{\partial}{\partial \xi_i}\left(\frac{1}{2} e_{kj}\xi_k\xi_j\right). \tag{6.4.12}$$

Consider the quadric

$$\frac{1}{2} e_{kj}\xi_k\xi_j = \text{const.} \tag{6.4.13}$$

Since this quadric is invariant with respect to an orthogonal transformation, we can choose the coordinate system with its origin at P, and having the axes along the axes of the quadric. In this case, the quadric (6.4.13) can be written in the canonical form:

$$e_{11}\xi_1'^2 + e_{22}\xi_2'^2 + e_{33}\xi_3'^2 = \text{const.} \tag{6.4.14}$$

This formula expresses the fact that the elementary deformations

$$d\xi_1 = e_{11}\xi_1, \quad d\xi_2 = e_{22}\xi_2, \quad d\xi_3 = e_{33}\xi_3 \tag{6.4.15}$$

are performed along the axes of the quadric (6.4.13). The quadric (6.4.13) is the *deformation quadric* and its axes are the *principal axes of deformation.*

If we take the value 1 for the constant in (6.4.13), then we may encounter one of the following situations. If all quantities e_{11}, e_{22}, e_{33} are positive, our quadric is an ellipsoid and the deformation is a *dilation*. In particular, if $e_{11} = e_{22} = e_{33}$, the ellipsoid becomes a sphere. If e_{11}, e_{22}, e_{33} have different signs, the quadric is a hyperboloid and the deformation can be either a *contraction*, or a *shear stress*. When the elements of the rotation matrix are zero, we have a *pure deformation* or a *pure strain*.

Concluding our discussion, during the motion of a CDM we can have both *rigid-body motion*, i.e. translations and rotations, and *pure deformations*, produced by the mutual displacements of the particles of the medium. In view of (6.4.4) and (6.4.11), we thus can write

$$\boldsymbol{\xi}' = \boldsymbol{\xi} + \boldsymbol{\omega} \times \boldsymbol{\xi} + \nabla\beta, \tag{6.4.16}$$

where

$$\beta = \frac{1}{2} e_{kj}\xi_k\xi_j.$$

6.4.2 Saint-Venant Compatibility Conditions

If the strain distribution in a CDM is known, then the displacement functions $u_i(x_1, x_2, x_3)$ can be found by solving the differential equations which, by virtue of the definition (6.4.7), represent the six independent components of the symmetric tensor e_{ik}:

$$e_{11} = \frac{\partial u_1}{\partial x_1}, \quad e_{22} = \frac{\partial u_2}{\partial x_2}, \quad e_{33} = \frac{\partial u_3}{\partial x_3},$$

$$e_{12} = \frac{1}{2}\left(\frac{\partial u_2}{\partial x_1} + \frac{\partial u_1}{\partial x_2}\right), \quad e_{23} = \frac{1}{2}\left(\frac{\partial u_3}{\partial x_2} + \frac{\partial u_2}{\partial x_3}\right), \quad e_{31} = \frac{1}{2}\left(\frac{\partial u_1}{\partial x_3} + \frac{\partial u_3}{\partial x_1}\right). \tag{6.4.17}$$

The system of six differential equations (6.4.17) in general overdetermines the three unknowns u_1, u_2 and u_3. However, due to the continuity of the displacement field in a CDM, there exists a functional dependence between the quantities e_{ij}. These extra conditions remove the overdetermination, rendering the system (6.4.17) integrable. Indeed, taking suitable derivatives in (6.4.17), we arrive at the following system of six second-order partial differential equations:

$$\begin{gathered}
\frac{\partial^2 e_{11}}{\partial x_2^2} + \frac{\partial^2 e_{22}}{\partial x_1^2} = 2\frac{\partial^2 e_{12}}{\partial x_1 \partial x_2}, \\
\frac{\partial^2 e_{22}}{\partial x_3^2} + \frac{\partial^2 e_{33}}{\partial x_2^2} = 2\frac{\partial^2 e_{23}}{\partial x_2 \partial x_3}, \\
\frac{\partial^2 e_{33}}{\partial x_1^2} + \frac{\partial^2 e_{11}}{\partial x_3^2} = 2\frac{\partial^2 e_{31}}{\partial x_3 \partial x_1}, \\
\frac{\partial^2 e_{11}}{\partial x_2 \partial x_3} = \frac{\partial}{\partial x_1}\left(\frac{\partial e_{31}}{\partial x_2} + \frac{\partial e_{12}}{\partial x_3} - \frac{\partial e_{23}}{\partial x_1}\right), \\
\frac{\partial^2 e_{22}}{\partial x_3 \partial x_1} = \frac{\partial}{\partial x_2}\left(\frac{\partial e_{12}}{\partial x_3} + \frac{\partial e_{23}}{\partial x_1} - \frac{\partial e_{31}}{\partial x_2}\right), \\
\frac{\partial^2 e_{33}}{\partial x_1 \partial x_2} = \frac{\partial}{\partial x_3}\left(\frac{\partial e_{23}}{\partial x_1} + \frac{\partial e_{31}}{\partial x_2} - \frac{\partial e_{12}}{\partial x_3}\right).
\end{gathered} \tag{6.4.18}$$

These equations were first obtained by *Barré de Saint-Venant* and they represent *compatibility conditions*. They can also be written in the condensed form

$$\epsilon_{ijk}\epsilon_{mpq}\frac{\partial^2 e_{kp}}{\partial x_j \partial x_q} = 0. \tag{6.4.18$'$}$$

The role of the conditions (6.4.18) is to ensure the compatibility of the deformations with the displacements. This can be intuitively seen by dividing the undeformed CDM into small volumes, which are then strained. The CDM can be re-composed after the deformation only if the strains of the individual small volumes are compatible with each other, what is ensured by Eqs. (6.4.18).

6.4.3 Finite-Strain Tensor

Let x_i and x'_i $(i = 1, 2, 3)$ be the coordinates of a particle before and after deformation, respectively. Assuming that the displacement $\mathbf{u}(\mathbf{r})$ is finite, we can write

$$x'_i = x_i + u_i(\mathbf{r}), \tag{6.4.19}$$

which, by differentiation, yields:

$$dx'_i = dx_i + \frac{\partial u_i}{\partial x_k} dx_k.$$

If ds^2 and ds'^2 are the squared distances between any two points before and after deformation, using the last relation we find:

$$\begin{aligned} dx'_i dx'_i &= \left(dx_i + \frac{\partial u_i}{\partial x_j} dx_j \right) \left(dx_i + \frac{\partial u_i}{\partial x_k} dx_k \right) \\ &= dx_i dx_i + \left(\frac{\partial u_k}{\partial x_j} + \frac{\partial u_j}{\partial x_k} + \frac{\partial u_i}{\partial x_j} \frac{\partial u_i}{\partial x_k} \right) dx_j dx_k, \end{aligned}$$

or

$$ds'^2 - ds^2 = 2E_{jk} dx_j dx_k,$$

where

$$E_{jk} = \frac{1}{2} \left(\frac{\partial u_k}{\partial x_j} + \frac{\partial u_j}{\partial x_k} + \frac{\partial u_i}{\partial x_j} \frac{\partial u_i}{\partial x_k} \right) \tag{6.4.20}$$

are the components of a second-order symmetric tensor, which is the *finite-strain tensor*, expressing the change in the squared length of the vector $d\mathbf{r}$ upon the deformation.

On the other hand, since

$$ds'^2 = dx'_i dx'_i = \frac{\partial x'_i}{\partial x_j} \frac{\partial x'_i}{\partial x_k} dx_j dx_k = g_{jk} dx_j dx_k,$$

where g_{ik} is the metric tensor of the transformation $x'_i = x'_i(x_1, x_2, x_3)$, we can also write:

$$ds'^2 - ds^2 = (g_{jk} - \delta_{jk})dx_j dx_k,$$

leading to the general expression for the finite-strain tensor:

$$E_{jk} = \frac{1}{2}(g_{jk} - \delta_{jk}).$$

Equations (6.4.20) show that the finite-strain tensor contains only linear and quadratic terms in the displacement-gradient components. This is an exact result and not a second-order approximation. If the deformations u_j are small enough so as to neglect the product $\frac{\partial u_i}{\partial x_j}\frac{\partial u_i}{\partial x_k}$ with respect to the rest of the terms in (6.4.20), then $E_{jk} \to e_{jk}$, which has been defined earlier for the *linear theory* of deformations. Noting that our study will be done within the framework of this linear approximation, we shall next discuss some of the basic models of CDM, like the elastic medium, the perfect fluid, and the viscous fluid.

6.5 Elastic Medium

6.5.1 Hooke's Generalized Law

By an *elastic medium* we mean a body which recovers its initial form and position when the deforming force stops its action.

The connection between applied forces and the deformations produced by them has been for a long time the object of experimental studies, leading to the so-called *stress-strain relations* or *constitutive equations*. In 1678, *Robert Hooke* revealed his discovery of the proportionality between forces and deformations (*ut tensio, sic vis*), and gave the law of elasticity:

$$\frac{\Delta l}{l_0} = \frac{1}{E}\frac{F}{S}. \tag{6.5.1}$$

Here, l_0 is the initial length of a thin and long wire, S – the area of its transverse section, F – the deforming force, Δl – the strain (elongation) produced by it and E – the *Young modulus*. This relation is in good agreement with the experimental observation if the force F is uniformly distributed on the surface S, which does not vary during the deformation. In the case of massive elastic bodies, we must consider both longitudinal and transversal deformations. In other words, Hooke's law (6.5.1) should be generalized so as to include all possible deformations.

Assuming that our model is a homogeneous and isotropic elastic medium, we delimit a rectangular parallelepiped $ABCDA'B'C'D'$ of the body, and choose a

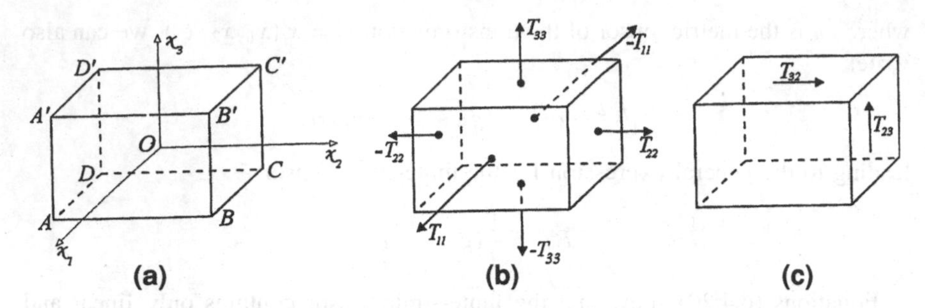

Fig. 6.7 Normal and tangential tensions in an elastic medium.

Cartesian orthogonal system of coordinates $Ox_1x_2x_3$ with its origin O at the centre of the parallelepiped (Fig. 6.7a).

Denote by T_{11}, T_{22}, T_{33} the normal stresses on the three pairs of faces, oriented along the axes x_1, x_2 and x_3, respectively. Under the action of these tensions, the parallelepiped is deformed infinitesimally, so that a point P, whose coordinates before deformation were x_1, x_2, x_3, will have after the deformation the coordinates

$$x_i' = x_i + \epsilon_i x_i \quad \text{(no summation; } i = 1,2,3).$$

The medium being homogeneous and isotropic, it follows that all the points will be shifted similarly, such that the change in relative distance between two arbitrary points will be:

$$\xi_i' = \xi_i + \epsilon_i \xi_i \quad \text{(no summation; } i = 1,2,3) \tag{6.5.2}$$

Comparing (6.5.2) with (6.4.15), we identify

$$e_{ii} = \epsilon_i \quad \text{(no summation; } i = 1,2,3). \tag{6.5.3}$$

If we have only normal tensions (i.e. the stress tensor T_{ik} is diagonal), the deformations are produced only along the coordinate axes (Fig. 6.7b), meaning that

$$e_{ij} = 0 \quad (i \neq j). \tag{6.5.4}$$

By virtue of (6.4.4), (6.5.3) and (6.5.4), for a pure deformation we therefore have

$$\frac{d\xi_i}{\xi_i} = e_{ii} = \epsilon_i \quad \text{(no summation; } i = 1,2,3). \tag{6.5.5}$$

Comparing now (6.5.5) with Hooke's law (6.5.1), we observe that under the action of normal tension (traction) T_{ii}, the medium elongates in the i direction by the relative rate T_{ii}/E, while in the orthogonal directions it contracts by Poisson effect by the rate $\sigma \frac{T_{ii}}{E}$. Here, σ is a material constant which depends on the nature of the medium, called *coefficient of transverse contraction* or *Poisson's ratio*. Since, by hypothesis, the medium is homogeneous and isotropic, we have:

$$e_{11} = \frac{T_{11}}{E} - \sigma\frac{T_{22}}{E} - \sigma\frac{T_{33}}{E},$$

$$e_{22} = \frac{T_{22}}{E} - \sigma\frac{T_{11}}{E} - \sigma\frac{T_{33}}{E},$$

$$e_{33} = \frac{T_{33}}{E} - \sigma\frac{T_{11}}{E} - \sigma\frac{T_{22}}{E}.$$

These relations can be written in a compressed form as

$$e_{ii} = \frac{1+\sigma}{E}T_{ii} - \frac{\sigma}{E}s \quad \text{(no summation)},$$

where we have used the notation

$$s = T_{11} + T_{22} + T_{33}. \tag{6.5.6}$$

Since $e_{ij} = 0$ for $i \neq j$, we have also:

$$e_{ij} = \frac{1+\sigma}{E}T_{ij} - \frac{\sigma}{E}s\delta_{ij}. \tag{6.5.7}$$

This relation has been deduced on the assumption that there are only normal tensions acting on the elastic medium. But the experiment shows that this is true in general. Indeed, if we had only tangential (shear) tensions $T_{12} = T_{21}$, $T_{23} = T_{32}$, $T_{31} = T_{13}$ (Fig. 6.7c), we would get

$$e_{ij} = \frac{1+\sigma}{E}T_{ij} \quad (i \neq j) \tag{6.5.8}$$

which proves to be true.

Formula (6.5.7) can be put in a different form, which displays the relation between the stress tensor T_{ij}, on the one hand, and the small-strain tensor e_{ij}, on the other. To this end, we define the *volume specific dilation*:

$$\theta = e_{11} + e_{22} + e_{33} = \frac{\partial u_i}{\partial x_i} = \operatorname{div}\mathbf{u}. \tag{6.5.9}$$

Taking $i = j$ in (6.5.7) and performing the summation, we have:

$$\theta = \frac{1-2\sigma}{E}s.$$

Introducing this result into (6.5.7), we obtain:

$$T_{ij} = \frac{E}{1+\sigma}e_{ij} + \frac{\sigma E}{(1+\sigma)(1-2\sigma)}\theta\delta_{ij}.$$

Denoting

$$\lambda = \frac{\sigma E}{(1+\sigma)(1-2\sigma)}, \quad \mu = \frac{E}{2(1+\sigma)}, \tag{6.5.10}$$

we finally arrive at *Hooke's generalized law* for homogeneous and isotropic elastic media:

$$T_{ij} = \lambda\theta\delta_{ij} + 2\mu e_{ij}. \tag{6.5.11}$$

The coefficients λ and μ are called *elastic constants* or *Lamé coefficients*, being determined by experimental methods. Eliminating σ between relations (6.5.10), we find Young's modulus E in terms of λ and μ:

$$E = \frac{\mu(2\mu + 3\lambda)}{\lambda + \mu}. \tag{6.5.12}$$

In the case of *anisotropic media* (crystals), Hooke's law has a more general form, namely

$$T_{ij} = C_{ijkm} e_{km}, \tag{6.5.13}$$

where C_{ijkm} are the components of a fourth rank tensor, called *elasticity tensor* or *stiffness tensor*. Its components satisfy the following symmetry relations:

$$C_{ijkm} = C_{jikm} = C_{ijmk}, \quad C_{ijkm} = C_{kmij}. \tag{6.5.14}$$

The first group of relations is trivially obtained from the symmetry of the tensors T_{ij} and e_{km}, and represent the so-called minor symmetries. The latter relation is obtained from thermodynamical considerations which are beyond the scope of this book. Here we have $45 + 15 = 60$ relations, therefore the maximum number of the independent components of C_{ijkm} is: $3^4 - 60 = 21$. This result can be also obtained by speculating the symmetry of the pairs of indices, which suggests to choose the following mapping of tensor indices (so-called *Voigt notation*):

1	2	3	4	5	6
11	22	33	23	31	12

The tensor C_{ijkm} becomes C_{IJ}, with the property

$$C_{IJ} = C_{JI}. \tag{6.5.15}$$

Therefore C_{IJ} is a second-rank symmetric tensor, defined on a six-dimensional space. The number of its independent components is

$$\binom{6}{2} = \frac{6(6+1)}{2} = 21.$$

If the anisotropic elastic medium possesses symmetry axes, the number of the independent components of the elasticity tensor is less than 21. Let a, b, c be the edges of an elementary cell of a crystal, and α, β, γ the angles at any intersection point of three edges. A cubic crystal ($a = b = c$, $\alpha = \beta = \gamma$) is the simplest anisotropic case and it has three independent components of C_{IJ}, while a crystal characterized by $a \neq b \neq c$, $a \neq c$, $\alpha \neq \beta \neq \gamma$, $\alpha \neq \gamma$ has the maximum number of components, namely 21.

Using our convention, Hooke's law (6.5.13) can be written as

$$T_I = C_{IK} e_K \quad (I, K = \overline{1,6}). \tag{6.5.16}$$

Comparing (6.5.11) and (6.5.13), we find the components of the tensor C_{ijkm} for a homogeneous and isotropic elastic medium:

$$C_{11} \equiv C_{1111} = \lambda + 2\mu,$$

$$C_{12} \equiv C_{1122} = \lambda,$$

$$C_{66} \equiv C_{1212} = 2\mu.$$

It is seen that the three components of the elasticity tensor C_{JK} are not independent, but obey the relation $C_{11} = C_{12} + C_{66}$, which means that only two components are distinct (the Lamé coefficients), as we expected.

We give here, as an example, the matrix of the elasticity tensor for a crystal with cubic symmetry:

$$\begin{pmatrix} C_{11} & C_{12} & C_{12} & 0 & 0 & 0 \\ C_{12} & C_{11} & C_{12} & 0 & 0 & 0 \\ C_{12} & C_{12} & C_{11} & 0 & 0 & 0 \\ 0 & 0 & 0 & C_{66} & 0 & 0 \\ 0 & 0 & 0 & 0 & C_{66} & 0 \\ 0 & 0 & 0 & 0 & 0 & C_{66} \end{pmatrix}. \tag{6.5.17}$$

Observation: If the temperature of the medium varies during the elastic deformation, Hooke's law (6.5.13) must be corrected by a term expressing the temperature changes, namely:

$$T_{ij} = C_{ijkm} e_{km} - \lambda_{ij}(T - T_0), \tag{6.5.18}$$

where $\Delta T = T - T_0$ is the variation of the temperature and λ_{ij} are some constants called *coefficients of thermo-elasticity*.

6.5.2 Equations of Motion of an Elastic Medium

To obtain the equations of motion of a homogeneous and isotropic elastic medium, we introduce T_{ik} given by Hooke's law (6.5.11) into Cauchy's equations (6.3.23). Recalling that the stress tensor T_{ik} is symmetric, these equations read:

$$\rho a_i = \rho F_i + \frac{\partial T_{ik}}{\partial x_k} \quad (i,k = 1,2,3). \tag{6.5.19}$$

We have from (6.5.11):

$$\frac{\partial T_{ik}}{\partial x_k} = \lambda \frac{\partial \theta}{\partial x_k} \delta_{ik} + 2\mu \frac{\partial e_{ik}}{\partial x_k}.$$

But

$$\frac{\partial e_{ik}}{\partial x_k} = \frac{1}{2}\frac{\partial}{\partial x_k}\left(\frac{\partial u_k}{\partial x_i} + \frac{\partial u_i}{\partial x_k}\right) = \frac{1}{2}\left[\frac{\partial}{\partial x_i}\left(\frac{\partial u_k}{\partial x_k}\right) + \frac{\partial^2 u_i}{\partial x_k \partial x_k}\right] = \frac{1}{2}\left(\frac{\partial \theta}{\partial x_i} + \Delta u_i\right),$$

hence:

$$\frac{\partial T_{ik}}{\partial x_k} = \lambda \frac{\partial \theta}{\partial x_i} + \mu\left(\frac{\partial \theta}{\partial x_i} + \Delta u_i\right) = (\lambda + \mu)\frac{\partial \theta}{\partial x_i} + \mu \Delta u_i,$$

which leads to the desired equations of motion:

$$\rho a_i = \rho F_i + (\lambda + \mu)\frac{\partial \theta}{\partial x_i} + \mu \Delta u_i \quad (i = 1,2,3). \tag{6.5.20}$$

The relations (6.5.20) form a system of three *linear partial differential equations* in the variables u_i ($i = 1, 2, 3$). They were first deduced by *Gabriel Lamé*. Let us write these equations in terms of Lagrange variables.

Recalling that our study is performed in the framework of the linear approximation, we rewrite (6.5.2) in the form:

$$x'_i = x_i + u_i(\mathbf{r}, t) \quad (i = 1,2,3), \tag{6.5.21}$$

where this time by x_i we mean the coordinates of some particle P at the initial time t_0, and by x'_i the coordinates of the same particle at the time t. In other words, x_1, x_2, x_3, t are Lagrange variables. Since x_i remain fixed while following the particle in its motion, by virtue of (6.2.3) and (6.2.4), we have:

$$v_i = \frac{\partial x'_i}{\partial t} = \frac{\partial u_i}{\partial t}, \tag{6.5.22}$$

$$a_i = \frac{\partial v_i}{\partial t} = \frac{\partial^2 u_i}{\partial t^2} \quad (i = 1,2,3). \tag{6.5.23}$$

The body force $\mathbf{F}$ is a function of the form

$$\mathbf{F}(\mathbf{r}', t) = \mathbf{F}(\mathbf{r} + \mathbf{u}, t) = \mathbf{F}(\mathbf{r}, t) + u_j \frac{\partial \mathbf{F}}{\partial x_j} + \cdots \simeq \mathbf{F}(\mathbf{r}, t). \tag{6.5.24}$$

The equation of continuity (6.3.10), with the notation adopted in this section, is

$$J\rho(\mathbf{r}',t) = \rho(\mathbf{r},t_0).$$

But

$$J = \frac{\partial(x'_j)}{\partial(x_j)} = 1 + \frac{\partial u_i}{\partial x_i} + \cdots \simeq 1.$$

If we denote $\rho(\mathbf{r}, t_0)$ by $\rho(\mathbf{r})$, the equation of continuity yields:

$$\rho(\mathbf{r}',t) \simeq \rho(\mathbf{r}). \tag{6.5.25}$$

Then Lamé's equations (6.5.20) read:

$$\rho\frac{\partial^2 u_i}{\partial t^2} = \rho F_i + (\lambda + \mu)\frac{\partial\theta}{\partial x_i} + \mu\,\Delta\,u_i, \tag{6.5.26}$$

or, in vector form,

$$\rho\frac{\partial^2 \mathbf{u}}{\partial t^2} = \rho\mathbf{F} + (\lambda + \mu)\mathrm{grad}\,\theta + \mu\,\Delta\,\mathbf{u}. \tag{6.5.26$'$}$$

An alternative form of this equation is obtained if we use the vector relation

$$\Delta\mathbf{u} = \mathrm{grad}\,\theta - \mathrm{curl}(\mathrm{curl}\,\mathbf{u}).$$

The result is:

$$\rho\frac{\partial^2 \mathbf{u}}{\partial t^2} = \rho\mathbf{F} + (\lambda + 2\mu)\mathrm{grad}\,\theta - \mu\,\mathrm{curl}(\mathrm{curl}\,\mathbf{u}). \tag{6.5.27}$$

To integrate Lamé's equations, one must know both the initial conditions (the elementary displacements and the velocity fields at the initial time) and the boundary conditions (e.g. the components of the stress tensor on the surface of the elastic body).

The equations of *elastic equilibrium*, called *Navier–Cauchy* equations, are obtained from (6.5.26) by taking $\mathbf{a} = 0$:

$$\rho F_i + (\lambda + \mu)\frac{\partial\theta}{\partial x_i} + \mu\Delta u_i = 0 \quad (i = 1,2,3). \tag{6.5.28}$$

6.5.3 Plane Waves in Isotropic Elastic Media

Assuming that our elastic medium is large enough, so as to obtain at least several wave-lengths in any direction of propagation, let us suppose that the elementary displacement $\mathbf{u}$ depends on a single space variable x_1 and the time t: $\mathbf{u} = \mathbf{u}(x_1,t)$.

If at the moment of observation the force **F** has stopped to act upon the medium, Lamé's equations (6.5.26) yield:

$$\rho\frac{\partial^2 u_i}{\partial t^2} = (\lambda+\mu)\frac{\partial\theta}{\partial x_i} + \mu\,\Delta\,u_i. \tag{6.5.29}$$

We also have:

$$\theta = \operatorname{div}\mathbf{u} = \frac{\partial u_1}{\partial x_1},$$

$$\Delta u_1 = \frac{\partial^2 u_1}{\partial x_1^2}, \quad \Delta u_2 = \frac{\partial^2 u_2}{\partial x_1^2}, \quad \Delta u_3 = \frac{\partial^2 u_3}{\partial x_1^2}.$$

Thus, the three components of equation (6.5.29) read:

$$(\lambda+2\mu)\frac{\partial^2 u_1}{\partial x_1^2} - \rho\frac{\partial^2 u_1}{\partial t^2} = 0, \tag{6.5.30}$$

$$\mu\frac{\partial^2 u_2}{\partial x_1^2} - \rho\frac{\partial^2 u_2}{\partial t^2} = 0, \tag{6.5.31}$$

$$\mu\frac{\partial^2 u_3}{\partial x_1^2} - \rho\frac{\partial^2 u_3}{\partial t^2} = 0. \tag{6.5.32}$$

We arrive at three partial differential equations of hyperbolic type, similar to the D'Alembert homogeneous wave equation:

$$\frac{\partial^2\Psi}{\partial x^2} - \frac{1}{v^2}\frac{\partial^2\Psi}{\partial t^2} = 0. \tag{6.5.33}$$

If we choose the displacement **u** so as to have $u_1 \neq 0$, $u_2 = 0, u_3 = 0$, the propagation is described by Eq. (6.5.30). Since the displacement is along the direction of propagation, we have a *longitudinal wave*. The speed of this wave is found by comparing (6.5.33) with (6.5.30):

$$v_l = \sqrt{\frac{\lambda+2\mu}{\rho}}. \tag{6.5.34}$$

If, on the contrary, we choose $u_1 = 0$, $u_2 \neq 0$, $u_3 \neq 0$, the propagation is given by Eqs. (6.5.31) and (6.5.32). In this case the direction of displacements is orthogonal on that of propagation of the oscillation, therefore we have a *transverse wave* which propagates with the speed

$$v_t = \sqrt{\frac{\mu}{\rho}}. \tag{6.5.35}$$

Observation: The longitudinal oscillations produced in an elastic rod are described by Eq. (6.5.30), while the transverse oscillations of a vibrating rope satisfy equations of the type (6.5.31), or (6.5.32).

6.6 Perfect Fluid

The *perfect* or *ideal* fluid is a model of CDM having the property that the tensions acting on any interior surface, which separates two arbitrary parts of the medium, are normal to this surface. In other words, there are no forces of friction between particles. For example, a fluid in a very slow motion is a good approximation for this model.

Since we have only normal tensions, we can write

$$\mathbf{T} = -p\mathbf{n}, \tag{6.6.1}$$

where $p(x_1, x_2, x_3, t)$ is a scalar called *pressure*. Here, x_1, x_2, x_3, t are Euler variables. Their use is recommended by the fact that a fluid in motion represents considerably large displacements of substance (convection). Because in fluid mechanics we deal only with pressures, the scalar p is positive. Projecting (6.6.1) on the x_i-axis and using Cauchy's formula (6.3.22), we have:

$$T_i = -pn_i = -pn_k\delta_{ik} = n_k T_{ik},$$

which yields:

$$T_{ik} = -p\delta_{ik}. \tag{6.6.2}$$

6.6.1 Equation of Motion of a Perfect Fluid

Utilizing the already known procedure, we introduce T_{ik} given by (6.6.2) in Cauchy's equations (6.3.23). The result is:

$$\rho a_i = \rho F_i - \frac{\partial p}{\partial x_i} \quad (i = 1, 2, 3). \tag{6.6.3}$$

These partial-derivative, non-linear equations were first obtained by *Leonhard Euler*. The equation of continuity (6.3.4), together with Euler's equations (6.6.3), form a system of four partial-derivative equations with five unknowns: v_1, v_2, v_3, p, ρ. We need one more equation. Since it cannot be furnished by the principles of mechanics, we 'borrow' from thermodynamics the equation of state

$$F(p, \rho, T) = 0, \tag{6.6.4}$$

where T is the absolute temperature. If $T = \text{const.}$, the equation of state takes the form

$$\rho = f(p), \tag{6.6.5}$$

and characterizes *barotropic fluids*. In this case, we can define the function

$$P(p) = \int \frac{dp}{\rho(p)} \tag{6.6.6}$$

to recast Euler's equation (6.6.3) in the form:

$$a_i = F_i - \frac{\partial P}{\partial x_i}. \tag{6.6.7}$$

Observations:

(a) Euler's equation can be written in different forms, required in practical applications. Thus, if we use the substantial derivative rule (6.2.14), we obtain:

$$\rho\left[\frac{\partial \mathbf{v}}{\partial t} + (\mathbf{v}\cdot\nabla)\mathbf{v}\right] = \rho\mathbf{F} - \nabla p. \tag{6.6.8}$$

Since

$$(\mathbf{v}\cdot\nabla)\mathbf{v} = \frac{1}{2}\nabla(\mathbf{v}\cdot\mathbf{v}) - \mathbf{v}\times(\nabla\times\mathbf{v}),$$

Euler's equation takes the form:

$$\frac{\partial \mathbf{v}}{\partial t} + \frac{1}{2}\nabla(\mathbf{v}\cdot\mathbf{v}) - \mathbf{v}\times(\nabla\times\mathbf{v}) = \mathbf{F} - \frac{1}{\rho}\nabla p. \tag{6.6.9}$$

This last form was given by *Hermann von Helmholtz*.

(b) To integrate the system of equations (6.3.4), (6.6.3) and (6.6.5), we must know both the initial conditions (quantities ρ, p, v_1, v_2, v_3 at the initial time) and the corresponding boundary conditions.

6.6.2 Particular Types of Motion of an Ideal Fluid

6.6.2.1 Irrotational Motion

Using a procedure similar to that applied in Sect. 6.4, we shall first define the *velocity of rotation* and *velocity of deformation* tensors. Let P and P' be any two infinitely closed points of the medium, and let $\mathbf{v}_P$, $\mathbf{v}'_{P'}$ be the velocities at these points at the times t and $t' > t$, respectively. The variation of $\mathbf{v}$ between these two points, at time t, is

$$\delta \mathbf{v} = \mathbf{v}_{P'} - \mathbf{v}_P = \frac{\partial \mathbf{v}}{\partial x_k} \delta x_k, \tag{6.6.10}$$

where the derivatives $\frac{\partial \mathbf{v}}{\partial x_k}$ are calculated at the point P. Projecting (6.6.10) on axes and denoting $\delta x_k = (PP')_k = X_k$, we have:

$$\begin{aligned} \delta v_i &= \frac{\partial v_i}{\partial x_k} X_k = \frac{1}{2}\left(\frac{\partial v_i}{\partial x_k} - \frac{\partial v_k}{\partial v_i}\right) X_k + \frac{1}{2}\left(\frac{\partial v_i}{\partial x_k} + \frac{\partial v_k}{\partial x_i}\right) X_k \\ &= \Omega_{ki} X_k + e'_{ki} X_k \quad (i,\ k = 1,2,3). \end{aligned} \tag{6.6.11}$$

Here,

$$\Omega_{ki} = \frac{1}{2}\left(\frac{\partial v_i}{\partial x_k} - \frac{\partial v_k}{\partial x_i}\right) \tag{6.6.12}$$

is the *velocity of rotation* tensor, while the corresponding axial vector

$$\boldsymbol{\Omega} = \frac{1}{2} \operatorname{curl} \mathbf{v} \tag{6.6.13}$$

is known as the *vorticity*. We therefore can write:

$$\Omega_{ki} X_k = \epsilon_{kij} \Omega_j X_k = (\boldsymbol{\Omega} \times \mathbf{X})_i. \tag{6.6.14}$$

The second term in (6.6.11) is

$$e'_{ki} X_k = \frac{\partial}{\partial X_i}\left(\frac{1}{2} e'_{kj} X_k X_j\right) = \frac{\partial \alpha}{\partial X_i}, \tag{6.6.15}$$

where e'_{ki} is the *velocity of deformation* tensor and

$$\alpha = \frac{1}{2} e'_{kj} X_k X_j = \text{const.} \tag{6.6.16}$$

is the *velocity of deformation quadric*. Then,

$$\mathbf{v}_{P'} = \mathbf{v}_P + \boldsymbol{\Omega} \times \mathbf{X} + \nabla \alpha,$$

i.e. the velocity of the fluid, at any point, is a vector sum of three terms: a rigid velocity of translation, one of rotation and one of deformation, orthogonal to the quadric (6.6.16).

If

$$\operatorname{curl} \mathbf{v} = 0 \tag{6.6.17}$$

at any point of the fluid and at any time, the motion is called *irrotational*. The relation (6.6.17) expresses the necessary and sufficient condition for the existence of a function $\varphi(\mathbf{r}, t)$, called *velocity potential*, such that

$$\mathbf{v} = \text{grad}\,\varphi(\mathbf{r},\ t). \tag{6.6.18}$$

The *streamlines* are defined as the lines which are tangent to the velocity at any point, being determined by the differential equations:

$$\frac{dx}{v_x} = \frac{dy}{v_y} = \frac{dz}{v_z}.$$

Note that the relation (6.6.18) defines the velocity potential up to a term which is a function of time only. Indeed, if we take

$$\varphi' = \varphi + F(t), \tag{6.6.19}$$

we obtain the same velocity field **v**.

We also have (see (6.6.9)):

$$a_i = \frac{\partial v_i}{\partial t} + \frac{1}{2}\frac{\partial}{\partial x_i}(\mathbf{v}\cdot\mathbf{v}) = \frac{\partial}{\partial x_i}\left[\frac{\partial\varphi}{\partial t} + \frac{1}{2}(\nabla\phi)^2\right] = \frac{\partial\psi}{\partial x_i},$$

or

$$\mathbf{a} = \text{grad}\,\psi(\mathbf{r},\ t), \tag{6.6.20}$$

where

$$\psi = \frac{\partial\varphi}{\partial t} + \frac{1}{2}(\nabla\varphi)^2 \tag{6.6.21}$$

is the *acceleration potential.* One observes that ψ has, in its turn, the property (6.6.19) of the velocity potential.

Assume that our fluid is barotropic and that there exists a function $V^*(\mathbf{r}, t)$, such that

$$\mathbf{F} = -\text{grad}\,V^*(\mathbf{r},\ t). \tag{6.6.22}$$

Substituting the last two relations into (6.6.7), we have:

$$\text{grad}\left[\frac{\partial\varphi}{\partial t} + \frac{1}{2}(\text{grad}\,\varphi)^2 + V^* + P\right] = 0,$$

which yields:

$$\frac{\partial\varphi}{\partial t} + \frac{1}{2}(\nabla\varphi)^2 + V^* + P = f_1(t), \tag{6.6.23}$$

where $f_1(t)$ is a function of time only. Thus, Euler's equations have been reduced to a single relation (6.6.23), named the *Lagrange–Bernoulli equation*, in which the velocity field **v** has been replaced by the scalar field $\varphi(\mathbf{r},\ t)$. In this case, the motion of the fluid is determined by three equations: the Lagrange–Bernoulli equation (6.6.23), the equation of continuity (6.3.4) and the equation of state (6.6.5).

In view of (6.6.19), we may take

$$\varphi' = \varphi - \int f_1(t)\, dt,$$

in which case the Lagrange–Bernoulli equation (6.6.23) reads:

$$\frac{\partial \varphi'}{\partial t} + \frac{1}{2}(\nabla \varphi')^2 + V^* + P = 0. \tag{6.6.24}$$

6.6.2.2 Fundamental Equation of Acoustics

Consider a volume of gas at rest ($\mathbf{v} = 0$). Assuming that at the moment of observation there are no body forces acting on the medium ($\mathbf{F} = 0$), Euler's equation (6.6.8) yields $p = p_0 =$ const., while (6.6.5) leads to $\rho = \rho_0 =$ const.

Suppose that in a certain point of the medium appears a small perturbation, expressed by:

$$p = p_0 + p', \quad \rho = \rho_0 + \rho' \quad (p' \ll p_0,\ \rho' \ll \rho_0),$$

whose velocity of propagation is small enough as to neglect the term $\frac{1}{2}(\nabla \varphi)^2$ in (6.6.24). On the other hand, the definition (6.6.6) yields:

$$P(p) = \int_{p_0}^{p} \frac{dp}{\rho(p)} \simeq \frac{p - p_0}{\rho_0} = \frac{p'}{\rho_0},$$

and consequently the Lagrange–Bernoulli equation reduces to

$$\frac{\partial \varphi}{\partial t} + \frac{p'}{\rho_0} = 0. \tag{6.6.25}$$

Due to the equation of continuity (6.3.4), we also have

$$\frac{\partial \rho'}{\partial t} + \rho_0 \mathrm{div}\, \mathbf{v} = \frac{\partial \rho'}{\partial t} + \rho_0 \Delta \varphi = 0. \tag{6.6.26}$$

But

$$\rho_0 + \rho' = \rho(p_0 + p') = \rho(p_0) + \left(\frac{\partial \rho}{\partial p}\right)_{p_0} p' + \cdots$$

and, since $\rho_0 = \rho(p_0)$,

$$p' = \left(\frac{\partial p}{\partial \rho}\right)_{\rho_0} \rho'. \tag{6.6.27}$$

Utilizing these results, let us now take the partial derivative with respect to time of (6.6.25). The result is:

$$\frac{\partial^2 \varphi}{\partial t^2} = -\frac{1}{\rho_0}\frac{\partial p'}{\partial t} = -\frac{1}{\rho_0}\left(\frac{\partial p'}{\partial \rho'}\right)\frac{\partial \rho'}{\partial t} = \left(\frac{\partial p}{\partial \rho}\right)_{\rho_0} \Delta\varphi,$$

or

$$\Delta\varphi - \frac{1}{c_0^2}\frac{\partial^2 \varphi}{\partial t^2} = 0, \tag{6.6.28}$$

where

$$c_0 = \sqrt{\left(\frac{\partial p}{\partial \rho}\right)_{\rho_0}} \tag{6.6.29}$$

is the *speed of sound*. In conclusion, in a gaseous medium the small perturbations propagate as *sound waves*, Eq. (6.6.28) being the *fundamental equation of acoustics*.

6.6.2.3 Stationary Motion. Bernoulli's Equation

If all particles of an ideal fluid which pass through a certain point follow the same trajectory, the motion of the fluid is called *permanent* or *stationary*. Mathematically, this is expressed by the fact that all quantities appearing in the equation of motion do not explicitly depend on time. Supposing that the motion is irrotational, Euler's equation (6.6.7) yields

$$\frac{1}{2}\mathbf{v}^2 + V^* + P = \text{const.}, \tag{6.6.30}$$

called *Bernoulli's equation*. If the fluid is incompressible, then $P = p/\rho$ and (6.6.30) becomes

$$\frac{1}{2}\rho\mathbf{v}^2 + \rho V^* + p = \text{const.} \tag{6.6.31}$$

This equation expresses the conservation of mechanical energy per unit volume of the fluid. The first term in (6.6.31) is the *dynamic pressure*, the second – the *potential pressure* and the third – the *static pressure*. In other words, *along a streamline, the total pressure is constant*. Note that the constant in (6.6.31) keeps its value at any point of the fluid, while in case when curl $\mathbf{v} \neq 0$, the constant varies from one streamline to another.

Taking the x_3-axis along the ascendent vertical, we have $V^* = gz$, and Bernoulli's equation reads:

$$\frac{1}{2}\mathbf{v}^2 + \rho g z + p = \text{const.} \tag{6.6.32}$$

The constant in (6.6.32) is determined from the boundary conditions: at $z = z_0$, $p = p_0$ and $v = v_0$ are known.

6.6.2.4 Plane, Irrotational, Stationary Motion of a Homogeneous, Incompressible Fluid

Assume a homogeneous and incompressible ideal fluid which moves in such a way that the velocities of its particles are permanently parallel to a fixed plane, say Ox_1x_2 (Oxy) and *do not depend on* x_3. Then the equation of continuity (6.3.11) reads:

$$\operatorname{div}\mathbf{v} = \frac{\partial v_x}{\partial x} + \frac{\partial v_y}{\partial y} = 0. \tag{6.6.33}$$

Since the motion is irrotational, we also have:

$$(\operatorname{curl}\mathbf{v})_z = \frac{\partial v_y}{\partial x} - \frac{\partial v_x}{\partial y} = 0. \tag{6.6.34}$$

Let now $z = x + iy$ be a complex variable (not to be confused with x_3!) and

$$f(z) = \varphi(x, y) + i\psi(x, y) \tag{6.6.35}$$

be a function of z, where φ and ψ are the real and imaginary parts, respectively. If the functions φ and ψ, assumed to be of class C^2, satisfy the *Cauchy–Riemann conditions*:

$$\frac{\partial \varphi}{\partial x} = \frac{\partial \psi}{\partial y}, \quad \frac{\partial \varphi}{\partial y} = -\frac{\partial \psi}{\partial x}, \tag{6.6.36}$$

then $f(z)$ is *analytical* or *holomorphic* at the point z.

Comparing (6.6.33) and (6.6.34) with the Cauchy–Riemann conditions (6.6.36), we realize that v_x and $-v_y$ can be taken as the real and imaginary parts of the complex function

$$w = v_x - iv_y, \tag{6.6.37}$$

or, by virtue of (6.6.18),

$$w = \frac{\partial \varphi}{\partial x} - i\frac{\partial \varphi}{\partial y}, \tag{6.6.38}$$

where $\varphi(x, y)$ is the velocity potential. The function w is called the *complex velocity*. Applying the Cauchy–Riemann conditions, we also have:

$$\frac{df}{dz} = \frac{\frac{\partial\varphi}{\partial x}dx + \frac{\partial\varphi}{\partial y}dy + i\left(\frac{\partial\psi}{\partial x}dx + \frac{\partial\psi}{\partial y}dy\right)}{dx + idy} = \frac{\partial\varphi}{\partial x} - i\frac{\partial\varphi}{\partial y},$$

which yields:

$$\frac{df}{dz} = w. \tag{6.6.39}$$

The function $f(z)$ is called the *complex potential* and $\psi(x, y)$ is the *stream function*.

The Cauchy–Riemann conditions also yield

$$\frac{\partial\varphi}{\partial x}\frac{\partial\psi}{\partial x} + \frac{\partial\varphi}{\partial y}\frac{\partial\psi}{\partial y} = (\nabla\varphi)\cdot(\nabla\psi) = 0, \tag{6.6.40}$$

which says that the two families of curves $\varphi(x, y) = \text{const.}$, $\psi(x, y) = \text{const.}$ are *orthogonal*.

Actually, by symmetry arguments, the role of velocity potential can be played either by $\varphi(x, y)$, or by $\psi(x, y)$. If we choose φ as the velocity potential, then the curves $\psi = \text{const.}$ are the *streamlines* while $\varphi = \text{const.}$ are the *equipotential lines*. If, on the contrary, one chooses ψ as the velocity potential, then $\varphi = \text{const.}$ are the streamlines, and $\psi = \text{const.}$ – the equipotential lines.

To illustrate the use of the complex potential method, let us discuss an example. Assume that the velocity potential is

$$f(z) = \varphi + i\psi = Az + \frac{B}{z}, \tag{6.6.41}$$

where A and B are two real constants. Separating the real and imaginary parts, we obtain:

$$\varphi = Ax + \frac{Bx}{x^2 + y^2}, \quad \psi = Ay - \frac{By}{x^2 + y^2}. \tag{6.6.42}$$

The complex velocity is found from Eq. (6.6.39):

$$w = v_x - iv_y = A - \frac{B}{z^2},$$

which yields:

$$v_x = A + B\frac{y^2 - x^2}{(x^2 + y^2)^2}, \quad v_y = -\frac{2Bxy}{(x^2 + y^2)^2}. \tag{6.6.43}$$

The form of the two last relations suggests the use of polar plane coordinates r, θ, instead of x, y. In this representation, we find:

Fig. 6.8 Image of *streamlines* and *equipotential lines* around a fixed, rigid and cylindrical obstacle.

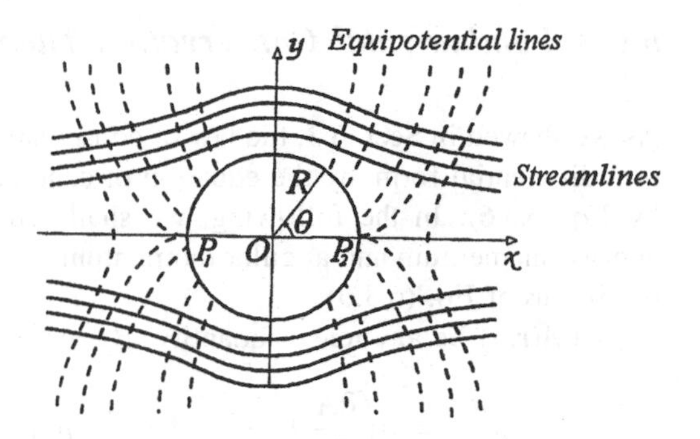

$$\varphi = \left(Ar + \frac{B}{r}\right)\cos\theta, \quad \psi = \left(Ar - \frac{B}{r}\right)\sin\theta, \tag{6.6.44}$$

$$v_x = A + \frac{B}{r^2}(\sin^2\theta - \cos^2\theta), \quad v_y = -2B\frac{\sin\theta\cos\theta}{r^2}. \tag{6.6.45}$$

Let us analyze the streamline

$$\psi(r,\theta) = \left(Ar - \frac{B}{r}\right)\sin\theta = 0.$$

This equation admits as solutions either $r^2 = B/A = \text{const.}$ $(\theta \neq 0, \pi)$, or $\theta = 0, \pi$ $(r^2 \neq B/A)$. The first solution represents a circle of radius $R = \sqrt{B/A}$, with its centre at the origin of the coordinate axes, and the second solution represents the x-axis (Fig. 6.8). On the other hand, (6.6.45) shows that far from the origin $(r \to \infty)$ we have $v_x = A$, $v_y = 0$, which indicates that the fluid performs a uniform motion of translation, parallel to the x-axis.

The same relations lead to the conclusion that at the point O both v_x and v_y become infinite, while at the points P_1 $(\theta = \pi)$ and P_2 $(\theta = 0)$ on the circle they vanish. The points P_1 and P_2 are called *stagnation points*. These considerations show that the circle of radius R represents the orthogonal cross section of a fixed, rigid, cylindrical obstacle.

From (6.6.45), we find the magnitude of velocity as

$$v^2 = \left(A - \frac{B}{r^2}\right)^2 + \frac{4AB}{r^2}\sin^2\theta,$$

i.e. the velocity attains its maximum values at the points with $\theta = \pm\frac{\pi}{2}$.

Observation: From $\mathbf{v} = \nabla\varphi$ and $\operatorname{div}\mathbf{v} = 0$, it follows that

$$\operatorname{div}(\operatorname{grad}\varphi) = \Delta\varphi = 0,$$

which is the *Laplace equation* for φ, i.e. the velocity potential is a *harmonic function*.

6.6.3 Fundamental Conservation Theorems

As we showed in Sect. 6.3, the law of conservation of mass can be expressed either in a differential form, by the equation of continuity (6.3.4), or in an integral form, by Eq. (6.3.6). In the following, we shall deduce the equations of variation of energy, momentum and angular momentum of an ideal fluid, in a form which will remind us of Eq. (6.3.6).

Let $A(\mathbf{r}, t)$ be any tensor quantity. By virtue of (6.3.4), we have:

$$\rho \frac{dA}{dt} = \rho\left(\frac{\partial A}{\partial t} + v_k \frac{\partial A}{\partial x_k}\right) = \frac{\partial}{\partial t}(\rho\, A) + \frac{\partial}{\partial x_k}(\rho\, A\, v_k). \tag{6.6.46}$$

6.6.3.1 Energy Conservation Theorem

By definition, the ideal fluid is free of both internal frictions and thermic conductivity, or, in short, there are no *dissipative* phenomena. From the thermodynamical point of view, the fact that there is no heat exchange between the system and the surrounding medium is expressed by the conservation of entropy. This process is called *isentropic*. Any reversible transformation of an isolated system is an isentropic transformation.

If we denote by s the unit mass entropy, then the constancy of entropy is given by

$$\frac{ds}{dt} = \frac{\partial s}{\partial t} + (\mathbf{v} \cdot \nabla)s = 0. \tag{6.6.47}$$

The energy of the fluid contained in an elementary volume consists of both kinetic and internal energies. If ε is the internal energy per unit mass, the energy per unit volume of the fluid is:

$$\tilde{E} = \frac{1}{2}\rho v^2 + \rho\varepsilon. \tag{6.6.48}$$

Assume now that the quantity A in (6.6.46) is the zero-rank tensor $A = \frac{1}{2}v^2 + \varepsilon$. Then, we have:

$$\rho \frac{d}{dt}\left(\frac{1}{2}v^2 + \varepsilon\right) = \frac{\partial}{\partial t}\left(\rho\frac{v^2}{2} + \rho\varepsilon\right) + \frac{\partial}{\partial x_k}\left[\rho v_k\left(\frac{v^2}{2} + \varepsilon\right)\right]. \tag{6.6.49}$$

On the other hand, the fundamental equation of the thermodynamics of reversible processes reads:

$$T\, ds = d\varepsilon + p\, d\left(\frac{1}{\rho}\right). \tag{6.6.50}$$

Using Euler's equations (6.6.3) (with $\mathbf{F} = 0$), the equation of continuity (6.3.3), the thermodynamical equation (6.6.50) and the fact that the process is isentropic ($ds = 0$), we can write:

$$\rho \frac{d}{dt}\left(\frac{1}{2}v^2 + \varepsilon\right) = \rho v_k \frac{dv_k}{dt} + \rho \frac{d\varepsilon}{dt} = -v_k \frac{\partial p}{\partial x_k} + \frac{p}{\rho}\frac{d\rho}{dt}$$
$$= -\frac{\partial}{\partial x_k}(p v_k) + \frac{p}{\rho}\left(\frac{d\rho}{dt} + \rho \frac{\partial v_k}{\partial x_k}\right) = -\frac{\partial}{\partial x_k}(p v_k).$$

A comparison between this relation and (6.6.49) yields

$$\frac{\partial}{\partial t}\left(\frac{1}{2}\rho v^2 + \rho\varepsilon\right) = -\frac{\partial}{\partial x_k}\left[\rho v_k \left(\frac{v^2}{2} + w\right)\right], \tag{6.6.51}$$

where

$$w = \varepsilon + \frac{p}{\rho} \tag{6.6.52}$$

is the *enthalpy* per unit mass. Integrating (6.6.51) on a fixed domain of volume V, bounded by the surface S, and applying the Green–Gauss formula, we finally obtain:

$$\frac{\partial}{\partial t}\int_V \left(\frac{1}{2}\rho v^2 + \rho\varepsilon\right) d\tau = -\oint_S \rho v_k \left(\frac{v^2}{2} + w\right) dS_k. \tag{6.6.53}$$

The l.h.s. gives the variation of energy in the volume V per unit time, while the r.h.s. expresses the density flux of this energy which flows through the surface bounding the domain. The quantity

$$\rho \mathbf{v}\left(\frac{v^2}{2} + w\right) \tag{6.6.54}$$

is the vector of *energy flux density*. Its magnitude is the energy which flows per unit time through the unit surface, orthogonal to the direction of the velocity.

6.6.3.2 Momentum Conservation Theorem

Assume now that A is the first-rank tensor v_i. Then,

$$\rho \frac{dv_i}{dt} = \frac{\partial}{\partial t}(\rho v_i) + \frac{\partial}{\partial x_k}(\rho v_i v_k).$$

On the other hand, Euler's equations (6.6.8) with $F_i = 0$ read:

$$\rho\frac{dv_i}{dt} = -\frac{\partial p}{\partial x_i} = -\frac{\partial}{\partial x_k}(p\delta_{ik}).$$

The last two relations yield:

$$\frac{\partial}{\partial t}(\rho v_i) = -\frac{\partial}{\partial x_k}\Pi_{ik}, \tag{6.6.55}$$

where the quantities

$$\Pi_{ik} = \rho v_i v_k + p\delta_{ik} \quad (i,k = 1,2,3) \tag{6.6.56}$$

are the components of a symmetric tensor. Integrating (6.6.55) on a fixed domain of volume V and applying the Green–Gauss formula, we obtain:

$$\frac{\partial}{\partial t}\int_V \tilde{P}_i\, d\tau = -\oint_S \Pi_{ik}\, dS_k, \tag{6.6.57}$$

which is the momentum theorem. Here, $\tilde{P}_i = \rho v_i$ and S is the closed surface bounding the integration volume. The integral on the l.h.s. of (6.6.57) is the i-component of the fluid momentum, and that in the r.h.s. is the momentum flux density through the surface S. The tensor Π_{ik} represents the i-component of the momentum passing in unit time through a unit area orthogonal to the x_k-axis. Equation (6.6.57) can also by written in a vector form:

$$\frac{\partial}{\partial t}\int_V \tilde{\mathbf{P}}\, d\tau = -\oint \{\Pi\}\cdot d\mathbf{S}, \tag{6.6.58}$$

where $\tilde{\mathbf{P}} = \rho\mathbf{v}$ and $\{\Pi\} = \rho\mathbf{v}\mathbf{v} + p\mathbf{u}_k\mathbf{u}_k$ is the dyadic representation of the momentum flux density tensor Π_{ik} (see Appendix A).

6.6.3.3 Angular Momentum Conservation Theorem

Let us choose $A \equiv (\mathbf{r}\times\mathbf{v})_i$ in (6.6.46). Applying Euler's equation (6.6.3) with $\mathbf{F} = 0$, we have:

$$\rho\frac{d}{dt}(\mathbf{r}\times\mathbf{v})_i = \epsilon_{ijk}x_j\rho\frac{dv_k}{dt} = -\epsilon_{ijk}x_j\frac{\partial p}{\partial x_k} = -\frac{\partial}{\partial x_l}(\epsilon_{ijk}x_j p\delta_{kl}).$$

Introducing this result into (6.6.46), then making a convenient grouping of the terms and integrating over a fixed volume V bounded by the surface S, we obtain the angular momentum theorem:

$$\frac{\partial}{\partial t}\int_V \tilde{L}_i\, d\tau = -\oint_S \mathcal{M}_{il}\, dS_l, \tag{6.6.59}$$

where we denoted

$$\tilde{L}_i = \epsilon_{ijk}\rho x_j v_k, \tag{6.6.60}$$

$$\mathcal{M}_{il} = \epsilon_{ijk} x_j \Pi_{kl}. \tag{6.6.61}$$

The l.h.s. of Eq. (6.6.59) represents the time-variation of the i-component of the angular momentum of the volume of fluid, while the r.h.s. is the i-component of the flux of angular momentum density through the surface S. In vector form, the angular momentum theorem reads:

$$\frac{\partial}{\partial t}\int_V \tilde{\mathbf{L}}\, d\tau = -\oint_S \{\mathcal{M}\} \cdot d\mathbf{S}, \tag{6.6.62}$$

where $\tilde{\mathbf{L}} = \rho \mathbf{r} \times \mathbf{v}$, $\{\mathcal{M}\} = \mathbf{r} \times \{\Pi\}$.

6.6.3.4 Kelvin's Velocity Circulation Theorem

Consider an ideal fluid in motion and let C be some closed curve, made out of fluid particles. We want to show that in the case of isentropic motion the circulation of the velocity along any closed curve is constant:

$$\oint_C \mathbf{v} \cdot d\mathbf{l} = \text{const.} \tag{6.6.63}$$

To prove this, we take the total time-derivative of the circulation:

$$\frac{d}{dt}\oint_C \mathbf{v} \cdot d\mathbf{l} = \oint_C \frac{d\mathbf{v}}{dt} \cdot d\mathbf{l} + \oint_C \mathbf{v} \cdot \frac{d}{dt}(d\mathbf{l}). \tag{6.6.64}$$

Since the motion is isentropic (s = const.), by virtue of (6.6.50) and (6.6.52) we have:

$$\nabla w = \frac{1}{\rho}\nabla p. \tag{6.6.65}$$

Assuming that $\mathbf{F}$ is conservative,

$$\mathbf{F} = -\text{grad}\, V^*(\mathbf{r}),$$

Euler's equation (6.6.3) reads:

$$\frac{d\mathbf{v}}{dt} = -\text{grad}(V^* + w). \tag{6.6.66}$$

Using these results, the first term on the r.h.s. of (6.6.64) vanishes:

$$\oint_C \frac{d\mathbf{v}}{dt} \cdot d\mathbf{l} = -\oint_C \nabla(V^* + w) \cdot d\mathbf{l} = -\oint_C d(V^* + w) = 0.$$

Next, we note that the contour element $d\mathbf{l}$ can be written as the difference between the position vectors of its end points, while the operators d and d/dt are independent. Having this in mind, the second term on the r.h.s. of (6.6.64) is also zero:

$$\oint_C \mathbf{v} \cdot \frac{d}{dt}(d\mathbf{l}) = \oint_C \mathbf{v} \cdot d\left(\frac{d\mathbf{l}}{dt}\right) = \oint_C \mathbf{v} \cdot d\mathbf{v} = \oint_C d\left(\frac{1}{2}v^2\right) = 0,$$

which completes the proof of (6.6.63), since

$$\frac{d}{dt}\oint_C \mathbf{v} \cdot d\mathbf{l} = 0. \tag{6.6.67}$$

This theorem is known as *Kelvin's velocity circulation theorem*. According to it, if at the time t_1 a certain number of particles form a closed contour, then the same particles will form a closed contour at any time $t_2 > t_1$. As a consequence of Kelvin's theorem, if the motion of a fluid is irrotational at a moment of time, it will remain irrotational.

6.6.3.5 Vorticity Equation

Applying in (6.6.67) the Stokes theorem, we have:

$$\frac{d}{dt}\int_S \operatorname{curl} \mathbf{v} \cdot d\mathbf{S} = 0,$$

or, in view of (6.6.13),

$$\frac{d}{dt}\int_S \mathbf{\Omega} \cdot d\mathbf{S} = 0. \tag{6.6.68}$$

This shows that the flux of the vorticity $\mathbf{\Omega}$ through the open surface S which moves together with the fluid is constant. A curve tangent in any point and at any time to $\mathbf{\Omega}$ is a *vortex line*. The differential equations of the vortex lines are deduced from the obvious relation $\mathbf{\Omega} \times d\mathbf{r} = 0$ and can be written as

$$\frac{dx}{\Omega_x} = \frac{dy}{\Omega_y} = \frac{dz}{\Omega_z}.$$

According to Kelvin's circulation theorem, the vortex lines move together with the fluid particles, as if being 'frozen' in the fluid. This result can be obtained also in a different way. Let us write Eq. (6.6.66) in the form

$$\frac{\partial \mathbf{v}}{\partial t} - \mathbf{v} \times (\operatorname{curl} \mathbf{v}) = -\operatorname{grad}\left(\frac{1}{2}|\mathbf{v}|^2 + V^* + w\right).$$

Taking the curl of this equation, we arrive at

$$\frac{\partial \boldsymbol{\Omega}}{\partial t} = \operatorname{curl}(\mathbf{v} \times \boldsymbol{\Omega}), \tag{6.6.69}$$

called the *vorticity equation*, which is the differential form of Kelvin's theorem. An alternative form of this equation, which is more convenient for our purpose, is

$$\frac{d\boldsymbol{\Omega}}{dt} - (\mathbf{v} \cdot \nabla)\boldsymbol{\Omega} = \operatorname{curl}(\mathbf{v} \times \boldsymbol{\Omega}),$$

or, using the formulas from Appendix B,

$$\frac{d\boldsymbol{\Omega}}{dt} = (\boldsymbol{\Omega} \cdot \nabla)\mathbf{v} - \boldsymbol{\Omega} \operatorname{div} \mathbf{v}.$$

Multiplying this equation by $1/\rho$ and using the equation of continuity, we easily arrive at

$$\frac{d}{dt}\left(\frac{\boldsymbol{\Omega}}{\rho}\right) = \left(\frac{\boldsymbol{\Omega}}{\rho} \cdot \nabla\right)\mathbf{v}, \tag{6.6.70}$$

which is *Beltrami's vorticity diffusion equation.*

To extract the physical significance of this equation, let $\boldsymbol{\Phi}$ be an arbitrary vector field, and let Γ, Γ' be two field lines, taken in such a way that all particles lying on Γ at time t, lie on Γ' at time $t + dt$. Also, let $\delta \mathbf{l}$ be an arbitrary infinitesimal vector on Γ, and $\delta \mathbf{l}'$ the infinitesimal vector, made up by the same particles, but on the line Γ'. In order to be conserved during the motion of the fluid, the lines of the field $\boldsymbol{\Phi}$ must satisfy the relation

$$\delta \mathbf{l}' - \delta \mathbf{l} = \frac{d}{dt}(\delta \mathbf{l})\, dt.$$

On the other hand, if $\mathbf{v} = \mathbf{v}_A$ and $\mathbf{v}' = \mathbf{v}_B$ are the velocities of displacement of the points A and B (Fig. 6.9), we can write:

$$\begin{aligned}
\delta \mathbf{l}' &= \delta \mathbf{l} + \mathbf{v}_B\, dt - \mathbf{v}_A\, dt \\
&= \delta \mathbf{l} + \frac{d}{dt}(\mathbf{r}_A + \delta \mathbf{l})\, dt - \mathbf{v}_A\, dt \\
&= \delta \mathbf{l} + [\mathbf{v} + (\delta \mathbf{l} \cdot \nabla)\mathbf{v} + \cdots]\, dt - \mathbf{v}\, dt \\
&\simeq \delta \mathbf{l} + [(\delta \mathbf{l} \cdot \nabla)\mathbf{v}]\, dt.
\end{aligned}$$

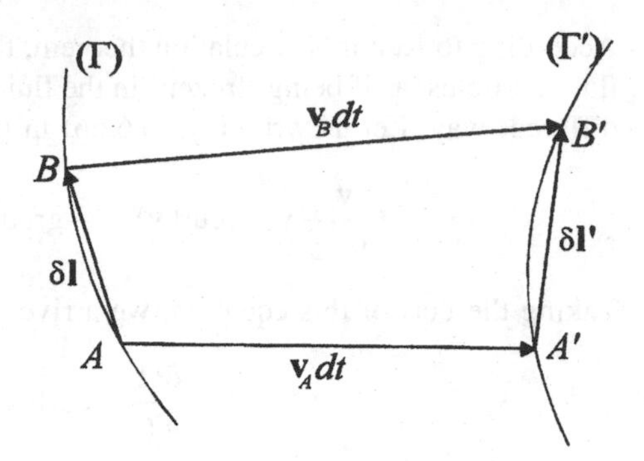

Fig. 6.9 Physical interpretation of *Beltrami's diffusion equation.*

The last two relations yield:

$$\frac{d}{dt}(\delta\mathbf{l}) = (\delta\mathbf{l} \cdot \nabla)\mathbf{v}. \tag{6.6.71}$$

Now, by comparing (6.6.71) with (6.6.70), we conclude that the lines of the field $\boldsymbol{\Omega}/\rho$ are conserved.

Observation: Beltrami's equation (6.6.70) has been deduced for an isentropic flow. If this condition is eliminated, this equation is written in a general form:

$$\frac{d}{dt}\left(\frac{\boldsymbol{\Omega}}{\rho}\right) = \left(\frac{\boldsymbol{\Omega}}{\rho} \cdot \nabla\right)\mathbf{v} + \frac{1}{2\rho}\operatorname{curl}\mathbf{a}, \tag{6.6.72}$$

where $\mathbf{a}$ is the acceleration vector. The derivation of this equation is straightforward and it is left to the reader. (Sometimes the vorticity is defined as $\boldsymbol{\Omega} = \operatorname{curl}\mathbf{v}$, which leads to the disappearance of the denominator '2' in (6.6.72).)

6.6.3.6 Clebsch's Theorem

Clebsch's theorem states that: *If the vector field* $\mathbf{A}$ *is given, then it is always possible to find three scalar functions* α, β, γ, *depending on coordinates and time, such that*

$$\mathbf{A} = -\nabla\alpha + \beta\nabla\gamma. \tag{6.6.73}$$

To prove the theorem, let us consider

$$\mathbf{B} = \operatorname{curl}\mathbf{A}, \tag{6.6.74}$$

which yields

$$\operatorname{div}\mathbf{B} = 0. \tag{6.6.75}$$

By integrating the differential equations of the solenoidal field $\mathbf{B}$,

$$\frac{dx}{B_x} = \frac{dy}{B_y} = \frac{dz}{B_z}, \tag{6.6.76}$$

we find two families of mobile surfaces,

$$f_1(x, y, z, t) = 0, \quad f_2(x, y, z, t) = 0, \tag{6.6.77}$$

which by intersection produce the lines of the field **B**. Then, Eq. (6.6.75) admits as a solution

$$\mathbf{B} = g(f_1, f_2, t)\nabla f_1 \times \nabla f_2. \tag{6.6.78}$$

Indeed, we have:

$$\begin{aligned}\operatorname{div}\mathbf{B} &= g\nabla \cdot (\nabla f_1 \times \nabla f_2) + (\nabla f_1 \times \nabla f_2) \cdot \nabla g \\ &= (\nabla f_1 \times \nabla f_2) \cdot \left(\frac{\partial g}{\partial f_1}\nabla f_1 + \frac{\partial g}{\partial f_2}\nabla f_2\right) = 0.\end{aligned}$$

Let β and γ be two other functions, so that:

$$\beta = \beta(f_1, f_2, t), \quad \gamma = \gamma(f_1, f_2, t). \tag{6.6.79}$$

Then, we have:

$$\nabla f_1 \times \nabla f_2 = J\nabla\beta \times \nabla\gamma,$$

where

$$J = \frac{\partial(f_1, f_2)}{\partial(\beta, \gamma)}$$

is the Jacobian of the transformation (6.6.79). If we choose β and γ so as to have $J = 1/g$, relation (6.6.78) becomes:

$$\mathbf{B} = \operatorname{curl}\mathbf{A} = \nabla\beta \times \nabla\gamma = \nabla \times (\beta\nabla\gamma),$$

or

$$\operatorname{curl}(\mathbf{A} - \beta\nabla\gamma) = 0,$$

which yields (6.6.73). This theorem was proved by *Alfred Clebsch*, and (6.6.73) expresses a *Clebsch transformation*. It is shown to be very useful in fluid mechanics and magnetofluid dynamics. The scalar functions α, β, γ are called *Clebsch potentials*. As in the case of the electrodynamic potentials, they are determined only up to a *gauge transformation*, meaning that for a given field **A**, they are not unique. Let α', β', γ' be another set of potentials, so that

$$\mathbf{A} = -\nabla\alpha' + \beta'\nabla\gamma'. \tag{6.6.80}$$

The transformations (6.6.73) and (6.6.80) yield

$$\beta'\nabla\gamma' - \beta\nabla\gamma + \nabla F = 0, \tag{6.6.81}$$

where $F = \alpha - \alpha'$. Equation (6.6.81) can be taken as a system of three algebraic equations with two unknown functions, β and β'. The solutions are non-trivial if the determinant

$$\frac{\partial(\gamma, \gamma', F)}{\partial(x, y, z)}$$

is zero. It follows that F is a function of γ, γ' and, possibly, of time: $F = F_1(\gamma, \gamma', t)$. Thus, (6.6.81) gives:

$$\beta = \frac{\partial F_1}{\partial \gamma}, \quad \beta' = -\frac{\partial F_1}{\partial \gamma'} \quad \alpha' = \alpha - F_1. \tag{6.6.82}$$

Consequently, we obtain:

$$\mathbf{A} = -\nabla(\alpha - F_1) - \frac{\partial F_1}{\partial \gamma'}\nabla\gamma' = -\nabla\alpha + \beta\nabla\gamma = \mathbf{A},$$

meaning that (6.6.82) is a suitable gauge transformation. Incidentally, this method was encountered while studying the canonical transformations, the generating function being $F_1(\gamma, \gamma', t)$. Similar formulas are obtained for the generating functions $F_2(\beta, \gamma', t)$, $F_3(\gamma, \beta', t)$, $F_4(\beta, \beta', t)$. The reader is encouraged to deduce them.

Application. Using Clebsch's representation (6.6.73), the velocity field $\mathbf{v}$ can be written as

$$\mathbf{v} = -\nabla\alpha + \beta\nabla\gamma, \tag{6.6.83}$$

which means that the vorticity $\boldsymbol{\Omega}$ is

$$\boldsymbol{\Omega} = \frac{1}{2}\operatorname{curl}\mathbf{v} = \frac{1}{2}\nabla\beta \times \nabla\gamma, \tag{6.6.84}$$

showing that the vortex lines lie at the intersection of the surfaces $\beta = \text{const.}$, $\gamma = \text{const.}$ (Fig. 6.10).

Assuming that the motion is isentropic, we introduce (6.6.83) into Euler's equation:

$$\frac{\partial \mathbf{v}}{\partial t} - \mathbf{v} \times \operatorname{curl}\mathbf{v} = -\operatorname{grad}\left(\frac{1}{2}|\mathbf{v}|^2 + V^* + w\right).$$

The result is:

$$\frac{\partial \beta}{\partial t}\nabla\gamma - \frac{\partial \gamma}{\partial t}\nabla\beta - \mathbf{v} \times (\nabla\beta \times \nabla\gamma) = -\nabla\left(\frac{1}{2}|\mathbf{v}|^2 + V^* + w - \frac{\partial \alpha}{\partial t} + \beta\frac{\partial \gamma}{\partial t}\right).$$

Denoting

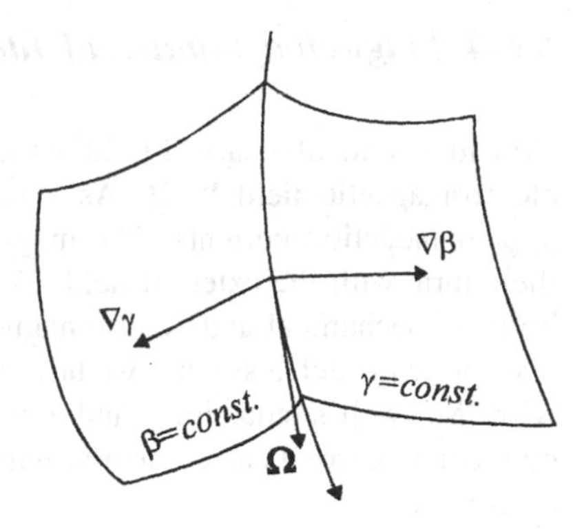

Fig. 6.10 Vortex lines lie at the intersection of the surfaces $\beta = \text{const.}$, $\gamma = \text{const.}$, where β and γ are the Clebsch potentials.

$$\Psi = \frac{1}{2}|\mathbf{v}|^2 + V^* + w - \frac{\partial\alpha}{\partial t} + \beta\frac{\partial\gamma}{\partial t}$$

and writting explicitly the double cross product, we have:

$$\left(\frac{\partial\beta}{\partial t} + \mathbf{v}\cdot\nabla\beta\right)\nabla\gamma - \left(\frac{\partial\gamma}{\partial t} + \mathbf{v}\cdot\nabla\gamma\right)\nabla\beta = \frac{d\beta}{dt}\nabla\gamma - \frac{d\gamma}{dt}\nabla\beta = -\nabla\Psi. \quad (6.6.85)$$

The vector equation (6.6.85) may be regarded as a system of three linear algebraic equations in the variables $d\beta/dt$ and $d\gamma/dt$. To have non-trivial solutions, the determinant

$$\partial(\Psi, \beta, \gamma)/\partial(x, y, z)$$

must vanish, which means that Ψ must be a function of the form: $\Psi = \Psi(\beta, \gamma, t)$. A solution of (6.6.84) is then

$$\frac{d\gamma}{dt} = \frac{d\Psi}{d\beta}, \quad \frac{d\beta}{dt} = -\frac{d\Psi}{d\gamma}. \quad (6.6.86)$$

As we can see, Eqs. (6.6.86) are similar to Hamilton's canonical equations (5.1.21). Here, the variables γ and β play the role of generalized coordinate and generalized momentum, respectively, while $\Psi(\beta, \gamma, t)$ stands for the Hamiltonian per unit mass.

If the motion is stationary, the partial derivatives with respect to time of all quantities are zero, and we have

$$\frac{d\Psi}{dt} = \dot{\beta}\frac{\partial\Psi}{\partial\beta} + \dot{\gamma}\frac{\partial\Psi}{\partial\gamma} = 0,$$

which expresses the conservation law

$$\frac{1}{2}v^2 + V^* + w = \text{const.},$$

i.e. *Bernoulli's equation* (6.6.30).

6.6.4 Magnetodynamics of Ideal Fluids

Consider an ideal, charged fluid, which undergoes isentropic motion in an external electromagnetic field **E**, **B**. As a result of the motion, in the conducting fluid appear induction currents. The magnetic fields of these currents will interact, in their turn, with the external field. Consequently, the charged particles are subject both to mechanical and electromagnetic forces. To describe the processes which take place in such a system, we have to use the equations of an ideal fluid, together with Maxwell's equations and Ohm's law. Neglecting the displacement and convection currents as compared with the conduction currents, Maxwell's system of equations reads:

$$\operatorname{curl}\mathbf{E} = -\frac{\partial \mathbf{B}}{\partial t}, \quad \operatorname{div}\mathbf{B} = 0, \tag{6.6.87}$$

$$\operatorname{curl}\mathbf{B} = \mu_0 \mathbf{j}, \quad \operatorname{div}\mathbf{E} = \frac{\rho^e}{\epsilon_0}, \tag{6.6.88}$$

while the differential form of Ohm's law is:

$$\mathbf{j} = \lambda(\mathbf{E} + \mathbf{v} \times \mathbf{B}). \tag{6.6.89}$$

In these formulas, ρ^e is the electric charge density and λ – the electric conductivity. It is assumed that $\mu \simeq \mu_0, \epsilon \simeq \epsilon_0$, where μ and ϵ are the permeability and the permittivity of the medium, respectively.

Using Maxwell's equations and Ohm's law, we can eliminate the field **E** and obtain a single vector equation for **B**:

$$\frac{\partial \mathbf{B}}{\partial t} = -\operatorname{curl}\mathbf{E} = \operatorname{curl}(\mathbf{v} \times \mathbf{B}) - \frac{1}{\lambda}\operatorname{curl}\mathbf{j} = \operatorname{curl}(\mathbf{v} \times \mathbf{B}) - \frac{1}{\lambda\mu_0}\operatorname{curl}\operatorname{curl}\mathbf{B},$$

or, by utilizing the results derived in Appendix B,

$$\frac{\partial \mathbf{B}}{\partial t} = \operatorname{curl}(\mathbf{v} \times \mathbf{B}) + \nu_m \Delta \mathbf{B}, \tag{6.6.90}$$

called *induction equation*. The quantity

$$\nu_m = \frac{1}{\lambda\mu_0}, \quad [\nu_m] = L^2 T^{-1}, \tag{6.6.91}$$

is known as the *magnetic viscosity*.

If the electric conductivity of the medium is very high ($\lambda \to \infty$), then $\nu_m \to 0$ and (6.6.90) reduces to

$$\frac{\partial \mathbf{B}}{\partial t} = \operatorname{curl}(\mathbf{v} \times \mathbf{B}), \tag{6.6.92}$$

which is similar to the vorticity equation (6.6.69). We also have:

$$\rho^e = \epsilon_0 \operatorname{div} \mathbf{E} = -\epsilon_0 \operatorname{div}(\mathbf{v} \times \mathbf{B}).$$

The equation of motion of the magnetofluid is obtained by considering the force of interaction between the magnetic field **B** and the currents **j** (Lorentz force):

$$\mathbf{F}_L = ne(\mathbf{v} \times \mathbf{B}) = \mathbf{j} \times \mathbf{B} = \frac{1}{\mu_0}(\operatorname{curl} \mathbf{B}) \times \mathbf{B}.$$

Denoting by $\mathbf{F}'$ the body forces which are not of magnetic nature, the equation of motion then reads:

$$\rho \frac{d\mathbf{v}}{dt} = \mathbf{j} \times \mathbf{B} - \nabla p + \rho \mathbf{F}'. \tag{6.6.93}$$

The complete system of equations describing the behaviour of an ideal, infinitely conducting fluid, performing isentropic motion in an external electromagnetic field, is therefore:

$$\operatorname{div} \mathbf{B} = 0, \quad \frac{\partial \mathbf{B}}{\partial t} = \operatorname{curl}(\mathbf{v} \times \mathbf{B}), \quad p = p(\rho),$$

$$\rho \frac{d\mathbf{v}}{dt} = \mathbf{j} \times \mathbf{B} - \nabla p, \quad \frac{ds}{dt} = 0, \quad \frac{\partial \rho}{\partial t} + \operatorname{div}(\rho \mathbf{v}) = 0. \tag{6.6.94}$$

Observation: Following a procedure similar to that leading to Beltrami's equation (6.6.70), one obtains

$$\frac{d}{dt}\left(\frac{\mathbf{B}}{\rho}\right) = \left(\frac{\mathbf{B}}{\rho} \cdot \nabla\right) \mathbf{v}.$$

Thus, if the electric conductivity is very high, the lines of the magnetic field are 'frozen' in the magnetofluid.

6.7 Viscous Fluid

In the study of the ideal fluid, as a model of CDM, we assumed that the stress tensor T_{ik} reduces to a single component p – the pressure. This model is a good approximation to the real flows in the case of slow motions, but if the velocity becomes high, the friction between the fluid particles cannot be neglected anymore. The existence of the interior friction is mathematically expressed by the appearance of non-diagonal components of the stress tensor. The resistance encountered by particles due to their mutual friction is called *viscosity*, while the fluids having this property are termed *viscous* or *real*.

A viscous fluid keeps its properties within certain pressure and temperature limits. Melted bitumen, for example, has properties of a viscous fluid, but as the temperature goes down, it becomes a *plastic medium*, and finally a rigid body.

Consider a viscous fluid at rest. In this state, the stress tensor has only diagonal (normal) components, as for the ideal fluids:

$$T_{ik} = -p\delta_{ik}.$$

If the fluid is put into motion, there appear both normal and tangent tensions, i.e.

$$T_{ik} = -p\delta_{ik} + T'_{ik}, \tag{6.7.1}$$

where T'_{ik} is the *viscous stress tensor*. The experimental data show that the tensions occurring in a viscous fluid depend on both the *velocity of deformation* θ',

$$\theta' = \frac{\partial v_i}{\partial x_i} = \operatorname{div} \mathbf{v}, \tag{6.7.2}$$

and the *velocity of deformation tensor* e'_{ik},

$$e'_{ik} = \frac{1}{2}\left(\frac{\partial v_k}{\partial x_i} + \frac{\partial v_i}{\partial x_k}\right). \tag{6.7.3}$$

We assume that, in the case of homogeneous and isotropic viscous fluids, the relation between T'_{ik} and e'_{ik} has the form:

$$T'_{ik} = \lambda'\theta'\delta_{ik} + 2\mu' e'_{ik},$$

where the quantities $\lambda' > 0, \mu' > 0$ are called *dynamic coefficients of viscosity*. Their dimension is $[\lambda', \mu'] = ML^{-1}T^{-1}$. Using (6.7.1), we then obtain:

$$T_{ik} = -(p - \lambda'\theta')\delta_{ik} + 2\mu' e'_{ik}. \tag{6.7.4}$$

The fluids obeying Eq. (6.7.4) are called *perfectly viscous* or *Newtonian*.

The equations of motion of a viscous fluid are found by an already known procedure. Introducing the tensor T_{ik} given by (6.7.4) into Cauchy's equations (6.3.23), we have:

$$\rho a_i = \rho F_i + \left(-\frac{\partial p}{\partial x_k} + \lambda' \frac{\partial \theta'}{\partial x_k}\right)\delta_{ik} + 2\mu' \frac{\partial e'_{ik}}{\partial x_k}.$$

But

$$\frac{\partial e'_{ik}}{\partial x_k} = \frac{1}{2}\frac{\partial}{\partial x_k}\left(\frac{\partial v_k}{\partial x_i} + \frac{\partial v_i}{\partial x_k}\right) = \frac{1}{2}\left(\frac{\partial \theta'}{\partial x_i} + \Delta v_i\right),$$

therefore

$$\rho a_i = \rho F_i - \frac{\partial p}{\partial x_i} + (\lambda' + \mu')\frac{\partial \theta'}{\partial x_i} + \mu' \Delta v_i, \tag{6.7.5}$$

or, in vector form,

$$\rho \mathbf{a} = \rho \mathbf{F} - \nabla p + (\lambda' + \mu')\operatorname{grad} \theta' + \mu' \Delta \mathbf{v}. \tag{6.7.6}$$

These are the equations of motion of our model of viscous fluid, called *Navier–Stokes equations*. They represent a system of *non-linear*, second-order partial differential equations.

If the homogeneous viscous fluid is incompressible ($\operatorname{div}\mathbf{v} = 0$), the last equation reduces to

$$\rho\mathbf{a} = \rho\mathbf{F} - \nabla p + \mu'\Delta\mathbf{v}, \tag{6.7.7}$$

which shows that in such a fluid the viscosity is determined by a single coefficient μ'. The quantity

$$\nu' = \frac{\mu'}{\rho}, \quad [\nu'] = L^2T^{-1} \tag{6.7.8}$$

is called the *kinematic coefficient of viscosity*. Comparing (6.7.8) with (6.6.91), we see that the coefficients ν' and ν_m have the same dimension. If the viscosity is absent ($\lambda' = \mu' = 0$), then the Navier–Stokes equations (6.7.6) yield Euler equations (6.6.3), as expected.

In order to determine the motion of the fluid, the Navier–Stokes equations (6.7.6) and the equation of continuity must be completed with one more equation. As we have shown, in the case of ideal fluids, this is the equation of conservation of entropy (6.6.47). Since viscous fluids are characterized by irreversible processes of dissipation of energy, the aforementioned equation is not valid any more. To obtain the corresponding equation for viscous fluids we shall proceed as follows.

The fact that the time-variation of the energy density of an ideal fluid is equal to the divergence of the energy flow which passes through the boundary of the domain occupied by the fluid, mathematically expressed by Eq. (6.6.51), is also valid for viscous fluids. Nevertheless, in this last case, two more terms must be added to the quantity $\rho v_k(v^2/2 + w)$, which is related to the displacements of the fluid. These two terms are due to viscosity, $(-v_i T'_{ik})$, and to heat transfer, $(-\chi \frac{\partial T}{\partial x_k})$, where χ is the *thermo-conductivity coefficient*. The equation of conservation of energy is then:

$$\frac{\partial}{\partial t}\left(\frac{\rho v^2}{2} + \rho\varepsilon\right) = -\frac{\partial}{\partial x_k}\left[\rho v_k\left(\frac{v^2}{2} + w\right) - v_i T'_{ik} - \chi\frac{\partial T}{\partial x_k}\right]. \tag{6.7.9}$$

Let us deduce this equation in a more rigorous way. Writing the Navier–Stokes equations (6.7.5) in the compressed form

$$\rho\frac{dv_i}{dt} = -\frac{\partial p}{\partial x_i} + \frac{\partial T'_{ik}}{\partial x_k}$$

and observing that, in view of (6.3.3) and (6.6.50),

$$\frac{d\varepsilon}{dt} = T\frac{ds}{dt} + \frac{p}{\rho^2}\frac{d\rho}{dt} = T\frac{ds}{dt} - \frac{p}{\rho}\frac{\partial v_k}{\partial x_k},$$

we obtain:

$$\rho \frac{d}{dt}\left(\frac{v^2}{2}+\varepsilon\right) = \rho v_k \frac{dv_k}{dt} + \rho \frac{d\varepsilon}{dt} = -\frac{\partial}{\partial x_k}(p v_k) + \rho T \frac{ds}{dt} + v_i \frac{\partial T'_{ik}}{\partial x_k}. \qquad (6.7.10)$$

Then Eqs. (6.6.49) and (6.7.10) yield:

$$\frac{\partial}{\partial t}\left(\rho \frac{v^2}{2} + \rho\varepsilon\right) = -\frac{\partial}{\partial x_k}\left[\rho v_k\left(\frac{v^2}{2}+w\right) - v_i T'_{ik}\right] + \rho T \frac{ds}{dt} - T'_{ik}\frac{p v_i}{\partial x_k},$$

or, after some rearrangements of terms,

$$\frac{\partial}{\partial t}\left(\frac{\rho v^2}{2} + \rho\varepsilon\right) = -\frac{\partial}{\partial x_k}\left[\rho v_k\left(\frac{v^2}{2}+w\right) - v_i T'_{ik} - \chi \frac{\partial T}{\partial x_k}\right]$$
$$+ \rho T \frac{ds}{dt} - T'_{ik}\frac{\partial v_i}{\partial x_k} - \frac{\partial}{\partial x_k}\left(\chi \frac{\partial T}{\partial x_k}\right).$$

We therefore obtain Eq. (6.7.9) if

$$\rho T \frac{ds}{dt} = T'_{ik}\frac{\partial v_i}{\partial x_k} + \frac{\partial}{\partial x_k}\left(\chi \frac{\partial T}{\partial x_k}\right). \qquad (6.7.11)$$

The last equation is equivalent to (6.7.9). It is called the *general equation of heat propagation* ($\rho T \frac{ds}{dt}$ is the heat gained by the unit volume of fluid per unit time). If the viscosity and the heat conduction are absent ($T'_{ik} = 0,\ \chi = 0$), we fall back on the equation of conservation of entropy for a perfect fluid (6.6.47), as expected.

Application. Let us determine the distribution of velocities in a viscous, incompressible fluid, contained between two infinite, coaxial cylinders, of radii R_1 and $R_2 > R_1$, which perform a uniform motion of rotation about their common axis, with the angular velocities ω_1 and ω_2, respectively. Choosing a cylindric system of coordinates (in which the radial coordinate is denoted by r, since ρ denotes the mass density in this chapter), with the z-axis along the cylinders' axis, by symmetry criteria we have: $v_r = v_z = 0,\ v_\varphi = v(r),\ p = p(r)$. The acceleration is constant and directed along the r-coordinate. Projecting Eq. (6.7.7) on cylindric coordinates, we obtain:

$$\frac{v^2}{r} = \frac{1}{\rho}\frac{dp}{dr}, \qquad (6.7.12)$$

$$\frac{d^2 v}{dr^2} + \frac{1}{r}\frac{dv}{dr} - \frac{v}{r^2} = 0. \qquad (6.7.13)$$

Equation (6.7.13) admits solutions of type r^n. Introducing this solution into the equation, we obtain $n = \pm 1$, therefore v is

$$v = Ar + \frac{B}{r}. \tag{6.7.14}$$

The constants A and B are determined by the condition that, on the surfaces of cylinders in contact with the fluid, the fluid velocity is equal to that of cylinders: $v_{R_1} = \omega_1 R_1,\ v_{R_2} = \omega_2 R_2$. The result is:

$$v = \frac{\omega_1 R_1^2 - \omega_2 R_2^2}{R_1^2 - R_2^2} r + \frac{(\omega_2 - \omega_1) R_1^2 R_2^2}{R_1^2 - R_2^2} \frac{1}{r}. \tag{6.7.15}$$

If $\omega_1 = \omega_2 = \omega$, we have $v = \omega r$, i.e. the fluid rotates together with the cylinders, like a rigid system. If the exterior cylinder is taken away $(R_2 = \infty,\ \omega_2 = 0)$, we obtain $v = \frac{\omega_1 R_1^2}{r}$.

Using Cauchy's formula (6.3.22) and the relation (6.7.1), we are able to determine the stress exerted on a solid surface which is in contact with a viscous fluid. Thus, we have:

$$T_i = n_k T_{ik} = -p n_i + n_k T'_{ik}. \tag{6.7.16}$$

The first term represents the normal tension, while the second – the force of friction per unit area, due to viscosity. If the fluid is incompressible (our case), Eq. (6.7.4) yields:

$$T'_{ik} = \mu' \left(\frac{\partial v_k}{\partial x_i} + \frac{\partial v_i}{\partial x_k} \right).$$

We are interested in finding the tangent tension $T'_{r\varphi}$ corresponding to the interior cylinder, namely:

$$T'_{r\varphi}|_{r=R_1} = \mu' \left(\frac{\partial v}{\partial r} - \frac{v}{r} \right)|_{r=R_1} = -2\mu' \frac{\omega_2 - \omega_1}{R_1^2 - R_2^2} R_2^2.$$

The problem of motion of a viscous fluid between two coaxial rotating cylinders lies at the basis of the hydrodynamical theory of lubrication.

Observation: To facilitate the study of different particular cases of motion of a viscous fluid, one defines the dimensionless ratio

$$Re = \frac{lv}{\nu'}, \tag{6.7.17}$$

called *Reynolds' number*, where l is the length of the macroscopic inhomogeneity in the fluid, v – the fluid velocity and ν' – the kinematic coefficient of viscosity. The value of Reynolds' number indicates the type of flow. For example, in a pipe of a circular section and smooth walls, for $Re < Re^c$, where Re^c is the *critical value* of Re $(Re^c = Dv/\nu' \simeq 2{,}400$, where D is the diameter of the pipe) the fluid flow is *laminar*. In such a flow the streamlines do not intersect each other. If $Re > Re^c$, the flow is *turbulent*, being characterized by a non-regular variation in space and time of the velocity. The turbulent flow usually occurs at high velocities of the fluid and for large dimensions of the leading pipes. The ideal fluids are characterized by $Re \to \infty$ ($\nu' \to 0$).

6.8 Lagrangian Formalism

The principles and methods used in the study of systems with a finite number of degrees of freedom are also valid in the mechanics of CDM. Since these systems are characterized by an infinite number of degrees of freedom, the number of the differential equations of motion should be infinite. In fact, these equations are similar to the Lagrange equations and their number is finite. We shall start by obtaining the Euler–Lagrange equations for continuous systems.

6.8.1 Euler–Lagrange Equations for Continuous Systems

Consider the functional

$$J(\varphi) = \int_{D_n} \mathcal{L}\left[x_1, .., x_n; \varphi(x_1, .., x_n), \frac{\partial \varphi}{\partial x_1}, .., \frac{\partial \varphi}{\partial x_n}\right] dx_1 \dots dx_n, \qquad (6.8.1)$$

defined on the bounded domain D_n of a n-dimensional space R_n, where $\mathcal{L}$ is a continuous and differentiable function, admitting as many partial derivatives as necessary, and φ is a function of class C^2. Assuming that the values of φ on the closed hypersurface S_{n-1} which bounds the domain D_n are given, we want to determine the function φ for which $J(\varphi)$ attains an extremum.

Suppose that $\varphi(x_1, .., x_n)$ realizes the stationary value of $J(\varphi)$. In this case, for any infinitesimal variation $\varphi + \delta\varphi$, where $\delta\varphi = \epsilon\eta(x_1, .., x_n)$, with

$$\eta(x_1, .., x_n)|_{S_{n-1}} = 0, \qquad (6.8.2)$$

the first variation $\delta J(\varphi)$ of the integral (6.8.1) must be zero. Since

$$\mathcal{L}\left(x, \varphi + \epsilon\eta, \frac{\partial \varphi}{\partial x} + \epsilon\frac{\partial \eta}{\partial x}\right) = \mathcal{L}\left(x, \varphi, \frac{\partial \varphi}{\partial x}\right) + \epsilon\eta\frac{\partial \mathcal{L}}{\partial \varphi} + \epsilon\sum_{i=1}^{n}\frac{\partial \eta}{\partial x_i}\frac{\partial \mathcal{L}}{\partial \varphi_{,i}} + \cdots,$$

where we used the notation $\varphi_{,i} = \partial\varphi/\partial x_i$, the first variation of $J(\varphi)$ is:

$$\delta J(\varphi) = \epsilon \int_{D_n}\left(\eta\frac{\partial \mathcal{L}}{\partial \varphi} + \sum_{i=1}^{n}\frac{\partial \eta}{\partial x_i}\frac{\partial \mathcal{L}}{\partial \varphi_{,i}}\right) d\Omega, \qquad (6.8.3)$$

with $d\Omega = dx_1 dx_2 \dots dx_n$.

Integrating by parts the second term in (6.8.3), we have:

$$\int_{D_n} \sum_{i=1}^{n} \frac{\partial \eta}{\partial x_i} \frac{\partial \mathcal{L}}{\partial \varphi_{,i}} d\Omega = \int_{D_n} \sum_{i=1}^{n} \frac{\partial}{\partial x_i} \left(\eta \frac{\partial \mathcal{L}}{\partial \varphi_{,i}} \right) d\Omega - \int_{D_n} \eta \sum_{i=1}^{n} \frac{\partial}{\partial x_i} \left(\frac{\partial \mathcal{L}}{\partial \varphi_{,i}} \right) d\Omega. \tag{6.8.4}$$

But

$$d\Omega = dx_1 \dots dx_k \dots dx_n = dx_k dS_k \quad \text{(no summation)}, \tag{6.8.5}$$

where $dS_k = dx_1 \dots dx_{k-1} dx_{k+1} \dots dx_n$ is the element of hypersurface orthogonal to dx_k. Using the Green–Gauss theorem and the boundary condition (6.8.2), we see that the first integral on the r.h.s. of (6.8.4) vanishes:

$$\int_{D_n} \sum_{i=1}^{n} \frac{\partial}{\partial x_i} \left(\eta \frac{\partial \mathcal{L}}{\partial \varphi_{,i}} \right) d\Omega = \int_{S_{n-1}} \sum_{i=1}^{n} \eta \frac{\partial \mathcal{L}}{\partial \varphi_{,i}} dS_i = 0.$$

Thus, the first variation of $J(\varphi)$ is:

$$\delta J(\varphi) = \epsilon \int_{D_n} \eta \left[\frac{\partial \mathcal{L}}{\partial \varphi} - \sum_{i=1}^{n} \frac{\partial}{\partial x_i} \left(\frac{\partial \mathcal{L}}{\partial \varphi_{,i}} \right) \right] d\Omega.$$

Since the function η is arbitrary (up to the condition (6.8.2)), the necessary and sufficient condition for a stationary value of $J(\varphi)$ is

$$\frac{\partial \mathcal{L}}{\partial \varphi} - \sum_{i=1}^{n} \frac{\partial}{\partial x_i} \left(\frac{\partial \mathcal{L}}{\partial \varphi_{,i}} \right) = 0. \tag{6.8.6}$$

Assuming now that $\mathcal{L}$ is a function of h variables $\varphi^{(1)}(x), \dots, \varphi^{(h)}(x)$, let us consider the functional

$$J[\varphi^{(s)}] = \int_{D_n} \mathcal{L}[x, \varphi^{(s)}(x), \varphi^{(s)}_{,x}]\, dx_1 \dots dx_n \quad (s = \overline{1,h}). \tag{6.8.7}$$

Using a similar procedure as for a single variable φ, the stationarity condition for the functional $J[\varphi^{(s)}]$ yields the following system of second-order partial differential equations:

$$\frac{\partial \mathcal{L}}{\partial \varphi^{(s)}} - \sum_{i=1}^{n} \frac{\partial}{\partial x_i} \left(\frac{\partial \mathcal{L}}{\partial \varphi^{(s)}_{,i}} \right) = 0 \quad (s = \overline{1,h}), \tag{6.8.8}$$

called the *Euler–Lagrange equations* of the continuous system.

In order to use these equations in CDM mechanics, we choose

$$x_1 = x, \quad x_2 = y, \quad x_3 = z, \quad x_4 = t. \tag{6.8.9}$$

With this choice, the functional (6.8.7) becomes:

$$J[\varphi^{(s)}] = \int_{t_1}^{t_2}\int\int\int \mathcal{L}[x, y, z, t, \varphi^{(s)}(x, y, z, t), \varphi^{(s)}_{,x}, \ldots, \varphi^{(s)}_{,t}]\, dx\, dy\, dz\, dt \quad (s = \overline{1, h}) \tag{6.8.10}$$

and the Euler–Lagrange equations read:

$$\frac{\partial \mathcal{L}}{\partial \varphi^{(s)}} - \frac{\partial}{\partial x_i}\left(\frac{\partial \mathcal{L}}{\partial \varphi^{(s)}_{,i}}\right) - \frac{\partial}{\partial t}\left(\frac{\partial \mathcal{L}}{\partial \varphi^{(s)}_{,t}}\right) = 0 \quad (s = \overline{1, h}), \tag{6.8.11}$$

where the summation convention has been used for the index $i = 1, 2, 3$.

Comparing (6.8.10) with the action integral (2.7.17), we realize the equivalence between them if we choose

$$L = \int_V \mathcal{L}\, dx\, dy\, dz = \int_V \mathcal{L}\, d\tau. \tag{6.8.12}$$

Therefore the function $\mathcal{L}$ stands for the Lagrangian per unit volume, i.e. the *Lagrangian density*.

With this notation, Hamilton's principle (2.7.16) reads:

$$\delta \int_{t_1}^{t_2}\int_V \mathcal{L}\, d\tau dt = 0, \tag{6.8.13}$$

and can be used as a fundamental postulate in the study of holonomic CDM, while the Euler–Lagrange equations (6.8.11) are the equations of motion of these systems.

Comparing Eqs. (6.8.8) with the Lagrange equations for systems with a finite number of degrees of freedom (2.5.17), we realize that they are different in certain respects. In case of CDM the role of generalized coordinates is played by the functions $\varphi^{(s)}$, called *dependent variables* or *variational parameters*, while $x_1, .., x_n$ play the role of *independent variables*. Our choice (6.8.9) shows that both the space coordinates x, y, z and the time t are now taken as independent parameters, while $\varphi^{(s)}$ are selected from the physical variables which characterize a given system. In view of these considerations, Eqs. (6.8.8) can be regarded as an infinite chain of Lagrange-type differential equations, each of them being obtained by a successive fixation of space variables x, y, z. Since $\mathcal{L}$ is a Lagrangian density, all quantities appearing in it must be represented by their densities, such as: mass density ρ, entropy density s, current density $\mathbf{j}$, etc.

Equations (6.8.8) are particularly useful in the analytical formalism of CDM, because they can be applied not only to the study of condensed media (solid, fluid), but also in the derivation of fundamental equations governing the *fields*.

Observation: The Euler–Lagrange equations (6.8.8) do not change their form if instead of $\mathcal{L}$ we choose

$$\mathcal{L}'(x, \varphi^{(s)}, \varphi^{(s)}_{,i}) = \mathcal{L}(x, \varphi^{(s)}, \varphi^{(s)}_{,i}) + \sum_{k=1}^{n} \frac{\partial}{\partial x_k} F_k(x, \varphi^{(s)}), \tag{6.8.14}$$

provided that the integration domain D_n remains unchanged and the field variables $\varphi^{(s)}$ take fixed values on the boundary S_{n-1} of D_n. To prove this, we integrate the last relation on D_n and, using the generalized Green–Gauss theorem, we obtain:

$$\int_{D_n} \mathcal{L}'\, d\Omega = \int_{D_n} \mathcal{L}\, d\Omega + \int_{D_n} \sum_{k=1}^{n} \frac{\partial F_k}{\partial x_k} d\Omega = \int_{D_n} \mathcal{L}\, d\Omega + \int_{S_{n-1}} \sum_{k=1}^{n} F_k\, dS_k.$$

Applying now the operator δ to this relation, we have:

$$\delta \int_{D_n} \mathcal{L}'\, d\Omega = \delta \int_{D_n} \mathcal{L}\, d\Omega + \int_{S_{n-1}} \sum_{i=1}^{n} \sum_{k=1}^{n} \left[\frac{\partial F_k}{\partial x_i} \delta x_i + \frac{\partial F_k}{\partial \varphi^{(s)}} \delta \varphi^{(s)} \right] dS_k.$$

But, by hypothesis, on the boundary S_{n-1} we have $\delta x_i = 0$, $\delta \varphi^{(s)} = 0$, therefore

$$\delta \int_{D_n} \mathcal{L}'\, d\Omega = \delta \int_{D_n} \mathcal{L}\, d\Omega,$$

which means that the condition of stationarity for $J[\varphi^{(s)}]$,

$$\delta J[\varphi^{(s)}] = \delta \int_{D_n} \mathcal{L}\, d\Omega = 0,$$

does not change upon the transformation (6.8.14). As a results, the Euler–Lagrange equations (6.8.8) do not change their form. In other words, two Lagrangian densities which differ from one another by a divergence term are *equivalent*.

6.8.2 Applications

6.8.2.1 Lamé's Equations

As a first example, let us obtain by means of the Lagrangian formalism the equations of motion of an isotropic and homogeneous elastic medium (6.5.26), within the frame of the linear approximation.

Recalling that in the study of an elastic medium the Lagrange variables are usually used, the velocity and acceleration fields are:

$$\mathbf{v} = \frac{\partial \mathbf{u}}{\partial t}, \quad \mathbf{a} = \frac{\partial \mathbf{v}}{\partial t} = \frac{\partial^2 \mathbf{u}}{\partial t^2}, \tag{6.8.15}$$

where $\mathbf{u}$ is the field of infinitesimal displacements. Since the displacements are small, the mass density ρ can be taken as a constant (see (6.5.25)).

Denoting by $\mathcal{T}$ and $\mathcal{V}$ the kinetic and the potential energy densities of the elastic medium, the Lagrangian density is $\mathcal{L} = \mathcal{T} - \mathcal{V}$.

Writing the kinetic energy density $\mathcal{T}$ is an easy matter. By means of (6.8.15), we obtain:

$$\mathcal{T} = \frac{1}{2}\rho \mathbf{v}^2 = \frac{1}{2}\rho \left|\frac{\partial \mathbf{u}}{\partial t}\right|^2 = \frac{1}{2}\rho u_{i,t} u_{i,t}, \tag{6.8.16}$$

where $u_{i,t}$ stands for $\partial u_i / \partial t$ and the summation convention has been used.

The potential energy density is formed by two parts: (a) the potential energy density of deformation $\mathcal{V}_1$; (b) the potential energy density of the external body forces $\mathcal{V}_2$. Let us deduce these quantities in a form suitable for an action principle.

(a) Consider an elastic body which, under the action of both body forces $\mathbf{F}\Delta m$ and superficial forces $\mathbf{T}\Delta S$, is brought to a deformed state. If we neglect the heat transformations, we can assume that the energy of deformation equals the sum of works done by body and superficial forces. Supposing that the displacement vector $\mathbf{u}$ performs an elementary variation $\delta \mathbf{u}$ compatible with the constraints, the variation of the deformation energy V_1 is:

$$\delta V_1 = \int_V \delta \mathcal{V}_1 \, d\tau = \int_V \mathbf{F} \cdot \delta \mathbf{u} \rho \, d\tau + \int_S \mathbf{T} \cdot \delta \mathbf{u} \, dS.$$

Using Cauchy's formula (6.3.22) and the Green–Gauss theorem, we obtain:

$$\begin{aligned} \int_V \delta \mathcal{V}_1 \, d\tau &= \int_V F_i \delta u_i \rho \, d\tau + \int_V \frac{\partial}{\partial x_k}(T_{ik}\delta u_i) \, d\tau \\ &= \int_V \left(\rho F_i + \frac{\partial T_{ik}}{\partial x_k}\right) \delta u_i \, d\tau + \int_V T_{ik} \delta\left(\frac{\partial u_i}{\partial x_k}\right) d\tau. \end{aligned} \tag{6.8.17}$$

Since the deformation is static, from (6.3.23), we have:

$$\rho F_i + \frac{\partial T_{ik}}{\partial x_k} = 0,$$

therefore (see (6.4.5))

$$\delta \mathcal{V}_1 = T_{ik} \delta\left(\frac{\partial u_i}{\partial x_k}\right) = T_{ik}(\delta \omega_{ik} + \delta e_{ik}).$$

But as the stress tensor is symmetric and the rotation tensor is antisymmetric, the product $T_{ik}\delta\omega_{ik}$ is zero. Using (6.5.13) and (6.5.14), we then obtain:

$$\delta\mathcal{V}_1 = T_{ik}\delta e_{ik} = C_{ikjm}e_{jm}\delta e_{ik} = C_{jmik}e_{ik}\delta e_{jm} = \frac{1}{2}C_{ikjm}\delta(e_{ik}e_{jm}).$$

If the components of the elasticity tensor C_{ikjm} are constants, we find:

$$\mathcal{V}_1 = \frac{1}{2}C_{ikjm}e_{ik}e_{jm} = \frac{1}{2}T_{ik}e_{ik}.$$

Recalling that the medium is isotropic, the components of the tensor T_{ik} are related to e_{ik} by Hooke's law (6.5.11), and finally we find the deformation energy density in the form:

$$\mathcal{V}_1 = \frac{1}{2}(\lambda\theta\delta_{ik} + 2\mu e_{ik})e_{ik} = \frac{1}{2}\lambda\theta^2 + \mu e_{ik}e_{ik}. \tag{6.8.18}$$

(b) If on the elastic medium acts a conservative and homogeneous force field (for example, the gravitational field), the potential energy V_2 of the body forces $\mathbf{F}$ is

$$V_2 = \int_V \mathcal{V}_2\,d\tau = -\int_V F_i u_i \rho\,d\tau,$$

which yields:

$$\mathcal{V}_2 = -\rho F_i u_i. \tag{6.8.19}$$

In view of (6.8.18) and (6.8.19), the potential energy density $\mathcal{V}$ is:

$$\mathcal{V} = \mathcal{V}_1 + \mathcal{V}_2 = \frac{1}{2}\lambda\theta^2 + \mu e_{ik}e_{ik} - \rho F_i u_i. \tag{6.8.20}$$

Now, we are able to write the Lagrangian density:

$$\mathcal{L} = \mathcal{T} - \mathcal{V} = \frac{1}{2}\rho u_{i,t}u_{i,t} - \frac{1}{2}\lambda\theta^2 - \mu e_{ik}e_{ik} + \rho F_i u_i.$$

But

$$\theta^2 = e_{ii}e_{kk} = \frac{\partial u_i}{\partial x_i}\frac{\partial u_k}{\partial x_k} = u_{i,i}u_{k,k},$$

$$e_{ik}e_{ik} = \frac{1}{4}\left(\frac{\partial u_k}{\partial x_i} + \frac{\partial u_i}{\partial x_k}\right)\left(\frac{\partial u_k}{\partial x_i} + \frac{\partial u_i}{\partial x_k}\right) = \frac{1}{2}u_{k,i}u_{k,i} + \frac{1}{2}u_{k,i}u_{i,k},$$

and the Lagrangian density finally acquires the form:

$$\mathcal{L}(x,u,u_{,x}) = \frac{1}{2}\rho u_{i,t}u_{i,t} - \frac{1}{2}\lambda u_{i,i}u_{k,k} - \frac{1}{2}\mu u_{k,i}u_{i,k} - \frac{1}{2}\mu u_{k,i}u_{k,i} + \rho F_i u_i. \quad (6.8.21)$$

Once the Lagrangian density is known, the next step is to use the Euler–Lagrange equations (6.8.11). Observing that in our case the variational parameters $\varphi^{(s)}$ are the components u_j $(j = 1,2,3)$ of the elementary displacement $\mathbf{u}$, the Euler–Lagrange equations read:

$$\frac{\partial \mathcal{L}}{\partial u_j} - \frac{\partial}{\partial x_m}\left(\frac{\partial \mathcal{L}}{\partial u_{j,m}}\right) - \frac{\partial}{\partial t}\left(\frac{\partial \mathcal{L}}{\partial u_{j,t}}\right) = 0 \qquad (j,m = 1,2,3). \qquad (6.8.22)$$

We obtain, successively:

$$\frac{\partial \mathcal{L}}{\partial u_j} = \rho F_i \delta_{ij} = \rho F_j,$$

$$\begin{aligned}\frac{\partial \mathcal{L}}{\partial u_{j,m}} &= -\frac{1}{2}\lambda(\delta_{ij}\delta_{im}u_{k,k} + \delta_{kj}\delta_{km}u_{i,i}) - \mu\delta_{kj}\delta_{im}u_{k,i} - \frac{1}{2}\mu(\delta_{kj}\delta_{im}u_{i,k} + \delta_{ij}\delta_{km}u_{k,i}) \\ &= -\lambda\theta\delta_{jm} - \mu(u_{j,m} + u_{m,j}),\end{aligned}$$

$$\frac{\partial}{\partial x_m}\left(\frac{\partial \mathcal{L}}{\partial u_{j,m}}\right) = -(\lambda+\mu)\frac{\partial\theta}{\partial x_j} - \mu\Delta u_j,$$

$$\frac{\partial \mathcal{L}}{\partial u_{j,t}} = \rho u_{i,t}\delta_{ij} = \rho u_{j,t},$$

$$\frac{\partial}{\partial t}\left(\frac{\partial \mathcal{L}}{\partial u_{j,t}}\right) = \rho u_{j,tt}.$$

Introducing these results into (6.8.22), we are led to:

$$\rho F_j + (\lambda+\mu)\frac{\partial\theta}{\partial x_j} + \mu\Delta u_j - \rho\frac{\partial^2 u_j}{\partial t^2} = 0 \quad (j = 1,2,3),$$

which are indeed Lamé's equations (6.5.26).

6.8.2.2 Euler's Equations

Consider an ideal, compressible fluid, which performs an isentropic motion in an external potential field.[1]

[1] Herivel, J.W.: The derivation of the equation of motion of an ideal fluid by Hamilton's principle. Proc. Camb. Phil. Soc. **51**, 344 (1955).

Hereafter we shall denote by $\mathcal{V}^*(\mathbf{r},t)$, $s(\mathbf{r},t)$ and $\varepsilon(\mathbf{r},t)$ the potential of the exterior field (e.g. the gravitational field), the entropy and the internal energy, taken per unit mass, respectively. Then the kinetic energy density is $\frac{1}{2}\rho|\mathbf{v}|^2$, where $\mathbf{v}(\mathbf{r}, t)$ is the velocity field, while the potential energy density is composed of two terms, $\rho\varepsilon$ and $\rho\mathcal{V}^*$, corresponding to the internal and external forces, respectively. Nevertheless, the expression

$$\mathcal{L}_0 = \frac{1}{2}\rho\mathbf{v}^2 - \rho(\varepsilon + \mathcal{V}^*) \tag{6.8.23}$$

cannot be used as a Lagrangian density, because it contains only some of the physical variables which define the system. In turn, this is due to the fact that we did not take into consideration the *constraints* acting on the fluid, which in our case are the equation of continuity (6.3.4) and the equation of conservation of entropy (6.6.47).

A suitable Lagrangian density is constructed by using the method of Lagrangian multipliers. To this end, we amplify the constraint Eqs. (6.3.4) and (6.6.47) by the multipliers $\alpha(\mathbf{r}, t)$ and $\beta(\mathbf{r}, t)$, respectively, and add the result to (6.8.23). This yields:

$$\mathcal{L} = \frac{1}{2}\rho\mathbf{v}^2 - \rho(\varepsilon + \mathcal{V}^*) - \alpha\left[\frac{\partial\rho}{\partial t} + \nabla\cdot(\rho\mathbf{v})\right] - \beta\rho\left(\frac{\partial s}{\partial t} + \mathbf{v}\cdot\nabla s\right). \tag{6.8.24}$$

It is more convenient for our purpose to use the Lagrangian density in a slightly modified form. Taking advantage of the property (6.8.14), we shall add to (6.8.24) the divergence

$$\frac{\partial}{\partial x_j}(\alpha\rho v_j) \quad (j = \overline{1,4}),$$

where we choose $x_1 = x,\ x_2 = y,\ x_3 = z,\ x_4 = t,\ v_1 = v_x,\ v_2 = v_y,\ v_3 = v_z,\ v_4 = 1.$ Since

$$-\alpha\left(\frac{\partial\rho}{\partial t} + \mathbf{v}\cdot\nabla\rho + \rho\nabla\cdot\mathbf{v}\right) + \nabla\cdot(\alpha\rho\mathbf{v}) + \frac{\partial}{\partial t}(\alpha\rho) = \rho\left(\frac{\partial\alpha}{\partial t} + \mathbf{v}\cdot\nabla\alpha\right),$$

we finally obtain:

$$\mathcal{L} = \frac{1}{2}\rho\mathbf{v}^2 - \rho(\varepsilon + \mathcal{V}^*) + \rho\left(\frac{\partial\alpha}{\partial t} + \mathbf{v}\cdot\nabla\alpha\right) - \beta\rho\left(\frac{\partial s}{\partial t} + \mathbf{v}\cdot\nabla s\right). \tag{6.8.25}$$

Choosing $s,\ \rho,\ v_x,\ v_y,\ v_z$ as the variational parameters $\varphi^{(i)}, i = 1,2,3,4,5$, in (6.8.11), we then have:

(i) $\varphi^{(1)} = s$. The corresponding Euler–Lagrange equation is

$$\frac{\partial\mathcal{L}}{\partial s} - \frac{\partial}{\partial x_i}\left(\frac{\partial\mathcal{L}}{\partial s_{,i}}\right) - \frac{\partial}{\partial t}\left(\frac{\partial\mathcal{L}}{\partial s_{,t}}\right) = 0. \tag{6.8.26}$$

Using the fundamental equation of the thermodynamics of equilibrium processes (6.6.50),

$$T\,ds = d\varepsilon(\rho, s) + p\,d\left(\frac{1}{\rho}\right),$$

we find:

$$\frac{\partial \mathcal{L}}{\partial s} = \frac{\partial \mathcal{L}}{\partial \varepsilon}\frac{\partial \varepsilon}{\partial s} = -\rho T, \qquad \frac{\partial \mathcal{L}}{\partial s_{,i}} = -\beta\rho v_i, \qquad \frac{\partial \mathcal{L}}{\partial s_{,t}} = -\beta\rho.$$

Introducing these results into (6.8.26) and using the equation of continuity, after simplifying by $\rho \neq 0$ we arrive at:

$$\frac{\partial \beta}{\partial t} + \mathbf{v}\cdot\nabla\beta = T. \tag{6.8.27}$$

(ii) $\varphi^{(2)} = \rho$. We have

$$\frac{\partial \mathcal{L}}{\partial \rho} - \frac{\partial}{\partial x_i}\left(\frac{\partial \mathcal{L}}{\partial \rho_{,i}}\right) - \frac{\partial}{\partial t}\left(\frac{\partial \mathcal{L}}{\partial \rho_{,t}}\right) = 0.$$

Performing the derivatives:

$$\frac{\partial \mathcal{L}}{\partial \rho} = \frac{1}{2}\mathbf{v}^2 - (\varepsilon + \mathcal{V}^*) - \frac{p}{\rho} + \frac{\partial \alpha}{\partial t} + \mathbf{v}\cdot\nabla\alpha,$$

$$\frac{\partial \mathcal{L}}{\partial \rho_{,i}} = 0, \quad \frac{\partial \mathcal{L}}{\partial \rho_{,t}} = 0,$$

we obtain:

$$\frac{1}{2}\mathbf{v}^2 - (\varepsilon + \mathcal{V}^*) - \frac{p}{\rho} + \frac{\partial \alpha}{\partial t} + \mathbf{v}\cdot\nabla\alpha = 0, \tag{6.8.28}$$

which is a Bernoulli-type equation.

(iii) $\varphi^{(3,4,5)} = v_k$ $(k = 1,2,3)$. In this case, we have three equations:

$$\frac{\partial \mathcal{L}}{\partial v_k} - \frac{\partial}{\partial x_i}\left(\frac{\partial \mathcal{L}}{\partial v_{k,i}}\right) - \frac{\partial}{\partial t}\left(\frac{\partial \mathcal{L}}{\partial v_{k,t}}\right) = 0.$$

Since

$$\frac{\partial \mathcal{L}}{\partial v_k} = \rho v_k + \rho\frac{\partial \alpha}{\partial x_k} - \beta\rho\frac{\partial s}{\partial x_k},$$

$$\frac{\partial \mathcal{L}}{\partial v_{k,i}} = 0, \quad \frac{\partial \mathcal{L}}{\partial v_{k,t}} = 0,$$

we arrive at:

$$\mathbf{v} = -\nabla\alpha + \beta\nabla s, \tag{6.8.29}$$

which is a *Clebsch transformation*. Therefore, the functions $\alpha(\mathbf{r}, t)$, $\beta(\mathbf{r}, t)$ and s play the role of *Clebsch potentials*.

The last step is now to eliminate the multipliers α and β from Eqs. (6.8.27)–(6.8.29). To do this, we shall first replace $\nabla\alpha = -\mathbf{v} + \beta\nabla s$ into (6.8.28):

$$-\frac{1}{2}|\mathbf{v}|^2 - (\varepsilon + \mathcal{V}^*) - \frac{p}{\rho} + \frac{\partial\alpha}{\partial t} + \beta\mathbf{v}\cdot\nabla s = 0.$$

Applying to this equation the operator *gradient*, we have:

$$-\mathbf{v}\times\operatorname{curl}\mathbf{v} - (\mathbf{v}\cdot\nabla)\mathbf{v} - \frac{p}{\rho^2}\nabla\rho - T\nabla s - \nabla\mathcal{V}^* - \frac{1}{\rho}\nabla p + \frac{p}{\rho^2}\nabla\rho$$
$$+\frac{\partial}{\partial t}(\beta\nabla s - \mathbf{v}) + \beta\nabla(\mathbf{v}\cdot\nabla s) + (\mathbf{v}\cdot\nabla s)\nabla\beta = 0.$$

But, by virtue of (6.6.47) and (6.8.27),

$$\begin{aligned}
&-\mathbf{v}\times\operatorname{curl}\mathbf{v} - T\nabla s + \frac{\partial\beta}{\partial t}\nabla s + \beta\nabla\left(\frac{\partial s}{\partial t}\right) + \beta\nabla(\mathbf{v}\cdot\nabla s) + (\mathbf{v}\cdot\nabla s)\nabla\beta\\
&= -\mathbf{v}\times\operatorname{curl}(\beta\nabla s) - T\nabla s - (\mathbf{v}\cdot\nabla\beta)\nabla s + T\nabla s + (\mathbf{v}\cdot\nabla s)\nabla\beta\\
&= -\mathbf{v}\times(\nabla\beta\times\nabla s) - (\mathbf{v}\cdot\nabla\beta)\nabla s + (\mathbf{v}\cdot\nabla s)\nabla\beta = 0,
\end{aligned}$$

therefore we obtain Euler's equation (6.6.8):

$$\frac{\partial\mathbf{v}}{\partial t} + (\mathbf{v}\cdot\nabla)\mathbf{v} = -\nabla\mathcal{V}^* - \frac{1}{\rho}\nabla p = \mathbf{F} - \frac{1}{\rho}\nabla p.$$

Observation: Before going further, we wish to make some remarks on the constraints used to construct the Lagrangian density (6.8.25). From the hydrodynamic point of view, there are motions consistent with the dynamic equations which are not included in this principle. Indeed, if the specific entropy is homogeneous in space, Eq. (6.8.29) leads to $\mathbf{v} = -\operatorname{grad}\alpha$, meaning that in this case the motion is restricted to irrotational flows. To remove this difficulty, an additional vector constraint was introduced[2] expressing the conservation of the identity of particles, in the form $d\mathbf{X}/dt = 0$. Later, it was shown[3] that a single component of $\mathbf{X}$ and, consequently, a single equation of this type is enough to avoid the aforementioned restriction. This component is one of the Lagrangian coordinates of the particle,

[2] Lin, C.C.: Liquid Helium. In: Proceedings of the International School of Physics, Course XXI. Academic, New York (1963).

[3] Selinger, R.L., Whitham, G.B.: Proc. Roy. Soc. A **305**, 1 (1968).

even if the description of the motion is Eulerian. We wanted to emphasize this point in order to draw attention to the fact that our description is not very general.

6.8.2.3 Maxwell's Equations

As we have mentioned earlier, the Euler–Lagrange equations (6.8.8) are also useful in field theory. As an example, in the following we shall deduce the fundamental system of equations describing electromagnetic phenomena in a homogeneous and isotropic medium.

Let the electromagnetic field be defined by the vectors **E**, **B**, while the field sources are given by the conduction current density **j** and by the electric charge density ρ^e. Then Maxwell's equations are:

$$\operatorname{curl} \mathbf{E} = -\frac{\partial \mathbf{B}}{\partial t}, \quad \operatorname{div} \mathbf{B} = 0, \tag{6.8.30}$$

$$\operatorname{curl} \frac{\mathbf{B}}{\mu_0} = \mathbf{j} + \epsilon_0 \frac{\partial \mathbf{E}}{\partial t}, \quad \operatorname{div}(\epsilon_0 \mathbf{E}) = \rho^e. \tag{6.8.31}$$

Here we separated the source-free equations (6.8.30) from the source equations (6.8.31). We assumed again that $\epsilon \simeq \epsilon_0$, $\mu \simeq \mu_0$, where ϵ and μ are the permittivity and the permeability of the medium, respectively.

The Lagrangian density of the system formed by the electromagnetic field and the sources is composed of two terms:

$$\mathcal{L} = \mathcal{L}_0 + \mathcal{L}_{int},$$

where

$$\mathcal{L}_0 = \frac{1}{2}\epsilon_0 |\mathbf{E}|^2 - \frac{1}{2\mu_0}|\mathbf{B}|^2 \tag{6.8.32}$$

is the Lagrangian density of the electromagnetic field when sources are absent, and

$$\mathcal{L}_{int} = -\rho^e \phi + \mathbf{j} \cdot \mathbf{A} \tag{6.8.33}$$

is the Lagrangian density which expresses the interaction between sources and the field. The first was derived by *Joseph Larmor*, while the second is obtained from the Lagrangian of interaction (see (2.5.29)) per unit volume. Writing the electromagnetic field **E**, **B** in terms of the vector and scalar potentials **A**, ϕ (see (2.5.26)), we obtain the following Lagrangian density:

$$\mathcal{L} = \frac{1}{2}\epsilon_0 \left| -\nabla\phi - \frac{\partial \mathbf{A}}{\partial t} \right|^2 - \frac{1}{2\mu_0}|\nabla \times \mathbf{A}|^2 - \rho^e \phi + \mathbf{j} \cdot \mathbf{A}. \tag{6.8.34}$$

Taking as variational parameters A_k $(k = 1, 2, 3)$ and ϕ, we have successively:

(i) $\varphi^{(1,2,3)} = A_k$. The Euler–Lagrange equations read:

$$\frac{\partial \mathcal{L}}{\partial A_k} - \frac{\partial}{\partial x_i}\left(\frac{\partial \mathcal{L}}{\partial A_{k,i}}\right) - \frac{\partial}{\partial t}\left(\frac{\partial \mathcal{L}}{\partial A_{k,t}}\right) = 0. \tag{6.8.35}$$

Since

$$E_m = -\phi_{,m} - A_{m,t}, \quad B_m = \epsilon_{msj} A_{j,s}, \tag{6.8.36}$$

we derive:

$$\frac{\partial \mathcal{L}}{\partial A_k} = j_k,$$

$$\frac{\partial \mathcal{L}}{\partial A_{k,i}} = \frac{\partial \mathcal{L}}{\partial B_m}\frac{\partial B_m}{\partial A_{k,i}} = -\frac{1}{\mu_0} B_m(\epsilon_{msj}\delta_{jk}\delta_{is}) = -\frac{1}{\mu_0}\epsilon_{ikm} B_m,$$

$$\frac{\partial}{\partial x_i}\left(\frac{\partial \mathcal{L}}{\partial A_{k,i}}\right) = -\frac{1}{\mu_0}\epsilon_{ikm} B_{m,i} = \frac{1}{\mu_0}(\operatorname{curl}\mathbf{B})_k,$$

$$\frac{\partial \mathcal{L}}{\partial A_{k,t}} = \frac{\partial \mathcal{L}}{\partial E_m}\frac{\partial E_m}{\partial A_{k,t}} = \epsilon_0 E_m(-\delta_{mk}) = -\epsilon_0 E_k,$$

$$\frac{\partial}{\partial t}\left(\frac{\partial \mathcal{L}}{\partial A_{k,t}}\right) = -\epsilon_0 \frac{\partial E_k}{\partial t}.$$

Introducing these results into (6.8.35), we obtain:

$$\frac{1}{\mu_0}(\operatorname{curl}\mathbf{B})_k = j_k + \epsilon_0 \frac{\partial E_k}{\partial t},$$

which is the k-component of Maxwell's equation $(6.8.31)_1$.

(ii) $\varphi^{(4)} = \phi$. We obtain a single equation:

$$\frac{\partial \mathcal{L}}{\partial \phi} - \frac{\partial}{\partial x_i}\left(\frac{\partial \mathcal{L}}{\partial \phi_{,i}}\right) - \frac{\partial}{\partial t}\left(\frac{\partial \mathcal{L}}{\partial \phi_{,t}}\right) = 0. \tag{6.8.37}$$

Performing the derivatives, we have:

$$\frac{\partial \mathcal{L}}{\partial \phi} = -\rho^e,$$

$$\frac{\partial \mathcal{L}}{\partial \phi_{,i}} = \frac{\partial \mathcal{L}}{\partial E_m}\frac{\partial E_m}{\partial \phi_{,i}} = \epsilon_0 E_m(-\delta_{im}) = -\epsilon_0\, E_i,$$

$$\frac{\partial}{\partial x_i}\left(\frac{\partial \mathcal{L}}{\partial \phi_{,i}}\right) = -\epsilon_0\, E_{i,i}$$

$$\frac{\partial \mathcal{L}}{\partial \phi_{,t}} = 0.$$

With these results, Eq. (6.8.37) becomes:

$$\epsilon_0 E_{i,i} = \rho^e,$$

i.e. Eq. $(6.8.31)_2$.

Observation: The variational problem can be put in a different way: given Maxwell's source equations (6.8.31), find a Lagrangian density leading to the relations (6.8.36) between the field and the potentials. To do this, we use the expression

$$\mathcal{L} = \frac{1}{2}\epsilon_0 |\mathbf{E}|^2 - \frac{1}{2\mu_0}|\mathbf{B}|^2 - \phi(\epsilon_0 \operatorname{div} \mathbf{E} - \rho^e) + \mathbf{A} \cdot \left(\frac{1}{\mu_0} \operatorname{curl} \mathbf{B} - \epsilon_0 \frac{\partial \mathbf{E}}{\partial t} - \mathbf{j} \right),$$

where $\mathbf{A}(\mathbf{r}, t)$ and $\phi(\mathbf{r}, t)$ are Lagrange multipliers. This time, the role of variational parameters is played by the field components E_i, B_i $(i = 1, 2, 3)$, while Maxwell's source equations are used as four *constraints* acting on the field. Choosing $\varphi^{(1,2,3)} = E_i$, $\varphi^{(4,5,6)} = B_i$, we find (6.8.36), and the source-free Maxwell's equations follow immediately. The calculation is left to the reader.

6.8.2.4 Schrödinger's Equation

As a last application of the Lagrangian formalism, let us find a suitable Lagrangian density leading to the well-known Schrödinger's equation for the wave associated to a microparticle, which is fundamental in quantum mechanics.

Denoting by ψ the wave function and by ψ^* its complex conjugate, we shall use the following Lagrangian density:

$$\mathcal{L} = \frac{\hbar^2}{2m} \nabla\psi \cdot \nabla\psi^* + V\psi\psi^* + \frac{\hbar}{2i}(\psi^*\psi_{,t} - \psi\psi^*_{,t}). \tag{6.8.38}$$

The Lagrangian density must be a *real* function, that is why the functions ψ, ψ^*, as well as their derivatives $\psi_{,i}$, $\psi^*_{,i}$, $\psi_{,t}$, $\psi^*_{,t}$ appear only in suitably chosen products.

Applying the Euler–Lagrange equations, we have:

$$\frac{\partial \mathcal{L}}{\partial \psi^*} = V\psi + \frac{\hbar}{2i}\psi_{,t}, \quad \frac{\partial \mathcal{L}}{\partial \psi^*_{,t}} = -\frac{\hbar}{2i}\psi, \quad \frac{\partial \mathcal{L}}{\partial (\nabla\psi)} = \frac{\hbar^2}{2m}\nabla\psi,$$

which yield indeed Schrödinger's equation:

$$\left(-\frac{\hbar^2}{2m}\Delta + V\right)\psi = -\frac{\hbar}{i}\frac{\partial \psi}{\partial t},$$

which can be written also as

$$\hat{H}\psi = \hat{E}\psi, \tag{6.8.39}$$

where $\hat{H} = -\frac{\hbar^2}{2m}\Delta + V$ is the *Hamiltonian operator* and $\hat{E} = -\frac{\hbar}{i}\frac{\partial}{\partial t}$ is the *energy operator*. In a similar way, performing the derivatives with respect to ψ, $\psi_{,i}$, $\psi_{,t}$, we obtain Schrödinger's equation for ψ^*:

$$\hat{H}\psi^* = \hat{E}^*\psi^*.$$

6.9 Hamiltonian Formalism

6.9.1 Hamilton's Canonical Equations for Continuous Systems

By analogy with Hamilton's function (5.1.20),

$$H(q, p, t) = \sum_{j=1}^{n} p_j \dot{q}_j - L(q, \dot{q}, t),$$

where $q_1, .., q_n$ are the generalized coordinates, $\dot{q}_1, .., \dot{q}_n$ – the generalized velocities, $p_1, .., p_n$ – the generalized momenta, and L – the Lagrangian, we define the *Hamiltonian density* (i.e. the Hamiltonian per unit volume) $\mathcal{H}$ by

$$\mathcal{H} = \sum_{s=1}^{h} \pi_{(s)} \varphi^{(s)}_{,t} - \mathcal{L}, \tag{6.9.1}$$

where $\mathcal{L}$ is the Lagrangian density, $\varphi^{(s)}_{,t}$ are the partial derivatives with respect to time of the variational parameters $\varphi^{(s)}$, and $\pi_{(s)}$ stand for the *momentum densities* conjugated with $\varphi^{(s)}$:

$$\pi_{(s)} = \frac{\partial \mathcal{L}}{\partial \varphi^{(s)}_{,t}}. \tag{6.9.2}$$

If we choose $x_1 = x$, $x_2 = y$, $x_3 = z$, $x_4 = t$, the functional dependence of the Lagrangian density $\mathcal{L}$ will be:

$$\mathcal{L} = \mathcal{L}[x_i, t, \varphi^{(s)}(x_i, t), \varphi^{(s)}_{,i}, \varphi^{(s)}_{,t}] \quad (i = 1, 2, 3;\ s = \overline{1, h}), \tag{6.9.3}$$

where $\varphi^{(s)}_{,i} = \partial \varphi^{(s)} / \partial x_i$. In this case, we have:

$$\mathcal{H} = \mathcal{H}[x_i, t, \varphi^{(s)}(x_i, t), \varphi^{(s)}_{,i}, \pi_{(s)}] \quad (i = 1, 2, 3;\ s = \overline{1, h}). \tag{6.9.4}$$

The Hamiltonian of a continuous system will be then:

$$H = \int_V \mathcal{H}[x_i, t, \varphi^{(s)}, \varphi^{(s)}_{,i}, \pi_{(s)}]\, d\tau. \tag{6.9.5}$$

Recalling that $\varphi^{(s)}$ are continuous and derivable functions of the independent variables x, y, z, t, let us perform an arbitrary variation δH of H for some fixed values of x, y, z ($\delta x_i = 0$). Using the expression (6.9.5) and the summation convention for the index i = 1, 2, 3, we obtain:

$$\begin{aligned} \delta H &= \int_V \delta \mathcal{H}\, d\tau \\ &= \int_V \left[\frac{\partial \mathcal{H}}{\partial t}\delta t + \sum_{s=1}^{h} \frac{\partial \mathcal{H}}{\partial \varphi^{(s)}}\delta\varphi^{(s)} + \sum_{s=1}^{h} \frac{\partial \mathcal{H}}{\partial \varphi^{(s)}_{,i}}\delta\varphi^{(s)}_{,i} + \sum_{s=1}^{h} \frac{\partial \mathcal{H}}{\partial \pi_{(s)}}\delta\pi_{(s)} \right] d\tau. \end{aligned} \tag{6.9.6}$$

The variation δH can also be written in an alternative form. Using (6.9.1) and (6.9.3), we find:

$$\begin{aligned} \delta H = \int_V \delta\left(\sum_{s=1}^{h} \pi_{(s)}\varphi^{(s)}_{,t} - \mathcal{L} \right) d\tau = \int_V \Bigg[\sum_{s=1}^{h} \left(\pi_{(s)}\delta\varphi^{(s)}_{,t} + \varphi^{(s)}_{,t}\delta\pi_{(s)} \right) - \frac{\partial \mathcal{L}}{\partial t}\delta t \\ - \sum_{s=1}^{h} \left(\frac{\partial \mathcal{L}}{\partial \varphi^{(s)}}\delta\varphi^{(s)} + \frac{\partial \mathcal{L}}{\partial \varphi^{(s)}_{,t}}\delta\varphi^{(s)}_{,t} + \frac{\partial \mathcal{L}}{\partial \varphi^{(s)}_{,i}}\delta\varphi^{(s)}_{,i} \right) \Bigg] d\tau. \end{aligned} \tag{6.9.7}$$

But

$$\int_V \sum_{s=1}^{h} \frac{\partial \mathcal{L}}{\partial \varphi^{(s)}_{,i}}\delta\varphi^{(s)}_{,i}\, d\tau = \int_V \frac{\partial}{\partial x_i}\left(\sum_{s=1}^{h} \frac{\partial \mathcal{L}}{\partial \varphi^{(s)}_{,i}}\delta\varphi^{(s)} \right) d\tau - \int_V \sum_{s=1}^{h} \frac{\partial}{\partial x_i}\left(\frac{\partial \mathcal{L}}{\partial \varphi^{(s)}_{,i}} \right)\delta\varphi^{(s)}\, d\tau. \tag{6.9.8}$$

The first integral on the r.h.s can be transformed into a surface integral on the boundary S of the domain of volume V. Since $\varphi^{(s)}$ have fixed values on S, we obtain

$$\int_V \frac{\partial}{\partial x_i}\left(\sum_{s=1}^{h} \frac{\partial \mathcal{L}}{\partial \varphi^{(s)}_{,i}}\delta\varphi^{(s)} \right) d\tau = \oint_S \sum_{s=1}^{h} \frac{\partial \mathcal{L}}{\partial \varphi^{(s)}_{,i}}\delta\varphi^{(s)}\, dS_i = 0. \tag{6.9.9}$$

Introducing these results in (6.9.7) and performing some reduction of terms, by means of the Euler–Lagrange equations (6.8.8), we arrive at:

$$\delta H = \int_V \left\{ \sum_{s=1}^{h} \varphi^{(s)}_{,t} \delta\pi_{(s)} - \frac{\partial \mathcal{L}}{\partial t}\delta t - \sum_{s=1}^{h}\left[\frac{\partial \mathcal{L}}{\partial \varphi^{(s)}} - \frac{\partial}{\partial x_i}\left(\frac{\partial \mathcal{L}}{\partial \varphi^{(s)}_{,i}}\right)\right]\delta\varphi^{(s)}\right\} d\tau$$

$$= \int_V \left[\sum_{s=1}^{h}\left(\varphi^{(s)}_{,t}\delta\pi_{(s)} - \pi_{(s),t}\delta\varphi^{(s)}\right) - \frac{\partial \mathcal{L}}{\partial t}\delta t\right] d\tau. \tag{6.9.10}$$

A similar integration by parts can be done in (6.9.6), yielding:

$$\delta H = \int_V \left\{ \sum_{s=1}^{h} \frac{\partial \mathcal{H}}{\partial \pi_{(s)}}\delta\pi_{(s)} + \frac{\partial \mathcal{H}}{\partial t}\delta t + \sum_{s=1}^{h}\left[\frac{\partial \mathcal{H}}{\partial \varphi^{(s)}} - \frac{\partial}{\partial x_i}\left(\frac{\partial \mathcal{H}}{\partial \varphi^{(s)}_{,i}}\right)\right]\delta\varphi^{(s)}\right\} d\tau. \tag{6.9.11}$$

Equating the coefficients of the same arbitrary variations $\delta\varphi^{(s)}$, $\delta\pi_{(s)}$ and δt in (6.9.10) and (6.9.11), we obtain the following system of equations:

$$\begin{aligned} \varphi^{(s)}_{,t} &= \frac{\partial \mathcal{H}}{\partial \pi_{(s)}}, \\ \pi_{(s),t} &= -\frac{\partial \mathcal{H}}{\partial \varphi^{(s)}} + \frac{\partial}{\partial x_i}\left(\frac{\partial \mathcal{H}}{\partial \varphi^{(s)}_{,i}}\right) \quad (i = 1,2,3;\ s = \overline{1,h}), \end{aligned} \tag{6.9.12}$$

as well as the identity:

$$\frac{\partial \mathcal{H}}{\partial t} = -\frac{\partial \mathcal{L}}{\partial t}. \tag{6.9.13}$$

The system of $2h$ partial derivative equations (6.9.12) is analogous to Hamilton's system of canonical equations (5.1.21), while the identity (6.9.13) is the local analogue of (5.1.22).

Equations (6.9.12) can be written in a symmetric form using the notion of *functional derivative*. To this end, we shall calculate the partial derivative of $L = \int_V \mathcal{L}\, d\tau$ with respect to $\varphi^{(s)}(x_j)$ $(j = \overline{1,4})$ for $x_1 = x$, $x_2 = y$, $x_3 = z$ fixed. Let us consider the family of functions $\varphi^{(s)}(x_j, \epsilon)$, where ϵ is a parameter chosen in such a way that

$$\varphi^{(s)}(x_j, \epsilon)|_S = \varphi^{(s)}(x_j, 0) \equiv \varphi^{(s)}(x_j), \tag{6.9.14}$$

where S is the closed surface bounding the domain of volume V. Consider the derivative:

$$\frac{dL}{d\epsilon} = \int_V \sum_{s=1}^{h}\left[\frac{\partial \mathcal{L}}{\partial \varphi^{(s)}}\frac{\partial \varphi^{(s)}}{\partial \epsilon} + \frac{\partial \mathcal{L}}{\partial \varphi^{(s)}_{,i}}\frac{\partial \varphi^{(s)}_{,i}}{\partial \epsilon}\right] d\tau.$$

But

$$\begin{aligned}\int_V \sum_{s=1}^h \frac{\partial \mathcal{L}}{\partial \varphi_{,i}^{(s)}} \frac{\partial \varphi_{,i}^{(s)}}{\partial \epsilon} d\tau &= \int_V \sum_{s=1}^h \frac{\partial \mathcal{L}}{\partial \varphi_{,i}^{(s)}} \frac{\partial}{\partial x_i}\left(\frac{\partial \varphi^{(s)}}{\partial \epsilon}\right) d\tau \\ &= \int_V \sum_{s=1}^h \frac{\partial}{\partial x_i}\left(\frac{\partial \mathcal{L}}{\partial \varphi_{,i}^{(s)}} \frac{\partial \varphi^{(s)}}{\partial \epsilon}\right) d\tau - \int_V \sum_{s=1}^h \frac{\partial}{\partial x_i}\left(\frac{\partial \mathcal{L}}{\partial \varphi_{,i}^{(s)}}\right) \frac{\partial \varphi^{(s)}}{\partial \epsilon} d\tau \\ &= \oint_S \sum_{s=1}^h \frac{\partial \mathcal{L}}{\partial \varphi_{,i}^{(s)}} \frac{\partial \varphi^{(s)}}{\partial \epsilon} dS_i - \int_V \sum_{s=1}^h \frac{\partial}{\partial x_i}\left(\frac{\partial \mathcal{L}}{\partial \varphi_{,i}^{(s)}}\right) \frac{\partial \varphi^{(s)}}{\partial \epsilon} d\tau.\end{aligned}$$

By condition (6.9.14), the surface integral vanishes, hence:

$$\frac{dL}{d\epsilon} = \int_V \sum_{s=1}^h \left[\frac{\partial \mathcal{L}}{\partial \varphi^{(s)}} - \frac{\partial}{\partial x_i}\left(\frac{\partial \mathcal{L}}{\partial \varphi_{,i}^{(s)}}\right)\right] \frac{\partial \varphi^{(s)}}{\partial \epsilon} d\tau. \qquad (6.9.15)$$

The choice of $\varphi^{(s)}(x_j, \epsilon)$ tells us that they are equal to $\varphi^{(s)}(x_j)$ everywhere, except for a vicinity Q (determined by ϵ) of the fixed point x, y, z. The same property holds for $\partial \varphi^{(s)}/\partial \epsilon$. In the limit $Q \to 0$, we can write

$$\lim_{Q \to 0} \frac{dL/d\epsilon}{\partial \varphi^{(s)}/\partial \epsilon} \equiv \frac{\delta L}{\delta \varphi^{(s)}} = \frac{\partial \mathcal{L}}{\partial \varphi^{(s)}} - \frac{\partial}{\partial x_i}\left(\frac{\partial \mathcal{L}}{\partial \varphi_{,i}^{(s)}}\right) \quad (i = \overline{1,4}). \qquad (6.9.16)$$

This expression is called the *functional derivative* or *variational derivative* of the Lagrangian L with respect to the field variables $\varphi^{(s)}$. Using this definition, the Euler–Lagrange equations (6.8.8) become:

$$\frac{\delta L}{\delta \varphi^{(s)}} = 0. \qquad (6.9.17)$$

Observing that $\mathcal{H}$ does not depend on $\partial \pi_{(s)}/\partial x_i$, the functional derivatives of the Hamiltonian H with respect to $\pi_{(s)}$ and $\varphi^{(s)}$ are:

$$\frac{\delta H}{\delta \pi_{(s)}} = \frac{\partial \mathcal{H}}{\partial \pi_{(s)}}, \quad \frac{\delta H}{\delta \varphi^{(s)}} = \frac{\partial \mathcal{H}}{\partial \varphi^{(s)}} - \frac{\partial}{\partial x_j}\left(\frac{\partial \mathcal{H}}{\partial \varphi_{,j}^{(s)}}\right) \quad (j = 1, 2, 3), \qquad (6.9.18)$$

allowing us to write equations (6.9.12) in a symmetric form:

$$\varphi_{,t}^{(s)} = \frac{\delta H}{\delta \pi_{(s)}}, \quad \pi_{(s),t} = -\frac{\delta H}{\delta \varphi^{(s)}}. \qquad (6.9.19)$$

The functional derivative serves, among other things, to define the *Poisson bracket for continuous systems*. To this end, let us consider the integral

$$F = \int_V \mathcal{F}[x_i, t, \varphi^{(s)}, \varphi^{(s)}_{,i}, \pi_{(s)}]\, d\tau \quad (i = 1, 2, 3), \tag{6.9.20}$$

where the *density* $\mathcal{F}$ of F is a function of the independent variables x_i, t, of the variational parameters $\varphi^{(s)}$ and their derivatives with respect to space coordinates, and of the momentum densities $\pi_{(s)}$. Assuming that the integration domain is fixed and that the variables $\varphi^{(s)}$, $\pi_{(s)}$ satisfy the system of canonical equations (6.9.19), we first take the total derivative with respect to time of (6.9.20):

$$\frac{dF}{dt} = \int_V \left[\frac{\partial \mathcal{F}}{\partial t} + \sum_{s=1}^{h}\left(\frac{\partial \mathcal{F}}{\partial \varphi^{(s)}}\varphi^{(s)}_{,t} + \frac{\partial \mathcal{F}}{\partial \varphi^{(s)}_{,i}}\varphi^{(s)}_{,it} + \frac{\partial \mathcal{F}}{\partial \pi_{(s)}}\pi_{(s),t}\right)\right] d\tau. \tag{6.9.21}$$

But

$$\int_V \sum_{s=1}^{h} \frac{\partial \mathcal{F}}{\partial \varphi^{(s)}_{,i}}\varphi^{(s)}_{,it}\, d\tau = \int_V \frac{\partial}{\partial x_i}\left(\sum_{s=1}^{h}\frac{\partial \mathcal{F}}{\partial \varphi^{(s)}_{,i}}\varphi^{(s)}_{,t}\right) d\tau - \int_V \sum_{s=1}^{h}\frac{\partial}{\partial x_i}\left(\frac{\partial \mathcal{F}}{\partial \varphi^{(s)}_{,i}}\right)\varphi^{(s)}_{,t}\, d\tau.$$

Using the Green–Gauss theorem, the first integral on the r.h.s. transforms into the surface integral

$$\oint_S \sum_{s=1}^{h}\frac{\partial \mathcal{F}}{\partial \varphi^{(s)}_{,i}}\varphi^{(s)}_{,t}\, dS_i,$$

which vanishes due to the boundary conditions. Using the definition of the functional derivative in the expression (6.9.21), we obtain:

$$\frac{dF}{dt} = \int_V \left[\frac{\partial \mathcal{F}}{\partial t} + \sum_{s=1}^{h}\left(\frac{\delta F}{\delta \varphi^{(s)}}\varphi^{(s)}_{,t} + \frac{\delta F}{\delta \pi_{(s)}}\pi_{(s),t}\right)\right] d\tau,$$

or, by virtue of the canonical equations (6.9.19),

$$\frac{dF}{dt} = \int_V \frac{\partial \mathcal{F}}{\partial t}\, d\tau + \int_V \sum_{s=1}^{h}\left[\frac{\delta F}{\delta \varphi^{(s)}}\frac{\delta H}{\delta \pi_{(s)}} - \frac{\delta F}{\delta \pi_{(s)}}\frac{\delta H}{\delta \varphi^{(s)}}\right] d\tau. \tag{6.9.22}$$

The expression

$$\{F, H\} = \int_V \sum_{s=1}^{h}\left[\frac{\delta F}{\delta \varphi^{(s)}}\frac{\delta H}{\delta \pi_{(s)}} - \frac{\delta F}{\delta \pi_{(s)}}\frac{\delta H}{\delta \varphi^{(s)}}\right] d\tau \tag{6.9.23}$$

is the definition of the *Poisson bracket* of the functions F and H for continuous systems. With this notation, we can write the time derivative of a global observable F of the continuous system as

$$\frac{dF}{dt} = \int_V \frac{\partial \mathcal{F}}{\partial t} d\tau + \{F, H\}. \qquad (6.9.24)$$

This relation is useful in the study of first integrals of canonical equations (6.9.19), in a way similar to that found for the systems with a finite number of degrees of freedom. The Poisson brackets (6.9.23) are widely applied in field theory.

6.9.2 Applications

6.9.2.1 Lamé's Equations

Let us consider again the problem of determining the fundamental equations of the linear theory of elasticity (6.5.20), but this time using the Hamiltonian formalism. The reader will follow more easily the calculation if we rewrite the Lagrangian density (6.8.21):

$$\mathcal{L} = \frac{1}{2}\rho u_{i,t} u_{i,t} - \frac{1}{2}\lambda u_{i,i} u_{k,k} - \frac{1}{2}\mu u_{k,i} u_{k,i} - \frac{1}{2}\mu u_{k,i} u_{i,k} + \rho F_i u_i. \qquad (6.9.25)$$

Recalling that the variational parameters are u_i, the momentum densities (6.9.2) are:

$$\pi_i = \frac{\partial \mathcal{L}}{\partial u_{i,t}} = \rho u_{i,t}, \qquad (6.9.26)$$

and the Hamiltonian density (6.9.1) is:

$$\mathcal{H} = \pi_i u_{i,t} - \mathcal{L} = \frac{1}{2}\rho u_{i,t} u_{i,t} + \frac{1}{2}\lambda u_{i,i} u_{k,k} + \frac{1}{2}\mu u_{k,i} u_{k,i} + \frac{1}{2}\mu u_{k,i} u_{i,k} - \rho F_i u_i. \qquad (6.9.27)$$

In this form, the Hamiltonian density is not yet suitable for the use of canonical equations (6.9.12), because our general formalism demands $\mathcal{H}$ to be expressed in terms of momentum densities $\pi_{(s)}$, and not of $\varphi^{(s)}_{,t}$. Using (6.9.26), we recast $\mathcal{H}$ in the form:

$$\mathcal{H} = \frac{1}{2\rho}\pi_i \pi_i + \frac{1}{2}\lambda u_{i,i} u_{k,k} + \frac{1}{2}\mu u_{k,i} u_{k,i} + \frac{1}{2}\mu u_{k,i} u_{i,k} - \rho F_i u_i. \qquad (6.9.28)$$

We have:

$$\frac{\partial \mathcal{H}}{\partial u_j} = -\rho F_i \delta_{ij} = -\rho F_j,$$

$$\frac{\partial \mathcal{H}}{\partial u_{j,m}} = \frac{1}{2}\lambda(\delta_{ij}\delta_{im}u_{k,k} + \delta_{kj}\delta_{km}u_{i,i}) + \mu\delta_{kj}\delta_{im}u_{k,i}$$
$$+ \frac{1}{2}\mu(\delta_{ij}\delta_{km}u_{k,i} + \delta_{kj}\delta_{im}u_{i,k}) = \lambda\theta\delta_{jm} + \mu(u_{j,m} + u_{m,j}),$$

and the canonical equations (6.9.12) yield:

$$u_{j,t} = \frac{1}{\rho}\pi_j,$$

$$\pi_{j,t} = (\lambda + \mu)\theta_{,j} + \mu\Delta u_j + \rho F_j,$$

leading straightforwardly to Lamé's equations (6.5.20):

$$\rho u_{j,tt} = \rho F_j + (\lambda + \mu)\theta_{,j} + \mu\Delta u_j. \tag{6.9.29}$$

As we have shown, the longitudinal oscillations produced in an elastic rod are described by a second-order partial differential equation of type (6.5.30). Denoting $x_1 = x$, $u_1 = u(x,t)$ $(u_2 = u_3 = 0)$, this equation reads:

$$Ku_{,xx} - \rho u_{,tt} = 0 \quad (K = \lambda + 2\mu). \tag{6.9.30}$$

Let us now apply the formalism presented above to the case of an elastic rod. First, we observe that Eq. (6.9.30) is obtained by using the Lagrangian density

$$\mathcal{L} = \frac{1}{2}\rho u_{,t}^2 - \frac{1}{2}\lambda u_{,x}^2 - \mu u_{,x}^2 = \frac{1}{2}\rho u_{,t}^2 - \frac{1}{2}K\, u_{,x}^2. \tag{6.9.31}$$

The momentum density π associated with u is given by (6.9.26):

$$\pi = \rho u_{,t},$$

and thus the Hamiltonian density reads:

$$\mathcal{H} = \frac{1}{2\rho}\pi^2 + \frac{1}{2}Ku_{,x}^2. \tag{6.9.32}$$

We shall prove that, if $\mathcal{F}$ occurring in (6.9.24) is chosen as

$$\mathcal{F} = \pi u_{,x} = \rho u_{,t}u_{,x}, \tag{6.9.33}$$

then its space integral F is a constant of motion. Since $\mathcal{F}$ does not explicitly depend on time, we have

$$\frac{dF}{dt} = \{F, H\} = \int_V \left(\frac{\delta F}{\delta u}\frac{\delta H}{\delta \pi} - \frac{\delta F}{\delta \pi}\frac{\delta H}{\delta u}\right) d\tau.$$

The variational derivatives appearing under the integral are:

$$\frac{\delta F}{\delta u} = -\frac{\partial}{\partial x}\left(\frac{\partial \mathcal{F}}{\partial u_{,x}}\right) = -\pi_{,x}, \qquad \frac{\delta F}{\delta \pi} = \frac{\partial \mathcal{F}}{\partial \pi} = u_{,x},$$

$$\frac{\delta H}{\delta u} = -\frac{\partial}{\partial x}\left(\frac{\partial \mathcal{H}}{\partial u_{,x}}\right) = -K\, u_{,xx}, \qquad \frac{\delta H}{\delta \pi} = \frac{\partial \mathcal{H}}{\partial \pi} = \frac{1}{\rho}\pi,$$

therefore

$$\{F, H\} = \int_V \frac{\partial}{\partial x}\left(\frac{1}{2}Ku_{,x}^2 - \frac{1}{2\rho}\pi^2\right) d\tau = \oint_S \left(\frac{1}{2}Ku_{,x}^2 - \frac{1}{2\rho}\pi^2\right) dS_x. \tag{6.9.34}$$

But u varies only in the x-direction, consequently we have $u = \text{const.}$ in any point of the cross section whose surface element is $dS_x = dy\,dz$, which means that the integral (6.9.34) vanishes. The fact that $F = \int_V \pi u_{,x}\, d\tau$ is a constant of motion is therefore expressed by the relation

$$\{F, H\} = 0. \tag{6.9.35}$$

6.9.2.2 Telegrapher's Equations

These equations are concerned with the transmission lines (telephone, telegraph, etc.) and describe space and time variation of some characteristic quantities (electric current, voltage) as well as of some electric constants (capacitance C, inductance L). Here by 'constants' we mean those quantities which do not depend on current intensity and voltage.

Even if in a transmission line the electric constants are distributed along the line, we may assume that an elementary segment dx of the line is equivalent to a circuit with concentrated constants, formed by a coil and a resistor, as shown in Fig. 6.11, with the electric resistance and the inductance per unit length of the line, denoted by R and L, respectively. The capacitance between the two conductors is represented by the shunt capacitor C, while the conductance G of the dielectric separating the signal wire from the return wire is represented by a shunt resistor with the resistance $1/G$. If we denote by $i(x, t)$ the electric current intensity on a line and by $u(x, t)$ the voltage between two lines at point x and at time t, according to Kirchhoff's rules (see Sect. 3.7), we have:

$$-du = \frac{\partial}{\partial t}(L\, i\, dx) + R\, i\, dx, \quad -di = \frac{\partial}{\partial t}(C\, u\, dx) + G\, u\, dx, \tag{6.9.36}$$

which can also be written as

$$-\frac{\partial u}{\partial x} = \frac{\partial}{\partial t}(L\, i) + Ri, \quad -\frac{\partial i}{\partial x} = \frac{\partial}{\partial t}(C\, u) + G\, u. \tag{6.9.37}$$

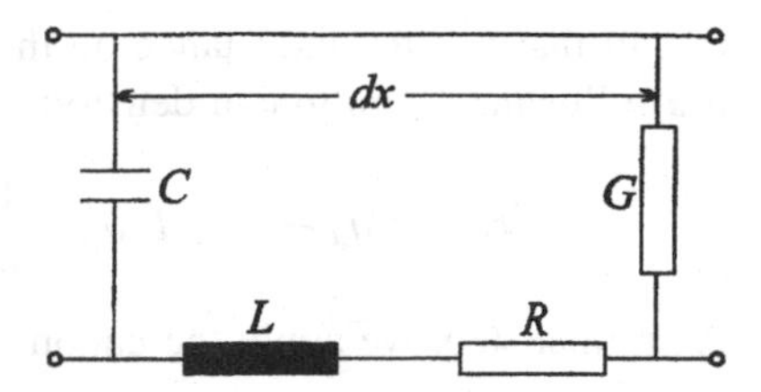

Fig. 6.11 Transmission line modeled by a circuit of concentrated constants *R,L,C,G*, used to obtain the *telegrapher's equations.*

These equations are called *telegrapher's equations.* If there are no losses $(R = 0,\ G = 0)$, these equations read:

$$-\frac{\partial u}{\partial x} = \frac{\partial}{\partial t}(Li), \quad -\frac{\partial i}{\partial x} = \frac{\partial}{\partial t}(C\,u). \tag{6.9.38}$$

Let us first show that this system of equations can be derived by using the following Lagrangian density:

$$\mathcal{L} = \frac{1}{2}L\,i^2 - \frac{1}{2}C\,u^2. \tag{6.9.39}$$

Observing that our Lagrangian density is similar in form with the Lagrangian density of the electromagnetic field (6.8.32), we shall express the quantities i and u in terms of the electric charge $q(x,\,t)$, chosen as a potential:

$$i = \frac{\partial q}{\partial t}, \quad u = -\frac{1}{C}\frac{\partial q}{\partial x}. \tag{6.9.40}$$

The Lagrangian density is then:

$$\mathcal{L} = \frac{1}{2}L\left(\frac{\partial q}{\partial t}\right)^2 - \frac{1}{2C}\left(\frac{\partial q}{\partial x}\right)^2. \tag{6.9.41}$$

Using (6.9.41), the Euler–Lagrange equations (6.8.11) yield a D'Alembert-type homogeneous wave equation:

$$q_{,xx} - \frac{1}{c^2}q_{,tt} = 0, \tag{6.9.42}$$

where $c = 1/\sqrt{LC}$ is the speed of propagation of the waves (electrical impulse). In the ideal case of perfect conductors forming a coaxial transmission line, with vacuum in between the conductors, this speed is the speed of light in empty space. Equation $(6.9.38)_1$ is then found by substituting (6.9.40) into (6.9.42), while $(6.9.38)_2$ is obtained by taking the partial derivative of $(6.9.40)_1$ with respect to x, then of $(6.9.40)_2$ with respect to t, and finally adding the results.

Let us now apply the Hamiltonian formalism. According to the definition (6.9.2), choosing $\varphi = q$, we have:

$$\pi = \frac{\partial \mathcal{L}}{\partial q_{,t}} = L\,q_{,t},$$

(recall that L is the inductance on the line, and not the Lagrangian!) which yields the following Hamiltonian density:

$$\mathcal{H} = \pi\, q_{,t} - \mathcal{L} = L\, q_{,t}^2 - \frac{1}{2} L\, q_{,t}^2 + \frac{1}{2C}\, q_{,x}^2 = \frac{1}{2L}\pi^2 + \frac{1}{2C} q_{,x}^2.$$

As a final step, we apply the canonical equations (6.9.12) and obtain:

$$q_{,t} = \frac{\partial \mathcal{H}}{\partial \pi} = \frac{\pi}{L}, \quad \pi_{,t} = \frac{\partial}{\partial x}\left(\frac{\partial \mathcal{H}}{\partial q_{,x}}\right) = \frac{1}{C}\, q_{,xx}. \tag{6.9.43}$$

If we now take the partial derivative with respect to time of $(6.9.43)_1$ and introduce the result into $(6.9.43)_2$, we find again equation (6.9.42), and therefore Eqs. (6.9.38) as well.

6.9.2.3 Equation of Motion of an Ideal Magnetofluid

Using the same formalism, let us deduce the equation of motion of an ideal one-component magnetofluid, undergoing isentropic motion in an external electromagnetic field (6.6.93), assuming that the non-electromagnetic (gravitational) field is also taken into consideration. The Lagrangian density of our problem should be formed by three terms: a term $\mathcal{L}_0^f$ corresponding to the fluid, a term $\mathcal{L}_0^{em}$ corresponding to the electromagnetic field, and a term $\mathcal{L}_{int}$ expressing the interaction between them: $\mathcal{L} = \mathcal{L}_0^f + \mathcal{L}_0^{em} + \mathcal{L}_{int}$. In this formulation, the usual electromagnetic potentials $\mathbf{A}$, ϕ are chosen as variational parameters.

But this choice is not unique. As an alternative approach to the variational formalism of our problem, we shall use a different representation of the electromagnetic field, which makes possible the simplification the Lagrangian density and, consequently, the solution of this application.

Since our model implies the existence of conduction, convection and displacement current densities, Maxwell's source equations read (see (6.6.88)):

$$\frac{1}{\mu_0} \operatorname{curl} \mathbf{B} = \mathbf{j} + \rho^e \mathbf{v} + \epsilon_0 \frac{\partial \mathbf{E}}{\partial t}, \quad \epsilon_0 \operatorname{div} \mathbf{E} = \rho^e. \tag{6.9.44}$$

These equations can be written in a symmetric form, similar to that of the source-free equations (6.6.87), by using the Lagrangian density

$$\begin{aligned} \mathcal{L}' =& \frac{1}{2}\epsilon_0 |\mathbf{E}|^2 - \frac{1}{2\mu_0} |\mathbf{B}|^2 + \mathbf{P} \cdot \left(\mathbf{E} + \mathbf{v} \times \mathbf{B} - \frac{1}{\lambda}\mathbf{j}\right) \\ & - \mathbf{M} \cdot \left(\operatorname{curl} \mathbf{E} + \frac{\partial \mathbf{B}}{\partial t}\right) - \psi \operatorname{div} \mathbf{B}, \end{aligned} \tag{6.9.45}$$

where the source-free Maxwell's equations (6.6.87) and Ohm's law (6.6.89) have been used as equations of constraint, while $\mathbf{P}(\mathbf{r}, t)$, $\mathbf{M}(\mathbf{r}, t)$ and $\psi(\mathbf{r}, t)$ are

Lagrange multipliers. If we choose as variational parameters E_i, B_i $(i = 1,2,3)$, the Euler–Lagrange equations (6.8.11) yield:

$$\mathbf{E} = \frac{1}{\epsilon_0}(\text{curl}\,\mathbf{M} - \mathbf{P}), \quad \mathbf{B} = \mu_0\left(\text{grad}\,\psi + \mathbf{P}\times\mathbf{v} + \frac{\partial \mathbf{M}}{\partial t}\right). \tag{6.9.46}$$

These relations define the electromagnetic field **E**, **B** in terms of the *generalized antipotentials* **M**, ψ (historically, the 'antipotential' designated the magnetic scalar potential). The appearance of the terms **P** and $\mathbf{P}\times\mathbf{v}$ generalizes the usual antipotentials, defined in the case $\mathbf{j} = 0$, $\rho^e = 0$. Taking the divergence of $(6.9.46)_1$ and the curl of $(6.9.46)_2$, we find the source Maxwell's equation in the symmetric form[4]

$$\text{curl}\left(\frac{1}{\mu_0}\mathbf{B} - \mathbf{P}\times\mathbf{v}\right) = \frac{\partial}{\partial t}(\epsilon_0\mathbf{E} + \mathbf{P}), \quad \text{div}(\epsilon_0\mathbf{E} + \mathbf{P}) = 0. \tag{6.9.47}$$

The vector field **P** is called the '*polarization*'. Comparing (6.9.47) with (6.9.44), we can write

$$\mathbf{j} = \mathbf{v}\,\text{div}\,\mathbf{P} + \text{curl}(\mathbf{P}\times\mathbf{v}) + \frac{\partial \mathbf{P}}{\partial t}, \quad \rho^e = -\text{div}\,\mathbf{P}, \tag{6.9.48}$$

which satisfy identically the equation of continuity

$$\frac{\partial \rho^e}{\partial t} + \text{div}(\mathbf{j} + \rho^e\mathbf{v}) = 0. \tag{6.9.49}$$

Like the usual electromagnetic potentials, the antipotentials **M**, ψ can be related by a Lorenz-type condition. Introducing (6.9.46) into Maxwell's source-free equations (6.6.87), we have:

$$\Delta\mathbf{M} - \epsilon_0\mu_0\frac{\partial^2\mathbf{M}}{\partial t^2} = \nabla\left(\nabla\cdot\mathbf{M} + \epsilon_0\mu_0\frac{\partial\psi}{\partial t}\right) + \epsilon_0\mu_0\frac{\partial}{\partial t}(\mathbf{P}\times\mathbf{v}) - \nabla\times\mathbf{P}.$$

In order that **M** satisfies the homogeneous D'Alembert wave equation, the following two conditions must be fulfilled:

$$\text{div}\,\mathbf{M} + \epsilon_0\mu_0\frac{\partial\psi}{\partial t} = 0, \tag{6.9.50}$$

$$\epsilon_0\mu_0\frac{\partial}{\partial t}(\mathbf{P}\times\mathbf{v}) = \text{curl}\,\mathbf{P}. \tag{6.9.51}$$

Relation (6.9.50) is the *Lorenz condition* for the antipotentials and we shall use it as a *constraint* in the Lagrangian density.

---

[4] Calkin, M.G.: An action principle for magnetohydrodynamics. Can. J. Phys. **41**, 2241 (1963).

The advantage of this representation is that the Lagrangian density of the electromagnetic field (6.8.32), in which **E**, **B** are given in terms of **M**, ψ, *includes* the interaction between field and particles. This is possible because the interaction between the electromagnetic field and the point sources has been replaced by an interaction between the electromagnetic and the 'polarization' fields. Starting from the Lagrangian density (6.8.25), we postulate the Lagrangian density for our model, in terms of antipotentials:

$$\mathcal{L} = \frac{1}{2\mu_0}|\mathbf{B}|^2 - \frac{1}{2}\epsilon_0|\mathbf{E}|^2 - \frac{1}{2\epsilon_0}\left(\nabla\cdot\mathbf{M} + \epsilon_0\mu_0\frac{\partial\psi}{\partial t}\right)^2 + \frac{1}{2}\rho|\mathbf{v}|^2 - \rho(\varepsilon + \mathcal{V}^*) + \rho\left(\frac{\partial\alpha}{\partial t} + \mathbf{v}\cdot\nabla\alpha\right) - \beta\rho\left(\frac{\partial s}{\partial t} + \mathbf{v}\cdot\nabla s\right), \quad (6.9.52)$$

where **E** and **B** are given by (6.9.46). Before going further, we need to define the explicit relation between velocity and the 'polarization' fields. Using Euler–Lagrange equations (6.8.11) with $\varphi^{(s)} \equiv v_i$ $(i = 1, 2, 3)$ as variational parameters, we have:

$$\frac{\partial\mathcal{L}}{\partial v_{i,k}} = 0, \qquad \frac{\partial\mathcal{L}}{\partial v_{i,t}} = 0,$$

$$\frac{\partial\mathcal{L}}{\partial v_i} = \frac{\partial\mathcal{L}}{\partial B_k}\frac{\partial B_k}{\partial v_i} + \rho v_i + \rho\alpha_{,i} - \beta\rho s_{,i} = 0.$$

Since

$$\frac{\partial\mathcal{L}}{\partial B_k} = \frac{1}{\mu_0}B_k,$$

$$\frac{\partial B_k}{\partial v_i} = \frac{\partial}{\partial v_i}[\mu_0(\psi_{,k} + \epsilon_{kjm}P_j v_m + M_{k,t})] = \mu_0\epsilon_{kjm}\delta_{im}P_j = \mu_0\epsilon_{kji}P_j,$$

we finally find:

$$v_j = -\alpha_{,j} + \beta s_{,j} - \frac{1}{\rho}\epsilon_{jkm}B_k P_m, \quad (6.9.53)$$

or, equivalently,

$$\mathbf{v} = -\nabla\alpha + \beta\nabla s - \frac{1}{\rho}\mathbf{B}\times\mathbf{P},$$

which is a *generalized Clebsch transformation.*[5]

According to (6.9.2), the *momentum densities* π_{M_j}, π_{P_j}, π_ψ, π_α, π_s, associated with the field variables M_j, P_j, ψ, α, s, are:

[5] Merches, I.: Variational principle in magnetohydrodynamics. Phys. Fluids **12** (10), 2225 (1969).

$$\pi_{M_j} = \frac{\partial \mathcal{L}}{\partial M_{j,t}} = B_j, \quad \pi_{P_j} = \frac{\partial \mathcal{L}}{\partial P_{j,t}} = 0, \quad \pi_\psi = \frac{\partial \mathcal{L}}{\partial \psi_{,t}} = -\mu_0(M_{i,i} + \epsilon_0\mu_0\psi_{,t}) = 0,$$

$$\pi_\alpha = \frac{\partial \mathcal{L}}{\partial \alpha_{,t}} = \rho, \quad \pi_s = \frac{\partial \mathcal{L}}{\partial s_{,t}} = -\beta\rho. \tag{6.9.54}$$

Using (6.9.1), we can now write the Hamiltonian density $\mathcal{H}$:

$$\begin{aligned}\mathcal{H} &= \sum_{s=1}^{6} \pi_{(s)}\dot{\varphi}^{(s)} - \mathcal{L} = B_j M_{j,t} - \mu_0(M_{j,j} + \epsilon_0\mu_0\psi_{,t})\psi_{,t} + \frac{1}{2\epsilon_0}(M_{j,j} + \epsilon_0\mu_0\psi_{,t})^2 \\ &\quad + \rho\alpha_{,t} - \beta\rho s_{,t} - \frac{1}{2\mu_0}B_jB_j + \frac{1}{2}\epsilon_0 E_jE_j - \frac{1}{2}\rho v_jv_j + \rho(\varepsilon + \mathcal{V}^*) - \rho(\alpha_{,t} + v_j\alpha_{,j}) \\ &\quad + \beta\rho(s_{,t} + v_js_{,j}).\end{aligned}$$

The form of the Hamiltonian density can be simplified if we observe that, using the cyclicity of the mixed product and by virtue of (6.9.53), we can write:

$$\begin{aligned}B_jM_{j,t} &= B_j\left[\frac{1}{\mu_0}B_j - (\mathbf{P}\times\mathbf{v})_j - \psi_{,j}\right] \\ &= \frac{1}{\mu_0}B_jB_j + v_j(\rho v_j + \rho\alpha_{,j} - \beta\rho s_{,j}) - B_j\psi_{,j}.\end{aligned}$$

Introducing this result into $\mathcal{H}$, we obtain the Hamiltonian density in terms of the field variables, and their partial derivatives with respect to coordinates and time:

$$\begin{aligned}\mathcal{H} &= \frac{1}{2\mu_0}B_jB_j + \frac{1}{2}\epsilon_0E_jE_j - B_j\psi_{,j} + \frac{1}{2\epsilon_0}\left[M_{j,j}M_{k,k} - \epsilon_0\mu_0^2(\psi_{,t})^2\right] \\ &\quad + \frac{1}{2}\rho v_jv_j + \rho(\varepsilon + \mathcal{V}^*).\end{aligned} \tag{6.9.55}$$

In order to apply the Hamiltonian technique, it is necessary to express $\mathcal{H}$ in terms of the field variables M_j, P_j, ψ, α, s and their conjugate momentum densities π_{M_j}, π_{P_j}, π_ψ, π_α, π_s. Using (6.9.54), we get the Hamiltonian density in the final form:

$$\begin{aligned}\mathcal{H} &= \frac{1}{2\mu_0}\pi_{M_j}\pi_{M_j} + \frac{1}{2\epsilon_0}(\epsilon_{jkm}M_{m,k} - P_j)(\epsilon_{jli}M_{i,l} - P_j) - \pi_{M_j}\psi_{,j} \\ &\quad - \frac{1}{2\epsilon\mu_0}\pi_\psi(M_{j,j} - \epsilon_0\mu_0\psi_{,t}) + \frac{1}{2}\pi_\alpha\left(-\alpha_{,j} - \frac{1}{\pi_\alpha}\pi_s s_{,j} - \frac{1}{\pi_\alpha}\epsilon_{jkm}\pi_{M_k}P_m\right) \\ &\quad \times\left(-\alpha_{,j} - \frac{1}{\pi_\alpha}\pi_s s_{,j} - \frac{1}{\pi_\alpha}\epsilon_{jli}\pi_{M_l}P_i\right) + \pi_\alpha\varepsilon(\pi_\alpha, s) + \pi_\alpha\mathcal{V}.\end{aligned} \tag{6.9.56}$$

Utilizing the calculation after formula (6.8.29) and applying the canonical equations (6.9.12), we obtain the following system:

$$\pi_{M_j,t} = -\epsilon_{jkm} E_{m,k},$$

$$M_{j,t} = \frac{1}{\mu_0}\pi_{M_j} - \psi_{,j} - \epsilon_{jkm} P_k v_m,$$

$$\pi_{P_j,t} = 0 = E_j + \epsilon_{jkm} v_k \pi_{M_m},$$

$$P_{j,t} = 0,$$

$$\pi_{\psi,t} = -\pi_{M_j,t},$$

$$\psi_{,t} = -\frac{1}{2\epsilon_0\mu_0}(M_{j,j} - \epsilon_0\mu_0\psi_{,t}),$$

$$\pi_{\alpha,t} = -(\pi_\alpha v_j)_{,j},$$

$$\alpha_{,t} = -\frac{1}{2}v_j v_j - v_j\alpha_{,j} + \varepsilon + \mathcal{V} + \frac{1}{\pi_\alpha}p,$$

$$\pi_{s,t} = -T\pi_\alpha - (\pi_s v_j)_{,t},$$

$$s_{,t} = -v_j s_{,j}.$$

Rearranging these equations and using the vector notation, we have:

$$\frac{\partial \mathbf{B}}{\partial t} = -\nabla\times\mathbf{E}, \qquad \nabla\cdot\mathbf{B} = 0,$$

$$\mathbf{E} + \mathbf{v}\times\mathbf{B} = 0,$$

$$\mathbf{B} = \mu_0\left(\nabla\psi + \mathbf{P}\times\mathbf{v} + \frac{\partial \mathbf{M}}{\partial t}\right),$$

$$\nabla\cdot\mathbf{M} + \epsilon_0\mu_0\frac{\partial\psi}{\partial t} = 0,$$

$$\frac{\partial s}{\partial t} + \mathbf{v}\cdot\nabla s = 0,$$

$$\frac{\partial \rho}{\partial t} + \nabla\cdot(\rho\mathbf{v}) = 0,$$

$$\frac{\partial \mathbf{P}}{\partial t} = 0,$$

$$\frac{\partial \beta}{\partial t} + \mathbf{v}\cdot\nabla\beta = T,$$

$$\frac{1}{2}|\mathbf{v}|^2 - \mathcal{V}^* - \varepsilon - \frac{p}{\rho} + \frac{\partial \alpha}{\partial t} + \mathbf{v} \cdot \nabla \alpha = 0.$$

Thus, we have found the source-free Maxwell's equations (6.6.87), Ohm's law for infinite conductivity, the field $\mathbf{B}$ in terms of the antipotentials $\mathbf{M}$, ψ, the Lorenz condition (6.9.50), the equation of conservation of entropy $(6.6.94)_5$, the equation of continuity $(6.6.94)_6$, as well as Eqs. (6.8.27) and (6.8.28), which we also obtained in the case of uncharged fluids.

Recalling that our final purpose is to obtain the equation of motion, the next step consists in eliminating the Lagrange multipliers from the equations:

$$\begin{aligned} &\mathbf{v} + \nabla \alpha - \beta \nabla s + \frac{1}{\rho}\mathbf{B} \times \mathbf{P} = 0, \\ &\frac{1}{2}|\mathbf{v}|^2 - \epsilon - \mathcal{V}^* - \frac{p}{\rho} + \frac{\partial \alpha}{\partial t} + \mathbf{v} \cdot \nabla \alpha = 0, \\ &\frac{\partial \beta}{\partial t} + \mathbf{v} \cdot \nabla \beta = T. \end{aligned} \tag{6.9.57}$$

Extracting $\nabla \alpha$ from $(6.9.57)_1$ and introducing this expression into $(6.9.57)_2$, then taking the gradient of the result, we have:

$$\nabla(\beta \mathbf{v} \cdot \nabla s) + \frac{\partial}{\partial t}(\nabla \alpha) - \frac{1}{2}\nabla|\mathbf{v}|^2 = \nabla\left(\varepsilon + \mathcal{V}^* + \frac{p}{\rho} + \frac{1}{\rho}\mathbf{v} \cdot (\mathbf{B} \times \mathbf{P})\right),$$

or, by using again $(6.9.57)_1$ to express $\nabla \alpha$,

$$\begin{aligned} &\frac{\partial}{\partial t}(\beta \nabla s) - \frac{\partial \mathbf{v}}{\partial t} - \frac{\partial}{\partial t}\left(\frac{1}{\rho}\mathbf{B} \times \mathbf{P}\right) - \frac{1}{2}\nabla|\mathbf{v}|^2 + \nabla(\beta \mathbf{v} \cdot \nabla s) \\ &\quad = \nabla\left(\varepsilon + \mathcal{V}^* + \frac{p}{\rho} + \frac{1}{\rho}\mathbf{v} \cdot (\mathbf{B} \times \mathbf{P})\right). \end{aligned}$$

Utilizing the vector formula (B.39) and the fundamental equation of thermodynamics (6.6.50), we obtain:

$$\begin{aligned} \frac{\partial \mathbf{v}}{\partial t} + (\mathbf{v} \cdot \nabla)\mathbf{v} = &-\nabla \mathcal{V}^* - \frac{1}{\rho}\nabla p + \frac{\partial}{\partial t}\left(\frac{1}{\rho}\mathbf{B} \times \mathbf{P}\right) + \mathbf{v} \times \left[\nabla \times \left(\frac{1}{\rho}\mathbf{B} \times \mathbf{P}\right)\right] \\ &+ \nabla\left(\frac{1}{\rho}\mathbf{v} \cdot (\mathbf{B} \times \mathbf{P})\right). \end{aligned} \tag{6.9.58}$$

If $\mathbf{a}$, $\mathbf{b}$, $\mathbf{c}$ are any three vector fields, it is not difficult to prove the following vector identity:

$$\begin{aligned} &\mathbf{a} \times [\nabla \times (\mathbf{b} \times \mathbf{c})] + \mathbf{b} \times [\nabla \times (\mathbf{c} \times \mathbf{a})] + \mathbf{c} \times [\nabla \times (\mathbf{a} \times \mathbf{b})] \\ &\quad = (\mathbf{b} \times \mathbf{c})\nabla \cdot \mathbf{a} + (\mathbf{c} \times \mathbf{a})\nabla \cdot \mathbf{b} + (\mathbf{a} \times \mathbf{b})\nabla \cdot \mathbf{c} + \nabla(\mathbf{c} \cdot (\mathbf{a} \times \mathbf{b})). \end{aligned} \tag{6.9.59}$$

Multiplying this relation by $\frac{1}{\rho}$ and performing some simple calculations, we arrive at

$$\nabla\left(\frac{1}{\rho}\mathbf{c}\cdot(\mathbf{a}\times\mathbf{b})\right)-(\mathbf{c}\cdot(\mathbf{a}\times\mathbf{b}))\nabla\frac{1}{\rho}+\left[\nabla\times\left(\frac{1}{\rho}\mathbf{a}\times\mathbf{b}\right)\right]\times\mathbf{c}$$
$$-\left[\nabla\left(\frac{1}{\rho}\right)\times(\mathbf{a}\times\mathbf{b})\right]\times\mathbf{c}+(\mathbf{a}\times\mathbf{b})\frac{1}{\rho}\nabla\cdot\mathbf{c}$$
$$=\frac{1}{\rho}[\nabla\times(\mathbf{a}\times\mathbf{c})+\mathbf{c}\nabla\cdot\mathbf{a}]\times\mathbf{b}+\frac{1}{\rho}\mathbf{a}\times[\nabla\times(\mathbf{b}\times\mathbf{c})+\mathbf{c}\nabla\cdot\mathbf{b}].$$

If in this expression we put $\mathbf{a}=\mathbf{P},\ \mathbf{b}=\mathbf{B},\ \ \mathbf{c}=\mathbf{v}$ and replace the term $\frac{1}{\rho}\nabla\cdot\mathbf{v}$ by means of the equation of continuity, we obtain:

$$\nabla\left(\frac{1}{\rho}\mathbf{v}\cdot(\mathbf{P}\times\mathbf{B})\right)+\mathbf{v}\times\left[\nabla\times\left(\frac{1}{\rho}\mathbf{B}\times\mathbf{P}\right)\right]+\frac{\partial}{\partial t}\left(\frac{1}{\rho}\mathbf{P}\times\mathbf{B}\right)$$
$$+\mathbf{v}\times\left[(\mathbf{P}\times\mathbf{B})\times\nabla\left(\frac{1}{\rho}\right)\right]-\left[\nabla\left(\frac{1}{\rho}\right)\times(\mathbf{P}\times\mathbf{B})\right]\times\mathbf{v}$$
$$=\frac{1}{\rho}\frac{\partial}{\partial t}(\mathbf{P}\times\mathbf{B})+\frac{1}{\rho}[\nabla\times(\mathbf{P}\times\mathbf{v})+\mathbf{v}\nabla\cdot\mathbf{P}]\times\mathbf{B}+\frac{1}{\rho}\mathbf{P}\times[\nabla\times(\mathbf{B}\times\mathbf{v})+\mathbf{v}\nabla\cdot\mathbf{B}],$$

or, after some reduction of terms,

$$\nabla\left(\frac{1}{\rho}\mathbf{v}\cdot(\mathbf{P}\times\mathbf{B})\right)+\mathbf{v}\times\left[\nabla\times\left(\frac{1}{\rho}\mathbf{B}\times\mathbf{P}\right)\right]+\frac{\partial}{\partial t}\left(\frac{1}{\rho}\mathbf{P}\times\mathbf{B}\right)$$
$$=\frac{1}{\rho}\left[\frac{\partial\mathbf{P}}{\partial t}+\nabla\times(\mathbf{P}\times\mathbf{v})+\mathbf{v}\nabla\cdot\mathbf{P}\right]\times\mathbf{B}+\frac{1}{\rho}\mathbf{P}\times\left[\frac{\partial\mathbf{B}}{\partial t}+\nabla\times(\mathbf{B}\times\mathbf{v})+\mathbf{v}\nabla\cdot\mathbf{B}\right].$$

In view of Maxwell's equation $(6.6.94)_1$, the induction equation for infinite conductivity $(6.6.94)_2$, and the relation $(6.9.48)_1$, the r.h.s. of the last equation reduces to $\frac{1}{\rho}\mathbf{j}\times\mathbf{B}$. Introducing this result into (6.9.58), we finally arrive at:

$$\rho\left[\frac{\partial\mathbf{v}}{\partial t}+(\mathbf{v}\cdot\nabla)\mathbf{v}\right]=\mathbf{j}\times\mathbf{B}-\nabla p+\rho\mathbf{F}',$$

which is the expected equation of motion (6.6.93).

Observation: Assume that, apart from the Lorenz condition (6.9.50), the generalized antipotentials satisfy the boundary conditions $\mathbf{M}|_S=0,\ \psi|_S=0$, where S is the closed surface which bounds the volume V of the magnetofluid. Integrating the Hamiltonian density (6.9.55) over V, we obtain the total Hamiltonian. By volume integration, the term $\mathbf{B}\cdot\nabla\psi$ gives:

$$\int_V\mathbf{B}\cdot\nabla\psi\,d\tau=\int_V\nabla\cdot(\psi\mathbf{B})\,d\tau-\int_V\psi\nabla\cdot\mathbf{B}\,d\tau.$$

The first integral on the r.h.s. can be transformed in a surface integral by virtue of the Green–Gauss theorem and it vanishes because of our assumption on the potentials:

$$\int_V \nabla \cdot (\psi \mathbf{B})\, d\tau = \oint_S \psi \mathbf{B} \cdot d\mathbf{S} = 0,$$

while the second integral is also zero in view of Maxwell's equation $\operatorname{div} \mathbf{B} = 0$. It then follows that

$$H = \int_V \mathcal{H}\, d\tau = \int_V \left(\frac{1}{2\mu_0} |\mathbf{B}|^2 + \frac{1}{2}\epsilon_0 |\mathbf{E}|^2 + \frac{1}{2}\rho |\mathbf{v}|^2 + \rho\varepsilon + \mathcal{V}^* \right) d\tau, \qquad (6.9.60)$$

which is the total energy of the system contained in the volume V. This result is one more proof that our formalism is a useful tool of investigation in a boundary physical discipline which, in this case, is magnetofluid dynamics.

6.10 Noether's Theorem for Continuous Systems

6.10.1 Hamilton's Principle and the Equations of Motion

In Chap. 2 we have seen that there is an intimate connection between the symmetry properties of a mechanical system and the equations of conservation of its characteristic physical quantities, such as momentum, angular momentum and energy. We have also proved that these equations of conservation follow from a general theorem due to *Emmy Noether*, which gives a method of derivation of the equations of conservation from Hamilton's principle. Since Hamilton's principle can be used to obtain the equations of motion of both discrete and continuous systems, it follows that practically the entire field of Physics falls under the incidence of Noether's theorem.

In this section we give a compressed proof of Noether's theorem for continuous systems.[6] Maintaining the notations used in the last two sections, we recall that by $\varphi^{(s)}(x)$ $(s = \overline{1,h})$ we denoted the C^2-class *dependent functions* or *field variables* $\varphi : \{\varphi^{(1)}, .., \varphi^{(h)}\}$, while $x : \{x_1, .., x_n\}$ stand for the *independent variables*.

As we know, the differential equation of motion can be derived from Hamilton's principle

[6] Hill, E.L.: Hamilton's principle and the conservation theorems of mathematical physics. Rev. Mod. Phys. **23**, 253 (1951).

$$\delta \mathcal{J}(\varphi) = 0, \tag{6.10.1}$$

where

$$\mathcal{J}(\varphi) \equiv \mathcal{J}[\varphi^{(1)}, .., \varphi^{(h)}] = \int_{D_n} \mathcal{L}(x, \varphi, \varphi_{,x})\, d\Omega. \tag{6.10.2}$$

Here, $\varphi_{,x} : \{\varphi^{(s)}_{,x_1}, \ldots, \varphi^{(s)}_{,x_n}\}$, while the integral is extended over an arbitrary domain Ω of the n-dimensional closed manifold D_n of coordinates $x_1, \ldots, x_n$.

Let us consider an infinitesimal transformation of coordinates:

$$x'_k = x_k + \delta x_k \quad (k = \overline{1, n}). \tag{6.10.3}$$

This will produce an elementary variation of the fields and their derivatives of the form:

$$\varphi^{(s)'}(x') = \varphi^{(s)}(x) + \delta \varphi^{(s)}(x), \tag{6.10.4}$$

$$\varphi^{(s)'}_{,k}(x') = \varphi^{(s)}_{,k}(x) + \delta \varphi^{(s)}_{,k}(x) \quad (k = \overline{1, n};\ s = \overline{1, h}). \tag{6.10.5}$$

The relation

$$\delta \mathcal{J} = \int_{D'_n} \mathcal{L}(x', \varphi', \varphi'_{,x})\, d\Omega' - \int_{D_n} \mathcal{L}(x, \varphi, \varphi_{,x})\, d\Omega, \tag{6.10.6}$$

where D'_n is the image of the domain of integration by the transformation (6.10.3), is called the *functional variation* of the integral $\mathcal{J}$, if between the 'volume' elements $d\Omega$ and $d\Omega'$ there exists a point-to-point correspondence, which means

$$d\Omega' = J\, d\Omega. \tag{6.10.7}$$

In view of (6.10.3), we have:

$$J = 1 + \frac{\partial}{\partial x_k}(\delta x_k). \tag{6.10.8}$$

Because of the occurrence of so many indices, in this section we shall use the summation convention for *all* repeated indices. Utilizing the Taylor series expansion in the first integral on the r.h.s. of relation (6.10.6) and keeping only the terms linear in δx, $\delta \varphi$, $\delta \varphi_{,x}$, we obtain:

$$\delta \mathcal{J} = \int_{D_n} \left[\mathcal{L} \frac{\partial}{\partial x_k}(\delta x_k) + \frac{\partial \mathcal{L}}{\partial x_k} \delta x_k + \frac{\partial \mathcal{L}}{\partial \varphi^{(s)}} \delta \varphi^{(s)} + \frac{\partial \mathcal{L}}{\partial \varphi^{(s)}_{,k}} \delta \varphi^{(s)}_{,k} \right] d\Omega. \tag{6.10.9}$$

Here we cannot perform an integration by parts, as we have usually done in the previous sections, because the relations (6.10.4) and (6.10.5) express the connection between the field variables and their transformed values at *different* points. Nevertheless, this can be done by introducing some new variations, $\delta_* \varphi^{(s)}$ and $\delta_* \varphi^{(s)}_{,k}$, defined by

$$\varphi^{(s)'}(x') = \varphi^{(s)}(x') + \delta_* \varphi^{(s)}(x'), \tag{6.10.10}$$

$$\varphi^{(s)'}_{,k}(x') = \varphi^{(s)}_{,k}(x') + \delta_* \varphi^{(s)}_{,k}(x'). \tag{6.10.11}$$

Since

$$\delta_* \varphi^{(s)}(x') = \delta_* \varphi^{(s)}(x + \delta x) = \delta_* \varphi^{(s)}(x) + \delta x_k \frac{\partial}{\partial x_k}(\delta \varphi_*{}^{(s)}) + \cdots$$

we may neglect the products of infinitesimal variations $(\delta x_k)(\delta \varphi_*{}^{(s)})$, so that

$$\delta_* \varphi^{(s)}(x') = \delta_* \varphi^{(s)}(x), \tag{6.10.12}$$

and, consequently,

$$\delta_* \varphi^{(s)}_{,k} = \frac{\partial}{\partial x_k}(\delta_* \varphi^{(s)}). \tag{6.10.13}$$

Comparing (6.10.10) and (6.10.11) with (6.10.4) and (6.10.5), we have also:

$$\delta \varphi^{(s)} = \delta_* \varphi^{(s)} + \varphi^{(s)}_{,l} \delta x_l, \tag{6.10.14}$$

$$\delta \varphi^{(s)}_{,k} = \delta_* \varphi^{(s)}_{,k} + \varphi^{(s)}_{,kl} \delta x_l. \tag{6.10.15}$$

Let us now define the operator

$$\frac{D}{Dx_k} = \frac{\partial}{\partial x_k} + \varphi^{(s)}_{,k} \frac{\partial}{\partial \varphi^{(s)}} + \varphi^{(s)}_{,kl} \frac{\partial}{\partial \varphi^{(s)}_{,l}}. \tag{6.10.16}$$

With this notation, the integral (6.10.9) reads:

$$\delta \mathcal{J} = \int_{D_n} \left[\frac{D}{Dx_k}(\mathcal{L} \delta x_k) + \frac{\partial \mathcal{L}}{\partial \varphi^{(s)}} \delta_* \varphi^{(s)} + \frac{\partial \mathcal{L}}{\partial \varphi^{(s)}_{,k}} \delta_* \varphi^{(s)}_{,k} \right] d\Omega.$$

Observing that

$$\frac{\partial \mathcal{L}}{\partial \varphi^{(s)}} \delta_* \varphi^{(s)}_{,k} = \frac{D}{Dx_k} \left[\frac{\partial \mathcal{L}}{\partial \varphi^{(s)}_{,k}} \delta_* \varphi^{(s)} \right] - \frac{D}{Dx_k} \left(\frac{\partial \mathcal{L}}{\partial \varphi^{(s)}_{,k}} \right) \delta_* \varphi^{(s)},$$

we obtain:

$$\delta\mathcal{J} = \int_{D_n} \left\{ \frac{D}{Dx_k} \left(\mathcal{L}\delta x_k + \frac{\partial \mathcal{L}}{\partial \varphi^{(s)}_{,k}} \delta_* \varphi^{(s)} \right) + [\mathcal{L}]_{(s)} \delta_* \varphi^{(s)} \right\} d\Omega, \tag{6.10.17}$$

where the expression

$$[\mathcal{L}]_{(s)} = \frac{\partial \mathcal{L}}{\partial \varphi^{(s)}} - \frac{D}{Dx_k} \left(\frac{\partial \mathcal{L}}{\partial \varphi^{(s)}_{,k}} \right) \tag{6.10.18}$$

is called the *Lagrangian derivative* of $\mathcal{L}$ with respect to $\varphi^{(s)}$. Going back to the initial variations $\delta\varphi^{(s)}$, we finally obtain for $\delta\mathcal{J}$ the following relation:

$$\delta\mathcal{J} = \int_{D_n} \left\{ \frac{D}{Dx_k} \left[\left(\mathcal{L}\delta_{kl} - \frac{\partial \mathcal{L}}{\partial \varphi^{(s)}_{,k}} \varphi^{(s)}_{,l} \right) \delta x_l + \frac{\partial \mathcal{L}}{\partial \varphi^{(s)}_{,k}} \delta\varphi^{(s)} \right] \right.$$
$$\left. + [\mathcal{L}]_{(s)} (\delta\varphi^{(s)} - \varphi^{(s)}_{,k} \delta x_k) \right\} d\Omega, \tag{6.10.19}$$

where δ_{kl} is the Kronecker symbol.

The calculation carried out so far was formal. Let us now assume that the function $\mathcal{L}$ is the Lagrangian density of a physical continuous system. In order to apply Hamilton's principle, we make the following two assumptions:

(a) The integration domain D_n is fixed ($\delta x_k = 0$ in D_n);
(b) The field variables take fixed values on the hypersurface S_{n-1} which bounds the domain ($\delta\varphi^{(s)}|_{S_{n-1}} = 0$).

Applying the Green–Gauss theorem to (6.10.19), we arrive at

$$\delta\mathcal{J} = \int_{D_n} [\mathcal{L}]_{(s)} \delta\varphi^{(s)} \, d\Omega. \tag{6.10.20}$$

According to Hamilton's principle, this integral vanishes for infinitesimal arbitrary variations $\delta\varphi^{(s)}$, subject to the aforementioned boundary conditions. Thus, we obtain:

$$[\mathcal{L}]_{(s)} = \frac{\partial \mathcal{L}}{\partial \varphi^{(s)}} - \frac{D}{Dx_k} \left(\frac{\partial \mathcal{L}}{\partial \varphi^{(s)}_{,k}} \right) = 0, \tag{6.10.21}$$

which are the *differential equations of motion* of the system. If the operator D/Dx_k reduces to $\partial/\partial x_k$, these equations lead to the Euler–Lagrange equations in the form (6.8.8).

Observation: Let us show that the form of Eqs. (6.10.21) does not change upon a *divergence transformation*:

$$\mathcal{L}' = \mathcal{L} + \frac{D\Theta_k}{Dx_k}, \tag{6.10.22}$$

where $\Theta_k(x_1, .., x_n, \varphi^{(1)}, .., \varphi^{(h)})$ are n arbitrary functions of class C^2. Indeed, we have:

$$\frac{\partial}{\partial \varphi^{(s)}}\left(\frac{D\Theta_k}{Dx_k}\right) = \frac{\partial}{\partial \varphi^{(s)}}\left[\frac{\partial \Theta_k}{\partial x_k} + \varphi^{(m)}_{,k}\frac{\partial \Theta_k}{\partial \varphi^{(m)}}\right]$$
$$= \frac{\partial}{\partial x_k}\left(\frac{\partial \Theta_k}{\partial \varphi^{(s)}}\right) + \varphi^{(m)}_{,k}\frac{\partial}{\partial \varphi^{(m)}}\left(\frac{\partial \Theta_k}{\partial \varphi^{(s)}}\right) = \frac{D}{Dx_k}\left(\frac{\partial \Theta_k}{\partial \varphi^{(s)}}\right),$$

$$\frac{\partial}{\partial \varphi^{(s)}_{,k}}\left(\frac{D\Theta_i}{Dx_i}\right) = \frac{\partial}{\partial \varphi^{(s)}_{,k}}\left(\varphi^{(m)}_{,i}\frac{\partial \Theta_i}{\partial \varphi^{(m)}}\right) = \delta_{sm}\delta_{ik}\frac{\partial \Theta_i}{\partial \varphi^{(m)}} = \frac{\partial \Theta_k}{\partial \varphi^{(s)}},$$

yielding:

$$\left[\frac{D\Theta_k}{Dx_k}\right]_{(s)} = 0,$$

which completes the proof.

6.10.2 Symmetry Transformations

As we have already learned in Chap. 2, by *symmetry transformations* we mean a class of transformations of variables which leaves unchanged the form of the equations of motion. From the physical point of view, this means that by such a transformation one passes from one possible motion of the system to another one.

Consider the transformation

$$x_k \to x'_k(x), \tag{6.10.23}$$

$$\varphi^{(s)} \to \varphi^{(s)'}[x, \varphi(x)]. \tag{6.10.24}$$

In order that this transformation represents a symmetry transformation, we must have, on the one hand, the invariance of the functional $\mathcal{J}$, i.e.

$$\mathcal{L}'(x', \varphi', \varphi'_{,x})\, d\Omega' = \mathcal{L}(x, \varphi, \varphi_{,x})\, d\Omega,$$

and on the other, by virtue of the property (6.10.22),

$$\mathcal{L}'(x', \varphi', \varphi'_{,x}) = \mathcal{L}(x', \varphi', \varphi'_{,x}) + \frac{D\Theta_k}{Dx_k}.$$

The last two relations define the *class of symmetry transformations*.

The most important type of symmetry transformations in the study of conservation theorems is that obtained by an iteration of a succession of infinitesimal transformations. Assume, then, the infinitesimal transformations:

$$x_k \to x'_k = x_k + \delta x_k, \tag{6.10.25}$$

$$\varphi^{(s)} \to \varphi^{(s)'}(x') = \varphi^{(s)}(x) + \delta\varphi^{(s)}(x), \tag{6.10.26}$$

$$\varphi^{(s)}_{,k} \to \varphi^{(s)'}_{,k}(x') = \varphi^{(s)}_{,k}(x) + \delta\varphi^{(s)}_{,k}(x). \tag{6.10.27}$$

Under this transformation, the quantities Θ_k will transform as

$$\Theta_k \to \Theta'_k(x', \varphi') = \Theta_k(x', \varphi') + \delta\Theta_k(x', \varphi'). \tag{6.10.28}$$

Since the variations δx_k, $\delta\varphi^{(s)}$ are infinitesimal, we may keep only the linear terms in the series expansion:

$$\delta\Theta(x', \varphi') = \delta\Theta(x + \delta x,\ \varphi + \delta\varphi) \simeq \delta\Theta(x, \varphi).$$

Going now back to the infinitesimal transformation (6.10.25)–(6.10.27), we realize that in order to be a symmetry transformation we must have:

$$\mathcal{L}'(x + \delta x,\ \varphi + \delta\varphi,\ \varphi_{,x} + \delta\varphi_{,x})\, d\Omega' = \mathcal{L}(x, \varphi, \varphi_{,x})\, d\Omega, \tag{6.10.29}$$

$$\mathcal{L}'(x + \delta x,\ \varphi + \delta\varphi,\ \varphi_{,x} + \delta\varphi_{,x})\, d\Omega' = \mathcal{L}(x + \delta x,\ \varphi + \delta\varphi,\ \varphi_{,x} + \delta\varphi_{,x})\, d\Omega' + \frac{D}{Dx_k}(\delta\Theta_k)\, d\Omega'. \tag{6.10.30}$$

Comparing the last two relations, we get:

$$\mathcal{L}(x + \delta x,\ \varphi + \delta\varphi,\ \varphi_{,x} + \delta\varphi_{,x}) = \left[\mathcal{L}(x, \varphi, \varphi_{,x}) - \frac{D}{Dx_k}(\delta\Theta_k)\right]\left(1 - \frac{\partial}{\partial x_j}(\delta x_j)\right),$$

or, by keeping only the terms linear in δ,

$$\left[\delta x_k \frac{\partial}{\partial x_k} + \delta\varphi^{(s)} \frac{\partial}{\partial \varphi^{(s)}} + \delta\varphi^{(s)}_{,k} \frac{\partial}{\partial \varphi^{(s)}_{,k}} + \frac{\partial}{\partial x_k}(\delta x_k)\right]\mathcal{L} = -\frac{D}{Dx_k}(\delta\Theta_k). \tag{6.10.31}$$

If, for a given $\mathcal{L}$, we can find the functions $\delta\Theta_k$ so as to satisfy (6.10.31), then (6.10.25)–(6.10.27) is a symmetry transformation. In particular, if the square bracket in the l.h.s. of (6.10.31) is identically zero and the Jacobian of the transformation equals one, one says that $\mathcal{L}$ is *form-invariant*.

Let us now integrate (6.10.31) over a fixed and bounded (but otherwise arbitrary) domain D_n. The result is:

$$\delta\mathcal{J} + \int\limits_{D_n} \frac{D}{Dx_k}(\delta\Theta_k)\, d\Omega = 0,$$

or, in view of (6.10.19),

$$\int\limits_{D_n} \left\{ \frac{D}{Dx_k}\left[\left(\mathcal{L}\delta_{kl} - \frac{\partial\mathcal{L}}{\partial\varphi^{(s)}_{,k}}\varphi^{(s)}_{,l}\right)\delta x_l + \frac{\partial\mathcal{L}}{\partial\varphi^{(s)}_{,k}}\delta\varphi^{(s)} + \delta\Theta_k\right] \right.$$
$$\left. + [\mathcal{L}]_{(s)}[\delta\varphi^{(s)} - \varphi^{(s)}_{,k}\delta x_k] \right\} d\Omega = 0. \qquad (6.10.32)$$

Since this equality holds for any domain of integration, if the equations of motion $[\mathcal{L}]_{(s)} = 0$ are satisfied, it follows that

$$\frac{D}{Dx_k}\left[\left(\mathcal{L}\delta_{kl} - \frac{\partial\mathcal{L}}{\partial\varphi^{(s)}_{,k}}\varphi^{(s)}_{,l}\right)\delta x_l + \frac{\partial\mathcal{L}}{\partial\varphi^{(s)}_{,k}}\delta\varphi^{(s)} + \delta\Theta_k\right] = 0. \qquad (6.10.33)$$

We conclude that with the infinitesimal symmetry transformation (6.10.25)–(6.10.27) one can associate the equation of conservation (6.10.33). This is a particular case of *Noether's theorem: any invariance with respect to a continuous transformation leads to an equation of conservation.* An application of this theorem in the mechanics of discrete systems has been described in Chap. 2.

If the operator D/Dx_k reduces to $\partial/\partial x_k$, and one chooses x_1, x_2, x_3 as the space coordinates and $x_4 = t$ as the time, then the equation of conservation (6.10.33) can be written as

$$\frac{\partial\gamma}{\partial t} + \operatorname{div}\mathbf{G} = 0, \qquad (6.10.34)$$

where

$$\gamma = \left(\mathcal{L} - \frac{\partial\mathcal{L}}{\partial\varphi^{(s)}_{,t}}\varphi^{(s)}_{,t}\right)\delta t - \frac{\partial\mathcal{L}}{\partial\varphi^{(s)}_{,t}}(\delta\mathbf{r}\cdot\nabla)\varphi^{(s)} + \frac{\partial\mathcal{L}}{\partial\varphi^{(s)}_{,t}}\delta\varphi^{(s)} + \delta\Theta^t, \qquad (6.10.35)$$

$$\mathbf{G} = \left[\mathcal{L}\delta\mathbf{r} - \frac{\partial\mathcal{L}}{\partial(\nabla\varphi^{(s)})}(\delta\mathbf{r}\cdot\nabla)\varphi^{(s)}\right] - \frac{\partial\mathcal{L}}{\partial(\nabla\varphi^{(s)})}\varphi^{(s)}_{,t}\delta t + \frac{\partial\mathcal{L}}{\partial(\nabla\varphi^{(s)})}\delta\varphi^{(s)} + \delta\boldsymbol{\Theta}. \qquad (6.10.36)$$

We observe that Eq. (6.10.34) is very similar in form with an equation of continuity, written for the *densities* γ and $\mathbf{G}$. Integrating over a fixed domain of volume V of the three-dimensional physical space and utilizing the Green–Gauss theorem, we obtain:

$$\frac{\partial}{\partial t}\int\limits_V \gamma\, d\tau = -\oint\limits_S \mathbf{G}\cdot d\mathbf{S}. \qquad (6.10.37)$$

Let us now apply Noether's theorem in order to derive the fundamental theorems which govern the motion of an ideal fluid, obtained in a different way in Sect. 6.6. Choosing again (6.8.24) as a suitable Lagrangian density (with $\mathcal{V}^* = 0$):

$$\mathcal{L} = \frac{1}{2}\rho|\mathbf{v}|^2 - \rho\varepsilon - \alpha\left[\frac{\partial\rho}{\partial t} + \nabla\cdot(\rho\mathbf{v})\right] - \beta\rho\left(\frac{\partial s}{\partial t} + \mathbf{v}\cdot\nabla s\right), \tag{6.10.38}$$

we shall deduce the equations of transformation and conservation associated with the *space–time symmetry transformations.*

6.10.3 Energy Conservation Theorem

We require that the action principle $\delta\mathcal{J} = 0$ with $\mathcal{L}$ given by (6.10.38) must be invariant with respect to an infinitesimal displacement of the time origin:

$$t \to t' = t + \delta t, \quad \delta\mathbf{r} = 0, \quad \delta\varphi^{(s)} = 0, \quad \delta\Theta = 0, \tag{6.10.39}$$

where δt is an infinitesimal constant (which obviously has the dimension of time). In this case, the relations (6.10.35) and (6.10.36) reduce to

$$\gamma^{(t)} = \left(\mathcal{L} - \frac{\partial\mathcal{L}}{\partial\varphi^{(s)}_{,t}}\varphi^{(s)}_{,t}\right)\delta t, \tag{6.10.40}$$

$$\mathbf{G}^{(t)} = -\frac{\partial\mathcal{L}}{\partial(\nabla\varphi^{(s)})}\varphi^{(s)}_{,t}\,\delta t, \tag{6.10.41}$$

where the superscript (t) indicates the type of symmetry transformation. Using (6.8.28) and (6.8.29), we have:

$$\gamma^{(t)} = \left(\mathcal{L} + \alpha\frac{\partial\rho}{\partial t} + \beta\rho\frac{\partial s}{\partial t}\right)\delta t = -\left[\frac{1}{2}\rho|\mathbf{v}|^2 + \rho\varepsilon + \nabla\cdot(\alpha\rho\mathbf{v})\right],$$

$$\mathbf{G}^{(t)} = \left(\alpha\mathbf{v}\frac{\partial\rho}{\partial t} + \beta\rho\mathbf{v}\frac{\partial s}{\partial t} + \alpha\rho\frac{\partial\mathbf{v}}{\partial t}\right)\delta t = \left[\frac{\partial}{\partial t}(\alpha\rho\mathbf{v}) - \rho\mathbf{v}\frac{\partial\alpha}{\partial t} - \beta\rho\mathbf{v}(\mathbf{v}\cdot\nabla s)\right]\delta t$$
$$= \left[\frac{\partial}{\partial t}(\alpha\rho\mathbf{v}) - \rho\mathbf{v}\left(\frac{1}{2}|\mathbf{v}|^2 + w\right)\right]\delta t.$$

Introducing these last two relations into (6.10.37) and simplifying by the arbitrary constant δt, we arrive at the energy conservation equation (6.6.53):

$$\frac{\partial}{\partial t}\int_V\left(\frac{1}{2}\rho|\mathbf{v}|^2 + \rho\varepsilon\right)d\tau = -\oint_S\left(\frac{1}{2}\rho|\mathbf{v}|^2 + \rho w\right)\mathbf{v}\cdot d\mathbf{S}.$$

6.10.4 Momentum Conservation Theorem

The action principle is required to be invariant also with respect to an infinitesimal displacement of the origin of the coordinate system:

$$\mathbf{r} \to \mathbf{r}' = \mathbf{r} + \delta\mathbf{r}, \quad \delta t = 0, \quad \delta\varphi^{(s)} = 0, \quad \delta\Theta = 0, \tag{6.10.42}$$

where $\delta\mathbf{r}$ is an infinitesimal constant vector oriented along the displacement. Then (6.10.35) yields

$$\gamma^{(r)} = -\frac{\partial\mathcal{L}}{\partial\varphi^{(s)}_{,t}}(\delta\mathbf{r}\cdot\nabla)\varphi^{(s)} = \alpha\delta\mathbf{r}\cdot\nabla\rho + \beta\rho\delta\mathbf{r}\cdot\nabla s = [\rho\mathbf{v} + \nabla(\alpha\rho)]\cdot\delta\mathbf{r},$$

while (6.10.36) leads to

$$\begin{aligned}\mathbf{G}^{(r)} &= \mathcal{L}\delta\mathbf{r} - \frac{\partial\mathcal{L}}{\partial(\nabla\varphi^{(s)})}(\delta\mathbf{r}\cdot\nabla)\varphi^{(s)} \\ &= \left[\frac{1}{2}\rho|\mathbf{v}|^2 - \rho\varepsilon + \rho\frac{\partial\alpha}{\partial t} - \frac{\partial}{\partial t}(\alpha\rho) - \alpha\nabla\cdot(\rho\mathbf{v}) - \beta\rho\left(\frac{\partial s}{\partial t} + \mathbf{v}\cdot\nabla s\right)\right]\delta\mathbf{r} \\ &\quad + \alpha\mathbf{v}\,\delta\mathbf{r}\cdot\nabla\rho + \beta\rho\mathbf{v}\,\delta\mathbf{r}\cdot\nabla s + \alpha\rho(\delta\mathbf{r}\cdot\nabla)\mathbf{v}.\end{aligned}$$

Since

$$\frac{1}{2}\rho|\mathbf{v}|^2 - \rho\varepsilon + \rho\frac{\partial\alpha}{\partial t} = p - \rho\mathbf{v}\cdot\nabla\alpha,$$

$$\alpha\mathbf{v}\,\delta\mathbf{r}\cdot\nabla\rho + \beta\rho\mathbf{v}\delta\mathbf{r}\cdot\nabla s = \mathbf{v}\delta\mathbf{r}\cdot\nabla(\alpha\rho) + \rho\mathbf{v}\mathbf{v}\cdot\delta\mathbf{r},$$

$$\begin{aligned}&\mathbf{v}\delta\mathbf{r}\cdot\nabla(\alpha\rho) - \rho(\mathbf{v}\cdot\nabla\alpha)\delta\mathbf{r} - \alpha(\mathbf{v}\cdot\nabla\rho)\delta\mathbf{r} \\ &\quad = \mathbf{v}\delta\mathbf{r}\cdot\nabla(\alpha\rho) - \mathbf{v}\cdot\nabla(\alpha\rho)\delta\mathbf{r} = \nabla(\alpha\rho)\times(\mathbf{v}\times\delta\mathbf{r}),\end{aligned}$$

we obtain:

$$\begin{aligned}\mathbf{G}^{(r)} &= p\delta\mathbf{r} + \rho\mathbf{v}\mathbf{v}\cdot\delta\mathbf{r} - \delta\mathbf{r}\frac{\partial}{\partial t}(\alpha\rho) + \nabla(\alpha\rho)\times(\mathbf{v}\times\delta\mathbf{r}) \\ &\quad + \alpha\rho(\delta\mathbf{r}\cdot\nabla)\mathbf{v} - \alpha\rho\delta\mathbf{r}\,\nabla\cdot\mathbf{v}.\end{aligned}$$

The divergence of the last three terms in $\mathbf{G}^{(r)}$ is zero. Indeed,

$$\begin{aligned}&\nabla\cdot[\nabla(\alpha\rho)\times(\mathbf{v}\times\delta\mathbf{r}) + \alpha\rho(\delta\mathbf{r}\cdot\nabla)\mathbf{v} - \alpha\rho\delta\mathbf{r}\nabla\cdot\mathbf{v}] \\ &\quad = -\nabla(\alpha\rho)\cdot\nabla\times(\mathbf{v}\times\delta\mathbf{r}) + \nabla\cdot[\alpha\rho(\delta\mathbf{r}\cdot\nabla)\mathbf{v}] - \nabla\cdot(\alpha\rho\delta\mathbf{r}\nabla\cdot\mathbf{v}) \\ &\quad = [\nabla(\alpha\rho)\cdot\delta\mathbf{r}]\nabla\cdot\mathbf{v} - \nabla(\alpha\rho)\cdot[(\delta\mathbf{r}\cdot\nabla)\mathbf{v}] + \nabla\cdot[\alpha\rho(\delta\mathbf{r}\cdot\nabla)\mathbf{v}] - \nabla\cdot(\alpha\rho\delta\mathbf{r}\nabla\cdot\mathbf{v}) \\ &\quad = \alpha\rho\nabla\cdot[\nabla\times(\mathbf{v}\times\delta\mathbf{r})] = 0.\end{aligned}$$

Using these results in (6.10.37), denoting $\Pi_{ik} = \rho v_i v_k + p\delta_{ik}$, $\tilde{P}_i = \rho v_i$, and dropping the infinitesimal vector constant $\delta\mathbf{r}$, we obtain:

$$\frac{\partial}{\partial t}\int_V \tilde{P}_i \, d\tau = -\oint_S \Pi_{ik} \, dS_k,$$

which is the momentum conservation theorem (6.6.57).

6.10.5 Angular Momentum Conservation Theorem

The action principle must be invariant with respect to the infinitesimal rotation of the coordinate axes:

$$\mathbf{r} \to \mathbf{r}' = \mathbf{r} + \mathbf{r} \times \delta\boldsymbol{\theta}, \quad \delta t = 0, \quad \delta\varphi^{(s)} = 0, \quad \delta\Theta = 0, \tag{6.10.43}$$

where the infinitesimal vector constant $\delta\boldsymbol{\theta}$ is oriented along the axis of rotation, its magnitude being equal to the angle of rotation. Denoting by the superscript (θ) the type of symmetry transformation and proceeding in the same manner as above, we have:

$$\begin{aligned}
\gamma^{(\theta)} &= -\frac{\partial \mathcal{L}}{\partial \varphi^{(s)}_{,t}}(\mathbf{r} \times \delta\boldsymbol{\theta}) \cdot \nabla\varphi^{(s)} \\
&= \alpha(\mathbf{r} \times \delta\boldsymbol{\theta}) \cdot \nabla\rho + \beta\rho(\mathbf{r} \times \delta\boldsymbol{\theta}) \cdot \nabla s = [\rho\mathbf{v} + \nabla(\alpha\rho)] \cdot (\mathbf{r} \times \delta\boldsymbol{\theta}),
\end{aligned}$$

$$\begin{aligned}
\mathbf{G}^{(\theta)} &= \mathcal{L}\mathbf{r} \times \delta\boldsymbol{\theta} - \frac{\partial \mathcal{L}}{\partial(\nabla\varphi^{(s)})}(\mathbf{r} \times \delta\boldsymbol{\theta}) \cdot \nabla\varphi^{(s)} \\
&= \left[\frac{1}{2}\rho|\mathbf{v}|^2 - \rho\varepsilon + \rho\frac{\partial\alpha}{\partial t} - \frac{\partial}{\partial t}(\alpha\rho) - \alpha\nabla\cdot(\rho\mathbf{v}) - \beta\rho\left(\frac{\partial s}{\partial t} + \mathbf{v}\cdot\nabla s\right)\right](\mathbf{r} \times \delta\boldsymbol{\theta}) \\
&\quad + \alpha\mathbf{v}(\mathbf{r} \times \delta\boldsymbol{\theta}) \cdot \nabla\rho + \beta\rho\mathbf{v}(\mathbf{r} \times \delta\boldsymbol{\theta}) \cdot \nabla s + \alpha\mathbf{v}(\mathbf{r} \times \delta\boldsymbol{\theta}) \cdot \nabla\rho \\
&= p\mathbf{r} \times \delta\boldsymbol{\theta} - \mathbf{r} \times \delta\boldsymbol{\theta}\frac{\partial}{\partial t}(\alpha\rho) + \nabla(\alpha\rho) \times [\mathbf{v} \times (\mathbf{r} \times \delta\boldsymbol{\theta})] \\
&\quad - \alpha\rho(\mathbf{r} \times \delta\boldsymbol{\theta})\nabla\cdot\mathbf{v} + \rho\mathbf{v}\,\mathbf{v}\cdot(\mathbf{r} \times \delta\boldsymbol{\theta}) + \alpha\rho(\mathbf{r} \times \delta\boldsymbol{\theta}) \cdot \nabla\mathbf{v}.
\end{aligned}$$

Introducing these results into (6.10.34), we have:

$$\frac{\partial}{\partial t}\rho\mathbf{v} \cdot (\mathbf{r} \times \delta\boldsymbol{\theta}) + \nabla \cdot [\rho\mathbf{v}(\mathbf{v} \cdot (\mathbf{r} \times \delta\boldsymbol{\theta})) + p\mathbf{r} \times \delta\boldsymbol{\theta}] = 0,$$

or, in projection on the x_i-axis,

$$\frac{\partial}{\partial t}(\rho\epsilon_{ijk}v_i x_j \delta\theta_k) + \frac{\partial}{\partial x_l}(\rho\epsilon_{ijk}v_i v_l x_j \delta\theta_k + p\epsilon_{ljk}x_j\delta\theta_k) = 0.$$

Observing that $\epsilon_{ljk} = \epsilon_{ijk}\delta_{il}$ and dropping the infinitesimal constant $\delta\theta_k$, we obtain:

$$\frac{\partial}{\partial t}(\rho\epsilon_{kij}x_j v_i) + \frac{\partial}{\partial x_l}[\epsilon_{kji}(\rho x_j v_i v_l + p x_j \delta_{il})] = 0.$$

Using the notation

$$\tilde{L}_k = \rho\epsilon_{kji}x_j v_i,$$

$$\mathcal{M}_{kl} = \epsilon_{kji}(\rho x_j v_i v_l + p x_j \delta_{il})$$

and integrating over the volume V, we arrive at the angular momentum conservation theorem (6.6.59):

$$\frac{\partial}{\partial t}\int_V \tilde{L}_k\, d\tau = -\oint_S \mathcal{M}_{kl}\, dS_l.$$

6.10.6 Centre of Mass Theorem

Let us resume the discussion carried out in Sect. 2.8 on Galilean transformations, but this time for the case of continuous systems. Starting from the postulate that two inertial frames are equivalent in describing the motion of a mechanical system, let us study the invariance of the action principle relative to the infinitesimal Galilean transformation

$$\mathbf{r} \to \mathbf{r}' = \mathbf{r} + \delta\mathbf{v}_0\, t, \quad \delta t = 0, \quad \delta\varphi^{(s)} \neq 0, \quad \delta\Theta \neq 0, \tag{6.10.44}$$

where the infinitesimal constant vector $\delta\mathbf{v}_0$ represents the relative velocity of the two frames. Since $\delta t = 0$, we have $t' = t$, and consequently $d/dt = d/dt'$. Taking the time-derivative of (6.10.44), we then obtain:

$$\mathbf{v}' = \mathbf{v} + \delta\mathbf{v}_0. \tag{6.10.45}$$

Using again the condition $d/dt = d/dt'$, we can write also:

$$\frac{\partial}{\partial t} + \mathbf{v}\cdot\nabla = \frac{\partial}{\partial t'} + \mathbf{v}'\cdot\nabla'. \tag{6.10.46}$$

The frames S and S' are inertial, therefore $\mathbf{u}_i = \mathbf{u}'_i$ $(i = 1, 2, 3)$. Since $\mathbf{r} = \mathbf{r}'$, we conclude that $\nabla = \nabla'$ and (6.10.46) yields:

$$\frac{\partial}{\partial t} = \frac{\partial}{\partial t'} + \delta\mathbf{v}_0\cdot\nabla'. \tag{6.10.47}$$

On the other hand we observe that, for s and β invariants relative to the transformation (6.10.44), the velocity field $\mathbf{v}$ remains unchanged with respect to the transformation

$$\alpha' = \alpha - \mathbf{r} \cdot \delta \mathbf{v}_0,$$

because, according to the Clebsch transformation (6.8.29), we have:

$$\mathbf{v}' - \mathbf{v} = -\nabla(\alpha' - \alpha) = \nabla(\mathbf{r} \cdot \delta \mathbf{v}_0) = \delta \mathbf{v}_0.$$

In this application we shall use the Lagrangian density in the equivalent form (6.8.25). Denoting by the superscript (g) the type of transformation, we obtain:

$$\begin{aligned}
\gamma^{(g)} &= -\frac{\partial \mathcal{L}}{\partial \varphi^{(s)}_{,t}}(\delta \mathbf{r} \cdot \nabla)\varphi^{(s)} + \frac{\partial \mathcal{L}}{\partial \varphi^{(s)}_{,t}} \delta \varphi^{(s)} + \delta \Theta^{(t)} \\
&= -\rho \delta \mathbf{r} \cdot \nabla \alpha + \beta \rho \delta \mathbf{r} \cdot - \rho \mathbf{r} \cdot \delta \mathbf{v}_0 + \delta \Theta^{(t)} \\
&= \rho \mathbf{v} \cdot \delta \mathbf{r} - \rho \mathbf{r} \cdot \delta \mathbf{v}_0 + \delta \Theta^{(t)},
\end{aligned}$$

as well as

$$\begin{aligned}
\mathbf{G}^{(g)} &= \mathcal{L} \delta \mathbf{r} - \frac{\partial \mathcal{L}}{\partial (\nabla \varphi^{(s)})}(\delta \mathbf{r} \cdot \nabla)\varphi^{(s)} + \frac{\partial \mathcal{L}}{\partial (\nabla \varphi^{(s)})} \delta \varphi^{(s)} + \delta \boldsymbol{\Theta} \\
&= \left[\frac{1}{2}\rho|\mathbf{v}|^2 - \rho \varepsilon + \rho\left(\frac{\partial \alpha}{\partial t} + \mathbf{v} \cdot \nabla \alpha\right) - \beta \rho\left(\frac{\partial s}{\partial t} + \mathbf{v} \cdot \nabla s\right)\right] \delta \mathbf{r} - \rho \mathbf{v}(\delta \mathbf{r} \cdot \nabla \alpha) \\
&\quad + \beta \rho \mathbf{v}(\delta \mathbf{r} \cdot \nabla s) - \rho \mathbf{v}(\mathbf{r} \cdot \delta \mathbf{v}_0) + \delta \boldsymbol{\Theta} \\
&= p \delta \mathbf{r} + \rho \mathbf{v}(\mathbf{v} \cdot \delta \mathbf{r}) - \rho \mathbf{v}(\mathbf{r} \cdot \delta \mathbf{v}_0) + \delta \boldsymbol{\Theta}.
\end{aligned}$$

Utilizing (6.10.34), we obtain the differential form of the equation of conservation associated with the symmetry transformation (6.10.44):

$$\frac{\partial}{\partial t}[(\tilde{P}_i t - \rho x_i)\delta v_{0i} + \delta \Theta^{(t)}] + \frac{\partial}{\partial x_k}\{[(\rho v_i v_k + p \delta_{ik})t - \rho x_i v_k]\delta v_{0i} + \delta \Theta_k\} = 0$$
$$(k = 1, 2, 3). \tag{6.10.48}$$

Keeping in mind that $\delta \mathbf{r} = \delta \mathbf{v}_0 t$, the last step is achieved by introducing these expressions into (6.10.37). Assuming that $\delta \Theta^{(g)} = 0$, $\delta \boldsymbol{\Theta} = 0$ and dropping the infinitesimal constant $\delta \mathbf{v}_0$, we arrive at the centre of mass theorem in the form

$$\frac{\partial}{\partial t}\int_V (\tilde{P}_i\, t - \rho x_i)\, d\tau = -\oint_S [(\rho v_i v_k + p \delta_{ik})t - \rho x_i v_k]\, dS_k. \tag{6.10.49}$$

If in (6.10.48) we choose the infinitesimal functions $\delta \Theta^{(t)}$ and $\delta \boldsymbol{\Theta}$ such that

$$\frac{\partial}{\partial t}(\delta \Theta^{(t)}) + \frac{\partial}{\partial x_k}(\delta \Theta_k) = -\frac{\partial}{\partial x_k}[(\rho v_i v_k + p \delta_{ik})t - \rho x_i v_k]\delta v_{0i},$$

then the centre of mass theorem reads:

$$\int_V (\tilde{P}_i t - \rho x_i)\, d\tau = \text{const.},$$

or in the equivalent, but more intuitive form,

$$t\, P_i - M \frac{\int_V \rho x_i\, d\tau}{M} = t\, P_i - M x_i^G = \text{const.} \qquad (6.10.50)$$

Here, $P_i = \int_V \rho v_i\, d\tau$ are the momentum components and x_i^G are the coordinates of the centre of mass of the fluid contained in the domain of volume V. We therefore conclude that *the centre of mass of the continuous system moves uniformly in a straight line*.

6.11 Problems

1. The velocity field in a fluid, expressed in Euler variables, is:

   $$v_1 = kx_2, \quad v_2 = kx_1, \quad v_3 = 0 \quad (k = \text{const.})$$

   (pure sliding). Find the velocity of deformation tensor, the velocity vortex vector, the density variation, the displacement vector and the deformation tensor.
2. Show that in an equilibrium state, if the body forces are absent, the components u_i of the elastic displacement are biharmonic functions, i.e. $\nabla^4 u_i = 0$.
3. A bead of radius R has been introduced in an incompressible perfect fluid. Study the potential flow of the fluid around the bead.
4. Determine the potential motion of a fluid moving inside a dihedral angle.
5. Determine the equation of motion of a sphere performing a motion of vibration in a perfect fluid, as well that of a sphere put into motion by a vibrating fluid.
6. Determine the motion of a fluid in the vicinity of the critical point.
7. Given the complex potential in the form

   $$f(z) = (2 - 3i)\ln(z^2 + 1) + \frac{2}{z},$$

   find the flow rate of an incompressible ideal fluid flowing through the circle $|z| = 2$, as well as the velocity circulation on the circle.
8. Find the shape of an incompressible fluid in the gravitational field, situated in a cylinder uniformly rotating about its axis.
9. Study the potential flow of an incompressible perfect fluid, contained in an ellipsoidal container uniformly rotating about its principal axis. Find the total angular momentum of the fluid in the container.

10. Consider the harmonic functions $\varphi_i(x_1, x_2, x_3)$ $(i = 1, 2, 3)$ and $\psi(x_1, x_2, x_3)$. Find the conditions under which the formulas

$$u_i = \varphi_i + (r^2 - a^2)\frac{\partial \psi}{\partial x_i} \quad (r^2 = x_1^2 + x_2^2 + x_3^2,\ a = \text{const.}),$$

where u_i are elementary displacements, determine a solution of Lamé's homogeneous equations.
11. From an incompressible fluid filling up the whole space is instantly removed a spherical volume of radius a. After how long time does the spherical cavity disappear?
12. A sphere immersed in an incompressible fluid dilates according to the law $R = R(t)$. Determine the pressure of the fluid on the surface of the sphere.
13. Determine the shape of a jet of fluid through an infinitely long opening performed in a plane wall.
14. Determine the motion of a fluid between two cylindrical pipes of radii R_1 and $R_2 (> R_1)$.
15. Solve the same problem for two elliptic-shaped pipes.
16. Once again, solve the same problem for a pipe whose cross section is a triangle with equal sides.
17. Determine the motion of a fluid filling up the space between two concentric spheres of radii R_1 and $R_2 (< R_1)$, rotating about two different diameters, the angular velocities being ω_1 and ω_2. (The Reynolds numbers satisfy the property: $\frac{\omega_1 R_1^2}{\nu} \ll 1, \frac{\omega_2 R_2^2}{\nu} \ll 1$).
18. Determine the velocity of a spherical drop of fluid of viscosity η', moving under the action of gravity in a fluid of viscosity η.
19. A film of viscous fluid is bounded by two parallel solid planes. If one of the planes performs oscillations (parallel to itself), determine the force of friction acting on the other plane.
20. A plane discus of large radius R performs small oscillations about its axis, the angle of rotation being $\theta = \theta_0 \sin \omega t$. Determine the moment of the forces of friction acting on the discus.

Addenda

Post-Classical Mechanics

In the following three addenda we very briefly present as some examples three different directions, into which the classical mechanics, as described in the previous chapters, has evolved. These addenda are intended for those readers who are interested to get acquainted with the new subjects and learn the basic ideas used nowadays in modern physics, where the classical mechanics though standing as their foundation, is valid only in specific situations and in certain approximations. For further study of these subjects an appropriate literature is given.

M. Chaichian et al., *Mechanics*, DOI: 10.1007/978-3-642-17234-2,

Addendum I
Special Theory of Relativity

When the speed of a particle or an object, such as a proton, electron or a nucleus, approaches the speed of light (the so-called relativistic particle), the usual classical mechanics is no more valid. Instead one has the relativistic mechanics, which is based on the special theory of relativity. In this theory, the speed of light denoted by c, is always constant (actually, it is an invariant) – when one goes from one moving system of reference to another, it does not change. The addition of velocities is not the same as in the usual (nonrelativistic) classical mechanics. All the basic quantities, such as the Lagrangian or the action and the corresponding equations of motion derived from it, are covariant under the so-called Lorentz transformations, while in the (nonrelativistic) classical mechanics those quantities are covariant under the Galilean transformations. What is the most drastic is that space–time becomes a four-dimensional manifold, in which the space and time are tightly connected and time is no more an absolute, universal concept but changes when we go from one system of reference to another. The usual classical mechanics is obtained as a limit when the velocities of the particles are small as compared to the speed of light. Here, the speed of light c is the new fundamental parameter which enters the special theory of relativity.

I.1 Introduction

We recall the importance of inertial frames of reference in the classical mechanics and the validity of Galilei's principle of relativity, which states that the laws of mechanics are the same in all inertial frames. This means that the oscillations of a pendulum, for instance, in an inertial frame, are produced in similar manner in any other frame moving at constant velocity with respect to it. A consequence of the principle of relativity of Galilei and the notions of absolute time and space were Galilei's transformations. The equations of mechanics, as that of the motion of a planet around the Sun under the action of Newton's gravitational force, do not change in form, i.e. they are said to be covariant under such transformations. From

M. Chaichian et al., *Mechanics*, DOI: 10.1007/978-3-642-17234-2,

this it results that if an object moves with velocity V with regard to an inertial frame, and this frame in turn with velocity V with regard to another inertial frame, the velocity of the object with regard to this second frame satisfies the law of addition of velocities $V'' = V + V'$, that is, the principle of relativity of Galilei leads to the additivity of the velocities when the motions are considered as referred to several inertial frames. On the other hand, when electromagnetic phenomena are concerned, the principle of relativity of Galilei is not valid. In particular, it is not satisfied by the light propagation (in general, for electromagnetic waves the Maxwell equations and the electromagnetic wave equation are not covariant under Galilean transformations). The existence of an absolute frame of reference was admitted, that of the *luminiferous æther* in which the electromagnetic waves would move at the speed of 300,000 km/s and it was expected that the light would have different velocities if measured in a frame at rest or in motion with regard to the æther, and also it would yield a different result if the velocity of light were measured moving in the sense of the Earth rotation or along a perpendicular direction. Experiments to verify these hypotheses were performed at the end of the nineteenth century, the most famous of them being the so-called Michelson–Morley experiment, performed by *Albert Abraham Michelson* and *Edward Williams Morley* in the year 1887, but leading to negative results: the effect due to the supposed difference of velocities of light along and perpendicular to the Earth rotation direction did not appear. In summary, the scientific community was facing the following facts:

1. Newtonian mechanics and the principle of relativity of Galilei were valid (verified in mechanical experiments and in astronomic observations);
2. The laws that govern electromagnetic phenomena, described by the Maxwell equations, were also valid, and verified experimentally. But these equations did not satisfied the Galilean relativity principle and it was expected from that reason that the speed of light would be different for an observer at rest, as compared with the value measured by an observer in motion;
3. The experiments carried out in order to measure such difference of velocities gave negative answers, as if the velocity of light were the same for both observers.

Apparently the statements 1, 2, and 3 could not be all valid simultaneously in the theoretical framework of that time and *Albert Einstein* proposed to solve this contradiction in 1905, by formulating two principles or basic postulates:

1. The speed of the light emitted by a source is the same for all observers, whatever would be their state of motion.
2. The laws of physics (including the electromagnetic phenomena) are valid in all inertial frames.

Thus, Einstein generalized the principle of relativity of Galilei to all physical phenomena, including the electromagnetic ones, and demonstrated that assuming the validity of his two postulates, all the previously mentioned contradictions would disappear. The essential differences between the consequences of the principles of relativity of Einstein and Galilei had enormous transcendence: not only the controversial luminiferous æther was not necessary, but there were no reasons to suppose its existence. The validity of his postulates implied also the disappearance of the absolute space and absolute time of Newtonian mechanics as independent entities: space and time formed now a joint entity, being intimately related among themselves; the space–time and the fundamental laws of physics could be written as mathematical expressions in a four-dimensional space, nowadays called the Minkowski space. Einstein created a new mechanics such that, when the velocities of the particles are small compared with the velocity of light, it coincides with the Newtonian mechanics, but differs greatly from it for velocities near to that of light.

I.2 Lorentz Transformations

We have already seen in Chap. 1 that a frame of reference is determined by a system of three coordinate axes to fix the position of the objects with regard to them, and a clock in order to measure the time at which the events occur. In classical mechanics, a unique clock serves for all frames of reference. In relativistic mechanics, each frame requires its own clock. The clocks of several frames of reference match in different manner. Suppose we are given a frame of reference S and we consider two events: the departure of a light signal from a point A and the arrival of that signal to another point B. The coordinates of the first event in such a reference frame would be (by including the time as a fourth coordinate):

$$x_1, y_1, z_1, t_1$$

and those of the second event:

$$x_2, y_2, z_2, t_2.$$

Since the signal propagates with the velocity of light, we have

$$\Delta r = c\Delta t, \tag{I.1}$$

where the distance between A and B is:

$$\Delta r = \sqrt{(x_2 - x_1)^2 + (y_2 - y_1)^2 + (z_2 - z_1)^2} \tag{I.2}$$

and the interval of time between the two events is:

$$\Delta t = t_2 - t_1. \tag{I.3}$$

Then, according to (I.1), the coordinates of the events satisfy:

$$(x_2 - x_1)^2 + (y_2 - y_1)^2 + (z_2 - z_1)^2 - c^2\ (t_2 - t_1)^2 = 0. \tag{I.4}$$

If now the events are studied in another frame of reference S' moving with velocity V with respect to S, in this new frame the two events will have the coordinates:

$$x'_1, y'_1, z'_1, t'_1 \quad \text{and} \quad x'_2, y'_2, z'_2, t'_2,$$

and they should again satisfy the equation

$$(x'_2 - x'_1)^2 + (y'_2 - y'_1)^2 + (z'_2 - z'_1)^2 - c^2(t'_2 - t'_1)^2 = 0. \tag{I.5}$$

Let us assume that S' moves parallel to the x-axis. The relations (I.4) and (I.5) will be satisfied by the coordinates of the events in the frames S and S' if they are related by a linear transformation – a so-called Lorentz transformation (or FitzGerald–Lorentz transformation):

$$\begin{aligned} x' &= \frac{x - Vt}{\sqrt{1 - V^2/c^2}}, \\ y' &= y, \\ z' &= z, \\ t' &= \frac{t - \frac{V}{c^2}x}{\sqrt{1 - V^2/c^2}}. \end{aligned} \tag{I.6}$$

These are the transformations which replace those of Galilei in Einstein's relativity (Fig. I.1). The initial formulation was proposed by *George Francis FitzGerald* in 1889, and developed by *Hendrik Lorentz* in 1892, in an attempt of interpreting the Michelson–Morley experiment as a contraction of all bodies along their direction of motion.

If $x'_1, y'_1, z'_1, t'_1, x'_2, y'_2, z'_2, t'_2$ are substituted by their transformed in terms of $x_1, y_1, z_1, t_1, z_2, y_2, z_2, t_2$ according to (I.6), the expression (I.5) is converted into

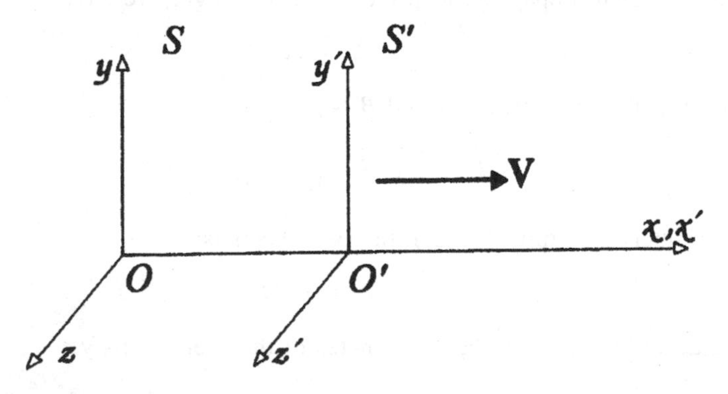

Fig. I.1 Two inertial reference frames S and S' oriented so that their axes x, x' coincide. The frame S' moves with respect to S with a constant velocity V.

(I.4). This means that the expression (I.4) is invariant with respect to the transformations (I.6) that, as we see, depend on the velocity V. For small velocities compared to that of the light, (I.6) turns into the Galilean transformations:

$$x' = x - Vt,$$
$$y' = y,$$
$$z' = z,$$
$$t' = t.$$

The Lorentz transformations are a consequence of the constancy of the speed of light for all the inertial frames and of the linearity of the coordinate transformations. If two events that we shall call 1 and 2 are not related by the departure and arrival of a light signal, then their coordinates would not satisfy the equality (I.1) and we can have one of the two possibilities:

$$\text{either } s_{12}^2 > 0, \quad \text{or} \quad s_{12}^2 < 0, \tag{I.7}$$

where $s_{12}^2 = [(x_2 - x_1)^2 + (y_2 - y_1)^2 + (z_2 - z_1)^2 - c^2(t_2 - t_1)^2]$ is named *space–time interval*. If the interval between the two events observed from the frame S' is calculated, its value is the same as the one calculated from the frame S. By applying the Lorentz transformations to the coordinates of the events 1 and 2 one can check that the space–time interval does not change. The interval between two events is the same in all inertial frames. It is *relativistically invariant*.

If $s_{12}^2 < 0$, which means $s_{12}^2/\Delta t^2 \equiv v^2 - c^2 < 0$, the interval between the two events is called *timelike* and the two events can be causally connected; they can be related to each other by means of a signal traveling at lower velocity v than that of the light. In particular, it is always possible to find a reference frame in which both events occur at the same point of space. As an example, let us suppose that a traveler throws some object through a window of a train and, five seconds later, throws another object through the same window. For an external observer the two events occurred at different points of space and at different times. For the traveler, both events occurred at the same point of space, but at different times.

If $s_{12}^2 > 0$, the interval between the events 1 and 2 is called *spacelike*. In this case, the two events cannot be related causally, since the spatial distance between the two points at which they occur is greater than the product of the velocity of light by the difference of time between them:

$$(x_2 - x_1)^2 + (y_2 - y_1)^2 + (z_2 - z_1)^2 > c^2(t_2 - t_1)^2. \tag{I.8}$$

When the interval between two events is spacelike, it is always possible to find a frame of reference in which the two events occur at the same instant of time, although at different space points. The interval between two events in space–time is a generalization of the distance between two points in ordinary space.

If $s_{12}^2 = 0$, the two events are related by a light signal (since we obtain $v^2 = c^2$) and the interval is called *lightlike*.

I.3 Addition of Velocities

The impossibility of exceeding the speed of light is a consequence of Einstein's postulates. This is easily derived from the law of summation of velocities in relativistic mechanics. By taking the relations (I.6),

$$x = \frac{x' + Vt'}{\sqrt{1 - V^2/c^2}}, \quad t = \frac{t' + (V/c^2)x'}{\sqrt{1 - V^2/c^2}}, \tag{I.9}$$

and after differentiating them with respect to (x', t'), let us divide the first equation by the second. We obtain:

$$v_x = \frac{v'_x + V}{1 + Vv'_x/c^2}, \tag{I.10}$$

where $v_x = \frac{dx}{dt}$ represents the velocity of a particle with respect to the rest frame S, while $v'_x = \frac{dx'}{dt'}$ represents the same velocity, but measured from the moving frame S'. For $V \ll c$, one can take the denominator equal to unity and get approximately

$$v_x \approx v'_x + V, \tag{I.11}$$

i.e. the law of composition of velocities of classical mechanics. But from (I.10), if the particle moves with respect to S' at large speeds, for instance, $v'_x = c/2$, and the frame S', in turn, moves with respect to S also at the same velocity $V = c/2$, it would result:

$$v_x = \frac{c/2 + c/2}{1 + 1/4} = \frac{4}{5}c, \tag{I.12}$$

which is smaller than c. Even by taking $v'_x = c$, $V = c$, it would not be possible to exceed the speed of light:

$$v_x = \frac{c + c}{1 + c^2/c^2} = c. \tag{I.13}$$

In other words, although S' would move with respect to S with the speed of light, and the particle moves with respect to S' also with the speed of light, its speed with respect to S would be precisely the speed of light. We see that it is not possible, by means of the relativistic law of composition of velocities, to exceed the velocity of light c, by summing velocities that are smaller than or equal to c.

I.4 Relativistic Four-Vectors

One of the most interesting geometrical consequences of the Lorentz transformations is that the simultaneous transformations of the space and time coordinates are geometrically equivalent to a rotation in a four-dimensional space, the *space–time*, which is called the *Minkowski space*. *Hermann Minkowski* elaborated on *Henri Poincaré*'s observation that in Einstein's theory of relativity the time could be treated as a fourth dimension.

In two-dimensional Euclidean space, a vector **a** has the components (a_1, a_2) in a frame of coordinates (x, y). In another frame of coordinates (x', y') forming an angle ϕ with the first, it will have the coordinates (a'_1, a'_2):

$$\begin{aligned} a'_1 &= a_1 \cos\phi + a_2 \sin\phi, \\ a'_2 &= a_1 \sin\phi + a_2 \cos\phi, \end{aligned} \tag{I.14}$$

and the relation ${a'_1}^2 + {a'_2}^2 = a_1^2 + a_2^2 = \mathbf{a}^2 \equiv a^2$ is satisfied.

In the theory of the relativity, to change the description of an interval between two events from a frame of reference S to another S', means to change an interval of components $(x_2 - x_1, y_2 - y_1, z_2 - z_1, t_2 - t_1)$ to another one of components $(x'_2 - x'_1, y'_2 - y'_1, z'_2 - z'_1, t'_2 - t'_1)$ by means of a transformation similar to (I.14). To do it, let us take $\tau = ict$, $\phi \to i\phi$ and choose the angle ϕ such that

$$\begin{aligned} \cos i\phi &= \frac{1}{\sqrt{1 - V^2/c^2}}, \\ \sin i\phi &= \frac{iV/c}{\sqrt{1 - V^2/c^2}}. \end{aligned} \tag{I.15}$$

Then,

$$\begin{aligned} x_2 - x_1 &= (x'_2 - x'_1) \cos i\phi + (\tau'_2 - \tau'_1) \sin i\phi, \\ y_2 - y_1 &= y'_2 - y'_1, \\ z_2 - z_1 &= z'_2 - z'_1, \\ \tau_2 - \tau_1 &= -(x'_2 - x'_1) \sin i\phi + (\tau'_2 - \tau'_1) \cos i\phi. \end{aligned} \tag{I.16}$$

The transformations for $x_2 - x_1, t_2 - t_1$ in (I.16) are similar to those in (I.6). The difference lies in the imaginary character of the variable $\tau = ict$ and in the fact that $\sin i\phi$, $\cos i\phi$ are not actually trigonometric, but hyperbolic functions, defined as $\sin i\phi = i \sinh\phi, \cos i\phi = \cosh\phi$. Then we may write the equation

$$\cos^2 i\phi + \sin^2 i\phi = 1 \tag{I.17}$$

in an equivalent form, in terms of the hyperbolic functions, as

$$\cosh^2\phi - \sinh^2\phi = 1.$$

We recall the definitions: $\cosh\phi = (e^{\phi} + e^{-\phi})/2, \sinh\phi = (e^{\phi} - e^{-\phi})/2$. But the transformations (I.16), which represent another way of writing the Lorentz transformation (I.6), as

$$\cosh\phi = \frac{1}{\sqrt{1 - V^2/c^2}},$$
$$\sinh\phi = \frac{V/c}{\sqrt{1 - V^2/c^2}}, \tag{I.18}$$

leave invariant the interval s_{12} and are to be viewed as a rotation by an imaginary angle in Minkowski space. Due to this, one can state that two events in space–time determine a vector whose components are the differences between their coordinates $[x_2 - x_1, y_2 - y_1, z_2 - z_1, c(t_2 - t_1)]$. All the physical quantities in the theory of relativity must be scalars, vectors, tensors, etc., under Lorentz transformations. This means new relations between quantities apparently independent in non-relativistic physics, in similar form as the new relations of dependence between space and time, which were not present in the mechanics of Galilei and Newton.

I.5 Energy and Momentum

The momentum of a free particle of mass m moving with a velocity v is defined in the special theory of relativity as

$$\mathbf{p} = \frac{m\,\mathbf{v}}{\sqrt{1 - v^2/c^2}}, \tag{I.19}$$

and its energy as

$$E = \frac{mc^2}{\sqrt{1 - v^2/c^2}}. \tag{I.20}$$

The two quantities form a four-vector $(p_x, p_y, p_z, i\frac{E}{c})$ whose modulus is

$$p^2 - \frac{E^2}{c^2} = -m^2c^2, \tag{I.21}$$

from which we obtain:

$$E = c\sqrt{p^2 + m^2c^2}. \tag{I.22}$$

For low velocities $v \ll c$, we have $\sqrt{1 - v^2/c^2} \approx 1$, and from the expressions (I.19) and (I.20) we get the nonrelativistic momentum $\mathbf{p} = m\mathbf{v}$ and the energy

$$E = mc^2 + mv^2/2, \tag{I.23}$$

that is, the term mc^2 plus the expression for the kinetic energy of Newtonian mechanics. For $v = 0$, we obtain the expression

$$E = mc^2, \tag{I.24}$$

which relates the mass of a body at rest with its energy content. This expression is probably the most popular consequence of the theory of the relativity. The largest amount of energy that a body is able to produce (for example, when transforming completely into radiation) is equal to the product of its mass by the square of the speed of light. This relation explains the production of enormous amounts of energy in nuclear fission processes (division of an atomic nucleus), in which a certain excess of the initial mass of the nucleus when compared to the sum of the masses of the final nuclei is totally converted into radiation energy.

From the above short description of the main ideas leading to the special theory of relativity, the reader can now realize that, while the usual (nonrelativistic) mechanics is based on Galilei's principle of relativity, its generalization to the so-called relativistic mechanics (and, in general, to any relativistic theory) is based on Einstein's principle of relativity. Thus, if a theory is invariant under the group of Galilei transformations, its relativistic version should be invariant under the group of Lorentz transformations. In particular, in relativistic mechanics the theory is described by a Lagrangian or action which are invariant under the Lorentz transformations and the corresponding equations of motion are covariant under those transformations. For readers interested in further study of the subject, we mention a partial list of literature, in which additional references can be found.

Literature

1. Bergmann, P.G.: Introduction to the Theory of Relativity. Prentice-Hall, New Jersey (1947)
2. Chaichian, M., Merches, I., Tureanu, A.: Electrodynamics: An Intensive Course. Springer-Verlag, Berlin Heidelberg (2012)
3. Kittel, C., Knight, W.D., Ruderman, M.A., Helmholz, A.C., Moyer, B.J.: Mechanics, Berkeley Physics Course, vol. 1. McGraw-Hill, New York (1973)
4. Landau, L.D., Lifschitz, E.M.: The Classical Theory of Fields. Pergamon Press, Oxford (1975)
5. Rindler, W.: Introduction to Special Relativity. Clarendon Press, Oxford (1991)
6. Taylor, E.F., Wheeler, J.A.: Spacetime Physics. W.A. Freeman and Co., San Francisco (1966)
7. Ugarov, V.A.: Special Theory of Relativity. Mir Publishers, Moscow (1979)

Addendum II
Quantum Theory and the Atom

While for macroscopic systems specified by their action being large with respect to some basic quantity called the Planck constant, classical mechanics is an accurate theory, for atomic and still smaller, subatomic systems, classical mechanics is no more applicable. Instead, a new theory, called quantum mechanics, formulated in its final form during the years 1925–1926, was developed, in which many of our usual, everyday-life intuitions have to be altered. For instance, we can no more tell along which trajectory a certain particle such as an electron moves. Newton's equations of motion are no more valid and instead one has to use the so-called Schrödinger equation, according to which only with some probability one can predict the events. In such a theory the new fundamental parameter which enters is the Planck constant, introduced by Max Planck in 1900, in his successful attempt to explain the law of the black body radiation. The description of such systems can be achieved by using operators associated with the observables (canonical formalism), or, equivalently, by using the so-called Feynman path integral approach. The predictions of classical mechanics are obtained in the limit when the action S of the system becomes much larger than the Planck constant, i.e. when the system is no more a microscopic, but a macroscopic one

II.1 Introduction

The 20th century commenced with modifications of the established physical ideas of the previous centuries, by the drastic changes in the concepts of space and time introduced with the formulation of special relativity. But also other new deep modifications of the ideas of classical physics were required for understanding the microscopic world. The investigations of the black body radiation and the photoelectric effect opened a way for understanding the quantum nature of the atomic world, which started to be revealed by studying the emission and absorption of the electromagnetic radiation.

The light, whose wave nature was demonstrated with no doubt in a large number of experiments, in some new phenomena appeared as if having a

M. Chaichian et al., *Mechanics*, DOI: 10.1007/978-3-642-17234-2,

corpuscular structure. The situation turned still more paradoxical when it became evident that the particles composing the atomic structure, as the electrons, showed manifestly wave properties.

Later it became clear that it was not possible to determine simultaneously the momentum and position of an atomic particle, as the electron. Thus it was necessary to invent a new mechanics, called *quantum mechanics*.

II.2 Motion of a Particle

Classically, the motion of a particle is described by giving its position and its velocity (or its momentum) and, in principle, one can know at each instant where the particle is and towards where it is moving. For a particle of the atomic world this is not possible. Different experiments on interference phenomena with electrons brought *Werner Heisenberg* to his famous uncertainty relation.

This relation can be expressed as follows: if Δx is the uncertainty (understood as *standard deviation*) in the position along x and Δp_x is the uncertainty in the momentum, then the relation $\Delta x \Delta p_x \geq \hbar/2$, where $\hbar$ is the reduced Planck constant, i.e. $h/2\pi$, is valid.

The phenomenon of interference of electrons shows that it has wave properties. In fact, in order to describe a particle in quantum mechanics, a wave function $\Psi(x,t)$ is introduced. The square of the modulus of the wave function, $|\Psi|^2$ (Ψ in general is a complex function), gives the probability density of localizing a particle at the point x and at the time t. The following sections will be devoted to a brief historical review about the atom and quantum mechanics.

II.3 Evolution of the Concept of Atom

Democritus of Abdera (470–380 BCE), the Greek philosopher, suggested the hypothesis that the Universe consists of empty space and an enormous number of indivisible particles, and that by joining and separating them we get the creation and disappearance of bodies.

Approximately a century later, another Greek philosopher, *Epicurus* (341–270 BCE), named *atoms* these particles. In the 18th century, *Daniel Bernoulli* was the first who attempted to construct a theory of gases based on the atomic structure model and using the calculus of probabilities. At the beginning of the 19th century, *John Dalton* introduced again the hypothesis of the atomic structure, and *Amedeo Avogadro* was the first to clearly distinguish between atoms and molecules (which are composed of atoms). Starting from the middle of nineteenth century, the kinetic theory of gases was developed by *James Prescott Joule, Rudolf Clausius* and *James Clerk Maxwell*, and subsequently by *Ludwig Boltzmann*, who based it on his statistical interpretation of the second law of thermodynamics. In 1881, *Hermann von Helmholtz*, as a result of the analysis of the works done by *Michael Faraday* on electrolysis, suggested the atomic nature of electricity and later, in 1891, *George Johnstone Stoney* proposed the term *electron* for the unit of electric charge.

In 1897, *Joseph John Thomson*, as a consequence of his experimental studies with cathode rays, stated again the atomic nature of electricity, and he also used the term *electrons* for the electric corpuscles.

Later, Thomson proposed a model of the atom named later *plum pudding*, since he supposed the atoms as positively charged lumps in which the electrons were embedded like the plums in a pudding. The electrons were supposed to oscillate around their mean positions when emitting or absorbing radiation.

II.4 Rutherford's Experiment

In 1884, the Swiss mathematician *Johann Balmer* published the result of his investigations on the hydrogen spectrum. From the spectroscopic measurements of *Anders Ångström*, it was known that when the radiation emitted by this gas is studied (for example, by producing electric arch sparks inside a bubble containing it), the spectrum is formed by a series of lines beginning in the visible zone and ending in the ultraviolet. Balmer gave an empirical formula for the frequencies of the different lines:

$$\nu = cR\left(\frac{1}{2^2} - \frac{1}{n^2}\right), \quad n = 3, 4, 5, \ldots \tag{II.1}$$

where R is the Rydberg constant with the value $1.09677 \times 10^5\,\mathrm{cm}^{-1}$ and c is the speed of light.

In 1911, *Ernest Rutherford* bombarded a thin sheet of gold with α particles (which have positive charge, being helium nuclei) and he concluded that the atoms are formed by a small massive positively-charged nucleus around which the electrons moved similarly to a planetary system, the nucleus playing the role of the Sun, and the electrons moving around it as the planets (see Sect. 3.3.4 for more details).

Rutherford counted the α particles scattered at different angles in his experiment. He found that most of the α particles passed almost without being deflected, but a very small number of them were deflected at very large angles. Rutherford concluded that the "plum pudding" model could not be correct, since if it had been, the large deflection angles could not be explained. Instead, he proposed the planetary model for the atom. The smallness of the positively charged nucleus accounted for the small number of strongly repelled α particles. This was the first experiment on the scattering of particles reported in physics.

II.5 Bohr's Atom

At this point a contradiction appeared with the electromagnetic theory. The planetary model suggested that the electrons moved around the nucleus on elliptical orbits. But in such a case, the electrons would be accelerated continuously and, according to the laws of electrodynamics, an accelerated charge should emit radiation, leading to a continuous loss of energy and the electron would fall finally unto the nucleus. This emission of energy would give a continuous spectrum.

However, the spectroscopists had shown that the atoms do not emit energy with a continuous spectrum, but discretely in the form of spectral lines.

It was *Niels Bohr* who found a way to resolve the crisis with the suggestion of using the quantum ideas introduced earlier by Planck and Einstein. Thus, Bohr proposed two fundamental postulates:

1. *Of all the electron orbits, only those are permissible, for which the angular momentum of the electron is an integer multiple of* $\hbar$ *and no energy is radiated while the electron remains on any one of these permissible orbits.* These orbits are called stationary;
2. *Whenever radiation energy is emitted or absorbed by an atom, this energy is emitted or absorbed in quanta which are integer multiples of* $2\pi\hbar\nu (= \hbar\omega)$, *where* ν *is the frequency of the radiation, and the energy of the atom is changed by this amount.*

In other words, if E_i, E_f are respectively the initial and the final energies of the atom emitting radiation, the following relation is satisfied:

$$E_i - E_f = 2\pi\hbar\nu. \tag{II.2}$$

A simple calculation leads to the expression (where m is the electron's mass and e – its charge) for the energy of the electron in the hydrogen atom:

$$E_n = -\frac{2\pi^2 m\, e^4}{h^2}\frac{1}{n^2}, \tag{II.3}$$

with $n = 1, 2, 3, \ldots$ We observe that the constant coefficient in (II.3) multiplied and divided by c^2 results in $mc^2\alpha^2/2$, where mc^2 is the rest energy of the electron according to Addendum I, and the dimensionless constant $\alpha = e^2/4\pi\hbar\, c \simeq 1/137$ denotes the so-called *fine structure constant*, characterizing the electromagnetic interactions in the atom. After substituting (II.3) into (II.2), we get for the frequency ν the expression

$$\nu = \frac{2\pi^2 m e^4}{h^3}\left(\frac{1}{n_f^2} - \frac{1}{n_i^2}\right). \tag{II.4}$$

Here, $n_f = 2$ for the Balmer series (n_i is greater than n_f), while for $n_f = 1, 3, 4, 5$, we have respectively the Lyman, Paschen, Brackett and Pfund series. The Balmer series lies in the visible and near the ultraviolet region. The Lyman series is in the ultraviolet, whereas the last three are in the infrared region. The number

$$\frac{me^4}{4\pi^2 c\,\hbar^3} = 109740\ \mathrm{cm}^{-1} \tag{II.5}$$

corresponds to the value of the Rydberg constant R for the hydrogen atom. The reader may compare it with (II.1), and observe that the value predicted by Bohr's theory agrees very well with the experimental results.

The distinct spectral series result from the jumps of the electron from diverse excited states to a final fixed state. For instance, Balmer series is produced from the jumps of the electron from the initial levels $n_i = 3, 4, 5, \ldots$, to the final level $n_f = 2$.

Bohr's postulates led to a suitable model to explain the spectra of the hydrogen atom, but finally they were substituted by a more complete quantum theory.

There is some historical analogy between the role of Bohr's quantum mechanics for the atom and the Newtonian mechanics with regard to the planetary motion. We know that at first by starting from the results of the observation, some empirical laws were formulated (the Kepler laws) and later Newton constructed the theory: the second law of motion and the gravitational interaction law. This time there was also an experimental result (discrete character of the emission spectra) and empirical laws (the Balmer series), and then a physical theory was formulated (based on Bohr's postulates), from which the empirical laws could be deduced.

The Newtonian mechanics and the theory of gravitation remained valid during more than two centuries, until their status as the limiting cases of more general theories – Einstein's relativistic mechanics and theory of gravitation – was demonstrated.

Bohr's quantum postulates, however, became obsolete in a very short time. The theory was unsatisfactory for describing more complicated atomic systems, as for instance the helium atom. It ignored the electron spin and the Pauli exclusion principle and it contradicted the uncertainty principle since it assumed classical orbits where position and momentum could be known simultaneously. Thus, in only some twelve years, Bohr's theory was substituted by the new quantum mechanics due to *Erwin Schrödinger, Werner Heisenberg, Max Born, Paul Adrien Maurice Dirac, Pascual Jordan* and others, and Bohr himself had the privilege of not only following this evolution of the quantum theory, but also of strongly participating in its development.

II.6 Schrödinger Equation: Quantum Mechanics

We owe to *Louis de Broglie* the idea that if the radiation has dual behaviour, as waves and particles, the atomic particles like electrons, should manifest also wave properties. That is, if the relation between energy and frequency,

$$E = 2\pi\hbar\nu = \frac{2\pi c\,\hbar}{\lambda}, \tag{II.6}$$

holds for a wave, there must also exist a relation between the momentum and the wavelength of a particle, as:

$$p = \frac{2\pi\hbar}{\lambda}. \tag{II.7}$$

This was a mere speculation by de Broglie in 1923–1924, based on Einstein's idea of the photon, but it was confirmed experimentally by *Clinton Davisson* and *Lester Germer* in 1927, while studying the phenomenon of diffraction of electrons on crystals.

It can be argued that de Broglie's hypothesis gives rise to Bohr's stationary states, since for the electrons having stable orbits around the nucleus, it is necessary that the closed orbit contains an integer number of wavelengths, otherwise the waves would interfere and cancel. Then, if r is the radius of the orbit, we should have

$$2\pi r = n\lambda. \tag{II.8}$$

But

$$\lambda = \frac{2\pi\hbar}{mv}, \tag{II.9}$$

leading to

$$rmv = n\hbar, \tag{II.10}$$

which is Bohr's first postulate. But Bohr's theory, developed subsequently, among others by *Arnold Sommerfeld*, could not account for new atomic phenomena.

Around 1925, Heisenberg initiated the matrix mechanics, which he developed together with Jordan and Born. Matrix mechanics differed from the Bohr model but gave, however, results compatible with the experiment. In 1926, Schrödinger made a crucial step with his famous equation that was the beginning of the new quantum mechanics.

The fundamental assumptions made by Schrödinger which led him to his final equation can be outlined as follows: there exists an analogy between the basic equations of classical mechanics and those of geometrical optics (we recall the analogy between the Hamilton's principle of least action and Fermat's principle in optics). Then, if the atomic particles have wave properties, they should be governed by a wave mechanics, that must bear with regard to classical mechanics a similar relation that wave optics has with regard to geometrical optics.

In essence, the mathematical way to derive the Schrödinger equation is the following:

(a) Write down the classical expression for the energy of the studied system, with the kinetic energy in terms of momentum:

$$\frac{1}{2m}[p_x^2 + p_y^2 + p_z^2] + U(r) = E, \tag{II.11}$$

where $p^2/2m, U(r)$ and E are the kinetic, potential and total energies, respectively. As an example, for the electron in the hydrogen atom $U(r) = -e^2/r$.

(b) The classical quantities are substituted by operators, according to the following rules:

$$\begin{aligned} p_x &\to \hat{p}_x = -i\hbar \frac{\partial}{\partial x}, \\ p_y &\to \hat{p}_y = -i\hbar \frac{\partial}{\partial y}, \\ p_z &\to \hat{p}_z = -i\hbar \frac{\partial}{\partial z}, \\ E &\to \hat{H} = i\hbar \frac{\partial}{\partial t}. \end{aligned} \tag{II.12}$$

(c) A differential equation is built for the *wave function*, using the substitutions (II.12) in (II.11) and applying the obtained operator identity to Ψ:

$$\left[-\frac{\hbar^2}{2m} \left(\frac{\partial^2}{\partial x^2} + \frac{\partial^2}{\partial y^2} + \frac{\partial^2}{\partial x^2} \right) + U(r) \right] \Psi = i\hbar \frac{\partial \Psi}{\partial t}. \tag{II.13}$$

(d) In general, (II.13) is solved by imposing some simple conditions: Ψ is periodic in time (as any wave motion), vanishes at infinity and is also normalized, $\int \Psi^* \Psi \, d^3x = 1$, where Ψ^* is the complex conjugate of Ψ. For the hydrogen atom, it leads as an immediate consequence to Bohr's postulates and to the energy of the stationary states:

$$E_n = \frac{me^4}{2\hbar^2} \frac{1}{n^2}. \tag{II.14}$$

But even more, it follows that the angular momentum is also quantized, that is the inclination of the orbit of the electron can take only some discrete set of values, depending on the value of n.

The Schrödinger equation (II.13) is the basic equation in quantum mechanics, the analog of the Newton equation in classical mechanics.

II.7 Wave Function

Schrödinger interpreted $\Psi(x, y, z, t)$ as a wave field and from it one could assume that the particles such as an electron would be something like a *wavepacket*, similar to the pulse of radiation. But this idea was not a convincing one, among other reasons because the wavepacket would be dispersed and disappear in a very short time. However, frequently the term wavepacket is used when referring to the particle–wave function system.

Max Born was the first to interpret the wave function as a quantity associated to the probability of localization of the particle. That is, if Ψ^* is the complex conjugate of the wave function, the square of the modulus of Ψ, that is $\Psi^*\Psi = |\Psi|^2$ describes the probability density of finding the particle in a given point. We wrote previously the uncertainty relations between position and momentum as

$$\Delta x \, \Delta p \geq \hbar/2. \tag{II.15}$$

The relation (II.15) is typical of a wave motion, due to the correspondence $p = \hbar k$. For instance, the wavefunction of a free particle with momentum p and energy E, moving along the x axis, is $\Psi = Ae^{i(px+Et)/\hbar}$.

In quantum mechanics, observable quantities are associated to quantum mechanical operators. A quantum measurement of one of these observables leads to the knowledge of one of the eigenvalues of these operators. Two quantities q and p can be known simultaneously if the corresponding quantum operators $\hat{q}$ and $\hat{p}$ commute, i.e.

$$[\hat{q}, \hat{p}] = \hat{q}\hat{p} - \hat{p}\hat{q} = 0. \tag{II.16}$$

But if $[\hat{q}, \hat{p}] \neq 0$, it is not possible to know the values of q and p simultaneously. For instance, if q is the position x and p – the momentum p_x, the two corresponding operators would be the position operator $\hat{q} = x$ and the momentum operator $\hat{p}_x = -i\hbar\frac{\partial}{\partial x}$. Applied to a function of the coordinates $f(x)$, one can verify that:

$$[\hat{q}, \hat{p}]f = i\hbar f. \tag{II.17}$$

Thus, since x and $\hat{p}_x$ do not commute, the corresponding position and momentum cannot be simultaneously measured accurately.

In quantum mechanics, the expectation values of quantities such as x or r (with r being, e.g., the distance of the electron from proton in a hydrogen atom) on given quantum states are defined as

$$\langle x \rangle = \int \Psi^* x \Psi \, d^3x, \tag{II.18}$$

and

$$\langle r \rangle = \int \Psi^* r \Psi \, d^3x, \tag{II.19}$$

respectively. Here, Ψ is the wave function, which is obtained as the solution of the Schrödinger equation (II.13) for the system under study (e.g. the hydrogen atom) with a given potential $U(\mathbf{r})$. Similarly, the expectation values of other quantum mechanical operators, for instance of the momentum operator $\mathbf{p}$, are obtained as:

$$\langle \hat{\mathbf{p}} \rangle = \int \Psi^* \hat{\mathbf{p}} \Psi \, d^3x = \int \Psi^* \left(-i\hbar \frac{\partial}{\partial \mathbf{r}} \right) \Psi \, d^3x. \tag{II.20}$$

It can be shown that these expectation values satisfy the classical equation of motion (the so-called *Ehrenfest theorem*).

We hope that from this short addendum the reader has grasped some of the main ideas and facts, which have caused the classical mechanics, so successful in our ordinary *macroscopic* world, to evolve into quantum mechanics. For further reading and study of the subject, an arbitrarily chosen partial list of books from a vast literature, is given below.

Literature

1. Bohm, D.: Quantum Theory. Prentice Hall Publications, New York (1951)
2. Chaichian, M., Hagedorn, R.: Symmetries in Quantum Mechanics: From Angular Momentum to Supersymmetry. Institute of Physics Press, Bristol (1998)
3. Davidov, A.S.: Quantum Mechanics. Pergamon Press, Oxford (1965)
4. Dicke, R.H., Wittke, J.P.: Introduction to Quantum Mechanics. Addison Wesley, Reading, Mass (1965)
5. Feynman, R.P.: The Feynman Lectures on Physics, vol. 3. Addison Wesley, Reading, Mass (1965)
6. Gasiorowicz, S.: Quantum Physics. Wiley, New York (2003)
7. Landau, L.D., Lifschitz, E.M.: Quantum Mechanics. Non-Relativistic Theory. Pergamon Press, London (1981)
8. Rae, A.M.: Quantum Physics: Illusion or Reality? Cambridge University Press, Cambridge (1986)
9. Messiah, A.: Quantum Mechanics. Dover Publications, New York (1999)

Addendum III Stochastic Processes and the Langevin Equation

Finally, let us very briefly mention also about the basics of the so-called *stochastic processes* and the *Langevin equations*. In the usual classical mechanics we have been dealing with in this book until now, the Hamiltonian or the Lagrangian and the dynamics of a system of particles have been given by a set of potentials or by *certain* (so-called *deterministic*) forces corresponding to them. Thus all the predictions obtained from the equations of motion are certain – deterministic. However, there exist systems in which the forces acting on the particles cannot be known exactly, due to the complexity of the systems, and these are called indeterministic. In such cases, the Newton equations of motion also become indeterministic and are replaced by the so-called *Langevin equations* of motion.

The motion of a Brownian particle in a medium with temperature T can be considered as a prototype example of a *stochastic process*. In this case, the forces acting on the Brownian particle are random ones and as a consequence the corresponding Newton equation of motion is replaced by the stochastic Langevin equation of motion, the probability distribution of which satisfies the so-called *diffusion equation*.

The solution of the diffusion equation, which describes the motion of a Brownian particle with no forces acting on it, except the stochastic forces coming from the medium with temperature T, can be given in terms of Wiener's path integral, developed by *Norbert Wiener* already in the early 1920s, before the birth of quantum mechanics. By Wiener path integrals can be represented also the solutions for other more general cases, where additional deterministic forces are also present. The fundamental parameter in this case, which characterizes such stochastic processes, is the temperature T of the medium, describing the intensity of the stochastic forces. In the limit of T going to zero, the stochastic equations of motion turn to the usual deterministic equations of classical mechanics, described in this book until now. Let us mention that quantum mechanics can also be described equivalently using the so-called Feynman path integral, developed by *Richard Feynman* in 1948. For a partial selection of literature, in which long lists of references on quantum mechanics, stochastic processes and other related fields

M. Chaichian et al., *Mechanics*, DOI: 10.1007/978-3-642-17234-2,

described by path integral, are given, the reader can consult the monographs listed below.

Literature

1. Feynman, R.P., Hibbs, A.: Quantum Mechanics and Path Integrals. McGraw Hill, New York (1965)
2. Chaichian, M., Nelipa, N.F.: Introduction to Gauge Field Theories. Springer-Verlag, Berlin Heidelberg (1984)
3. Chaichian, M., Demichev, A.: Path Integrals in Physics, vols. I and II. Institute of Physics Publishing, Bristol and Philadelphia (2001)
4. Gardiner, C.W.: Handbook of Stochastic Methods for Physics, Chemistry and the Natural Sciences. Springer-Verlag, Berlin Heidelberg (2004)

Appendix A
Elements of Vector and Tensor Algebra

A.1 Orthogonal Transformations

The position in space of a point P relative to a given reference frame can be determined by its Cartesian coordinates (x_1, x_2, x_3) with respect to a system of orthogonal axes, having its origin at some point O. This is a Cartesian system of coordinates $S(Ox_1x_2x_3)$ (Fig. A.1). Since the choice of both the origin O and the direction of the axes is arbitrary, it is necessary to define the law of transformation of the coordinates of the point P, when passing to another coordinate system $S'(O'x_1'x_2'x_3')$. It is obvious that the coordinates x_i $(i = 1, 2, 3)$ of the point P relative to S' are functions of the coordinates x_i' $(i = 1, 2, 3)$ of the same point relative to S:

$$x_i' = f_i(x_1, x_2, x_3) \quad (i = 1, 2, 3).$$

In order that $x_i' = \text{const.}$ be the equation of a plane relative to S, this transformation has to be linear:

$$x_i' = a_{ij}x_j + X_i \quad (i = 1, 2, 3), \tag{A.1}$$

where X_i are the coordinates of O' relative to S. Here, the summation convention for repeated indices running from 1 to 3 has been used. If the origin O' is displaced, but the axes of S' remain parallel to the axes of S, we have $a_{ij} = \delta_{ij}$, where δ_{ij} is the *Kronecker delta symbol*, while if the axes of S' are only rotated about the fixed origin O' relative to the axes of S, we have $X_i = 0$. The first case corresponds to a *translation* of axes and the second to a *rotation*.

The distance between any two points P_1 and P_2 must be independent of the reference frame, therefore it must be invariant with respect to the transformation (A.1), i.e.

$$(x_i^1 - x_i^2)(x_i^1 - x_i^2) = (x_i'^1 - x_i'^2)(x_i'^1 - x_i'^2), \tag{A.2}$$

where the indices 1 and 2 correspond to the two points P_1 and P_2. Using (A.1), we obtain the *orthogonality condition*:

M. Chaichian et al., *Mechanics*, DOI: 10.1007/978-3-642-17234-2,

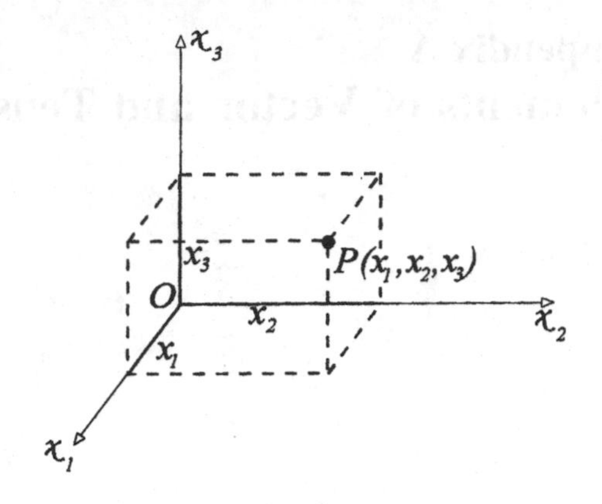

Fig. A.1 Cartesian system of coordinates.

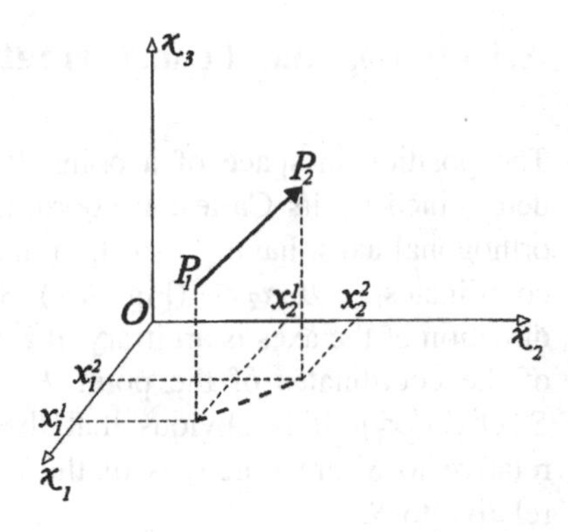

Fig. A.2 Vector distance between two arbitrary points P_1 and P_2.

$$a_{ij}a_{ik} = \delta_{jk} \quad (j,k = 1,2,3), \tag{A.3}$$

which means that out of nine parameters a_{ij}, only three are independent.

The linear transformation (A.1) which satisfies the condition (A.3) is called a *non-homogeneous orthogonal transformation* of coordinates. If $X_i = 0$, we have a *homogeneous orthogonal transformation*.

Multiplying (A.1) by a_{ik} and performing summation over the index i, in view of (A.3) we obtain the *inverse transformation*:

$$x_i = a_{ji}x'_j + X'_i \quad (i = 1,2,3), \tag{A.1$'$}$$

where $X'_i = -a_{ji}X_j$ are the coordinates of O relative to S'. Introducing (A.1′) into (A.2), we obtain

$$a_{ji}a_{ki} = \delta_{jk} \quad (j,k = 1,2,3). \tag{A.4}$$

This last relation is a consequence of (A.3) and therefore does not imply any supplementary condition on the coefficients a_{ij}.

The parameters a_{ij} stand for the elements of a matrix (a_{ij}), called the *transformation matrix* $\hat{A}$. If we also define the one-column matrices $\mathbf{x}, \mathbf{x}', \mathbf{X}, \mathbf{X}'$

having as elements x_i, x'_i, X_i, X'_i $(i = 1, 2, 3)$, then the relations (A.1) and (A.1) become:

$$\mathbf{x}' = \hat{A}\mathbf{x} + \mathbf{X}, \quad \mathbf{x} = \hat{A}^T\mathbf{x}' + \mathbf{X}', \tag{A.5}$$

where $\mathbf{X}' = -\hat{A}^T\mathbf{X}$. Relations (A.3) and (A.4) also yield

$$\hat{A}^T\hat{A} = \hat{A}\hat{A}^T = \hat{I}, \tag{A.6}$$

in which $\hat{I}$ is the *unit matrix* $(I_{ij} = \delta_{ij})$. From (A.6) it results that $\hat{A}^{-1} = \hat{A}^T$, i.e. in the case of orthogonal transformations the inverse and the transpose of a matrix are identical. We also have

$$\det(\hat{A}^T\hat{A}) = (\det\hat{A}^T)(\det\hat{A}) = (\det\hat{A})^2 = \det\hat{I} = 1,$$

hence:

$$\det\hat{A} = \pm 1. \tag{A.7}$$

Transformations with $\det\hat{A} = +1$ are called *proper transformations* (proper rotations or, simply, rotations), while those with $\det\hat{A} = -1$ are said to be *improper transformations*. In each of these two categories there is an important particular transformation, namely the *identity transformation* (proper transformation):

$$x'_i = x_i \quad (i = 1, 2, 3) \tag{A.8}$$

and the *space inversion* (improper transformation):

$$x'_i = -x_i \quad (i = 1, 2, 3). \tag{A.9}$$

In view of these definitions, the translations and rotations of the axes (Fig. A.3) belong to the proper transformations, while the mirror reflection (Fig. A.4) is an improper transformation. For example, the matrix of the transformation which gives the mirror reflection presented in Fig. A.4 is

$$A = \begin{pmatrix} -1 & 0 & 0 \\ 0 & 1 & 0 \\ 0 & 0 & 1 \end{pmatrix}.$$

The set of orthogonal transformations form a group, called *the group of non-homogeneous orthogonal transformations*. To prove this, we must show that it satisfies the axioms of the group, i.e. it is closed, associative, contains an identity element and contains an inverse of each of its elements. Let

$$x'_i = a_{ij}x_j + X_i, \quad x''_i = b_{ij}x'_j + X'_i \quad (i = 1, 2, 3)$$

be two successive orthogonal transformations from S to S' and then from S' to S''. It is obvious that

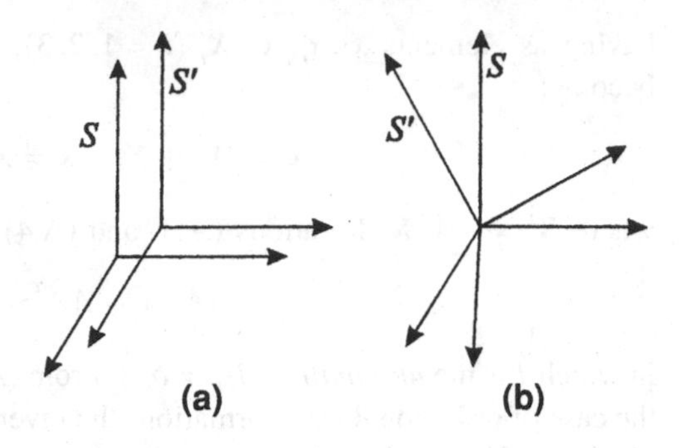

Fig. A.3 Proper transformations: *translation* and *rotation* of axes.

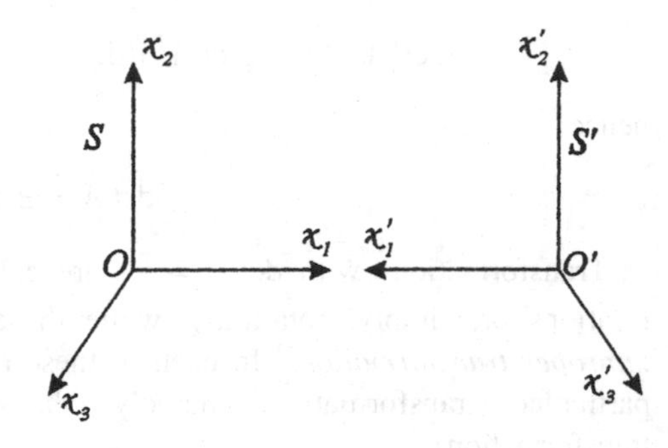

Fig. A.4 Improper transformation: *mirror reflection*.

$$x_i'' = b_{ij}a_{jk}x_k + b_{ij}X_j + X_i' = c_{ik}x_k + X_i'' \qquad (i = 1, 2, 3),$$

where

$$c_{ik} = b_{ij}a_{jk}, \quad X_i'' = b_{ij}X_j + X_i'.$$

By virtue of (A.6), it follows that the matrix $\hat{C} = \hat{B}\hat{A}$ is orthogonal:

$$\hat{C}\hat{C}^T = \hat{B}\hat{A}\hat{A}^T\hat{B}^T = \hat{B}\hat{B}^T = \hat{I},$$

which shows that the transformation from S to S'' is also orthogonal.

The identity element is the identity transformation, while the inverse element is the inverse transformation (A.1) which, in view of (A.4), belongs to the group. Consequently, all axioms are satisfied.

Some of the most important groups in physics are:

O(3): the *group of rotations*, which is formed by the set of the homogeneous orthogonal transformations, $x_i' = a_{ij}x_j$. This group is isomorphic to the group of 3×3 real and orthogonal matrices.

SO(3): the *special orthogonal group* or the *group of proper rotations*, which is composed of the subset of the homogeneous proper orthogonal transformations: $x_i' = a_{ij}x_j$, with $\det \hat{A} = 1$.

T(3): the *group of translations*, $x'_i = x_i + X_i$, which is a *commutative* or *Abelian* group, meaning that the result of two successive translations does not depend on their order.

The set of improper orthogonal transformations does not have the group property, because it does not contain the identity element.

Adding to these properties the fact that $\hat{A}$ is a continuous function of the elements a_{ij}, it results that any transformation belonging to $SO(3)$ can be obtained by a continuous rotation (a succession of homogeneous proper orthogonal transformations, which are infinitesimally different from the identity transformation). We also note that by no continuous rotation can we superpose two coordinate systems which are obtained one from the other by an improper transformation.

A coordinate system (x_1, x_2, x_3) is *right-handed* if a right screw directed along the x_3-axis produces through a rotation of angle $\pi/2$ the superposition of x_1 and x_2 axes. A *left-handed* coordinate system is obtained from a right-handed one by a space inversion. It results that any right-handed coordinate system can be obtained from another right-handed coordinate system by a proper rotation and/or a translation, i.e. by a transformation which belongs to the group of *non-homogeneous proper orthogonal transformations* given as the *semi-direct product* $SO(3) \ltimes T(3)$, which is the group of isometries of the three-dimensional Euclidean space.

A.2 Scalars, Vectors and Tensors

A.2.1 Algebraic Definition of Tensor Quantities

In the definition of tensor quantities we shall use orthonormal transformations with Cartesian coordinates.

A *scalar* is a quantity characterized by a real number which remains unchanged upon any transformation of the coordinate system. A scalar whose algebraic form is the same in all coordinate systems is called an *invariant*.

A *vector* $\mathbf{v}$ is a quantity characterized by an ordered system of three real numbers $\{v_1, v_2, v_3\}$, called its *components*, which under an orthogonal transformation transforms according to the rule

$$v'_i = a_{ij} v_j \quad (i = 1, 2, 3), \tag{A.10}$$

where v'_i are the components of the vector in the new coordinate system, while a_{ij} are the elements of the transformation matrix.

In general, a n-th rank tensor in the Euclidean space E_3 is a quantity with 3^n components $t_{i_1 i_2 \dots i_n} (i_1, i_2, \dots, i_n = 1, 2, 3)$ which transforms according to

$$t'_{i_1 i_2 \dots i_n} = a_{i_1 j_1} a_{i_2 j_2} \dots a_{i_n j_n} t_{j_1 j_2 \dots j_n} \tag{A.11}$$

under orthogonal transformations.

Note that scalars and vectors can be discussed within the frame of this general definition as 0th and first rank tensors, respectively. Nevertheless, we shall discuss them separately, in order to emphasize some useful particular properties.

A n-th order *pseudo-tensor* in E_3 is a quantity having 3^n components $t^*_{i_1 i_2 \dots i_n} (i_1, i_2, \dots, i_n = 1, 2, 3)$ which transform according to the rule

$$t^{*\prime}_{i_1 i_2 \dots i_n} = (\det \hat{A}) a_{i_1 j_1} a_{i_2 j_2} \dots a_{i_n j_n} t^*_{j_1 \dots j_n} \tag{A.12}$$

under orthogonal transformations. In other words, under a proper orthogonal transformation pseudo-tensors transform like tensors, while under an improper transformation appears a change of sign. A pseudo-vector is also called an *axial vector*, while an ordinary vector is called a *polar vector*. For example, under an inversion of axes $x'_i = -x_i$, a *vector* $\mathbf{v}$ transforms according to $v'_i = -v_i$ (as examples can serve the position vector $\mathbf{r}$ or the linear momentum $\mathbf{p}$), while a *pseudo-vector* obeys the rule $v^{*\prime} = v^*_i$ $(i = 1, 2, 3)$ (an example is the angular momentum, $\mathbf{L} = \mathbf{r} \times \mathbf{p}$).

A.2.2 General Properties of Tensors

We shall use a boldface letter to denote a tensor of specified rank, for example $\mathbf{t}$ stands for all its 3^n components $\{t_{i_1 \dots i_n},\ i_1, \dots, i_n = 1, 2, 3\}$.

Two tensors $\mathbf{t}^{(1)}$ and $\mathbf{t}^{(2)}$ of the same order are *equal* if all their components in some coordinate system are equal:

$$t^{(1)}_{i_1 \dots i_n} = t^{(2)}_{i_1 \dots i_n}, \qquad i_1, i_2, \dots, i_n = 1, 2, 3. \tag{A.13}$$

Using (A.11) or (A.12) it is easily seen that this property is *intrinsic*, i.e. it remains valid in any coordinate system obtained by an orthogonal transformation.

A *null tensor (pseudo-tensor)* has all its components equal to zero. From (A.11) and (A.12) it results that a null tensor (pseudo-tensor) is an invariant.

A tensor is termed *symmetric* or *antisymmetric* relative to a pair of indices, say i_r and i_s, if by their interchange we obtain:

$$t_{i_1 \dots i_r \dots i_s \dots i_n} = t_{i_1 \dots i_s \dots i_r \dots i_n}, \tag{A.14}$$

and, respectively,

$$t_{i_1 \dots i_r \dots i_s \dots i_n} = -t_{i_1 \dots i_s \dots i_r \dots i_n}. \tag{A.15}$$

The properties of symmetry and antisymmetry are intrinsic. A tensor which is symmetric (antisymmetric) relative to all indices of its components is called *completely symmetric* (*completely antisymmetric*).

Linear operations with tensors. The *sum* (*difference*) of two tensors makes sense only if the tensors are of the same rank. The result of summation (subtraction) is a tensor whose components are equal to the sum (difference) of the corresponding components of the two tensors.

Multiplying a tensor by a scalar, we obtain a tensor of the same rank whose components are the components of the initial tensor multiplied by that scalar.

Let us consider two tensors, $\mathbf{t}$ of rank m and $\tilde{\mathbf{t}}$ of rank n. The quantity

$$T_{i_1\ldots i_{m+n}} = t_{i_1\ldots i_m}\tilde{t}_{i_{m+1}\ldots i_{m+n}} \quad (i_1,\ldots,i_{m+n} = 1,2,3) \tag{A.16}$$

is a $m+n$-rank tensor, with 3^{m+n} components, called the *tensor product* of the given tensors. Indeed, in view of (A.11), we have:

$$\begin{aligned} T'_{i_1\ldots i_{m+n}} &= t'_{i_1\ldots i_m}\tilde{t}'_{i_{m+1}\ldots i_{m+n}} \\ &= a_{i_1 j_1}\ldots a_{i_m j_m} a_{i_{m+1} j_{m+1}}\ldots a_{i_{m+n} j_{m+n}} T_{j_1\ldots j_{m+n}}. \end{aligned}$$

In a similar way, using (A.11) and (A.12) it can be shown that the tensor product between a tensor and a pseudo-tensor is a pseudo-tensor whose rank equals the sum of ranks of tensor and pseudo-tensor, while the tensor product of two pseudo-tensors is a tensor whose rank is equal to the sum of ranks of the two pseudo-tensors.

By a *contraction* of a tensor relative to a pair of indices we mean the operation of equalizing the two indices and then performing the summation over the common value. By a contraction, the rank of a tensor reduces by two. For example, performing a contraction over the indices (i_{n-1}, i_n) of the tensor $t_{i_1\ldots i_n}$, one obtains:

$$\sum_{i=1}^{3} t_{i_1\ldots i_{n-2}ii} = t_{i_1\ldots i_{n-2}ii},$$

which is a tensor quantity with 3^{n-2} components. By virtue of (A.11) and (A.4), under an orthogonal transformation, we have:

$$\begin{aligned} t_{i_1\ldots i_{n-2}ii} &= a_{j_1 i_1}\ldots a_{j_{n-2} i_{n-2}} a_{j_{n-1} i} a_{j_n i} t'_{j_1\ldots j_{n-2} j_{n-1} j_n} \\ &= a_{j_1 i_1}\ldots a_{j_{n-2} i_{n-2}} \delta_{j_{n-1} j_n} t'_{j_1\ldots j_{n-2} j_{n-1} j_n} = a_{j_1 i_1}\ldots a_{j_{n-2} i_{n-2}} t'_{j_1\ldots j_{n-2} j_n j_n}, \end{aligned}$$

or

$$t'_{i_1\ldots i_{n-2}ii} = a_{i_1 j_1}\ldots a_{i_{n-2} j_{n-2}} t_{j_1\ldots j_{n-2}jj},$$

which is the law of transformation of a rank $(n-2)$ tensor. The operation of contraction can be performed or repeated relative to any pair of indices.

A special role is played by the *Kronecker delta* and *Levi-Civita* symbols.

The second-rank tensor $t_{ij} = \delta_{ij}$, where δ_{ij} is the Kronecker symbol, is an invariant under an orthogonal transformation. Indeed, using (A.4), we have:

$$t'_{ij} = a_{ik}a_{jl}t_{kl} = a_{ik}a_{jl}\delta_{kl} = a_{ik}a_{jk} = \delta_{ij}.$$

Note that, except for the null tensor, the Kronecker tensor is the only second-rank invariant tensor.

Let us now show that the tensor quantity having its components defined by the Levi-Civita symbol

$$\epsilon_{ijk} = \begin{cases} +1, & \text{if } i, j \text{ and } k \text{ are an even permutation of } 1, 2, \text{ and } 3 \\ -1, & \text{if } i, j \text{ and } k \text{ are an odd permutation of } 1, 2, \text{ and } 3 \\ 0, & \text{if any of the indices are equal} \end{cases}$$

is an invariant pseudo-tensor of the third rank, called the *Levi-Civita pseudo-tensor*. To do this, we first observe that the determinant of an arbitrary 3×3 matrix can be written as

$$\det \hat{A} = \epsilon_{ijk} A_{1i} A_{2j} A_{3k}, \tag{A.17}$$

which yields:

$$\epsilon_{lmn} \det \hat{A} = \epsilon_{ijk} A_{li} A_{mj} A_{nk}. \tag{A.18}$$

The last two relations can be easily proved by a straightforward calculation.

Assume that the Levi-Civita pseudo-tensor is defined in some coordinate system by $t^{*}_{ijk} = \epsilon_{ijk}$ $(i, j, k = 1, 2, 3)$. Then, we have

$$t^{*\prime}_{ijk} = (\det \hat{A}) A_{il} A_{jm} A_{kn} \epsilon_{lmn},$$

or, in view of (A.18),

$$t^{*\prime}_{ijk} = (\det \hat{A})^2 \epsilon_{ijk} = \epsilon_{ijk},$$

which shows that the Levi-Civita pseudo-tensor is an invariant.

It can be proven that there are neither invariant tensors of odd rank, nor invariant pseudo-tensors of even rank. Any even rank invariant tensor can be written as a linear combination with scalar coefficients (numbers) of tensor products of the Kronecker tensor by itself. For example, the fourth rank invariant tensor t_{ijkl} has three distinct terms, as follows:

$$t_{ijkl} = C_1 \delta_{ij} \delta_{kl} + C_2 \delta_{ik} \delta_{jl} + C_3 \delta_{il} \delta_{jk}, \tag{A.19}$$

where C_1, C_2, C_3 are some numbers. In particular, since the tensor product of two Levi-Civita pseudo-tensors is a sixth rank tensor, it can be shown that

$$\epsilon_{ijk} \epsilon_{lmn} = \det \begin{pmatrix} \delta_{il} & \delta_{im} & \delta_{in} \\ \delta_{jl} & \delta_{jm} & \delta_{jn} \\ \delta_{kl} & \delta_{km} & \delta_{kn} \end{pmatrix}. \tag{A.20}$$

The reader can prove this relation by setting first $i = 1$, $j = 2$, $k = 3$ and then making a generalization for any $i, j, k = 1, 2, 3$. After some contractions, we obtain two useful relations:

$$\epsilon_{ijk} \epsilon_{lmk} = \delta_{il} \delta_{jm} - \delta_{im} \delta_{jl}, \tag{A.21}$$

$$\epsilon_{ijk} \epsilon_{ljk} = 2 \delta_{il}. \tag{A.22}$$

A.2.3 Vectors

Consider the oriented segment of a straight line $\vec{P_1P_2}$ whose projections on axes are $x_i^2 - x_i^1$ $(i = 1, 2, 3)$ (Fig. A.2). Passing to another coordinate system, by virtue of (A.1), we obtain:

$$x_i'^2 - x_i'^1 = a_{ij}\,(x_j^2 - x_j^1) \quad (i = 1, 2, 3), \tag{A.23}$$

which is similar to (A.10). Consequently, we can define a vector as a quantity which transforms as the difference of coordinates under an orthogonal transformation. In our case, we have a correspondence between a (free) vector and the set of oriented segments parallel to P_1P_2, which justifies the notation $\mathbf{v}$ for the set of three real numbers that characterize the vector $\mathbf{v}$: $\{v_1, v_2, v_3\}$. In particular, the vector quantity $\vec{OP}$, known as *radius-vector* or *position vector* of the point P, is the vector denoted by $\mathbf{r}$, of components x_1, x_2, x_3, i.e. $\mathbf{r}$: $\{x_1, x_2, x_3\}$.

Two vectors $\mathbf{A}$ and $\mathbf{B}$ are *equal* ($\mathbf{A} = \mathbf{B}$) if their components are equal, $A_i = B_i$ $(i = 1, 2, 3)$. It results from (A.10) that this property is intrinsic.

The *multiplication* of a vector $\mathbf{A}$: $\{A_1, A_2, A_3\}$ by a scalar α is a vector $\mathbf{B}$, denoted by $\alpha\,\mathbf{A}$, of components αA_i $(i = 1, 2, 3)$:

$$B_i' = \alpha' A_i' = \alpha\, a_{ij}A_j = a_{ij}(\alpha\, A_j) = a_{ij}B_j.$$

The vector $\mathbf{B}$ is parallel $(\alpha > 0)$ or antiparallel $(\alpha < 0)$ to $\mathbf{A}$. In particular, $-\mathbf{A} = (-1)\mathbf{A}$ is the *opposite* of $\mathbf{A}$.

The *sum* of two vectors $\mathbf{A}$: $\{A_1, A_2, A_3\}$ and $\mathbf{B}$: $\{B_1, B_2, B_3\}$ is a vector $\mathbf{C}$, denoted by $\mathbf{A} + \mathbf{B}$, of components $A_i + B_i$ $(i = 1, 2, 3)$. Indeed, by virtue of (A.10), we have:

$$C_i' = A_i' + B_i' = a_{ij}A_j + a_{ij}B_j = a_{ij}(A_j + B_j) = a_{ij}C_j.$$

The *difference* between the vectors $\mathbf{A}$ and $\mathbf{B}$ is defined as the sum of $\mathbf{A}$ and the opposite of $\mathbf{B}$, i.e. $\mathbf{A} - \mathbf{B} = \mathbf{A} + (-\mathbf{B})$.

The *null vector*, denoted by 0, is the vector with all components equal to zero. It is obvious that the sum of a vector with its opposite equals the null vector.

It is easy to verify that all these definitions are geometrically equivalent to the parallelogram rule.

The properties of summation and multiplication by scalars show that the set of all vectors form a linear space, namely the three-dimensional linear space L_3.

The non-zero vectors $\mathbf{A}_1, \ldots, \mathbf{A}_n$ are *linearly independent* if the equality $\alpha_1 A_1 + \cdots + \alpha_n A_n = 0$ is satisfied if and only if all scalars $\alpha_1, \ldots, \alpha_n$ are zero. If not, they are *linearly dependent.*

A system of three linearly independent vectors, say $\mathbf{A}_1$, $\mathbf{A}_2$, $\mathbf{A}_3$, forms a *basis* in L_3, because the maximum number of linearly independent vectors is equal to the dimension of the space, which is three. In order to form a basis, it is necessary and sufficient that the three vectors are not coplanar. Any vector $\mathbf{B}$ of the space L_3

can be written as a linear combination of the basis vectors. Indeed, the non-zero vectors $\mathbf{A_1}$, $\mathbf{A_2}$, $\mathbf{A_3}$ and $\mathbf{B}$ are linearly dependent, meaning that in the equality

$$\alpha_i \mathbf{A}_i + \beta \mathbf{B} = 0$$

we can choose $\beta \neq 0$, which yields

$$\mathbf{B} = \left(-\frac{\alpha_i}{\beta} \right) \mathbf{A}_i.$$

If the three vectors are orthogonal to each other, they form an *orthogonal basis*. An orthogonal basis of unit vectors is termed as an *orthonormal basis*.

Most frequently is used the orthonormal basis $\{\mathbf{u}_1, \mathbf{u}_2, \mathbf{u}_3\}$, where $\mathbf{u}_i$ $(i = 1, 2, 3)$ are unit vectors of the orthogonal coordinate system $Ox_1x_2x_3$, i.e. the vectors whose components are $(\mathbf{u}_i)_j = \delta_{ij}$. With the aforementioned definitions, for any vector $\mathbf{A}$ we can write the expansion:

$$\mathbf{A} = A_i \mathbf{u}_i. \tag{A.24}$$

If we denote by $\mathbf{u}_i'$ the unit vectors of the coordinate system $O'x_1'x_2'x_3'$, obtained from $Oxyz$ by an orthogonal transformation, we have:

$$\mathbf{u}_i' = (\mathbf{u}_i')_j \mathbf{u}_j = a_{ij} \mathbf{u}_j. \tag{A.25}$$

Scalar product. Being given the vectors $\mathbf{A}$: $\{A_1, A_2, A_3\}$ and $\mathbf{B}$: $\{B_1, B_2, B_3\}$, it is easy to verify that the quantity $A_i B_i$ is invariant under an orthogonal transformation. It is called the *scalar product* or *inner product* of the two vectors $\mathbf{A}$ and $\mathbf{B}$. Denoting by $\mathbf{A} \cdot \mathbf{B}$ the scalar product, we have:

$$\mathbf{A} \cdot \mathbf{B} = A_i B_i. \tag{A.26}$$

In view of the invariance of the inner product, a suitable choice of axes (x_1 along $\mathbf{A}$ and x_3 orthogonal to the plane defined by the vectors $\mathbf{A}$ and $\mathbf{B}$), yields the well-known formula of the scalar product:

$$\mathbf{A} \cdot \mathbf{B} = AB \cos \varphi, \tag{A.27}$$

where $\varphi \in [0, 2\pi]$ is the angle between the two vectors and A and B are their magnitudes. In particular, for the orthogonal unit vectors we obtain

$$\mathbf{u}_i \cdot \mathbf{u}_j = \delta_{ij}. \tag{A.28}$$

The scalar product has the following properties:

(a) $\mathbf{A} \cdot \mathbf{B} = \mathbf{B} \cdot \mathbf{A}$ (commutative law of multiplication);
(b) $(\mathbf{A} + \mathbf{B}) \cdot \mathbf{C} = \mathbf{A} \cdot \mathbf{C} + \mathbf{B} \cdot \mathbf{C}$ (distributive law);
(c) $(\alpha \mathbf{A}) \cdot (\beta \mathbf{B}) = \alpha \beta \mathbf{A} \cdot \mathbf{B}$ (α, β scalars);
(d) $\mathbf{A} \cdot \mathbf{A} = |\mathbf{A}|^2 \geq 0$; the equality sign appears only for $\mathbf{A} = 0$.

The *magnitude* or *length* of a vector **A** is defined by means of the scalar product as

$$A = |\mathbf{A}| = \sqrt{\mathbf{A} \cdot \mathbf{A}} = \sqrt{A_i A_i}. \tag{A.29}$$

Cross product. The *cross product* of any two polar vectors **A** and **B**, in a right-handed coordinate system, is the pseudo-vector

$$C_i^* = \epsilon_{ijk} A_j B_k. \tag{A.30}$$

It can be conventionally represented by an oriented segment using notation $\mathbf{A} \times \mathbf{B}$, where

$$\mathbf{A} \times \mathbf{B} = C_i^* \mathbf{u}_i = \epsilon_{ijk} A_j B_k \mathbf{u}_i, \tag{A.31}$$

or

$$\mathbf{A} \times \mathbf{B} = \begin{vmatrix} \mathbf{u}_1 & \mathbf{u}_2 & \mathbf{u}_3 \\ A_1 & A_2 & A_3 \\ B_1 & B_2 & B_3 \end{vmatrix}.$$

To find the geometric significance of the cross product, we shall use the same choice of axes as for the scalar product. The only non-zero component is then $C_3^* = AB_2 = AB \sin \varphi$, and the magnitude is

$$|\mathbf{A} \times \mathbf{B}| = AB \sin \varphi. \tag{A.32}$$

Thus, the magnitude of the cross product $\mathbf{A} \times \mathbf{B}$ is given by the area of the parallelogram determined by the two vectors, while its direction is determined by the right-hand screw rule: if **A** turns unto **B** around the $\mathbf{u}_3$-axis, then a right-hand screw will advance in the direction of $\mathbf{A} \times \mathbf{B}$. If **A** and **B** are parallel (or antiparallel), the cross product is zero.

The cross product of the unit vectors of coordinate axes is

$$\mathbf{u}_i \times \mathbf{u}_j = \epsilon_{ijk} \mathbf{u}_k, \tag{A.33}$$

while the cross product of a vector **A** by a unit vector $\mathbf{u}_i$ reads

$$\mathbf{A} \times \mathbf{u}_i = \epsilon_{ijk} A_k \mathbf{u}_j. \tag{A.34}$$

The cross product has the following properties:

(a) $\mathbf{A} \times \mathbf{B} = -\mathbf{B} \times \mathbf{A}$ (antisymmetry or anticommutative law of multiplication);
(b) $(\mathbf{A} + \mathbf{B}) \times \mathbf{C} = \mathbf{A} \times \mathbf{C} + \mathbf{B} \times \mathbf{C}$ (distributive law);
(c) $(\alpha \mathbf{A}) \times (\beta \mathbf{B}) = \alpha\beta \mathbf{A} \times \mathbf{B}$, for any scalars α, β.

The *mixed product* of three polar vectors **A**, **B**, **C** is the pseudo-scalar

$$\mathbf{A} \cdot (\mathbf{B} \times \mathbf{C}) = \epsilon_{ijk} A_i B_j C_k = \begin{vmatrix} A_1 & A_2 & A_3 \\ B_1 & B_2 & B_3 \\ C_1 & C_2 & C_3 \end{vmatrix}. \tag{A.35}$$

In view of (A.26) and (A.31), the magnitude of the mixed product is the volume of the parallelepiped formed by the three vectors. Using the definitions of scalar and cross products, we obtain:

$$\begin{aligned}\mathbf{A}\cdot(\mathbf{B}\times\mathbf{C}) &= \mathbf{B}\cdot(\mathbf{C}\times\mathbf{A}) = \mathbf{C}\cdot(\mathbf{A}\times\mathbf{B})\\ &= -\mathbf{B}\cdot(\mathbf{A}\times\mathbf{C}) = -\mathbf{A}\cdot(\mathbf{C}\times\mathbf{B}) = -\mathbf{C}\cdot(\mathbf{B}\times\mathbf{A}).\end{aligned} \tag{A.36}$$

If the three vectors are coplanar, the mixed product is zero.

The *double cross product* of the three vectors **A**, **B**, **C** is a vector defined as

$$[\mathbf{A}\times(\mathbf{B}\times\mathbf{C})]_i = (\mathbf{A}\cdot\mathbf{C})B_i - (\mathbf{A}\cdot\mathbf{B})C_i \quad (i = 1,2,3),$$

or

$$\mathbf{A}\times(\mathbf{B}\times\mathbf{C}) = (\mathbf{A}\cdot\mathbf{C})\mathbf{B} - (\mathbf{A}\cdot\mathbf{B})\mathbf{C}. \tag{A.37}$$

A.2.4 Second-Rank Tensors

Let $\mathcal{V}$ be the set of all vectors of E_3. Then, given the vector $\mathbf{v}\in\mathcal{V}$ and the second-rank tensor $\hat{\mathbf{t}}$, the quantity

$$w_i = t_{ij}v_j \quad (i = 1,2,3) \tag{A.38}$$

is also a vector. Obviously, the following statement is also true: if $\mathbf{v},\ \mathbf{w}\in\mathcal{V}$, then the nine-component quantity t_{ij} $(i, j = 1,2,3)$ is a second-rank tensor. The matrix form of (A.38) is

$$\mathbf{w} = \hat{\mathbf{t}}\mathbf{v}, \tag{A.39}$$

called *right scalar product* between the tensor and the vector. In a similar way we can define the *left scalar product* by

$$w'_i = t_{ji}v_j \quad (i = 1,2,3), \tag{A.40}$$

whose matrix form is

$$\mathbf{w}' = \hat{\mathbf{t}}^T\mathbf{v}, \tag{A.41}$$

where $\hat{\mathbf{t}}^T$ is the transposed matrix associated with the tensor.

These considerations show that a second-rank tensor defines a linear application of $\mathcal{V}$ onto itself, i.e. $\hat{\mathbf{t}}$ is a linear operator on $\mathcal{V}$. In particular, if we take $\mathbf{v} = \mathbf{u}_k$ and observe that $(\mathbf{u}_k)_j = \delta_{kj}$, we obtain:

$$\mathbf{T}_i = t_{ik}\mathbf{u}_k \quad (i = 1,2,3). \tag{A.42}$$

These quantities are the *vector components* of the tensor $\hat{\mathbf{t}}$.

The algebraic operations involving tensors are considerably simplified by introducing the notion of *dyadic*. We call *dyadic product* or *Kronecker product* of an Euclidean space E_3 by itself a space denoted by $E_3 \otimes E_3$, defined in such a way that a pair of vectors $\mathbf{A}$, $\mathbf{B} \in E_3$ is associated with one and only one element of $E_3 \otimes E_3$, denoted by $\mathbf{AB}$ and having the following properties:

1°. $(\mathbf{A}_1 + \mathbf{A}_2)\mathbf{B} = \mathbf{A}_1\mathbf{B} + \mathbf{A}_2\mathbf{B}$, $\mathbf{A}(\mathbf{B}_1 + \mathbf{B}_2) = \mathbf{AB}_1 + \mathbf{AB}_2$,
2°. $(\alpha\mathbf{A})\mathbf{B} = \mathbf{A}(\alpha\mathbf{B}) = \alpha\mathbf{AB}$, where α is a scalar. It results that $E_3 \otimes E_3$ is a vector space of $3 \times 3 = 9$ dimensions. If in E_3 we have an orthonormal basis $\mathbf{u}_i$ $(i = 1, 2, 3)$, then in $E_3 \otimes E_3$ we have the basis $\mathbf{u}_i\mathbf{u}_j$, and we can write

$$\mathbf{AB} = (A_i\mathbf{u}_i)(B_j\mathbf{u}_j) = A_iB_j\mathbf{u}_i\mathbf{u}_j.$$

The quantity $\mathbf{AB}$ is called a *dyadic vector*, or simply a *dyadic*. One can define the following operations:

(a) $\mathbf{C} \cdot (\mathbf{AB}) = (\mathbf{C} \cdot \mathbf{A})\mathbf{B}$ (semiscalar left product);
$(\mathbf{AB}) \cdot \mathbf{C} = \mathbf{A}(\mathbf{B} \cdot \mathbf{C})$ (semiscalar right product);
(b) $\mathbf{C} \times (\mathbf{AB}) = (\mathbf{C} \times \mathbf{A})\mathbf{B}$ (semicross left product);
$(\mathbf{AB}) \times \mathbf{C} = \mathbf{A}(\mathbf{B} \times \mathbf{C})$ (semicross right product);
(c) $(\mathbf{AB}) \cdot (\mathbf{CD}) = (\mathbf{B} \cdot \mathbf{C})\mathbf{AD}$ (inner product of two dyadics).

By virtue of (A.42), a second-rank tensor can be associated with the dyadic

$$\{t\} = \mathbf{u}_i\mathbf{T}_i = t_{ij}\mathbf{u}_i\mathbf{u}_j \tag{A.43}$$

of $E_3 \otimes E_3$. In particular, the Kronecker tensor δ_{ij} is associated with the dyadic

$$\{1\} = \mathbf{u}_i\mathbf{u}_i \tag{A.44}$$

called the *dyadic unit vector*.

The use of dyadics has the advantage of employing the usual vector operations. For example, in view of (a), (b) and (A.42), we have:

$$\mathbf{A} \cdot \{t\} = \mathbf{A} \cdot (\mathbf{u}_i\mathbf{T}_i) = (\mathbf{A} \cdot \mathbf{u}_i)\mathbf{T}_i = A_i\mathbf{T}_i, \tag{A.45}$$

$$\mathbf{A} \times \{t\} = (\mathbf{A} \times \mathbf{u}_i)\mathbf{T}_i = \epsilon_{ijk}A_k\mathbf{u}_j\mathbf{T}_i, \tag{A.46}$$

$$\mathbf{A} \cdot \{t\} \cdot \mathbf{B} = (A_i\mathbf{T}_i) \cdot (B_k\mathbf{u}_k) = A_it_{ij}B_j. \tag{A.47}$$

The dyadic unit vector has the following properties:

$$\mathbf{A} \cdot \{1\} = \{1\} \cdot \mathbf{A} = \mathbf{A}, \tag{A.48}$$

$$\{t\} \cdot \{1\} = \{1\} \cdot \{t\} = \{t\}. \tag{A.49}$$

Any second-rank tensor can be expressed as the sum of a symmetric and an antisymmetric part:

$$t_{ij} = S_{ij} + A_{ij},$$

where

$$S_{ij} = \frac{1}{2}(t_{ij} + t_{ji}) \ , \quad A_{ij} = \frac{1}{2}(t_{ij} - t_{ji}).$$

Since the properties of symmetry and antisymmetry are intrinsic, the quantities S_{ij} and A_{ij} are tensors.

If $\hat{\mathbf{t}}$ is a second-rank tensor and $\mathbf{v}, \mathbf{w} \in \mathcal{V}$, we can define the bilinear form

$$L(\mathbf{v}, \mathbf{w}) \equiv \mathbf{v}(\hat{\mathbf{t}}\mathbf{w}) = v_i t_{ij} w_j. \tag{A.50}$$

It is seen that $L(\mathbf{v}, \mathbf{w})$ is a linear function in each of its arguments and, in view of (A.38), it has scalar values. If $\hat{\mathbf{t}}$ is symmetric, then $L(\mathbf{v}, \mathbf{w}) = L(\mathbf{w}, \mathbf{v})$. If $\mathbf{w} = \mathbf{v}$, we obtain a quadratic form:

$$L(\mathbf{v}) \equiv L(\mathbf{v}, \mathbf{v}) = \mathbf{v} \cdot (\hat{\mathbf{t}}\mathbf{v}) = t_{ij} v_i v_j. \tag{A.51}$$

If by $\mathbf{v}$ we mean the position vector $\mathbf{r}$ of a point in E_3, then equation $L(\mathbf{r}) = C$, where C is a constant, defines a quadric with its centre of symmetry in O, associated with the second-rank tensor $\hat{\mathbf{t}}$. Since a quadratic form is an invariant under an orthogonal transformation, the associated quadric does not depend on the choice of the coordinate axes.

If in (A.18) we set $\hat{A} = \hat{t}$ and use the tensor properties of ϵ_{ijk}, we arrive at the conclusion that the determinant of the coefficients of the quadric (A.51),

$$D = \det(\hat{\mathbf{t}}) = \det(t_{ij}),$$

is an invariant under orthogonal transformations. Another invariant of a second-rank tensor is the sum of its diagonal elements, called *trace* or *spur*:

$$\mathrm{Tr}\,\hat{\mathbf{t}} = t_{ii}. \tag{A.52}$$

Observing that $x_i x_i = \text{const.}$ defines a quadric (sphere), the expression

$$(t_{ij} - \lambda\delta_{ij}) x_i x_j = \text{const.}$$

is also a quadric, attached to the tensor $\tilde{t}_{ij} = t_{ij} - \lambda\delta_{ij}$, for any scalar λ. This means that $D_\lambda = \det(\tilde{t}_{ij})$ must be an invariant for any λ, and consequently the coefficient of each power of λ in the polynomial

$$D_\lambda = -\lambda^3 + U\lambda^2 - \Delta\lambda + D$$

must be an invariant. The quantities $D = \det\,\hat{\mathbf{t}}$, $U = \mathrm{Tr}\,\hat{\mathbf{t}}$ and

$$\Delta = t_{11}t_{22} + t_{11}t_{33} + t_{22}t_{33} - t_{12}t_{21} - t_{13}t_{31} - t_{23}t_{32} \tag{A.53}$$

are called the *principal invariants of the tensor* $\hat{\mathbf{t}}$.

Consider now the second-rank symmetric tensor t_{ij} $(i, j = 1, 2, 3)$ and let us find out if there exist unit vectors $\mathbf{f} \in \mathcal{V}$ and scalars λ, so that the vector defined by (A.38), with $\mathbf{v} = \mathbf{f}$, is equal to $\lambda\,\mathbf{f}$, i.e.

$$\hat{\mathbf{t}}\mathbf{f} = \lambda \mathbf{f}, \tag{A.54}$$

or

$$t_{ij} f_j = \lambda \, f_i \quad (i = 1, 2, 3). \tag{A.55}$$

The homogeneous system of equations (A.55) has a non-trivial solution if and only if the determinant of the matrix of coefficients, which is precisely D_λ, is zero. The equation $D_\lambda = 0$ is called the *characteristic equation*, and the three roots of this equation are the *proper values* or *eigenvalues* of the tensor $\hat{\mathbf{t}}$. The invariance of D_λ leads to the invariance of the proper values, which means that they are intrinsic.

Since $\hat{\mathbf{t}}$ is symmetric, all three roots of the characteristic equations are real. Indeed, if they were complex, then the vectors $\mathbf{f}$ should be also complex and then, by virtue of hypothesis of non-trivial solutions, we obtain from (A.55):

$$\lambda = (t_{ij} f_j f_i^*)(f_k f_k^*)^{-1},$$

which yields $t_{ij} = t_{ji}$, $\lambda^* = \lambda$. Therefore, we can consider only real unit vectors $\mathbf{f}$. Those vectors $\mathbf{f}$ which satisfy (A.54) are called the *proper vectors* of the symmetric tensor $\hat{\mathbf{t}}$. It can be shown that (A.54) admits three proper vectors $\mathbf{f}_k$ $(k = 1, 2, 3)$ which are (or can be made) orthonormal:

$$\mathbf{f}_k \cdot \mathbf{f}_l = \delta_{kl} \quad (k, l = 1, 2, 3). \tag{A.56}$$

Let $\lambda_k \mathbf{f}_k$ and $\lambda_l \mathbf{f}_l$ be two pairs of proper value – proper vector which satisfy (A.54). Then (A.54) yields $(\hat{\mathbf{t}}\mathbf{f}_k) \cdot \mathbf{f}_l = \lambda_k(\mathbf{f}_k \cdot \mathbf{f}_l)$ (no summation over k). Replacing k by l and subtracting, we obtain:

$$(\lambda_k - \lambda_l)(\mathbf{f}_k \cdot \mathbf{f}_l) = 0.$$

This shows that (A.56) can be satisfied if, for $k \neq l$, we have $\lambda_k \neq \lambda_l$. But if there is a double root, say $\lambda_1 = \lambda_2 \neq \lambda_3$, then in general $\mathbf{f}_1 \cdot \mathbf{f}_2 \neq 0$. It can be shown that the vector

$$\mathbf{g} = \alpha_1 \mathbf{f}_1 + \alpha_2 \mathbf{f}_2$$

is also a proper vector, corresponding to the proper value $\lambda_1 = \lambda_2$.

The relation (A.56) shows that the matrix $\hat{\mathcal{R}}$, of components $\mathcal{R}_{ij} = (\mathbf{f}_i)_j = f_{ij}$ is orthogonal, and therefore it defines an orthogonal transformation leading to a new coordinate system. Denoting by $\mathbf{u}_i'$ the new unit vectors, from (A.25) we obtain:

$$\mathbf{u}_i' = f_{ij} \mathbf{u}_j = \mathbf{f}_i,$$

which means that the axes of the new coordinate system are determined by the proper vectors $\mathbf{f}_i$. In the new frame, the components of the tensor $\hat{\mathbf{t}}$ transform according to

$$t'_{ij} = f_{ik} f_{jl} t_{kl} = \lambda_j f_{ik} f_{jk} = \lambda_j \delta_{ij}.$$

(Note that there is no summation over j in the last relation.) This shows that the matrix associated with the tensor $\hat{\mathbf{t}}$ in the basis $\{\mathbf{f}_1, \mathbf{f}_2, \mathbf{f}_3\}$ is *diagonal*, i.e. the only non-zero components lie on the principal diagonal, being equal to the proper values of the tensor. Thus, by a rotation of matrix $\hat{\mathcal{R}}$, we brought the tensor to the diagonal form. The axes defined by the unit vectors $\mathbf{f}_i$ are called the *principal axes* of the tensor $\hat{\mathbf{t}}$. In this frame the quadratic form (A.51) reduces to a sum of squared quantities, while the quadric $L(\mathbf{r}) = C$ is brought to the canonical form:

$$\sum_{i=1}^{3} \lambda_i x_i^2 = C. \tag{A.57}$$

This means that the principal axes of the symmetric tensor $\hat{\mathbf{t}}$ are also symmetry axes of the associated quadric.

As for second-rank antisymmetric tensors, from the definition $t_{ij} = -t_{ji}$ it results that such a tensor has only three distinct components, leading to a pseudo-vector $\mathbf{v}^*$ of components $v_1^* = t_{23},\ v_2^* = t_{31},\ v_3^* = t_{12}$. In a compact form, this can be written as

$$v_i^* = \frac{1}{2} \epsilon_{ijk} t_{jk} \quad (i = 1, 2, 3). \tag{A.58}$$

The pseudo-vector $\mathbf{v}^*$ whose components are given by (A.58) is the *dual pseudo-vector* associated with the antisymmetric tensor $\hat{\mathbf{t}}$. Using (A.21), we can also write

$$t_{ij} = \epsilon_{ijk}\, v_k^*, \tag{A.59}$$

which shows that any antisymmetric tensor of the second rank can be associated with a pseudo-vector. Thus, the relations (A.58) and (A.59) are equivalent.

Appendix B
Elements of Vector and Tensor Analysis

B.1 Fundamental Notions

B.1.1 Vector Functions on E_3

Let t be a variable parameter. If to any value of t corresponds a value of a vector $\mathbf{A}$, we say that $\mathbf{A}(t)$ is a *vector function* of t.

Let t_1 be a value of t. Then, if with any given infinitesimal ϵ we can associate a positive number η_ϵ, so that $|t - t_1| < \eta_\epsilon$ yields $|\mathbf{A}(t) - \mathbf{A}(t_1)| < \epsilon$, we say that $\mathbf{A}(t)$ is a *continuous vector function* of t.

B.1.2 Derivative of a Vector Function

Let P be a point mass (particle) and $\mathbf{r}(t)$ – its radius-vector relative to an arbitrary point O. When t varies, the arrow of the vector describes a curve C, called the *hodograph* of the function $\mathbf{r}(t)$. To some variation Δt of t will correspond a displacement of the particle from P to P_1, i.e. a variation $\Delta\mathbf{r} = \mathbf{r}(t + \Delta t) - \mathbf{r}(t)$ of the function $\mathbf{r}(t)$. The ratio $\Delta\mathbf{r}/\Delta t$ is a vector collinear with $\Delta\mathbf{r}$. The limit

$$\lim_{\Delta t \to 0} \frac{\Delta\mathbf{r}}{\Delta t} = \frac{d\mathbf{r}}{dt} = \mathbf{r}'(t) \tag{B.1}$$

is called the *derivative* of the vector function $\mathbf{r}(t)$ with respect to t at the point P. The vector $d\mathbf{r}/dt$ is tangent to the curve C at the point P. If the parameter t is the time, the derivative is denoted by $\dot{\mathbf{r}}$.

By definition, the *differential* of $\mathbf{r}(t)$ is $d\mathbf{r} = \mathbf{r}'(t)\, dt$, where $\mathbf{r}'(t)$ denotes the derivative of $\mathbf{r}(t)$ and dt is an arbitrary elementary variation of t. If $\mathbf{r}$ is a function of n variables $t_1, \ldots, t_n$, its differential is

$$d\mathbf{r} = \frac{\partial\mathbf{r}}{\partial t_i} dt_i, \tag{B.2}$$

M. Chaichian et al., *Mechanics*, DOI: 10.1007/978-3-642-17234-2,

where the symbol $\partial/\partial t_i$ stands for the partial derivative with respect to t_i and the summation convention over index $i = \overline{1,n}$ has been used.

The differentiation of vector functions is performed according to the same rules as for the scalar functions, but taking into account the properties of the vector algebra.

Application. Let $\mathbf{R}(t)$ be a vector of constant modulus, with its origin at O, whose arrow traces the circle of radius R lying in the xy-plane (Fig. B.1). If $\mathbf{u}_r$ is the unit vector of $\mathbf{R}$ and t is the time, we can write

$$\frac{d\mathbf{R}}{dt} = R\frac{d\mathbf{u}_r}{dt}.$$

We denote the unit vectors of the x-, y- and z-axis by $\mathbf{i}$, $\mathbf{j}$ and $\mathbf{k}$, respectively. Since $\mathbf{u}_r = \mathbf{i}\cos\varphi + \mathbf{j}\sin\varphi$, we have:

$$\begin{aligned}\frac{d\mathbf{u}_r}{dt} &= \dot{\varphi}(-\mathbf{i}\sin\varphi + \mathbf{j}\cos\varphi)\\ &= \dot{\varphi}\left[\mathbf{i}\cos\left(\varphi + \frac{\pi}{2}\right) + \mathbf{j}\sin\left(\varphi + \frac{\pi}{2}\right)\right] = \dot{\varphi}\mathbf{u}_\varphi,\end{aligned}$$

where

$$\mathbf{u}_\varphi = \mathbf{i}\cos\left(\varphi + \frac{\pi}{2}\right) + \mathbf{j}\sin\left(\varphi + \frac{\pi}{2}\right)$$

is the unit vector obtained by a counterclockwise rotation of angle $\pi/2$ of $\mathbf{u}_r$ about its origin. Thus,

$$\frac{d\mathbf{R}}{dt} = R\dot{\varphi}\mathbf{u}_\varphi, \quad \left|\frac{d\mathbf{R}}{dt}\right| = R\,\dot{\varphi}. \tag{B.3}$$

B.1.3 Integral of a Vector Function

The definition of the integral of a vector function is analogous to that of a scalar function. Let $\mathbf{A}(\tau)$ be a vector function of the parameter τ in the interval $[t_0, t]$. Dividing this interval in partial intervals by a partition $t_0 = \tau_0 \le \tau_1 \le \ldots \le \tau_n = t$, let us take the sum:

$$\sum_{i=1}^{n} \mathbf{A}(\tilde{\tau}_i)[\tau_i - \tau_{i-1}], \quad \tilde{\tau}_i \in [\tau_{i-1}, \tau_i].$$

If, for any choice of points, there exists a limit of this sum for $n \to \infty$ and max $[\tau_i - \tau_{i-1}] \to 0$, the same for any division of the interval $[t_0, t]$, then this limit is denoted

$$\int_{t_0}^{t} \mathbf{A}(\tau)\,d\tau = \mathbf{a}(t) - \mathbf{a}(t_0), \tag{B.4}$$

being called the *integral* of the vector function $\mathbf{A}(\tau)$ between the limits t_0 and t. The function $\mathbf{a}(\tau)$ is the *primitive* of $\mathbf{A}(\tau)$.

In mechanics, most frequently are encountered the following special integrals:

(a) The line integral $\int_C \mathbf{A}(\mathbf{r}) \cdot d\mathbf{s} = \int_a^b \mathbf{A}(\mathbf{r}(t)) \cdot \mathbf{r}'(t)\, dt$ along a curve C, parametrized bijectively by $\mathbf{r} : [a, b] \to C$, such that $\mathbf{r}(a)$ and $\mathbf{r}(b)$ represent the end-points of C and $d\mathbf{s}$ is a vector element of C. This integral is called the *circulation* of the vector $\mathbf{A}$ on the curve C between a and b. If C is a closed curve, the circulation is denoted $\oint_C \mathbf{A} \cdot d\mathbf{s}$.
(b) The surface integral $\int_S \mathbf{A} \cdot d\mathbf{S}$, where $\mathbf{A}$ is a vector with its origin on the surface S and $d\mathbf{S}$ is a vector surface element. This integral is called the *flux* of the vector $\mathbf{A}$ through the surface S. Since a surface is parametrized by two variables, the surface integral is basically a double integral. If the surface is closed, one uses the notation $\oint_S \mathbf{A} \cdot d\mathbf{S}$.
(c) The volume integral $\int_V \mathbf{A}\, d\tau$, where V is the volume of a domain D of E_3, $d\tau$ is a volume element, while $\mathbf{A}$ has its origin somewhere in D.

B.1.4 Scalar and Vector Fields

If in a domain $D \subset E_3$ with any point $P \in D$ one can associate a value of a scalar $\varphi(P)$, we say that on D is defined a *scalar field*. For example, such a field exists around a radiator: in each point of the surrounding space can be defined a scalar called the *temperature*.

If with any point $P \in D$ one can associate a vector quantity $\mathbf{A}(P)$, then on D we have defined a *vector field*. Such a field is, for example, the electric field strength $\mathbf{E}$. The velocities of molecules of a moving fluid represent also a vector field.

A scalar field φ or a vector field $\mathbf{A}$ are called *non-stationary* if φ or $\mathbf{A}$ depends explicitly on time. If not, the field is termed *stationary*.

B.2 Applications

B.2.1 The Serret–Frenet Frame

Assume that the trajectory of a particle P is a skew-regular curve C. Let P_0 be a fixed point on the curve and denote by s the arc length P_0P, called the *curve abscissa* of P (Fig. B.2). Then, if $\mathbf{r}$ is the radius-vector of P relative to the origin of the Cartesian orthogonal frame $Ox_1x_2x_3$, the parametric equations of the curve C are $x_i = x_i(s)$ $(i = 1, 2, 3)$ or, in vector form,

$$\mathbf{r} = \mathbf{r}(s). \tag{B.5}$$

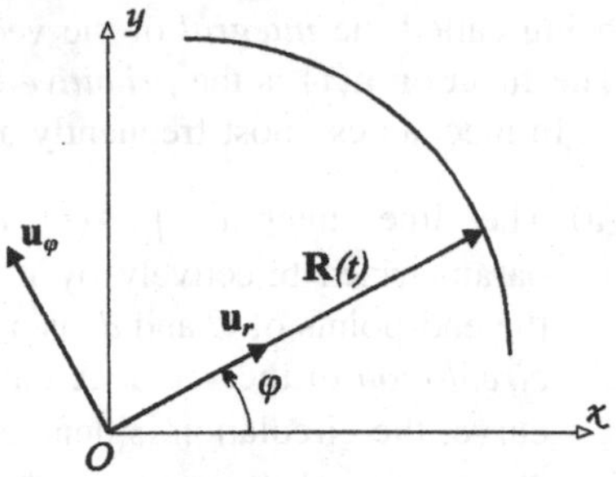

Fig. B.1 A vector $\mathbf{R}(t)$ of constant modulus, whose arrow traces a circle of radius R.

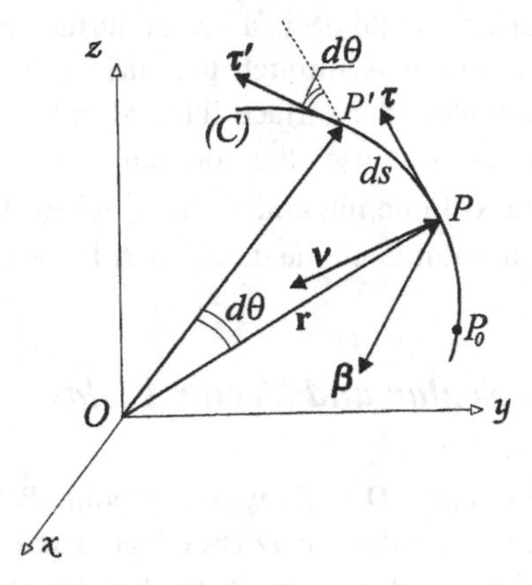

Fig. B.2 The Serret–Frenet reference frame.

Since $|d\mathbf{r}| = ds$, the quantity $d\mathbf{r}/ds$ is a unit vector $\boldsymbol{\tau}$ oriented along the tangent to the curve C at the point P:

$$\frac{d\mathbf{r}}{ds} = \boldsymbol{\tau}. \tag{B.6}$$

When the point P describes the curve C, the direction of $\boldsymbol{\tau}$ varies, which means that $\boldsymbol{\tau}$ is a function of s. The elementary arc ds depends, in its turn, on the angle $d\theta$ between the tangents at P and P' (Fig. B.2). Therefore

$$\frac{d\boldsymbol{\tau}}{ds} = \frac{d\theta}{ds}\frac{d\boldsymbol{\tau}}{d\theta} = \frac{1}{\rho}\boldsymbol{\nu}, \tag{B.7}$$

where $\boldsymbol{\nu}$ is the unit vector and $d\theta/ds = 1/\rho > 0$ is the magnitude of $d\boldsymbol{\tau}/ds$. The unit vector $\boldsymbol{\nu}$ is orthogonal to $\boldsymbol{\tau}$ at the point P and defines the *principal normal* to the curve C at P. The scalar ρ is the *radius of curvature* and $1/\rho$ is the *curvature* of the curve at the point P. The plane determined by the unit vectors $\boldsymbol{\tau}$ and $\boldsymbol{\nu}$ is called the *osculating* plane at the point P.

One can define a third unit vector $\boldsymbol{\beta} = \boldsymbol{\tau} \times \boldsymbol{\nu}$, known as the *unit bi-normal* to the curve C at the point P. The plane determined by $\boldsymbol{\nu}$ and $\boldsymbol{\beta}$ is called the *normal plane* to the curve at P. The orthogonal right-handed system of unit vectors $\boldsymbol{\tau}, \boldsymbol{\nu}$ and $\boldsymbol{\beta}$ is called the *Serret–Frenet trihedral system*.

Let us express the variation with respect to s of the unit vectors $\boldsymbol{\beta}$ and $\boldsymbol{\nu}$. Since $\boldsymbol{\beta} \cdot \boldsymbol{\beta} = 1$, we have

$$\boldsymbol{\beta} \cdot \frac{d\boldsymbol{\beta}}{ds} = 0,$$

i.e. the vectors $\boldsymbol{\beta}$ and $d\boldsymbol{\beta}/ds$ are orthogonal. In other words, the vectors $d\boldsymbol{\beta}/ds$, $\boldsymbol{\nu}$ and $\boldsymbol{\tau}$ are coplanar and, recalling the results of Appendix A, Sect. A.2.3, we can write:

$$\frac{d\boldsymbol{\beta}}{ds} = \lambda \boldsymbol{\tau} + \mu \boldsymbol{\nu}. \tag{B.8}$$

On the other hand, taking the derivative with respect to s of the relation $\boldsymbol{\tau} \cdot \boldsymbol{\beta} = 0$, we have:

$$\boldsymbol{\tau} \cdot \frac{d\boldsymbol{\beta}}{ds} + \boldsymbol{\beta} \cdot \frac{d\boldsymbol{\tau}}{ds} = 0.$$

In view of (B.7), we obtain:

$$\boldsymbol{\tau} \cdot \frac{d\boldsymbol{\beta}}{ds} = 0$$

and (B.8) yields $\lambda = 0$. Setting $\mu = -1/T$, we arrive at

$$\frac{d\boldsymbol{\beta}}{ds} = -\frac{1}{T}\boldsymbol{\nu}. \tag{B.9}$$

The scalar quantity T is the *torsion* of the curve C at the point P, while $1/T$ is the *radius of torsion*. The torsion indicates the deviation of a curve from the plane shape. Indeed, if C is a plane curve the direction of the unit vector $\boldsymbol{\beta}$ will remain unchanged, therefore $d\boldsymbol{\beta}/ds = 0$, consequently $1/T = 0$ (i.e. $T = \infty$).

To justify the choice of the minus sign in (B.9), let us obtain this formula in a different way. We observe that when P moves on the curve, the angle α between the unit bi-normal and some reference direction varies according to

$$\frac{d\boldsymbol{\beta}}{ds} = \frac{d\alpha}{ds}\frac{d\boldsymbol{\beta}}{d\alpha}. \tag{B.10}$$

Suppose that α is described by a positive (counterclockwise) rotation of the trihedral system about $\boldsymbol{\tau}$. In this case, the quantity $d\alpha/ds = 1/T > 0$ is the radius of torsion at P, while $d\boldsymbol{\beta}/d\alpha = -\boldsymbol{\nu}$ (see (B.7)). Thus, we fall back on the relation (B.9).

Taking the derivative with respect to s of $\boldsymbol{\nu} = \boldsymbol{\beta} \times \boldsymbol{\tau}$ and using (B.7) and (B.9), we have:

$$\frac{d\boldsymbol{\nu}}{ds} = \boldsymbol{\beta} \times \frac{d\boldsymbol{\tau}}{ds} + \frac{d\boldsymbol{\beta}}{ds} \times \boldsymbol{\tau} = \frac{1}{\rho}\boldsymbol{\beta} \times \boldsymbol{\nu} - \frac{1}{T}\boldsymbol{\nu} \times \boldsymbol{\tau},$$

and finally,

$$\frac{d\boldsymbol{\nu}}{ds} = -\frac{1}{\rho}\boldsymbol{\tau} + \frac{1}{T}\boldsymbol{\beta}. \tag{B.11}$$

The relations (B.7), (B.9) and (B.11) are known as the *Serret–Frenet* formulas. The reader is advised to check the validity of the following relations, giving the *curvature* and the *torsion* of a curve, as an application of the Serret–Frenet formulas:

$$\frac{1}{\rho} = \left(\frac{d^2x_1}{ds^2} + \frac{d^2x_2}{ds^2} + \frac{d^2x_3}{ds^2} \right)^{\frac{1}{2}}, \tag{B.12}$$

$$\frac{1}{T} = \rho^2 \frac{d\mathbf{r}}{ds} \cdot \left(\frac{d^2\mathbf{r}}{ds^2} \times \frac{d^3\mathbf{r}}{ds^3} \right). \tag{B.13}$$

The Serret–Frenet trihedral system, also called the *natural system of coordinates*, can be attached to any point of the trajectory of a particle, in order to study the elements which characterize its motion: velocity, acceleration, etc. This makes it a particularly useful instrument in mechanics.

B.2.2 Differential Vector Operators in Cartesian Coordinates

B.2.2.1 Gradient

Consider a scalar function $\varphi(x_1, x_2, x_3)$ of class C^2, defined on some domain $D \subset E_3$. Using the aforementioned definition, it results that we have in D a stationary scalar field. One observes that the differential $d\varphi = \frac{\partial \varphi}{\partial x_i} dx_i$ is the scalar product of the vectors $d\mathbf{r} = \mathbf{u}_i dx_i$ and $\mathbf{A} = \mathbf{u}_i(\partial\varphi/\partial x_i)$. The vector field $\mathbf{A}$ is called the *gradient* of the scalar function φ and is written grad φ. Therefore

$$d\varphi = \text{grad}\,\varphi \cdot d\mathbf{r}. \tag{B.14}$$

It is seen that the vector field $\mathbf{A} = \text{grad}\,\varphi$ has been obtained by applying the differential operator

$$\nabla = \mathbf{u}_i \frac{\partial}{\partial x_i} \tag{B.15}$$

to the scalar function $\varphi(x_1, x_2, x_3)$. This vector operator is called *del* or *nabla*. It was introduced by Hamilton. Therefore, we can write

$$\nabla\varphi = \text{grad}\,\varphi = \mathbf{u}_i \frac{\partial \varphi}{\partial x_i}. \tag{B.16}$$

The vector field $\mathbf{A} = \text{grad}\,\varphi(x_1, x_2, x_3)$ is called a *conservative field*. Such a field is, for example, the electrostatic field strength $\mathbf{E} = -\text{grad}\,\phi$, where $\phi(x_1, x_2, x_3)$ is the electrostatic *scalar potential*.

Equipotential surfaces. Consider the fixed surface

$$\varphi(x_1, x_2, x_3) = C(\text{const.}). \tag{B.17}$$

Then, from (B.14) we have $\nabla\varphi \cdot d\mathbf{r} = 0$. Since $d\mathbf{r}$ lies in the plane tangent to the surface (B.17), it follows that at any point of this surface the vector $\nabla\varphi$ is directed along the normal to the surface. Each value of the constant C gives another surface. These surfaces are called *equipotential surfaces*. In other words, given the scalar field $\varphi(x_1, x_2, x_3)$, the geometric locus of the points having the property $\nabla\varphi \cdot d\mathbf{r} = 0$ is an equipotential surface.

Field lines. Consider the stationary vector field $\mathbf{A}(x_1, x_2, x_3)$ and a curve C given by its parametric equations $x_i = x_i(s)$ $(i = 1, 2, 3)$. If the field $\mathbf{A}$ is tangent to the curve C at any point, then the curve C is a *line of the field* $\mathbf{A}$. The differential equations of the field lines are obtained from the obvious relation $\mathbf{A} \times d\mathbf{s} = 0$, where $d\mathbf{s}$ is a vector element of the field line. In projection on axes, this yields

$$\frac{dx_1}{A_1} = \frac{dx_2}{A_2} = \frac{dx_3}{A_3}. \tag{B.18}$$

Directional derivative. The component of grad φ along the x_i-axis is

$$(\operatorname{grad}\varphi)_i = \left(\mathbf{u}_k \frac{\partial\varphi}{\partial x_k}\right) \cdot \mathbf{u}_i = \frac{\partial\varphi}{\partial x_k}\delta_{ik} = \frac{\partial\varphi}{\partial x_i} \quad (i = 1, 2, 3). \tag{B.19}$$

Partial derivatives of φ with respect to x_i are denoted also by $\partial_i\varphi$ or $\varphi_{,i}$.

Let us project the vector grad φ on some direction defined by the unit vector $\mathbf{w}$. We can write $d\mathbf{r} = \mathbf{w}dr = \mathbf{w}ds$, where ds is the magnitude of the elementary displacement vector $d\mathbf{s}$. Dividing (B.14) by ds, we then obtain the *directional derivative* of φ

$$(\operatorname{grad}\varphi) \cdot \mathbf{w} = (\operatorname{grad}\varphi)_w = \frac{d\varphi}{dw}. \tag{B.20}$$

If, in particular, $\mathbf{w}$ coincides with the unit vector $\mathbf{n}$ of the normal to the surface (B.17), chosen as positive, then

$$(\operatorname{grad}\varphi) \cdot \mathbf{n} = \frac{d\varphi}{dn} \geq 0, \tag{B.21}$$

which shows that the gradient is oriented along the normal at the equipotential surface at any point and its direction indicates the maximum rate of variation of the function φ.

B.2.2.2 Divergence

Taking into account the properties of the vector operator ∇, let us apply it on an arbitrary vector field $\mathbf{A}$ by taking the scalar product

$$\operatorname{div}\mathbf{A} = \nabla \cdot \mathbf{A} = \left(\mathbf{u}_i \frac{\partial}{\partial x_i}\right) \cdot (A_k\mathbf{u}_k) = \delta_{ik}\frac{\partial A_k}{\partial x_i} = \frac{\partial A_i}{\partial x_i}. \tag{B.22}$$

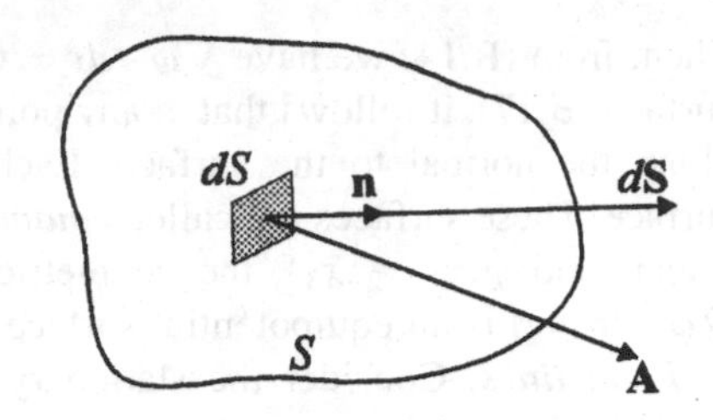

Fig. B.3 Vector representation of a surface element.

If **A** is a polar vector, the expression (B.22) is a scalar called the *divergence* of **A**. One also uses the notation $\partial_i A_i$ or $A_{i,i}$. A vector field with the property

$$\operatorname{div}\mathbf{A} = 0 \tag{B.23}$$

is called *source-free* or *solenoidal*. The lines of such a field are closed curves. This property characterizes, for example, the magnetic induction **B**.

B.2.2.3 Curl

Let us apply again the operator ∇ on a vector **A** field, but this time by taking the cross product $\nabla \times \mathbf{A}$. The expression

$$\operatorname{curl}\mathbf{A} = \nabla \times \mathbf{A} = \mathbf{u}_i \epsilon_{ijk} \frac{\partial A_k}{\partial x_j} \tag{B.24}$$

is called the curl of the vector **A**. The x_s-component of the curl **A** is

$$(\operatorname{curl}\mathbf{A})_s = (\mathbf{u}_s \cdot \mathbf{u}_i)\epsilon_{ijk} A_{k,j} = \epsilon_{sjk} A_{k,j}. \tag{B.25}$$

A vector field **A** which satisfies the relation

$$\operatorname{curl}\mathbf{A} = 0 \tag{B.26}$$

is called *irrotational* or *vorticity-free*. The electrostatic field strength **E** possesses this property.

B.2.3 Fundamental Theorems

B.2.3.1 Green–Gauss Theorem

Let S be a bounded surface and **A** a vector field with its origin on S (Fig. B.3). By definition, the *flux* of **A** through the surface S is the quantity

$$\Phi = \int_S \mathbf{A} \cdot d\mathbf{S} = \int_S \mathbf{A} \cdot \mathbf{n}\, dS, \tag{B.27}$$

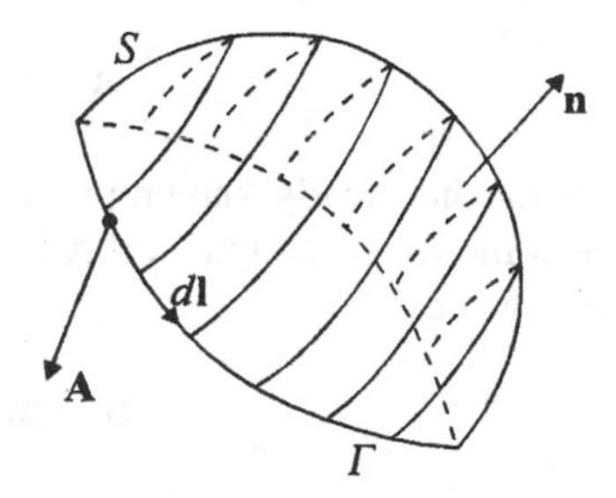

Fig. B.4 An open surface S bordered by a closed contour Γ. $\mathbf{A}$ is a vector field with its origin on Γ.

where $d\mathbf{S}$ is an oriented element of the surface S. If S is a closed surface and $\mathbf{n}$ is the unit vector of the outward normal, chosen as positive, it can be shown that

$$\oint_S \mathbf{A} \cdot d\mathbf{S} = \int_V \operatorname{div} \mathbf{A}\, d\tau, \tag{B.28}$$

where $d\tau$ is an element of the volume V bounded by the surface S. This formula expresses the *Green–Gauss theorem*: the flux of the vector field $\mathbf{A}$ through the closed surface S is equal to the integral over the volume V bounded by S of the divergence of the vector $\mathbf{A}$.

Let us contract the surface S so that the volume V becomes smaller and smaller. At the limit, the formula (B.28) yields:

$$\operatorname{div} \mathbf{A} = \lim_{\Delta\tau \to 0} \frac{1}{\Delta\tau} \oint_S \mathbf{A} \cdot d\mathbf{S}. \tag{B.29}$$

This relation can be considered as the definition formula of divergence at a point. If $\operatorname{div} \mathbf{A} > 0$, at that point there is a *positive* source, while if $\operatorname{div} \mathbf{A} < 0$, there is a *negative source* at that point.

B.2.3.2 Stokes' Theorem

Consider an open surface S bordered by the closed contour Γ and let $\mathbf{A}$ be a vector field with its origin on Γ. In this case, choosing a sense of integration, we can define $\oint_\Gamma \mathbf{A} \cdot d\mathbf{s}$ as being the *circulation* of $\mathbf{A}$ along the contour Γ (Fig. B.4). The positive sense of the normal unit vector is given by the right-screw rule. It can be shown that

$$\oint_\Gamma \mathbf{A} \cdot d\mathbf{s} = \int_S \operatorname{curl} \mathbf{A} \cdot d\mathbf{S}, \tag{B.30}$$

where $d\mathbf{S}$ is a vector element of the surface S. Relation (B.30) is known as *Stokes' theorem*: the circulation of the vector $\mathbf{A}$ along the closed path Γ is equal to the flux of $\operatorname{curl} \mathbf{A}$ through the open surface S which is bounded by the contour. In particular, if $\mathbf{A}$ is a conservative field $\mathbf{A} = \operatorname{grad} \varphi(x_1, x_2, x_3)$, we have:

$$\oint_\Gamma \mathbf{A}\cdot d\mathbf{s} = \oint_\Gamma \frac{\partial\varphi}{\partial x_i}dx_i = \oint_\Gamma d\varphi = 0. \tag{B.31}$$

Using the already known procedure, let us contract the surface element $d\mathbf{S}$ until it becomes so small that curl $\mathbf{A}$ has no significant variation within it. At the limit, we can write:

$$\mathbf{n}\cdot \operatorname{curl}\mathbf{A} = \lim_{\Delta S\to 0}\frac{1}{\Delta S}\oint_\Gamma \mathbf{A}\cdot d\mathbf{s}, \tag{B.32}$$

which can serve as the definition of curl $\mathbf{A}$.

Relations (B.29) and (B.32) are useful, because they stand for the definition of the divergence and curl of a vector, independently of the coordinate system. Therefore, these definitions are *intrinsic*.

Observation: Both the Green–Gauss and Stokes' theorems can be generalized in a space with n dimensions, S_n. Starting with formula (B.28), if we assume that $A_i\ (i=\overline{1,n})$ are n derivable functions of $x_1,..,x_n$ in some domain D of volume V in S_n, and extend the summation over the index i from 1 to n, we obtain:

$$\oint_{S_{n-1}}\sum_{i=1}^{n}A_i\,dx_i = \int_V\sum_{i=1}^{n}\frac{\partial A_i}{\partial x_i}\,d\Omega, \tag{B.33}$$

where S_{n-1} is the closed hypersurface which bounds the volume V of the manifold S_n.

In a similar way, we can extend the Stokes' theorem (B.30) to n dimensions. Using the analytic formulas of the vectors and the results derived in Appendix A, we have:

$$\oint_\Gamma A_i\,ds_i = \int_S \epsilon_{ijk}\frac{\partial A_k}{\partial x_j}\,dS_i = \frac{1}{2}\int_S\left(\frac{\partial A_k}{\partial x_j}-\frac{\partial A_j}{\partial x_k}\right)\epsilon_{jki}\,dS_i.$$

But $\epsilon_{jki}\,dS_i = dx_j\,dx_k$ and so, extending the summation for all indices from 1 to n, we have:

$$\oint_\Gamma\sum_{i=1}^{n}A_i\,ds_i = \frac{1}{2}\int_S\sum_{j,k=1}^{n}\left(\frac{\partial A_k}{\partial x_j}-\frac{\partial A_j}{\partial x_k}\right)dx_j\,dx_k, \tag{B.34}$$

where Γ is a generalized closed contour in S_n and S – an open hypersurface bounded by Γ.

B.2.4 Useful Formulas

The operator *nabla* is sometimes applied to products of two or more functions. In some other cases we meet repeated operations. In the following, we shall give some useful vector identities, frequently encountered in mechanical applications.

1. The *gradient* of a product of two scalar functions, $\varphi\psi$. Taking into account the properties of the operator *nabla*, with the observation that we first consider its differential character and then its vector behaviour, we have:

$$\text{grad}(\varphi\psi) = \nabla(\varphi\psi) = \varphi\nabla\psi + \psi\nabla\varphi = \varphi\,\text{grad}\,\psi + \psi\,\text{grad}\,\varphi. \tag{B.35}$$

2. The *divergence* of the product $\varphi\mathbf{A}$:

$$\text{div}(\varphi\mathbf{A}) = \nabla\cdot(\varphi\mathbf{A}) = \varphi\nabla\cdot\mathbf{A} + \mathbf{A}\cdot\nabla\varphi = \varphi\,\text{div}\,\mathbf{A} + \mathbf{A}\cdot\text{grad}\,\varphi. \tag{B.36}$$

3. The *curl* of the same product:

$$\begin{aligned}\text{curl}(\varphi\mathbf{A}) &= \nabla\times(\varphi\mathbf{A}) = \varphi\nabla\times\mathbf{A} + (\nabla\varphi)\times\mathbf{A}\\ &= \varphi\,\text{curl}\,\mathbf{A} + (\text{grad}\,\varphi)\times\mathbf{A}.\end{aligned} \tag{B.37}$$

4. The *divergence* of the cross product $\mathbf{A}\times\mathbf{B}$:

$$\begin{aligned}\text{div}(\mathbf{A}\times\mathbf{B}) &= \nabla\cdot(\mathbf{A}\times\mathbf{B}) = \partial_i(\mathbf{A}\times\mathbf{B})_i = \epsilon_{ijk}\partial_i(A_jB_k)\\ &= B_k(\epsilon_{kij}\partial_iA_j) - A_j(\epsilon_{jik}\partial_iB_k) = \mathbf{B}\cdot(\nabla\times\mathbf{A}) - \mathbf{A}\cdot(\nabla\times\mathbf{B})\\ &= \mathbf{B}\cdot\text{curl}\,\mathbf{A} - \mathbf{A}\cdot\text{curl}\,\mathbf{B}.\end{aligned} \tag{B.38}$$

5. The *gradient* of the *scalar product* $\mathbf{A}\cdot\mathbf{B}$. We first observe that a component of the gradient reads

$$\partial_j(\mathbf{A}\cdot\mathbf{B}) = \partial_j(A_kB_k) = A_k\partial_jB_k + B_k\partial_jA_k.$$

Next, let us multiply the relation $(\text{curl}\,\mathbf{B})_s = \epsilon_{slm}\partial_lB_m$ by ϵ_{sjk} and perform the summation over the indices l and m:

$$\epsilon_{sjk}(\text{curl}\,\mathbf{B})_s = (\delta_{lj}\delta_{mk} - \delta_{lk}\delta_{mj})\partial_lB_m = \partial_jB_k - \partial_kB_j.$$

Using this result, we have:

$$\partial_j(\mathbf{A}\cdot\mathbf{B}) = \epsilon_{jks}\left[A_k(\text{curl}\,\mathbf{B})_s + B_k(\text{curl}\,\mathbf{A})_s\right] + (A_k\partial_k)B_j + (B_k\partial_k)A_j,$$

or, in vector form,

$$\text{grad}(\mathbf{A}\cdot\mathbf{B}) = \mathbf{A}\times\text{curl}\,\mathbf{B} + \mathbf{B}\times\text{curl}\,\mathbf{A} + (\mathbf{A}\cdot\nabla)\mathbf{B} + (\mathbf{B}\cdot\nabla)\mathbf{A}. \tag{B.39}$$

6. The *curl* of the *cross product* $\mathbf{A}\times\mathbf{B}$:

$$\begin{aligned}\text{curl}(\mathbf{A}\times\mathbf{B}) &= \nabla\times(\mathbf{A}\times\mathbf{B}) = \mathbf{u}_i\epsilon_{ijk}\partial_j(\mathbf{A}\times\mathbf{B})_k = \mathbf{u}_i\epsilon_{kij}\epsilon_{klm}\partial_jA_lB_m\\ &= \mathbf{u}_i(\delta_{il}\delta_{jm} - \delta_{im}\delta_{jl})(A_l\partial_jB_m + B_m\partial_jA_l)\\ &= \mathbf{u}_i(A_i\partial_mB_m - B_i\partial_lA_l + B_m\partial_mA_i - A_l\partial_lB_i),\end{aligned}$$

or

$$\text{curl}(\mathbf{A}\times\mathbf{B}) = \mathbf{A}\,\text{div}\,\mathbf{B} - \mathbf{B}\,\text{div}\,\mathbf{A} + (\mathbf{B}\cdot\nabla)\mathbf{A} - (\mathbf{A}\cdot\nabla)\mathbf{B}. \tag{B.40}$$

Consider now some repeated operations with *nabla*.

7. The *divergence* of a *curl*:

$$\text{div}(\text{curl}\,\mathbf{A}) = \nabla\cdot(\nabla\times\mathbf{A}) = 0, \tag{B.41}$$

because the determinant has two identical lines.

8. The *curl* of a *gradient*:

$$\operatorname{curl}(\operatorname{grad}\varphi) = \nabla \times (\nabla\varphi) = 0, \tag{B.42}$$

for ∇ and $\nabla\varphi$ are collinear.

9. The *divergence* of a *gradient*:

$$\operatorname{div}(\operatorname{grad}\varphi) = \nabla \cdot (\nabla\varphi) = \nabla^2\varphi = \Delta\varphi, \tag{B.43}$$

where the operator

$$\nabla^2 = \Delta = \frac{\partial^2}{\partial x_i \partial x_i} \tag{B.44}$$

is the *Laplacian*. The equation

$$\Delta\varphi = 0 \tag{B.45}$$

is called the *Laplace equation*. The solutions of these equation are termed as *harmonic functions*.

10. The *curl* of a *curl*:

$$\begin{aligned} \operatorname{curl}(\operatorname{curl}\mathbf{A}) &= \nabla \times (\nabla \times \mathbf{A}) = \nabla(\nabla \cdot \mathbf{A}) - \nabla^2\mathbf{A} \\ &= \operatorname{grad}(\operatorname{div}\mathbf{A}) - \Delta\mathbf{A}. \end{aligned} \tag{B.46}$$

11. If $\mathbf{r}$ is the position vector of a particle relative to the origin of the frame $Ox_1x_2x_3$, we have:

$$\begin{aligned} &\operatorname{grad} r = \frac{\mathbf{r}}{r} = \mathbf{u}_r, \quad |\operatorname{grad} r| = 1, \quad \operatorname{div}\mathbf{r} = 3, \\ &\operatorname{curl}\mathbf{r} = 0, \quad \Delta\left(\frac{1}{r}\right) = 0 \quad (r \neq 0). \end{aligned} \tag{B.47}$$

12. Given the scalar field $\varphi(r)$ and the vector field $\mathbf{A}(r)$, where $r = |\mathbf{r}|$, it is easy to prove that

$$\operatorname{grad}\varphi(r) = \frac{\mathbf{r}}{r}\varphi', \quad \operatorname{div}\mathbf{A}(r) = \frac{1}{r}(\mathbf{r} \cdot \mathbf{A}'), \quad \operatorname{curl}\mathbf{A}(r) = \frac{1}{r}\mathbf{r} \times \mathbf{A}', \tag{B.48}$$

where $\varphi' = d\varphi/dr$, $\mathbf{A}' = d\mathbf{A}/dr$. The proof of the identities (B.47) and (B.48) is left to the reader.

B.3 Orthogonal Curvilinear Coordinates

B.3.1 Generalities

Let $\mathbf{r}$ be the position vector of a point P and x_1, x_2, x_3 – its Cartesian coordinates. Assume that x_1, x_2, x_3 are functions of class C^1 of three independent real parameters q_1, q_2, q_3:

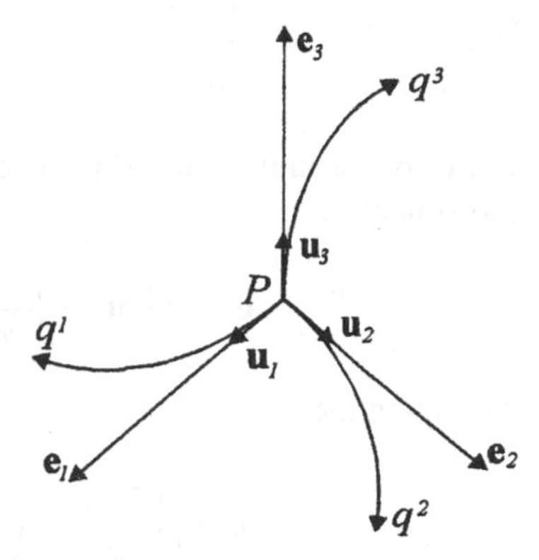

Fig. B.5 Orthogonal curvilinear coordinates.

$$x_i = x_i(q_1,\ q_2,\ q_3) \quad (i = 1, 2, 3). \tag{B.49}$$

We also assume that between the two representations there is a point-to-point correspondence, which means that to a given set (q) corresponds a single set (x) and the other way round. Therefore, the transformation (B.49) is (locally) reversible if

$$J = \frac{\partial(x_1, x_2, x_3)}{\partial(q_1, q_2, q_3)} \neq 0, \tag{B.50}$$

where J is the Jacobian of the transformation (B.49). If we give fixed values to two of the parameters $q_j\ (j = 1, 2, 3)$, say q_2 and q_3, we obtain the coordinate line $q_1 =$ *variable*. Similarly we can obtain the lines $q_2 =$ *variable* and $q_3 =$ *variable*. This shows that through any point in space pass three coordinate lines (Fig. B.5). The parameters q_1, q_2, q_3 are called the *general* or *curvilinear* coordinates of the point P.

Let $\mathbf{e}_i = \frac{\partial \mathbf{r}}{\partial q_i}$ be a vector tangent to the coordinate line $q_i =$ *variable*. In this case, the condition (B.50) expresses the fact that the system of vectors $\mathbf{e}_1, \mathbf{e}_2, \mathbf{e}_3$ are linearly independent, i.e. they form a *basis*. Indeed,

$$\mathbf{e}_1 \cdot (\mathbf{e}_2 \times \mathbf{e}_3) = \frac{\partial \mathbf{r}}{\partial q_1} \cdot \left(\frac{\partial \mathbf{r}}{\partial q_2} \times \frac{\partial \mathbf{r}}{\partial q_3} \right) = J \neq 0.$$

If at each point of the domain $D^{(q)}$ defined by the set of all possible values of the parameters (q) the vectors $\mathbf{e}_1, \mathbf{e}_2, \mathbf{e}_3$ form a right-handed orthogonal trihedral system, the coordinate lines q_1, q_2, q_3 form a system of *orthogonal curvilinear coordinates*. In the following, we shall refer to such coordinates only.

B.3.2 Line, Surface and Volume Elements in Orthogonal Curvilinear Coordinates

Suppose that the point P describes some curve C. An elementary displacement of the point P, by virtue of (B.49), is:

$$d\mathbf{s} \equiv d\mathbf{r} = \frac{\partial \mathbf{r}}{\partial q_i} dq_i = \mathbf{e}_i dq_i. \tag{B.51}$$

Let $\mathbf{u}_i$ be the unit vector of $\mathbf{e}_i$. Then we have, on the one hand, $\mathbf{u}_i \cdot \mathbf{u}_k = \delta_{ik}$, and on the other,

$$\mathbf{e}_i = |\mathbf{e}_i|\mathbf{u}_i = \left|\frac{\partial \mathbf{r}}{\partial q_i}\right| \mathbf{u}_i = h_i \mathbf{u}_i \quad \text{(no summation)}. \tag{B.52}$$

The quantities

$$h_i = \left|\frac{\partial \mathbf{r}}{\partial q_i}\right| = \sqrt{\left(\frac{\partial x_1}{\partial q_i}\right)^2 + \left(\frac{\partial x_2}{\partial q_i}\right)^2 + \left(\frac{\partial x_3}{\partial q_i}\right)^2} \tag{B.53}$$

are called *Lamé's coefficients*.

The *line element* then reads:

$$d\mathbf{s} = h_i dq_i \, \mathbf{u}_i. \tag{B.54}$$

To obtain the *surface element* we shall use the obvious relation

$$d\mathbf{S}_i = \frac{1}{2} \epsilon_{ijk} d\mathbf{s}_j \times d\mathbf{s}_k \quad (i = 1, 2, 3). \tag{B.55}$$

Performing summation over repeated indices, we arrive at

$$d\mathbf{S} = (h_2 h_3 dq_2 dq_3)\mathbf{u}_1 + (h_3 h_1 dq_3 dq_1)\mathbf{u}_2 + (h_1 h_2 dq_1 dq_2)\mathbf{u}_3. \tag{B.56}$$

The *volume element* is obtained by taking the mixed product of line elements, $d\tau = d\mathbf{s}_1 \cdot (d\mathbf{s}_2 \times d\mathbf{s}_3)$. Using (B.54), we immediately find:

$$d\tau = h_1 h_2 h_3 dq_1 dq_2 dq_3. \tag{B.57}$$

B.3.3 Differential Vector Operators in Orthogonal Curvilinear Coordinates

Consider a scalar field $\Psi(x_1, x_2, x_3)$, where x_1, x_2, x_3 are given in terms of q_1, q_2, q_3 (see (B.49)).

By virtue of (B.16) and (B.54), the *gradient* of Ψ reads

$$\operatorname{grad} \Psi = \frac{1}{h_i} \frac{\partial \Psi}{\partial q_i} \mathbf{u}_i. \tag{B.58}$$

If in (B.58) we choose $\Psi \equiv q_j$, we have $\operatorname{grad} q_j = \mathbf{u}_j / h_j$ (no summation) and thus we may write:

$$\mathrm{curl}(\mathrm{grad}\, q_j) = \mathrm{curl}\left(\frac{\mathbf{u}_j}{h_j}\right) = 0.$$

Performing the calculation according to (B.37), we find:

$$\mathrm{curl}\, \mathbf{u}_j = \frac{1}{h_j} \mathrm{grad}\, h_j \times \mathbf{u}_j \quad \text{(no summation)}. \tag{B.59}$$

The *divergence* of a vector field $\mathbf{A}(x_1, x_2, x_3)$, in view of (B.36), is

$$\mathrm{div}\, \mathbf{A} = \mathrm{div}(A_i \mathbf{u}_i) = A_i\, \mathrm{div}\, \mathbf{u}_i + \mathbf{u}_i \cdot \mathrm{grad}\, A_i, \tag{B.60}$$

or, by use of (B.38) and (B.59),

$$\begin{aligned} \mathrm{div}\, \mathbf{u}_i &= \frac{1}{2} \epsilon_{ijk} \mathrm{div}(\mathbf{u}_j \times \mathbf{u}_k) = \frac{1}{2} \epsilon_{ijk} (\mathbf{u}_k \cdot \mathrm{curl}\, \mathbf{u}_j - \mathbf{u}_j \cdot \mathrm{curl}\, \mathbf{u}_k) \\ &= \frac{1}{2} \epsilon_{ijk} \left(\frac{1}{h_j h_s} \frac{\partial h_j}{\partial q_s} \mathbf{u}_k \cdot (\mathbf{u}_s \times \mathbf{u}_j) - \frac{1}{h_k h_s} \frac{\partial h_k}{\partial q_s} \mathbf{u}_j \cdot (\mathbf{u}_s \times \mathbf{u}_k) \right). \end{aligned}$$

Since $\mathbf{u}_s \cdot (\mathbf{u}_j \times \mathbf{u}_k) = \epsilon_{sjk}$ and $\epsilon_{ijk}\epsilon_{sjk} = 2\delta_{is}$, we have:

$$\mathrm{div}\, \mathbf{u}_i = \frac{1}{h_i} \left(\frac{1}{h_j} \frac{\partial h_j}{\partial q_i} + \frac{1}{h_k} \frac{\partial h_k}{\partial q_i} \right). \tag{B.61}$$

We also obtain

$$\mathbf{u}_i \cdot \mathrm{grad}\, A_i = \mathbf{u}_i \cdot \mathbf{u}_k \frac{1}{h_k} \frac{\partial A_i}{\partial q_k} = \frac{1}{h_i} \frac{\partial A_i}{\partial q_i}.$$

Introducing these results into (B.60), we are led to

$$\mathrm{div}\, \mathbf{A} = \frac{1}{h_i h_j h_k} \frac{\partial}{\partial q_i} (A_i h_j h_k) \quad (i, j, k = \text{cyclic permutations of } 1, 2, 3), \tag{B.62}$$

or, if the summation is performed,

$$\mathrm{div}\, \mathbf{A} = \frac{1}{h_1 h_2 h_3} \left[\frac{\partial}{\partial q_1} (A_1 h_2 h_3) + \frac{\partial}{\partial q_2} (A_2 h_3 h_1) + \frac{\partial}{\partial q_3} (A_3 h_1 h_2) \right]. \tag{B.63}$$

Suppose now that $\mathbf{A} = \mathrm{grad}\, \Psi$. Then $\mathrm{div}\, \mathbf{A} = \Delta \Psi$ and (B.63) yields the expression for the *Laplacian*:

$$\Delta \Psi = \frac{1}{h_1 h_2 h_3} \left[\frac{\partial}{\partial q_1} \left(\frac{h_2 h_3}{h_1} \frac{\partial \Psi}{\partial q_1} \right) + \frac{\partial}{\partial q_2} \left(\frac{h_3 h_1}{h_2} \frac{\partial \Psi}{\partial q_2} \right) + \frac{\partial}{\partial q_3} \left(\frac{h_1 h_2}{h_3} \frac{\partial \Psi}{\partial q_3} \right) \right]. \tag{B.64}$$

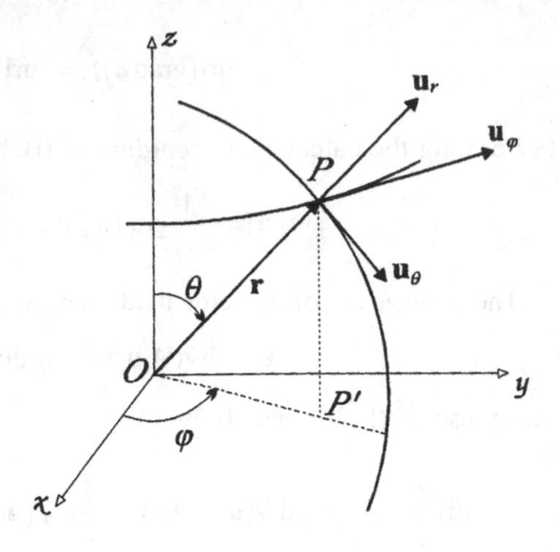

Fig. B.6 Spherical coordinates.

The *curl* of the field $\mathbf{A}$ is obtained according to (B.37) and (B.59):

$$\begin{aligned}\operatorname{curl}\mathbf{A} &= \operatorname{curl}(A_i\mathbf{u}_i) = A_i \operatorname{curl}\mathbf{u}_i + \operatorname{grad} A_i \times \mathbf{u}_i \\ &= \frac{A_i}{h_i}\frac{1}{h_k}\frac{\partial h_i}{\partial q_k}\mathbf{u}_k \times \mathbf{u}_i + \frac{1}{h_k}\frac{\partial A_i}{\partial q_k}\mathbf{u}_k \times \mathbf{u}_i.\end{aligned}$$

But $\mathbf{u}_k \times \mathbf{u}_i = \epsilon_{kis}\mathbf{u}_s$, hence

$$\operatorname{curl}\mathbf{A} = \epsilon_{ski}\left(\frac{A_i}{h_i h_k}\frac{\partial h_i}{\partial q_k} + \frac{1}{h_k}\frac{\partial A_i}{\partial q_k}\right)\mathbf{u}_s. \tag{B.65}$$

If the summation is performed, we finally arrive at:

$$\begin{aligned}\operatorname{curl}\mathbf{A} = &\frac{1}{h_2 h_3}\left[\frac{\partial}{\partial q_2}(A_3 h_3) - \frac{\partial}{\partial q_3}(A_2 h_2)\right]\mathbf{u}_1 \\ &+ \frac{1}{h_3 h_1}\left[\frac{\partial}{\partial q_3}(A_1 h_1) - \frac{\partial}{\partial q_1}(A_3 h_3)\right]\mathbf{u}_2 \\ &+ \frac{1}{h_1 h_2}\left[\frac{\partial}{\partial q_1}(A_2 h_2) - \frac{\partial}{\partial q_2}(A_1 h_1)\right]\mathbf{u}_3.\end{aligned} \tag{B.66}$$

B.3.4 Examples of Orthogonal Curvilinear Coordinates

B.3.4.1 Spherical Coordinates

Let $Ox_1x_2x_3$ be a Cartesian orthogonal system of coordinates. The position in space of a point (particle) P can be defined by: the distance $|\mathbf{r}| = OP$, the angle θ (latitude) between Ox_3 and $\mathbf{r}$ and the angle φ (longitude) between Ox_1 and the

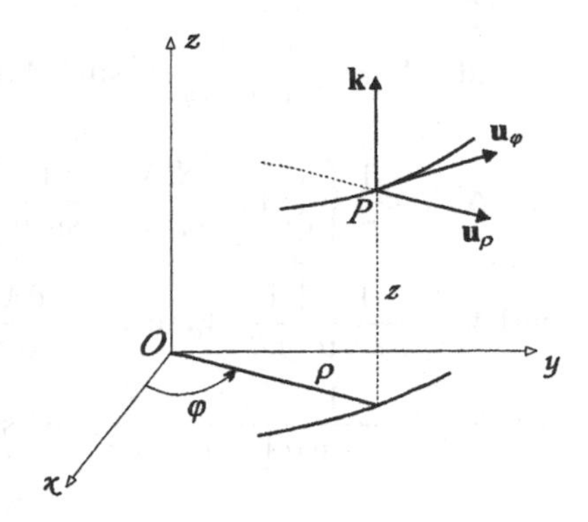

Fig. B.7 Cylindrical coordinates.

projection $\vec{OP'}$ of $\mathbf{r}$ on the Ox_1x_2-plane (Fig. B.6). The parameters r, θ, φ are called the *spherical coordinates* of the point P. The coordinate lines $q_1 = r, q_2 = \theta, q_3 = \varphi$ lie at the intersection of the surfaces $r = \lambda_1, \theta = \lambda_2, \varphi = \lambda_3$, which are orthogonal to each other. The surfaces $r = \lambda_1$ are concentric spheres with the centre at O; the surfaces $\theta = \lambda_2$ are right circular cones with their common top at O; finally, the surfaces $\varphi = \lambda_3$ are semi-meridian planes of the spheres $r = \lambda_1$. The variation intervals of the spherical coordinates are: $0 \leq r < \infty; 0 \leq \theta \leq \pi$; $0 \leq \varphi < 2\pi$.

The components of $\mathbf{r}$ on the axes Ox_1, Ox_2, Ox_3 are

$$\begin{aligned} x_1 &= r \sin\theta \cos\varphi, \\ x_2 &= r \sin\theta \sin\varphi, \\ x_3 &= r \cos\theta. \end{aligned} \tag{B.67}$$

By virtue of (B.53), the Lamé coefficients are:

$$h_1 = 1, \quad h_2 = r, \quad h_3 = r \sin\theta. \tag{B.68}$$

If we denote the unit vectors of the three mutually orthogonal directions by $\mathbf{u}_r = \mathbf{u}_1, \mathbf{u}_\theta = \mathbf{u}_2, \mathbf{u}_\varphi = \mathbf{u}_3$, according to the formulas deduced in Sect. B.3.3, we have:

$$d\mathbf{s} = \mathbf{u}_r \, dr + \mathbf{u}_\theta \, r \, d\theta + \mathbf{u}_\varphi \, r \, \sin\theta \, d\varphi, \tag{B.69}$$

$$d\mathbf{S} = \mathbf{u}_r r^2 \, \sin\theta \, d\theta \, d\varphi + \mathbf{u}_\theta \, r \, \sin\theta \, dr \, d\varphi + \mathbf{u}_\varphi \, r \, dr \, d\theta, \tag{B.70}$$

$$d\tau = r^2 \sin\theta \, dr \, d\theta \, d\varphi. \tag{B.71}$$

In particular, if $\theta = \pi/2$, the point P lies in the plane Ox_1x_2. Its coordinates are then r, φ and they are called *plane polar coordinates*.

Using the formulas obtained in Sect. B.3.4, we can write the *gradient*, the *divergence*, the *Laplacian* and the *curl* in spherical coordinates:

$$\text{grad}\,\Psi = \frac{\partial \Psi}{\partial r}\mathbf{u}_r + \frac{1}{r}\frac{\partial \Psi}{\partial \theta}\mathbf{u}_\theta + \frac{1}{r \sin\theta}\frac{\partial \Psi}{\partial \varphi}\mathbf{u}_\varphi, \tag{B.72}$$

$$\operatorname{div}\mathbf{A} = \frac{1}{r^2 \sin\theta}\left[\frac{\partial}{\partial r}(r^2 \sin\theta\, A_r) + \frac{\partial}{\partial\theta}(r \sin\theta\, A_\theta) + \frac{\partial}{\partial\varphi}(rA_\varphi)\right], \tag{B.73}$$

$$\Delta\Psi = \frac{1}{r^2}\left\{\frac{\partial}{\partial r}\left(r^2\frac{\partial\Psi}{\partial r}\right) + \frac{1}{\sin\theta}\left[\frac{\partial}{\partial\theta}\left(\sin\theta\frac{\partial\Psi}{\partial\theta}\right) + \frac{1}{\sin\theta}\frac{\partial^2\Psi}{\partial\varphi^2}\right]\right\}, \tag{B.74}$$

$$\begin{aligned}\operatorname{curl}\mathbf{A} &= \frac{1}{r\sin\theta}\left[\frac{\partial}{\partial\theta}(A_\varphi \sin\theta) - \frac{\partial A_\theta}{\partial\varphi}\right]\mathbf{u}_r \\ &+ \frac{1}{r\sin\theta}\left[\frac{\partial A_r}{\partial\varphi} - \frac{\partial}{\partial r}(r A_\varphi \sin\theta)\right]\mathbf{u}_\theta + \frac{1}{r}\left[\frac{\partial}{\partial r}(r A_\theta) - \frac{\partial A_r}{\partial\theta}\right]\mathbf{u}_\varphi.\end{aligned} \tag{B.75}$$

B.3.4.2 Cylindrical Coordinates

The position in space of the point P can be also defined by its *cylindrical coordinates*: the magnitude ρ of the projection of $\mathbf{r}$ on the plane Ox_1x_2, the angle φ between Ox_1 and ρ, and the Cartesian coordinate z. The coordinate lines lie at the intersection of the surfaces: $\rho = \lambda_1$ (cylinders of revolution about z-axis), $|\varphi| = \lambda_2$ (planes which contain the z-axis) and $z = \lambda_3$ (planes parallel to Ox_1x_2). The relations between the Cartesian and the cylindrical coordinates are given by

$$\begin{aligned}x_1 &= \rho\cos\varphi,\\ x_2 &= \rho\sin\varphi,\\ x_3 &= z.\end{aligned} \tag{B.76}$$

The Lamé coefficients are then

$$h_1 = 1, \quad h_2 = \rho, \quad h_3 = 1. \tag{B.77}$$

If we denote by $\mathbf{u}_\rho = \mathbf{u}_1, \mathbf{u}_\varphi = \mathbf{u}_2, \mathbf{k} = \mathbf{u}_3$ the unit vectors of the three orthogonal directions, we have:

$$d\mathbf{s} = \mathbf{u}_\rho\, d\rho + \mathbf{u}_\varphi\, \rho\, d\varphi + \mathbf{k}\, dz, \tag{B.78}$$

$$d\mathbf{S} = \mathbf{u}_\rho\, \rho d\varphi dz + \mathbf{u}_\varphi\, d\rho\, dz + \mathbf{k}\, \rho\, d\rho\, d\varphi, \tag{B.79}$$

$$d\tau = \rho\, d\rho\, d\varphi\, dz. \tag{B.80}$$

Finally, the relations deduced in Sect. B.3.4 yield:

$$\operatorname{grad}\Psi = \frac{\partial\Psi}{\partial\rho}\mathbf{u}_\rho + \frac{1}{\rho}\frac{\partial\Psi}{\partial\varphi}\mathbf{u}_\varphi + \frac{\partial\Psi}{\partial z}\mathbf{k}, \tag{B.81}$$

$$\operatorname{div}\mathbf{A} = \frac{1}{\rho}\left[\frac{\partial}{\partial\rho}(\rho A_\rho) + \frac{\partial A_\varphi}{\partial\varphi} + \frac{\partial}{\partial z}(\rho A_z)\right], \tag{B.82}$$

$$\Delta \Psi = \frac{1}{\rho}\left[\frac{\partial}{\partial \rho}\left(\rho \frac{\partial \Psi}{\partial \rho}\right) + \frac{1}{\rho}\frac{\partial^2 \Psi}{\partial \varphi^2} + \rho \frac{\partial^2 \Psi}{\partial z^2}\right], \tag{B.83}$$

$$\begin{aligned} \operatorname{curl} \mathbf{A} = & \left(\frac{1}{\rho}\frac{\partial A_z}{\partial \varphi} - \frac{\partial A_\varphi}{\partial z}\right)\mathbf{u}_\rho + \left(\frac{\partial A_\rho}{\partial z} - \frac{\partial A_z}{\partial \rho}\right)\mathbf{u}_\varphi \\ & + \left(\frac{\partial A_\varphi}{\partial \rho} + \frac{1}{\rho}A_\varphi - \frac{1}{\rho}\frac{\partial A_\rho}{\partial \varphi}\right)\mathbf{k}. \end{aligned} \tag{B.84}$$

References

1. Amenzade, Yu.A.: Theory of Elasticity. Mir, Moscow (1979)
2. Arnold, V.I.: Mathematical Methods of Classical Mechanics. Springer-Verlag, Berlin Heidelberg (1997)
3. Balian, R., Peube, J.L.: Fluid Dynamics. Macmillan Education Australia (1977)
4. Bartlett, W.H.C.: Elements of Analytical Mechanics. Scholarly Publishing Office, University of Michigan Library (2006)
5. Bose, S.K., Chattoraj, D.: Elementary Analytical Mechanics. Alpha Science International, Ltd. (2000)
6. Bradbury, T.C.: Theoretical Mechanics. Wiley, New York (1968)
7. Chaichian, M., Hagedorn, R.: Symmetries in Quantum Mechanics: From Angular Momentum to Supersymmetry. Taylor & Francis (1998)
8. Chaichian, M., Demichev, A.: Path Integrals in Physics, vols. I and II. Institute of Physics Publishing, Bristol and Philadelphia (2001)
9. Chaichian, M., Perez Rojas, H.C., Tureanu, A.: From the Cosmos to Quarks: Basic Concepts in Physics. Springer-Verlag, Berlin Heidelberg (2012)
10. Fassano, A., Marmi, S.: Analytical Mechanics: An Introduction. Oxford University Press (2006)
11. Forray, M.J.: Variational Calculus in Science and Engineering. McGraw Hill, New York (1968)
12. Fowles, G.R., Cassiday, G.L.: Analytical Mechanics. Cengage Learning (2005)
13. Fox, R.W., McDonald, A.T., Pritchard, P.J.: Introduction to Fluid Mechanics. Wiley, New York (2009)
14. Gantmacher, F.: Lectures in Analytical Mechanics. Mir, Moscow (1975)
15. Goldstein, H., Poole, C.P., Safko, J.L.: Classical Mechanics. Addison-Wesley (2002)
16. Grechko, L.G., et al.: Problems in Theoretical Physics. Mir, Moscow (1977)
17. Greenwood, D.T.: Principles of Dynamics. Dover, New York (1997)
18. Hand, L.N., Finch, J.D.: Analytical Mechanics. Cambridge University Press (1998)
19. Johns, O.D.: Analytical Mechanics for Relativity and Quantum Mechanics. Oxford University Press (2005)
20. Kibble, T.W.: Classical Mechanics. Addison-Wesley Longman (1986)
21. Kittel, C., Knight, W.D., Ruderman, M.A.: Mechanics. Berkeley Physics Course, vol. 1. McGraw-Hill, New York (1973)
22. Lagrange, J.L.: Analytical Mechanics. Kluwer Academic Publishers (2010)
23. Landau, L.D., Lifshitz, E.M.: Fluid Mechanics, 2nd edn. Pergamon (1987)
24. Landau, L.D., Lifshitz, E.M.: The Classical Theory of Fields. Pergamon (1987)
25. Landau, L.D., Lifshitz, E.M.: Theory of Elasticity. Pergamon (1975)

M. Chaichian et al., *Mechanics*, DOI: 10.1007/978-3-642-17234-2,

26. Landau, L.D., Lifshitz, E.M.: Mechanics, 3rd edn. Pergamon (1976)
27. Layton, R.A.: Principles of Analytical System Dynamics. Springer-Verlag, Berlin Heidelberg (1998)
28. Lurie, A.I.: Analytical Mechanics. Springer-Verlag, Berlin Heidelberg (2002)
29. Malvern, L.E.: Introduction to the Mechanics of a Continuous Medium. Prentice-Hall, New Jersey (1969)
30. Merches, I., Burlacu, L.: Applied Analytical Mechanics. The Voice of Bucovina Press, Iasi (1995)
31. Papastavridis, J.G.: Analytical Mechanics: A Comprehensive Treatise on the Dynamics of Constrained Systems for Engineers, Physicists and Mathematicians. Oxford University Press (2002)
32. Pellegrini, C., Cooper, R.K.: Modern Analytical Mechanics. Springer-Verlag, Berlin Heidelberg (1999)
33. Rossberg, K.: A First Course in Analytical Mechanics. Wiley, New York (1983)
34. Saletan, E.J., Cromer, A.H.: Theoretical Mechanics. Wiley, New York (1971)
35. Scheck, F.: Mechanics: From Newton's Laws to Deterministic Chaos. Springer-Verlag, Berlin Heidelberg (2010)
36. Serrin, J.: Mathematical Principles of Classical Fluid Mechanics. Handbuch der Physik. Springer-Verlag, Berlin Heidelberg, vol. VIII/1, Fluid Dynamics I (1959)
37. Spiegel, M.R.: Theory and Problems of Theoretical Mechanics. McGraw Hill, New York (1967)
38. Ter Haar, D.: Elements of Hamiltonian Mechanics. North-Holland, Amsterdam (1964)
39. Torok, J.S.: Analytical Mechanics: With an Introduction to Dynamical Systems. Wiley-Interscience (1999)
40. Wess, J.: Theoretische Mechanik. Springer-Verlag, Berlin Heidelberg (2007)
41. Ziwet, A., Field, P.: Introduction to Analytical Mechanics. Scholarly Publishing Office, University of Michigan (2005)

Author Index

A
Ångström, Anders, 401
Avogadro, Amedeo, 400

B
Balmer, Johann Jakob, 401–403
Beltrami, Eugenio, 333
Bernoulli, Daniel, 322, 324
Bernoulli, Johann, 67
Bertrand, Joseph Louis François, 105
Binet, Jacques Philippe Marie, 101
Bohr, Niels, 263, 287, 401–403
Boltzmann, Ludwig, 400
Born, Max, 404, 406
Brown, Robert (see *Brownian* in Subject Index), 409
Busch, Hans, 221

C
Cartan, Eli, 257
Cauchy, Augustin-Louis, 302–304, 317, 325
Christoffel, Elwin Bruno, 34, 72
Clausius, Rudolf, 400
Clebsch, Rudolf Friedrich Alfred, 334
Coriolis, Gaspard-Gustave, 62, 170
Coulomb, Charles-Augustin de, 108, 131

D
D'Alembert, Jean-Baptiste le Rond, 53, 299, 318
Dalton, John, 400
Davisson, Clinton, 404
De Broglie, Louis, 93, 403–404
Democritus of Abdera, 400
Descartes, René (*Renatus Cartesius*), (see *Cartesian* in Subject Index), 2
Dirac, Paul Adrien Maurice, 403
Dirichlet, Johann Peter Gustav Lejeune, 36

E
Ehrenfest, Paul, 287, 407
Einstein, Albert, 4, 390–397, 402, 404
Eötvös, Loránd, 8
Epicurus, 400
Euler, Leonhard, 66, 86, 175, 185, 295, 298, 319

F
Faraday, Michael, 400
Fermat, Pierre de, 93, 404
Feynman, Richard, 399, 409
FitzGerald, George Francis, 392
Foucault, Jean Bernard Léon, 172
Frenet, Jean Frédéric, 429

G
Galilei, Galileo, ix, 7, 389–390
Gauss, Johann Carl Friedrich, ix, 434
Germer, Lester, 404

M. Chaichian et al., *Mechanics*, DOI: 10.1007/978-3-642-17234-2,

G (*cont.*)
Gibbs, Josiah Willard, 214, 248–249
Green, George, 434

H
Hamilton, William Rowan, ix, 74, 211, 214, 265
Heisenberg, Werner, 232, 400, 404
Helmholtz, Hermann von, 213, 249, 320, 400
Hooke, Robert, 311

J
Jacobi, Carl, 86, 89, 226, 237, 265–266
Jordan, Pascual, 403
Joule, James Prescott, 400

K
Kelvin, William Thomson, Baron, 331
Kepler, Johannes, 13, 108, 114–115
Kirchhoff, Gustav Robert, 153, 157
König, Samuel, 21
Kronecker, Leopold, 411, 417

L
Lagrange, Joseph-Louis, ix, 36, 40, 43, 53, 66, 86, 234, 294, 322
Lamé, Gabriel Léon Jean Baptiste, 314, 316
Langevin, Paul, 409
Laplace, Pierre-Simon (see *Laplacian* in Subject Index), 327, 438
Larmor, Joseph, 205–206, 354
Legendre, Adrien-Marie, 212
Leibniz, Gottfried Wilhelm, 226
Lenz, Wilhelm, 116
Levi-Civita, Tullio, 417
Liouville, Joseph, 258
Lorentz, Hendrik, 7, 62, 389–397
Lorenz, Ludvig Valentin, 367

M
Maupertuis, Pierre-Louis Moreau de, ix, 86, 89
Maxwell, James Clerk, 153, 208, 338, 354, 390, 400
Michelson, Albert Abraham, 390
Minkowski, Hermann, 391, 395
Morley, Edward Williams, 390
Morse, Philip McCord, 95

N
Navier, Claude-Louis, 317, 341
Newton, Isaac (see *Newtonian* in Subject Index), ix, 1–23, 53, 274, 389, 403, 409
Noether, Emmy, 82, 84, 373

O
Ohm, Georg Simon, 154, 338

P
Pfaff, Johann Friedrich, 39, 261
Planck, Max, 233, 399–400, 402
Poincaré, Henri, 253, 257, 395
Poisson, Siméon Denis, 25, 167, 225, 227, 312

R
Rayleigh, John William Strutt, Baron, 61
Reynolds, Osborne, 343
Riemann, Georg Friedrich Bernhard (see *Riemannian* in Subject Index), 70, 325
Routh, Edward John, 223
Runge, Carl David Tolmé, 116
Rutherford, Ernest, 131, 401
Rydberg, Johannes Robert, 401, 403

S
Saint-Venant, Barré de, 309
Schrödinger, Erwin, 93, 356, 403
Serret, Joseph Alfred, 429
Sommerfeld, Arnold, 263, 287, 404
Stokes, George Gabriel, 341, 435
Stoney, George Johnstone, 401

T
Taylor, Brook, 63
Thomson, Joseph John, 156, 401

W
Wallis, John, 135
Wiener, Norbert, 409

Z
Zeeman, Pieter, 8

Subject Index

A
Abbreviated Hamilton–Jacobi equation, 269, 272, 274, 282
Absolute
absolute acceleration, 167
absolute integral invariant, 254
absolute motion, 1, 165
absolute reference frame, 1, 390
absolute time and space, 389–391
absolute velocity, 166
principle of absolute simultaneity, 7
Acceleration
absolute acceleration, 167
acceleration potential, 322
average acceleration, 4
centripetal acceleration, 167
Coriolis acceleration, 167
instantaneous acceleration, 4
relative acceleration, 167
transport acceleration, 167
Acoustics
fundamental equation of acoustics, 323
Action
action integral, 77, 346
action modulus, 263
action variable, 278
equal-action wave front, 90–93
Hamiltonian action, 90
Maupertuisian action, 89, 90
principle of action and reaction, 8
principle of least action, 86
stationary action, 83
Action–angle variables, 278–286
Adiabatic invariants, 287–289
Æther, 390–391
Amplitude
amplitude of integral, 135
amplitude of oscillation, 134, 141, 289
Analytical mechanics, ix–x, 1
principles of analytical mechanics, 27–93
Angle
action–angle variables, 278–286
angle of nutation, 176, 198
angle of precession, 176, 198
angle of self-rotation, 176, 198
angle variable, 278, 280
Euler's angles, 175–177, 190, 198, 200
imaginary angle, 396
polar angle, 114, 118, 127
scattering angle, 122, 124, 126, 131
Angular momentum, 11, 21, 81, 179
angular momentum and Bohr's quantization rule, 402
angular momentum conservation, 85, 126, 330, 382
angular momentum theorem, 12, 17
Poisson bracket of angular momentum, 228–231
Angular velocity, 168, 190
Anomaly
eccentric anomaly, 114–115
real anomaly, 114–115
Antipotentials, 367–368, 371–372
Antisymmetric tensor, 416, 423
and associated axial vector, 166, 229, 426
Aphelion, 102
Apocentre, 102–103, 111
Apogee, 102
Apparent force (see *force of inertia*)
Arbitrary displacements, 49
Areal velocity, 13, 222
Areas theorem, 13, 173, 231
Artificial satellite, 118–121

M. Chaichian et al., *Mechanics*, DOI: 10.1007/978-3-642-17234-2,

A (*cont.*)
Asymmetrical top, 187
Asynchronous variations, 87
Attraction force, 99, 108, 131
Attractive potential, 108, 113, 129, 140
Average
 average acceleration, 4
 average stress, 303
 average velocity, 3
Axial
 axial moments of inertia, 180
 axial symmetry, 127, 220
 axial vector, 166, 168, 321, 416
Axis
 axis of symmetry, 187, 190, 197
 instantaneous axis of rotation, 178

B
Ballistics, 10
Balmer series (see *spectrum of hydrogen atom*)
Barotropic fluid, 320, 322
Basis vector, 420
Beam, 126, 128, 131
 deflected beam, 126, 401
 incident beam, 125
 paraxial beam, 221
Beltrami's vorticity diffusion equation, 333–334, 339
Bernoulli
 Bernoulli's equation, 324–325, 337, 352
 Lagrange–Bernoulli equation, 322–323
Bertrand's theorem, 103, 105–108
Bilinear covariant, 261–263
Binet's equation, 101
Bi-normal, 32, 430, 431
Body forces, 300, 303, 316, 348
 specific body forces, 300
Bohr's postulates, 402–403, 405
Bohr–Sommerfeld quantization condition, 263, 287
Bound orbit, 102–104, 109, 110
 closed, 103, 105
 open, 103
Brachistochrone, 66–67
Brackets
 Lagrange brackets, 234–236
 Poisson brackets, 225–234, 360
Brackett series (see *spectrum of hydrogen atom*)
Brownian motion, 409
Busch's relation, 221

C
Calculus of variations, 62–74
Canonical
 canonical transformations, 236–249
  generating function for, 238–249
  group structure, 244
  infinitesimal, 249–253
  invariance of Poisson and Lagrange brackets, 242–244
  valence of, 238
 canonical variables, 214, 278
 Hamilton's canonical equations, 211–222, 360
Canonicity condition, 237–238, 240
Capacitance, 154, 155, 157, 158, 364
Cartan
 Poincaré–Cartan integral invariant, 257, 262
Cartesian
 orthogonal Cartesian coordinates, 2, 411
Catenary, 70
Cauchy
 Cauchy equations, 304
 Cauchy stress tensor, 302–306
 Cauchy's formula, 303–304
 Cauchy's tetrahedron, 303
 Navier–Cauchy equations, 317
Cauchy–Riemann conditions of analyticity, 325–326
CDM (see *continuous deformable medium*)
Central force field, 97–108
Centre
 centre of force, 102, 104, 125, 141
  falling on, 104–105
 centre of scattering, 125, 126
Centre of mass (CM), 20
 system of reference (CMS), 122
Centrifugal
 centrifugal force, 46, 100
 centrifugal momenta, 180
 centrifugal potential, 100, 104
Centripetal acceleration, 167
Characteristic
 characteristic equation, 146, 425
 characteristic function, 212–213, 224, 248
Christoffel symbols
 of the first kind, 34, 56, 72
 of the second kind, 34, 72
Circular orbit, 105–106, 108, 109
Circulation
 circulation as integral operator, 15, 429, 435

Kelvin's velocity circulation theorem, 331–333
Clebsch
Clebsch potentials, 335, 337, 353
Clebsch theorem, 334–337
Clebsch transformation, 335, 353, 368
CMS (see *centre of mass system of reference*)
Coefficient
coefficient of thermo-conductivity, 341
coefficient of thermo-elasticity, 315
coefficient of transverse contraction (Poisson's ratio), 312
dynamic coefficient of viscosity, 340
kinematic coefficient of viscosity, 341, 343
Collision
elastic collision, 121, 125
inelastic collision, 121
Commensurable frequencies, 284
Commutation relation, 234
Heisenberg's commutation relations, 233
Commutator of operators, 231–234
Compatibility conditions (Saint-Venant), 309–310
Complementary force (see *force of inertia*)
Complex
complex potential, 326
complex velocity, 326
Composite motion, 165
Compressible medium, 294, 350
Condition
"escape to infinity" condition, 104
canonicity condition, 237–238, 240
initial conditions, 9
Lorenz condition, 367, 371, 372
orthogonality condition, 164, 411
Saint-Venant compatibility conditions, 309–310
Cone, 140, 443
herpolhodic cone, 191
polhodic cone, 191
Configuration, 49
equilibrium configuration, 144–145, 149
Configuration space, 48–49, 236
generalized trajectory in, 49
Conic
conic pendulum, 140
eccentricity of a conic, 110, 113
equation of a conic, 110, 118, 276
Conjugate
conjugate momenta, 81, 214
conjugate momentum density, 357
conjugate variables, 214, 233
Conservation
conservation of entropy, 328
angular momentum conservation
for discrete particle systems, 12, 17, 85
for continuous systems, 302, 330–331, 382–383
linear momentum conservation
for discrete particle systems, 12, 17, 85
for continuous systems, 302, 329–330, 381
total energy conservation, 14–15, 328–329, 380
Conservation laws, 300, 337
and continuous symmetry transformations, 82–86, 373–380
and first integrals, 80–81
Conservative field, 14, 432, 435
and total energy, 15
Constant
fine structure constant, 402
gravitational constant, 118
Planck constant, 233, 263, 399, 400
Rydberg constant, 401, 403
Constants of motion, 80, 81, 216, 250
Constitutive equations (see *stress-strain relations*)
Constraint
bilateral constraint, 28, 30
constraint force components, 31–32
normal reaction, 31
force of friction, 31
constraint forces, 9, 16, 27, 30
constraints and degrees of freedom, 29–30
criteria of classification of constraints, 27–28
geometric (finite) constraint, 28
integrable (holonomic) constraint, 28
kinematic (differential) constraint, 28
non-integrable (non-holonomic) constraint, 29
particle subject to constraints, 37–38, 42–43
Pfaffian constraints, 39
rheonomous (non-stationary) constraint, 28, 41, 216
rigidity constraints, 44
scleronomous (stationary) constraint, 28, 40, 42
unilateral constraint, 28
Continuity equation (see *equation of continuity*)
Continuous
continuous symmetry transformations and conservation laws, 82–86, 373–380
continuous system of particles, 294

C (*cont.*)
Continuous deformable medium (CDM), 293–385
  Euler variables, 296
  Euler's kinematical method, 295–296
  Lagrange variables, 294, 297, 316, 347
  Lagrange's kinematical method, 294–295
Contraction, 308
  coefficient of transverse contraction, 312
  relativistic contraction of length, 392
  tensor contraction, 417
Contravariant components, 34, 71
Convention
  Einstein's summation convention, 4
  sign convention for stresses, 303
Coordinate operator, 233
Coordinates
  cyclic coordinates, 59, 81
  cylindrical coordinates, 2, 5, 443, 444
  generalized coordinates, 47–52, 346
  ignorable coordinates (see *cyclic coordinates*)
  natural coordinates, 5, 432
  normal coordinates, 146–149
    and normal frequencies, 147
  orthogonal Cartesian coordinates, 2, 4, 412
  plane polar coordinates, 5, 443
  spherical coordinates, 2, 4, 442–443
  system of coordinates, 4
    left-handed, 415
    right-handed, 415
Correspondence principle, 79
Cosmic velocities, 121
Coulomb field, 108, 131, 132
Coupled equations, 144, 146
Covariant
  bilinear covariant, 261–263
  covariant components of applied force, 33
  covariant equations, 389, 397
  covariant laws, 8
Cross product of vectors, 421
Cross section
  differential cross section, 128
  effective cross section, 125–129
  total cross section, 130–131
Current lines (see *streamlines*)
Curvature, 432
  radius of curvature, 5, 430
Curve
  brachistochrone, 66–67
  catenary, 69
  geodesic, 56, 69–74, 93
  ideal (perfectly smooth) curve, 31
  perfectly rough curve, 31
  stationary curve, 32
Curvilinear (general) coordinates
  expression for gradient, divergence, Laplacian, curl, 440–442
  expression for line, surface and volume elements, 439–440
Cyclic coordinate, 59
  conservation of momentum associated to a cyclic coordinate, 81
Cycloidal pendulum, 136–137
Cycloid, 68
Cylindrical coordinates, 2, 5, 443, 444
  expression for differential operators, 444–445
  expression for line, surface and volume elements, 444
Cylindrical symmetry, 270

D
D'Alembert's principle, 52, 53
Davisson–Germer experiment, 404
De Broglie's hypothesis, 403–404
Deflected beam, 126, 401
Deformation (strain), 7, 297, 306–311
  contraction, 308
  deformation quadric, 308
  dilation, 308
  linear theory of deformations, 306–308
  velocity of deformation quadric, 321
  velocity of deformation tensor, 321, 340
Degeneracy, 146, 286
Degenerate system, 284
Degree(s) of freedom, 15
  constraints and degrees of freedom, 29
  rotation degrees of freedom, 164
  translation degrees of freedom, 164
Density
  energy flux density, 329
  Hamiltonian density, 357
  Lagrangian density, 346
  mass current density, 299
  mass density, 15, 21, 293, 298
  momentum density, 357
Derivative
  directional derivative, 433
  functional (variational) derivative, 359–360
  Lagrangian derivative, 376

space (local) derivative, 297
substantial (total) derivative, 311
Determinant
determinant of a second-rank tensor, 424–425
functional (Wronskian) determinant, 37, 234
Hessian determinant, 212
Jacobian determinant, 244
Deterministic forces, 409
Differential
differential constraint (see *kinematic constraint*)
differential cross section, 128
differential equations of motion, 2, 11, 30, 52, 133, 373, 376
differential principles, 52
Differential operators
curl, 434, 435–436, 442, 444, 445
divergence, 433–436, 441, 444
gradient, 432–433, 440, 443
Laplacian, 233, 438, 441, 444
Diffusion equation, 409
Beltrami's vorticity diffusion equation, 333–334
Dilation, 308
volume specific dilation, 313
Dipole
electric dipole, 177
magnetic dipole, 222
Directional derivative, 433
Displacement(s)
arbitrary displacements, 49
atemporal displacement, 41
displacement consistent with the constraints, 30, 40, 42, 44
displacement of atoms in molecules
longitudinal, 150
transverse, 152
elementary displacements, 40–41
independent displacements, 44, 45, 63
possible displacement, 40–41
real displacement, 40
synchronic displacements, 40, 87
virtual displacement, 40–42
Displacement-gradient matrix, 307, 311
Dissipation, 341
Rayleigh dissipation function, 61, 155
Dissipative forces
dissipative forces as non-potential forces, 61, 62, 155
power of dissipative forces, 60
Distribution function, 259–260
Double cross product of vectors, 422
Dual
dual behaviour of radiation, 403
dual pseudo-vector, 426
Dyadic
dyadic product, 423
dyadic vector, 182–183, 330, 423
Dynamic
dynamic pressure, 324
dynamic coefficient of viscosity, 340

E

Eccentric anomaly, 114–115
Eccentricity of a conic, 110, 113
Effective
effective force, 100
effective potential, 100
effective scattering cross section, 125
Ehrenfest theorem, 407
Eigenvalues, 406, 425
Eikonal, 92–93
Einstein's summation convention, 4
Elastance, 154, 155
Elastic
elastic collision (scattering), 121, 125
elastic constant, 155, 314
elastic force, 140–144
elastic medium, 311–319
Elasticity
elasticity (stiffness) tensor, 314
law of elasticity (Hooke's law), 311
linear theory of elasticity, 79, 297, 362
Electric
electric charge, 153, 155, 231, 338, 354, 365, 401
electric circuit, 153–159
electric conductivity, 338, 339
electric dipole, 177
electric field, 205
electric lens, 222
Electromagnetic
electromagnetic antipotentials, 367–368, 371–372
electromagnetic field, 354, 367
Lagrangian density, 354, 365, 368
electromagnetic force, 57, 206, 338
electromagnetic potentials, 57, 366, 367
mechanical–electromagnetic analogies, 205–208
Elementary displacements, 40–41
Ellipse, 46, 109, 113, 115, 173, 264
Ellipsoid of inertia, 181, 182–183, 190

E (*cont.*)
canonical form, 182
central ellipsoid of inertia, 182
principal axes of the ellipsoid of inertia, 182, 185
Elongation, 141, 311
Energy
electrical energy, 20
energy conservation, 14–15
energy first integral, 15, 216
energy flux density, 329
energy operator, 357
free energy (Gibbs), 248
free energy (Helmholtz), 213, 248
internal energy, 213, 248, 328, 351
law of conservation and transformation of mechanical energy, 20
potential energy, 15
thermal energy, 20
Ensemble
statistical ensemble, 259
Entropy
conservation of entropy, 328
entropy as thermodynamical parameter, 213, 248
Equal-action wave front, 90–93
Equation(s) of motion
differential equation of motion, 2, 9
finite equation of motion, 2
fundamental equation of motion, 30
Equilibrium, 35
equilibrium and extremum of potential energy, 36
equilibrium conditions, 36
for one particle, 36
for rigid body, 44–45
geodetic form, 34
equilibrium configuration, 144, 145, 149, 152
equilibrium on a curve, 37
equilibrium on a surface, 37
equilibrium position of pendulum, 139
stable equilibrium, 36
Lagrange–Dirichlet theorem of absolute stability, 36–37
static equilibrium, 35
and principle of virtual work, 42
unstable equilibrium, 38, 139, 145, 279
Equipotential
equipotential lines, 326–327
equipotential surfaces, 432–433
"Escape to infinity" condition, 104
Essential constants (in Hamilton–Jacobi formalism), 266
Euclidean space, 48, 69, 73, 294, 395, 415, 423
Euler variables for a CDM, 296
Euler's angles, 175–177
angle of nutation, 176, 198
angle of self-rotation, 176, 198
precession angle, 176, 198
Euler's equations
equations of motion of an ideal fluid, 319
equations of motion of a rigid body, 185
Euler's theorem, 86, 255
Euler–Lagrange equations, 66
for continuous systems, 344–347
for discrete systems of particles, 77–79
Expectation value of operator, 406–407
Experiment
Davisson–Germer experiment, 404
Michelson–Morley experiment, 390, 392
Rutherford's experiment, 131, 401
Extended phase space (see *state space*)
Extremum
extremum of potential energy, 36
extremum problem, 62, 344

F
Fast top, 197
Fermat's principle, 93, 404
Feynman path integral, 406
Field
central force field, 97–108
conservative field, 432, 435
Coulomb field, 108, 131, 132
electric field, 205, 429
electromagnetic field, 354, 367
gravitational field, 72, 118, 191
irrotational (vorticity-free) field, 434
magnetic field, 205, 338
potential field, 14
solenoidal (source-free) field, 300, 434
Fine structure constant, 402
Finite
finite constraint (see *geometric constraint*)
finite equations of motion, 2
Finite-strain tensor, 310–311
First integral(s)
distinct first integrals, 11
energy first integral, 15, 216
first integrals and conservation laws, 80–81

first integrals as constants of motion, 80, 216
FitzGerald–Lorentz transformations (see *Lorentz transformations*)
Flow
laminar flow, 343
turbulent flow, 343
Fluid
barotropic fluid, 320, 322
ideal (perfect) fluid, 319–339
irrotational motion of fluid, 320–323
magnetodynamics of ideal fluid, 338–339
perfectly viscous (Newtonian) fluid, 340
phase fluid, 260
stationary motion of fluid, 324–327
viscous fluid, 339–344
Flux
angular momentum flux density, 331
energy flux density, 329
flux of a vector, 435
flux of phase fluid, 261
flux of vorticity, 333
magnetic flux, 153
momentum flux density, 330
Force
apparent force (see *force of inertia*)
attraction force, 99, 108, 131
body forces, 300, 303, 316, 348
central force field, 97–108
centre of force, 102, 104–105, 125, 141
centrifugal force, 46, 100
complementary force (see *force of inertia*)
conservative force field, 14
constraint force, 9, 16, 27, 30
deterministic forces, 409
effective force, 100
elastic force, 140–144
external force, 16
force of gravity, 59, 69, 132–140, 172, 189
force(s) of friction (see *constraint forces*)
force of inertia, 53, 168–175
generalized force, 49–51
gyroscopic force, 62
internal force, 16, 301
non-potential force, 60–62, 77, 78
principle of independence of forces, 8
reaction force, 8
as constraint force, 30–31, 202
repulsion force, 131
stochastic forces, 409
superficial forces, 300–301
Foucault pendulum, 172–175
Four-dimensional space, 389, 391, 394
Four-vector, 394–396
Frame
centre of mass frame (CMS), 23
frame of reference, 1
inertial frame, 6
laboratory frame, 122
non-inertial frame, 21, 23, 168–175
Free
free energy (Helmholtz), 213, 248
free enthalpy (see *Gibbs' free energy*)
free particle, 29
Frequency
angular frequency, 141
normal frequencies, 146
and normal coordinates, 147
Larmor frequency, 206
commensurable frequencies, 284
Friction
force(s) of friction, 31, 61, 319, 339
Function
characteristic function, 212–213, 224, 248
distribution function, 259–260
generating function for canonical transformations, 238–249
Hamilton's principal function, 266
harmonic function, 327, 438
potential function, 14
stream function, 326
wave function, 356, 400, 405–407
Functional derivative, 359–360
Fundamental equation
fundamental equation of acoustics, 323–324
fundamental equation of linear theory of elasticity, 362
fundamental equation of motion, 30
fundamental equation of Newtonian mechanics, 7
fundamental equation of thermodynamics, 213, 248, 328
Fundamental problem of mechanics, 9

G

Galilei
Galilei (Galilean) transformations, 7, 389, 390, 393, 397
infinitesimal, 85
Galilei group of transformations, 7, 8, 79
Galilei's principle of relativity, 6, 389, 390, 397
Gauge transformations, 82, 335

G (*cont.*)
Gauss
  Green–Gauss theorem, 434–435
General integral of motion, 9
General theory of relativity, 69
Generalized
  generalized coordinates, 47–52
    arbitrary choice of, 48
    kinetic energy in, 51
  generalized forces, 49–51
  generalized Hamilton's principle, 78
  generalized momenta, 81
  generalized potential (see *velocity-dependent potential*)
  generalized trajectory in configuration space, 49
  generalized velocities, 50
Generating function for canonical transformation, 238–249
Generator
  generator of infinitesimal canonical transformation, 250
  generator of rotations, 251
  generator of time-evolution, 252
  generator of translations, 251
Geodesic, 34, 56, 69–73
  geodesics of a sphere, 73
Geometric constraint, 28
Geometrical optics, 92, 93, 404
Gibbs' free energy, 248
Gibbs–Helmholtz relations, 249
Gradient, 432
  in cylindrical coordinates, 444
  in orthogonal curvilinear coordinates, 440
  in spherical coordinates, 443
Gravitational
  gravitational constant, 118
  gravitational field, 72, 118, 191
  gravitational force, 8
  gravitational mass, 8
Gravity
  motion under the influence of gravity, 132–140
  law of universal gravity, 25
Green–Gauss theorem, 434–435
Group
  Galilei transformations group, 7, 79
  group of proper rotations, 414
  group of rotations, 414
  group of translations, 415
  Lorentz group, 7, 79, 392
  orthogonal group, 413
  special orthogonal group, 414
Gyrocompass, 198–201
Gyroscope, 197–198, 199
Gyroscopic forces, 62

H
Hamilton's canonical equations, 211–222, 360
Hamilton's principal function, 266
Hamilton's principle, 74–77
  generalized Hamilton's principle, 78
  Hamilton's principle as variational principle, 78–79
Hamiltonian, 86, 214
  Hamiltonian action, 90
  Hamiltonian as generator of time-evolution, 252
  Hamiltonian density, 357
  Hamiltonian of conservative systems and total energy, 86
  Hamiltonian operator, 233, 357
Hamiltonian formalism
  for continuous systems, 357–373
  for discrete systems of particles, 211–289
Hamiltonian variables (see *canonical variables*)
Hamilton–Jacobi
  Hamilton–Jacobi equation, 265–268
    abbreviated (restricted), 269
    methods for solving, 268–272
  Hamilton–Jacobi formalism, 265–289
    action–angle variables, 278–286
    for free particle, 273
    for linear harmonic oscillator, 274
    for Newtonian central force field, 274–276
    for symmetrical top, 276–277
Harmonic function (see *Laplace equation*)
Harmonic linear oscillator
  classical harmonic oscillator, 140–141
  quantum mechanical harmonic oscillator, 263–264
Heat propagation
  general equation of heat propagation, 342
Heisenberg
  Heisenberg's commutation relation, 233
  Heisenberg's uncertainty principle, 232, 400, 403, 406
Helmholtz
  Gibbs–Helmholtz relations, 249
  Helmholtz free energy, 213, 248
Herpolhodic cone, 191
Hodograph, 427
Holonomic

holonomic constraint, 28
Lagrange equations for holonomic systems, 53–56
Hooke's generalized law, 311–315
Hyperbola, 111, 112, 113, 120

I
Ideal
ideal curve, 31
ideal fluid, 319–339
Euler's equations, 319, 350
magnetodynamics, 338–339
ideal mechanical model, 20
ideal surface, 35, 37, 42
Identity transformation, 241, 249, 413, 414, 415
Ignorable coordinates (see *cyclic coordinates*)
Imaginary angle, 396
Immobile reference frame, 1
Impact parameter, 126
Impedance, 157
Improper transformation, 413
Impulse, 7
Incident beam, 125
Incompressible fluid, 260, 294, 325–327, 341
Independence of forces
principle of independence of forces, 8
Index of refraction, 92, 130
Inductance, 155, 364
mutual inductance, 154
reactive inductance, 157
self-inductance, 154
Induction
induction currents, 338
induction equation, 338, 372
Inertia
centre of inertia (see *centre of mass*)
inertia products, 180
inertia tensor, 179
moment of inertia, 181
principal axes of inertia, 182
principal planes of inertia, 182
principle of inertia (Newton's first law), 6–7
Inertial
inertial force, 168–175
inertial frame, 6
inertial mass, 8
Infinitesimal
generator of transformation, 250
work, 14
Instantaneous
instantaneous acceleration, 4
instantaneous axis of rotation, 178
instantaneous linear velocity, 3
instantaneous vector of rotation, 168
Integrable constraint (see *holonomic constraint*)
Integral
action integral, 77, 346
elliptic integral, 135
first integral(s), 11, 80
general integral of motion, 9
integral invariants, 253–264
of the canonical equations, 255–256
Poincaré–Cartan, 257
relative, 254, 257
integral principles of analytical mechanics, 52
line integral, 429
surface integral, 429
volume integral, 429
Internal
internal energy, 213, 248, 328, 351
internal force, 16, 301
Interval
lightlike interval, 393
spacelike interval, 393
space–time interval, 393
invariant, 395
timelike interval, 393
Invariance principle, 79
Invariant
adiabatic invariant, 287–289
integral invariant, 253–264
principal invariants of a second-rank tensor, 424
relativistic invariant, 393
Inverse transformation, 48, 236, 244, 412, 414
Irrotational
irrotational field (see *vorticity-free field*)
irrotational motion of fluid, 320–323
Isentropic process, 328
Isoenergetic paths, 89
Isoperimetric problem, 93
Isotropic oscillator, 229
Isotropy of space, 82

J
Jacobi
Hamilton–Jacobi equation, 265–268
Hamilton–Jacobi formalism, 265–289
Jacobi identity for Poisson brackets, 226
Jacobi's formulation of principle of least action, 89
Jacobi's theorem, 266

K

Kelvin's velocity circulation theorem, 331–332
Kepler's equation, 114
Kepler's laws, 13, 115
Kepler's problem, 108–121
Kinematic
  kinematic coefficient of viscosity, 341
  kinematic constraint, 28
Kinetic energy
  expression in generalized coordinates, 51
  total kinetic energy, 18
Kinetic energy theorem, 14
Kirchhoff's rules, 153–158
König's theorems, 21–23
Kronecker
  Kronecker product (see *dyadic product*)
  Kronecker symbol, 411, 417

L

Laboratory frame (LS), 122
Lagrange multipliers, 44, 64, 356, 371
Lagrange brackets, 234–236
  fundamental Lagrange brackets, 235
  invariance under canonical transformations, 244
Lagrange equations, 39
  for holonomic systems, 53–56
  of the first kind, 39
  of the second kind, 55
  with multipliers, 203
Lagrange–Bernoulli equation, 322
Lagrange–Dirichlet theorem of absolute stability, 36
Lagrange variables for CDM, 294, 297, 347
Lagrangian, 55
  equivalent Lagrangians, 59
  Lagrangian density, 346
  Lagrangian derivative, 376
  Lagrangian function, 55
  Lagrangian variables, 55
Lagrangian formalism
  for continuous systems, 344–357
  for discrete systems of particles, 97–159
Lamé
  Lamé coefficients
    for curvilinear coordinates, 440
    in linear elasticity, 314
  Lamé's equations, 316, 347, 362
Laminar flow, 343
Langevin equation, 409
Laplace equation, 327, 438
Larmor
  Larmor frequency, 206
  Larmor precession, 205
  Larmor's theorem, 206
Law
  Hooke's generalized law, 311–315
  Kepler's laws, 13, 115
  law of conservation and transformation of mechanical energy, 20
  law of motion (see *equation of motion*)
  law of universal gravity, 25
  laws of Newtonian mechanics (see *Principles of Newtonian mechanics*)
Legendre transformation, 212, 213, 223, 238
Levi-Civita symbol, 417
Libration, 278
Light
  speed of light, 7, 389, 390, 393
  propagation of light waves, 91, 93
Lightlike interval, 393
Line of nodes, 176
Linear
  linear momentum, 7
    conservation, 85, 329, 381
    theorem, 12, 17
  linear theory of deformations, 306
Liouville's theorem, 258–261
Local (space) derivative, 297
Longitudinal wave, 318
Lorentz transformations, 7, 79, 389, 391–393, 397
Lorenz condition (gauge fixing), 367
LS (see *laboratory frame*)
Lyman series (see *spectrum of hydrogen atom*)

M

Magnetic
  magnetic dipole, 222
  magnetic field, 205, 338
  magnetic lens, 221
  magnetic viscosity, 338
Magnetodynamics of ideal fluids, 338–339
Magnetofluid, 339, 366
Mass
  centre of mass, 20
  gravitational mass, 8
  inertial mass, 8
  mass current density, 299
  mass density, 15, 21, 293, 298
  point mass, 1
  reduced mass, 98
  rest mass, 397
Mathematical pendulum, 132

Matrix
displacement-gradient matrix, 307, 311
transformation matrix, 165, 412
Maupertuis' principle, 52, 86
Maupertuisian action, 89, 90
Maxwell's equations, 338, 354–356, 390
Mechanical–electromagnetic analogies, 205–208
Mechanics
fundamental problem of mechanics, 9
Heisenberg's matrix mechanics, 404
Newtonian mechanics, 1–23
quantum mechanics, 79, 231–233, 264, 399–407
wave mechanics, 404
Medium
anisotropic medium, 314, 315
compressible medium, 294
continuous deformable medium (CDM), 293–385
incompressible medium, 294, 300, 325–327, 341
isotropic medium, 311, 314, 317, 340, 354
Metric tensor, 33, 69, 70, 311
contravariant components of, 71
covariant components of, 70
metric tensor components as gravitational potentials, 72
Michelson–Morley experiment, 390, 392
Minimal optical path, 93
Minkowski space, 391, 395, 396
Mixed product of vectors, 421–422
Modulus
action modulus, 263
modulus of integral, 135
modulus of wave function, 400
Young modulus, 311, 314
Moment of inertia, 181, 182
axial moment of inertia, 180
Momentum
angular momentum, 11, 21, 81, 179
conservation of, 85, 126, 330, 382
centrifugal momentum, 180
generalized momentum, 81
linear momentum, 7
conservation of, 85, 329, 381
momentum density, 357
momentum operator, 233, 406
Monotonic precession, 195
Morse potential, 95
Motion
Brownian motion, 409
composite motion, 165
equation(s) of motion, 2, 9, 30
general integral of motion, 9
irrotational motion of fluid, 320–323
law of motion (see *equation of motion*)
periodical motion, 132–153, 278–289
relative motion, 1, 165
transport motion, 165
uniform motion, 6
uniformly varied motion, 4
varied motion, 3
virtual motion, 75
Multipliers
Lagrange equations with, 203
Lagrange multipliers, 44, 64, 356, 371

N

Nabla symbol, 432
Natural
natural coordinates, 5
natural system of coordinates (see *Serret-Frenet frame*)
natural systems, 58, 211
Navier–Stokes equation, 341
Newtonian
Newtonian fluid (perfectly viscous), 340
Newtonian mechanics, 1–23
principles, 6–9
general theorems, 9–23
Nodes
line of nodes, 176
Noether's theorem
for continuous systems, 373–385
for discrete systems of particles, 82–86
Non-holonomic constraint, 29
Non-inertial frame, 21, 23, 168–175
Non-integrable constraint (see *non-holonomic constraint*)
Non-linear
non-linear differential equation, 134, 143, 144, 319
non-linear mechanics, 143
non-linear oscillations, 143–144
Non-potential forces, 60–62
dissipative forces, 61
gyroscopic forces, 62
power of, 60
Non-stationary constraint (see *rheonomous constraint*)
Normal
normal coordinates, 146–149
normal frequencies, 147
normal reaction force, 31, 202
normal stress (tension), 301, 304, 312

N (*cont.*)
pressure, 301
traction, 301
principal normal, 5, 32, 430
Nutation, 196–197
angle of nutation, 176, 198

O

Ohm's law, 154, 338, 366, 371
Operator
coordinate operator, 233
energy operator, 357
expectation values of operators, 406–407
Hamiltonian operator, 233, 357
momentum operator, 233, 406
Optical path, 92, 93
Optics
geometrical optics, 92, 93, 404
wave optics, 93, 404
Orbit
bound orbit, 102–104
circular orbit, 105
stable orbit, 404
unbound orbit, 104
Orthogonal
orthogonal basis, 420
orthogonal Cartesian coordinates, 4, 412
orthogonal transformations, 411–415
homogeneous, 412, 414
non-homogeneous, 412
special orthogonal group, 414
Orthonormal basis, 420
Oscillations
forced oscillations, 141, 156–157
non-linear oscillations, 143–144
small oscillations, 144–153
of molecules, 149–153
tautochronous oscillations, 136
Oscillator
harmonic linear oscillator, 140–141, 274
isotropic oscillator, 229
quantum mechanical harmonic oscillator, 263–264
space oscillator, 141–142
Osculating plane, 430

P

Parabola, 11, 113
Parameter
impact parameter, 126
variational parameters, 346, 350, 366
Parametric
parametric equations of trajectory, 2
parametric resonance, 289
Paraxial beam, 221
Particle
α particles, 401
electrically charged particle, 219–22
fictitious particle, 98
free particle, 29
particle subject to constraints, 37–38, 42–43
Paschen series (see *spectrum of hydrogen atom*)
Path
isoenergetic path, 89
optical path, 92, 93
Path integral
Feynman path integral, 399, 409
Wiener path integral, 409
Pendulum
conic pendulum, 140
cycloidal pendulum, 136–137
Foucault pendulum, 172–175
mathematical pendulum, 132
physical pendulum, 188–189
plane pendulum (see *simple pendulum*)
simple pendulum, 133–136
spherical pendulum, 137–140
Perfect fluid, 319–339
magnetodynamics, 338–339
Euler's equations, 319, 350
Pericentre, 102, 103
Perigee, 102
Perihelion, 102
Periodical motion
and action–angle variables, 278–286
and adiabatic invariance, 287–289
under the influence of elastic force, 140–144
under the influence of gravity, 132–140
Periodically conditioned system, 284
Pfaff forms, 261–263
Pfaffian constraints, 39
Pfund series (see *spectrum of hydrogen atom*)
Phase
extended phase space (see *state space*)
flux of phase fluid, 261
phase fluid, 260
phase space, 214
Physical pendulum, 188–189
Planck constant, 233, 263, 399, 400
Plane
osculating plane, 430
plane pendulum (see *simple pendulum*)

plane polar coordinates, 5, 443
plane trajectory, 99
Planetary model of atom, 401
Plasma, 287, 294
Poincaré–Cartan integral invariant, 257
Point
point mass, 1
representative point in configuration space, 48
stagnation points, 327
turning point for periodical motion, 102
Poisson
Poisson effect, 312
Poisson's differential equation, 25
Poisson's formula, 167
Poisson's ratio (see *coefficient of transverse contraction*)
Poisson's theorem, 227
Poisson brackets, 225–236
fundamental Poisson brackets, 226
invariance under canonical transformations, 242–244
Jacobi's identity for, 226
Poisson brackets and commutators, 231–234
Poisson brackets for angular momentum, 228–231
Poisson brackets of the Hamiltonian, 228, 230
properties of, 226
Polar
plane polar coordinates, 5, 443
polar angle, 114, 118, 127
polar vector, 8, 416
Polhodic cone, 191
Postulates, 52
Bohr's postulates (hydrogen atom), 402
Einstein's postulates (special relativity), 390
Potential
acceleration potential, 322
attractive potential, 108, 113, 129, 140
centrifugal potential, 100, 104
Clebsch potential, 335, 337, 353
effective potential, 100
generalized potential (see *velocity-dependent potential*)
Morse potential, 95
potential energy, 15
potential field, 14
potential function, 14
potential pressure, 324
repulsive potential, 113, 127
spherical potential well, 129
thermodynamic potential, 247–249
velocity potential, 321
velocity-dependent potential, 57–59
Precession
Larmor precession, 205–206
monotonic precession, 195
regular precession, 190
Pressure
dynamic pressure, 324
potential pressure, 324
pressure as normal stress (tension), 301, 319
pressure as thermodynamical parameter, 213, 248
static pressure, 324
Principal
Hamilton's principal function, 266
principal axes of inertia, 182
principal normal, 5, 32, 430
principal planes of inertia, 182
Principle
correspondence principle, 79
D'Alembert's principle, 52, 53
differential principles, 52
Fermat's principle, 93, 404
Galilei's principle of relativity, 6, 389, 390, 397
generalized Hamilton's principle, 78
Hamilton's principle, 74–79
Heisenberg's uncertainty principle, 232, 400, 403, 406
integral principles, 52
invariance principle, 79
Maupertuis' principle, 86
principle of absolute simultaneity, 7
principle of action and reaction, 8
principle of independence of forces, 8
principle of inertia, 6
principle of least action, 86
principle of physical symmetry, 79
principle of virtual work, 42
principles of analytical mechanics, 27–93
principles of Newtonian mechanics, 6–9
superposition principle, 79
variational principles, 52, 62, 79
Probabilistic interpretation of wave function, 406
Product
cross product, 421
double cross product, 422
dyadic (Kronecker) product, 423
inner product (see *scalar product*)
mixed product, 421–422

P (*cont.*)
products of inertia, 180
scalar product, 420
semi-direct product of groups, 415
Proper
proper transformation, 413
proper values of a second-rank tensor, 425
proper vectors, 425
Pseudo-tensor, 417, 418
Pseudo-vector, 426

Q
Quadratic form, 51, 424, 426
Quadric
deformation quadric, 308
ellipsoid of inertia quadric, 182
velocity of deformation quadric, 321
Quantization condition
Bohr–Sommerfeld quantization condition, 263, 287
Quantum mechanical harmonic oscillator, 263–264
Quantum mechanics, 79, 231–233, 264, 399–407

R
Radius of curvature, 5, 430
Radius-vector, 2
Rayleigh dissipation function, 61, 155
Reaction
force of reaction, 8, 30–31, 202
normal reaction, 31, 202
principle of action and reaction, 8
Reactive
reactive capacitance, 157
reactive inductance, 157
Reduced mass, 98
Refraction
index of refraction, 92, 130
Regular precession, 190
Relative
relative motion, 1
relative universal invariant of mechanics, 256–258
Relativity
Galilei's principle of relativity, 6, 389, 390, 397
special theory of relativity, 389–397
Repulsive potential, 113, 127
Resonance, 157
parametric resonance, 289
Restricted Hamilton–Jacobi equation, 269, 272, 274, 282
Reynold's number, 343
Rheonomous constraint, 28
Riemannian manifold, 70, 73
Rigid body, 19, 44, 45, 163–208
Rigidity constraints, 163
Rolling, 202
Rotation
instantaneous axis of rotation, 178
instantaneous vector of rotation, 168
rotation coordinates, 163
rotation degrees of freedom, 164
rotation tensor, 306–308
symmetry of rotation, 82
velocity of rotation tensor, 321
Rotator, 187
Routh's equations, 222–225
Routhian variables, 223
Runge–Lenz vector, 116–118
Rutherford
Rutherford's experiment, 131, 401
Rutherford's formula, 131
Rydberg constant, 401, 403

S
Saint-Venant compatibility conditions, 309–310
Satellite, 118
Scalar, 396, 415
magnetic scalar potential, 367
scalar potential, 354, 432
scalar product, 420
Scattering, 121–132
backwards scattering, 125
differential scattering cross section, 128
effective scattering cross section, 125–129
elastic scattering, 121, 122, 125
inelastic scattering, 121
scattering angle, 122, 124, 126, 131
scattering centre, 125, 126
Schrödinger's equation, 79, 356–357, 399, 403–405
Scleronomous constraint, 28, 40, 42
Serret–Frenet frame, 429–432
Shear stress, 301, 304, 308, 313
Simple pendulum, 133–136
Simultaneity
principle of absolute simultaneity, 7
Sliding, 202
Small-strain tensor, 306–308
Smooth
(perfectly) smooth curve, 31

(perfectly) smooth surface, 33, 37
Solenoidal (source-free) field, 300, 434
Sound
speed of sound, 324
Space
Euclidean space, 48, 69, 73, 294, 395, 415, 423
four-dimensional space, 389, 391, 394
Minkowski space, 391, 395, 396
space derivative (see *local derivative*)
space homogeneity and translational invariance, 82
space inversion, 413, 415
space isotropy and rotational invariance, 82
space oscillator, 141–142
space transformations, 85
state space, 257–258
Spacelike interval, 393
Space–time, 389, 395
space–time interval, 393
Special theory of relativity, 389–397
Spectrum of hydrogen atom, 402
Speed
speed of light, 7, 389
invariance in special relativity, 390
speed of sound, 324
Sphere
geodesics of a sphere, 73–74
Spherical
motion on a spherical surface, 59–60
spherical coordinates, 2, 4, 442–444
expression for differential operators, 443–444
expression for line, surface and volume elements, 443
spherical pendulum, 137–140
spherical potential well, 129
spherical symmetry, 99
spherical top, 187
Spur (see *trace*)
Stable
stable equilibrium, 36
Lagrange–Dirichlet theorem of absolute stability, 36–37
stable orbit, 404
Stagnation points, 327
State
state function
in quantum mechanics, 233
in thermodynamics, 213, 248
state of a quantum mechanical system, 403–406
state space, 257–258
Static
static equilibrium, 35
static pressure, 324
Stationary
stationary action, 77, 89
principle of, 77
stationary constraint (see *scleronomous constraint*)
stationary curve, 32
stationary field, 220
stationary motion of fluid, 324–327
stationary value in calculus of variations, 63
Statistical ensemble, 259
Stiffness tensor (see *elasticity tensor*)
Stochastic processes, 409–410
Stokes
Navier–Stokes equation, 341
Stokes' theorem, 435–436
Strain, 308, 311
finite-strain tensor, 310–311
small-strain tensor, 306–308
stress-strain relations, 311
Stream function, 326
Streamlines, 296–297, 322, 326–327
Stress
Cauchy stress tensor, 302–306
normal stress, 301, 304, 312
pressure, 301
traction, 301
tangential (shear) stress, 301, 304, 308, 312, 313
viscous stress tensor, 340
Stress-strain relations, 311
Substantial (total) derivative, 311
Summation convention, 4
Superficial forces, 300–301
Superposition principle, 79
Surface
equipotential surfaces, 432–433
ideal surface, 35, 37, 42
perfectly smooth surface, 33, 37
surface integral, 429
surface of revolution of minimum area, 68–69
Symbol
Kronecker symbol, 411, 417
Levi-Civita symbol, 417
nabla symbol, 432
Symmetric tensor, 424–426
Symmetrical top, 189–191
Symmetry, 79–86
axial (cylindrical) symmetry, 127, 220, 270

S (*cont.*)
continuous symmetry and conservation laws, 82–86
symmetry transformations, 377–380
Symplectic transformations (see *canonical transformations*)
Synchronic displacements, 40, 87
System
natural system, 58, 211
periodically conditioned system, 284
System (frame) of reference
centre of mass system (CMS), 122
laboratory system (LS), 122

T
Tangent (shear) stress, 301, 304, 308, 313
Target, 122
Tautochronous oscillations, 136
Telegrapher's equations, 364–366
Tensor
antisymmetric tensor, 166, 229, 416, 423, 426
Cauchy stress tensor, 302–306
eigenvalues of a second-rank tensor, 425
elasticity (stiffness) tensor, 314
finite-strain tensor, 310–311
general properties of tensors, 411–445
inertia tensor, 179
metric tensor, 33, 69, 70, 72, 311
pseudo-tensor, 417, 418
rotation tensor, 306–308
small-strain tensor, 306–308
symmetric tensor, 424–426
tensor contraction, 417
vector components of a tensor, 422
velocity of deformation tensor, 321, 340
velocity of rotation tensor, 321
Theorem
areas theorem, 13, 173, 231
Bertrand's theorem, 103, 105–108
Clebsch's theorem, 334–337
Ehrenfest theorem, 407
Green–Gauss theorem, 434–435
Jacobi's theorem, 266
Kelvin's velocity circulation theorem, 331–332
König's theorems, 21–23
Lagrange–Dirichlet theorem, 36
Larmor's theorem, 206
Liouville's theorem, 258–261
Noether's theorem, 82–86, 373–385
Poisson's theorem, 227
Stokes' theorem, 435–436
total angular momentum theorem, 17
total kinetic energy theorem, 18
total linear momentum theorem, 17
Theory of relativity
special theory of relativity, 389–397
general theory of relativity, 69
Thermo-conductivity coefficient, 341
Thermodynamic potentials, 247–249
Thermodynamics, 213
fundamental equation of thermodynamics, 213, 248, 328
Thermo-elasticity coefficient, 315
Timelike interval, 393
Top
asymmetrical top, 187
fast top, 197
spherical top, 187
symmetrical top, 189–191
Torsion, 431, 432
Total derivative (see *substantial derivative*)
Trace (of a second-rank tensor), 424
Trajectory
generalized trajectory in configuration space, 49
parametric equations of trajectory, 2
plane trajectory, 99
virtual trajectories, 75
Transformation
canonical transformations, 236–249
infinitesimal, 249–253
generating functions for, 238–249
Clebsch transformation, 335, 353, 368
Galilei transformations, 7, 8, 79, 85, 389, 390, 393, 397
identity transformation, 241, 249, 413–415
improper transformation, 413
inverse transformation, 48, 236, 244, 412, 414
Legendre transformation, 212, 213, 223, 238
Lorentz transformations, 7, 79, 389, 391–393, 397
orthogonal transformations, 411–415
homogeneous, 412, 414
non-homogeneous, 412
proper transformation, 413
transformation matrix, 165, 412
Translation, 6, 84
group of translations, 415
total momentum as generator of translations, 251
translation coordinates, 163
translation degrees of freedom, 164

translation symmetry, 82
translation transformation, 413, 414
Transport
transport acceleration, 167
transport motion, 165
transport velocity, 167
Transverse
coefficient of transverse contraction, 312
transverse oscillation, 153, 319
transverse wave, 318
Trihedral system (see *Serret–Frenet frame*)
Turbulent flow, 343–344
Turning points, 102, 134, 174, 194
apocentre, 102
pericentre, 102
Two-body problem, 97–99

U
Uncertainty principle of Heisenberg, 232, 400, 403, 406
Uniform motion, 6
Universal invariant, 256
relative universal invariant of mechanics, 256–258
Unstable equilibrium, 38, 139, 145, 279

V
Valence of a canonical transformation, 238
Variables
action–angle variables, 278–286
canonical variables, 214, 278
conjugate variables (see *canonical variables*)
Euler variables for CDM, 296, 297
Lagrange variables for CDM, 294, 297, 347
Lagrangian variables, 55
Routhian variables, 223
Variational
variational calculus (see *calculus of variations*)
variational derivative (see *functional derivative*)
variational parameters, 346, 350, 366
variational principles, 52, 62, 79
Variations
asynchronous variations, 87
calculus of variations, 62–74
Varied
uniformly varied motion, 4
varied motion, 3
Vector
axial vector, 166, 168, 321, 416
dyadic vector, 182–183, 330, 423
four-vector, 394–396
instantaneous vector of rotation, 168
polar vector, 8, 416
position vector (see *radius-vector*)
proper vectors, 425
pseudo-vector, 426
radius-vector, 2
Runge–Lenz vector, 116–118
vector components of a tensor, 422
Velocity
addition of velocities in special relativity, 394
areal velocity, 13, 222
average velocity, 3
cosmic velocities, 121
generalized velocities, 50
instantaneous linear velocity, 3
transport velocity, 167
velocity circulation
Kelvin's theorem, 331–332
velocity of deformation tensor, 321, 340
velocity of rotation tensor, 321
velocity potential, 321
velocity-dependent potential, 57–59
Vibration
molecule vibration, 149
vibration in action–angle variables, 278–284
Virtual
virtual displacements, 40–42
virtual motion, 75
virtual trajectories, 75
virtual work, 42, 49, 51, 53
infinitesimal, 60, 76
principle of, 42–47
Viscosity, 339
dynamic coefficient of viscosity, 340
kinematic coefficient of viscosity, 341
magnetic viscosity, 338
Viscous
perfectly viscous fluid, 340
viscous fluid, 339–344
viscous stress tensor, 340
Voigt notation, 314
Volume
volume integral, 429
volume specific dilation, 313
Vortex line, 332
and Clebsch potentials, 336–337

V (*cont.*)
Vorticity, 321
  Beltrami's vorticity diffusion equation, 333
  vorticity equation, 332–334
  vorticity in terms of Clebsch potentials, 336
  vorticity-free field, 434

W
Wallis' formula, 135
Wave
  electromagnetic waves, 390
  equal-action wave front, 90–93
  longitudinal wave, 318
  plane waves, 160
    in isotropic elastic media, 317–319
  propagation of light waves, 91, 93
  sound waves, 324
  speed of wave, 318
  transverse wave, 318
  wave associated to a microparticle, 356
    de Broglie's hypothesis, 403–404
  wave equation, 318, 365, 367
  wave function, 356, 400, 405–407
    probabilistic interpretation of, 406
  wave mechanics, 404
  wave number, 92
  wave optics, 93, 404
Wavelength, 91, 92, 403, 404
Wavepacket, 405
Well
  potential well, 129
Wiener path integral, 409
Work, 14
  infinitesimal work, 14
  principle of virtual work, 42–47
  virtual work, 42, 49, 51, 53
  work and total kinetic energy theorem, 18

Y
Young modulus, 311, 314